Proceedings of GeoShanghai 2018 International Conference: Multi-physics Processes in Soil Mechanics and Advances in Geotechnical Testing

Liangbo Hu · Xiaoqiang Gu
Junliang Tao · Annan Zhou
Editors

Proceedings of GeoShanghai 2018 International Conference: Multi-physics Processes in Soil Mechanics and Advances in Geotechnical Testing

Springer

Editors
Liangbo Hu
University of Toledo
Toledo, OH
USA

Junliang Tao
The University of Akron
Akron, OH
USA

Xiaoqiang Gu
Geotechnical Engineering
Tongji University
Shanghai
China

Annan Zhou
Royal Melbourne Institute of Technology
Melbourne, VIC
Australia

ISBN 978-981-13-0094-3 ISBN 978-981-13-0095-0 (eBook)
https://doi.org/10.1007/978-981-13-0095-0

Library of Congress Control Number: 2018939621

This Springer imprint is published by the registered company Springer Nature Singapore Pte Ltd.
part of Springer Nature
The registered company address is: 152 Beach Road, #21-01/04 Gateway East, Singapore 189721, Singapore

Preface

The 4th GeoShanghai International Conference was held on May 27–30, 2018, in Shanghai, China. GeoShanghai is a series of international conferences on geotechnical engineering held in Shanghai every four years. The conference was inaugurated in 2006 and was successfully held in 2010 and 2014, with more than 1200 participants in total. The conference offers a platform of sharing recent developments of the state-of-the-art and state-of-the-practice in geotechnical and geoenvironmental engineering. It has been organized by Tongji University in cooperation with the ASCE Geo-Institute, Transportation Research Board, and other cooperating organizations.

The proceedings of the 4th GeoShanghai International Conference include eight volumes of over 560 papers; all were peer-reviewed by at least two reviewers. The proceedings include Volumes 1: Fundamentals of Soil Behavior edited by Dr. Annan Zhou, Dr. Junliang Tao, Dr. Xiaoqiang Gu, and Dr. Liangbo Hu; Volume 2: Multi-physics Processes in Soil Mechanics and Advances in Geotechnical Testing edited by Dr. Liangbo Hu, Dr. Xiaoqiang Gu, Dr. Junliang Tao, and Dr. Annan Zhou; Volume 3: Rock Mechanics and Rock Engineering edited by Dr. Lianyang Zhang, Dr. Bruno Goncalves da Silva, and Dr. Cheng Zhao; Volume 4: Transportation Geotechnics and Pavement Engineering edited by Dr. Xianming Shi, Dr. Zhen Liu, and Dr. Jenny Liu; Volume 5: Tunneling and Underground Construction edited by Dr. Dongmei Zhang and Dr. Xin Huang; Volume 6: Advances in Soil Dynamics and Foundation Engineering edited by Dr. Tong Qiu, Dr. Binod Tiwari, and Dr. Zhen Zhang; Volume 7: Geoenvironment and Geohazards edited by Dr. Arvin Farid and Dr. Hongxin Chen; and Volume 8: Ground Improvement and Geosynthetics edited by Dr. Lin Li, Dr. Bora Cetin, and Dr. Xiaoming Yang. The proceedings also include six keynote papers presented at the conference, including "Tensile Strains in Geomembrane Landfill Liners" by Prof. Kerry Rowe, "Constitutive Modeling of the Cyclic Loading Response of Low Plasticity Fine-Grained Soils" by Prof. Ross Boulanger, "Induced Seismicity and Permeability Evolution in Gas Shales, CO_2 Storage and Deep Geothermal Energy" by Prof. Derek Elsworth, "Effects of Tunneling on Underground Infrastructures" by Prof. Maosong Huang, "Geotechnical Data Visualization and Modeling of Civil

Infrastructure Projects" by Prof. Anand Puppala, and "Probabilistic Assessment and Mapping of Liquefaction Hazard: from Site-specific Analysis to Regional Mapping" by Prof. Hsein Juang. The Technical Committee Chairs, Prof. Wenqi Ding and Prof. Xiong Zhang, the Conference General Secretary, Dr. Xiaoqiang Gu, the 20 editors of the 8 volumes and 422 reviewers, and all the authors contributed to the value and quality of the publications.

The Conference Organizing Committee thanks the members of the host organizations, Tongji University, Chinese Institution of Soil Mechanics and Geotechnical Engineering, and Shanghai Society of Civil Engineering, for their hard work and the members of International Advisory Committee, Conference Steering Committee, Technical Committee, Organizing Committee, and Local Organizing Committee for their strong support. We hope the proceedings will be valuable references to the geotechnical engineering community.

Shijin Feng
Conference Chair
Ming Xiao
Conference Co-chair

Organization

International Advisory Committee

Herve di Benedetto	University of Lyon, France
Antonio Bobet	Purdue University, USA
Jean-Louis Briaud	Texas A&M University, USA
Patrick Fox	Penn State University, USA
Edward Kavazanjian	Arizona State University, USA
Dov Leshchinsky	University of Illinois, USA
Wenhao Liang	China Railway Construction Corporation Limited, China
Robert L. Lytton	Texas A&M University, USA
Louay Mohammad	Louisiana State University, USA
Manfred Partle	KTH Royal Institute of Technology, Switzerland
Anand Puppala	University of Texas at Arlington, USA
Mark Randolph	University of Western Australia, Australia
Kenneth H. Stokoe	University of Texas at Austin, USA
Gioacchino (Cino) Viggiani	Université Joseph Fourier, France
Dennis T. Bergado	Asian Institute of Technology, Thailand
Malcolm Bolton	Cambridge University, UK
Yunmin Chen	Zhejiang University, China
Zuyu Chen	Tsinghua University, China
Jincai Gu	PLA, China
Yaoru Lu	Tongji University, China
Herbert Mang	Vienna University of Technology, Austria
Paul Mayne	Georgia Institute of Technology, USA
Stan Pietruszczak	McMaster University, Canada
Tom Papagiannakis	Washington State University, USA
Jun Sun	Tongji University, China

Scott Sloan	University of Newcastle, Australia
Hywel R. Thomas	Cardiff University, UK
Atsashi Yashima	Gifu University, Japan

Conference Steering Committee

Jie Han	University of Kansas, USA
Baoshan Huang	University of Tennessee, USA
Maosong Huang	Tongji University, China
Yongsheng Li	Tongji University, China
Linbin Wang	Virginia Tech, USA
Lianyang Zhang	University of Arizona, USA
Hehua Zhu	Tongji University, China

Technical Committee

Wenqi Ding (Chair)	Tongji University, China
Charles Aubeny	Texas A&M University, USA
Rifat Bulut	Oklahoma State University, USA
Geoff Chao	Asian Institute of Technology, Thailand
Jian Chu	Nanyang Technological University, Singapore
Eric Drumm	University of Tennessee, USA
Wen Deng	Missouri University of Science and Technology, USA
Arvin Farid	Boise State University, Idaho, USA
Xiaoming Huang	Southeast University, China
Woody Ju	University of California, Los Angeles, USA
Ben Leshchinsky	Oregon State University, Oregon, USA
Robert Liang	University of Dayton, Ohio, USA
Hoe I. Ling	Columbia University, USA
Guowei Ma	Hebei University of Technology, China
Roger W. Meier	University of Memphis, USA
Catherine O'Sullivan	Imperial College London, UK
Massimo Losa	University of Pisa, Italy
Angel Palomino	University of Tennessee, USA
Krishna Reddy	University of Illinois at Chicago, USA
Zhenyu Yin	Tongji University, China
ZhongqiYue	University of Hong Kong, China
Jianfu Shao	Université des Sciences et Technologies de Lille 1, France
Jonathan Stewart	University of California, Los Angeles, USA

Wei Wu	University of Natural Resources and Life Sciences, Austria
Jianhua Yin	The Hong Kong Polytechnic University, China
Guoping Zhang	University of Massachusetts, USA
Jianmin Zhang	Tsinghua University, China
Xiong Zhang (Co-chair)	Missouri University of Science and Technology, USA
Yun Bai	Tongji University, China
Jinchun Chai	Saga University, Japan
Cheng Chen	San Francisco State University, USA
Shengli Chen	Louisiana State University, USA
Yujun Cui	École Nationale des Ponts et Chaussees (ENPC), France
Mohammed Gabr	North Carolina State University, USA
Haiying Huang	Georgia Institute of Technology, USA
Laureano R. Hoyos	University of Texas at Arlington, USA
Liangbo Hu	University of Toledo, USA
Yang Hong	University of Oklahoma, USA
Minjing Jiang	Tongji University, China
Richard Kim	North Carolina State University, USA
Juanyu Liu	University of Alaska Fairbanks, USA
Matthew Mauldon	Virginia Tech., USA
Jianming Ling	Tongji University, China
Jorge Prozzi	University of Texas at Austin, USA
Daichao Sheng	University of Newcastle, Australia
Joseph Wartman	University of Washington, USA
Zhong Wu	Louisiana State University, USA
Dimitrios Zekkos	University of Michigan, USA
Feng Zhang	Nagoya Institute of Technology, Japan
Limin Zhang	Hong Kong University of Science and Technology, China
Zhongjie Zhang	Louisiana State University, USA
Annan Zhou	RMIT University, Australia
Fengshou Zhang	Tongji University, China

Organizing Committee

Shijin Feng (Chair)	Tongji University, China
Xiaoqiang Gu (Secretary General)	Tongji University, China
Wenqi Ding	Tongji University, China
Xiongyao Xie	Tongji University, China

Yujun Cui	École Nationale des Ponts et Chaussees (ENPC), France
Daichao Sheng	University of Newcastle, Australia
Kenichi Soga	University of California, Berkeley, USA
Weidong Wang	Shanghai Xian Dai Architectural Design (Group) Co., Ltd., China
Feng Zhang	Nagoya Institute of Technology, Japan
Yong Yuan	Tongji University, China
Weimin Ye	Tongji University, China
Ming Xiao (Co-chair)	Penn State University, USA
Yu Huang	Tongji University, China
Xiaojun Li	Tongji University, China
Xiong Zhang	Missouri University of Science and Technology, USA
Guenther Meschke	Ruhr-Universität Bochum, Germany
Erol Tutumluer	University of Illinois, Urbana—Champaign, USA
Jianming Zhang	Tsinghua University, China
Jianming Ling	Tongji University, China
Guowei Ma	Hebei University of Technology, Australia
Hongwei Huang	Tongji University, China

Local Organizing Committee

Shijin Feng (Chair)	Tongji University, China
Zixin Zhang	Tongji University, China
Jiangu Qian	Tongji University, China
Jianfeng Chen	Tongji University, China
Bao Chen	Tongji University, China
Yongchang Cai	Tongji University, China
Qianwei Xu	Tongji University, China
Qingzhao Zhang	Tongji University, China
Zhongyin Guo	Tongji University, China
Xin Huang	Tongji University, China
Fang Liu	Tongji University, China
Xiaoying Zhuang	Tongji University, China
Zhenming Shi	Tongji University, China
Zhiguo Yan	Tongji University, China
Dongming Zhang	Tongji University, China
Jie Zhang	Tongji University, China
Zhiyan Zhou	Tongji University, China
Xiaoqiang Gu (Secretary)	Tongji University, China
Lin Cong	Tongji University, China
Hongduo Zhao	Tongji University, China

Fayun Liang	Tongji University, China
Bin Ye	Tongji University, China
Zhen Zhang	Tongji University, China
Yong Tan	Tongji University, China
Liping Xu	Tongji University, China
Mengxi Zhang	Tongji University, China
Haitao Yu	Tongji University, China
Xian Liu	Tongji University, China
Shuilong Shen	Tongji University, China
Dongmei Zhang	Tongji University, China
Cheng Zhao	Tongji University, China
Hongxin Chen	Tongji University, China
Xilin Lu	Tongji University, China
Jie Zhou	Tongji University, China

Contents

Seepage and Porous Mechanics

About the Editors

Dr. Liangbo Hu received his Ph.D. degree from Duke University with a focus on Geomechanics and Geotechnical Engineering. He is currently an Associate Professor at the University of Toledo, Ohio, USA.

Dr. Xiaoqiang Gu obtained his Ph.D. degree in Geotechnical Engineering from the University of Hong Kong (Hong Kong SAR China). Currently he is working as an Associate Professor in the School of Civil Engineering, Tongji University (China).

Dr. Junliang Tao graduated with his Ph.D. degree from the Case Western Reserve University, USA. Currently he is an Assistant Professor, Department of Civil Engineering, the University of Akron (USA). He was awarded the NSF CAREER award in 2017.

Dr. Annan Zhou obtained his Ph.D. degree in Geotechnical Engineering from the University of Newcastle, Australia. Now, he works as a tenured Senior Lecturer at Royal Melbourne Institute of Technology, Australia. He was awarded the ARC DECRA Fellowship in 2012 and the AGS Hugh Trollope Medal in 2014.

Induced Seismicity and Permeability Evolution in Gas Shales, CO_2 Storage and Deep Geothermal Energy

Derek Elsworth[1(✉)], Kyunjae Im[1], Yi Fang[1,2], Takuya Ishibashi[1,3], and Chaoyi Wang[1]

[1] Energy and Mineral Engineering, G3 Center, EMS Energy Institute, Pennsylvania State University, University Park, USA
elsworth@psu.edu
[2] Institute for Geophysics, The University of Texas at Austin, Austin, USA
[3] Fukushima Renewable Energy Institute, National Institute for Advanced Industrial Science and Technology, Koriyama, Fukushima, Japan

Abstract. Contemporary methods of energy conversions that reduce carbon intensity include sequestering CO_2, fuel switching to lower-carbon sources, such as from gas shales, and recovering deep geothermal energy via EGS. In all of these endeavors, either maintaining the low permeability and integrity of caprocks or in controlling the growth of permeability in initially very-low-permeability shales and geothermal reservoirs represent key desires. At short-timescales of relevance, permeability is driven principally by deformations – in turn resulting from changes in total stresses, fluid pressure or thermal and chemical effects. These deformations may be intrinsically stable or unstable, result in aseismic or seismic deformation, with resulting changes in permeability conditioned by the deformational mode. We report observations, experiments and models to represent the respective roles of mineralogy, texture, scale and overpressures on the evolution of friction, stability and permeability in fractured rocks – and their interrelationships. The physics of these observed behaviors are explored via parametric studies and surface measurement of fractures, showing that both permeability and frictional strength are correlated to the fracture asperity evolution that is controlled in-turn by the sliding velocity and fracture material.

Keywords: Induced seismicity · Permeability evolution · Shale gas CO_2 sequestration · EGS

1 Introduction

The presence of pre-existing faults and fractures in the upper crust (Anderson and Zoback 1982; Curtis 2002) contribute to induced seismicity as a result of fluid injection in hydraulic fracturing, deep storage of CO_2, and stimulation of EGS reservoirs (Ellsworth 2013; Im et al. 2017; Guglielmi et al. 2015; Majer et al. 2007; McGarr et al. 2002; Walsh and Zoback 2015). In all of these, either maintaining the low permeability and integrity of caprocks or in controlling the growth of permeability in initially very-low-permeability shales and geothermal reservoirs are key desires. Hence, it is of

L. Hu et al. (Eds.): GSIC 2018, *Proceedings of GeoShanghai 2018 International Conference: Multi-physics Processes in Soil Mechanics and Advances in Geotechnical Testing*, pp. 1–20, 2018.
https://doi.org/10.1007/978-981-13-0095-0_1

particular interest to understand the seismicity-permeability interaction in caprocks and unconventional reservoirs.

Mechanically, the occurrence of induced seismicity depends on the shear strength and the frictional stability of a fault - which in-turn depends on its mineralogical composition (Ikari et al. 2011; Fang et al. 2017). The weakness of natural faults can be explained by the presence of frictionally weak minerals, including talc (Collettini et al. 2009; Moore and Rymer 2007). Early experiments using synthetic mixtures of salts and muscovite/kaolinite (Bos and Spiers 2002; Niemeijer and Spiers 2006) showed that weakening can occur with as low as 10 wt% (weight percentage) of frictionally weak minerals. Shear experiments using mixtures of talc and quartz sand (Carpenter et al. 2009) suggested that 30 wt%–50 wt% of frictionally weak minerals were required to weaken the composite gouge – much larger percentages than that observed in some natural weak faults. This difference can be explained by the presence of a structured through-going layer of weak minerals (e.g. talc), which weakens the fault. These observations pose the question of what proportion and in what arrangement frictionally weak minerals are needed what may be their impact on the evolution of rheology and permeability. Experiments (Moore and Lockner 2011; Niemeijer et al. 2010) suggest that the frictional strength of gouge decreases systematically with an increase in thickness of weak stable layers. Additionally, coupled shear-permeability experiments suggest that the permeability evolution of faults is likely linked to such mineralogical properties (Ishibashi et al. 2016).

Permeability is known to change during shear deformation (Elsworth and Goodman 1986). It has been widely observed that failure may occur stably (aseismically) at slow creep rates of long duration (order of 1–100 mm/yr) or unstably (seismically) at fast frictional sliding rates of short duration (order of 1 m/s) (Anderson et al. 1996; Peng and Gomberg 2010; Schmidt et al. 2005). The stability of sliding is governed by the frictional properties of faults and can be described with rate-and-state friction laws (Dieterich 1979; Marone 1998; Ruina 1983; Scholz 1998). These studies provide potential insights into the rheological response of caprocks and unconventional reservoirs with regard to the mode and timing of induced earthquakes. However, it is still unclear whether different styles of permeability evolve from unstable fast sliding of seismic events versus slow-slip aseismic events. In this study, we integrate both experimental and computational methods to explore how fracture permeability changes in response to fracture reactivation; identify hydraulic behavior for different rock types; and probe the relationship between frictional strength with respect to mineralogical composition. We explore these questions through (1) evaluating seismicity permeability relationships of fractures, (2) defining the role of healing and sealing on response, and in (3) representing these behaviors through coupled micromechanical models.

2 Seismicity-Permeability Observations

Induced seismic or aseismic events are primarily controlled by the frictional response of faults or fractures during dynamic shearing. The evolution of frictional strength during fracture slip may be generally interpreted in the framework of rate- and state-dependent friction theory (Dieterich 1979; Ruina 1983) as,

$$\mu = \mu_0 + a \ln\left(\frac{v}{v_0}\right) + b \ln\left(\frac{v\theta}{D_c}\right) \tag{1}$$

$$\frac{d\theta}{dt} = 1 - \frac{v\theta}{D_c} \tag{2}$$

where, upon a velocity increase from v_0 to v (Fig. 1a), the friction coefficient (μ) increases from a reference steady state (μ_0), by an instantaneous so-called direct effect, and then evolves to a new steady state (evolution effect) over a characteristic critical slip distance (D_c) (Fig. 1b). θ is a state variable. The parameters a and b are scaling factors that define the frictional stability. If friction increases with a velocity increase, the material is velocity-strengthening, leading to stable/aseismic slip (i.e., $(a - b) > 0$). Conversely, if ultimate friction decreases with a velocity increase, the material is velocity-weakening (i.e., $(a - b) < 0$), a necessary (though not sufficient) condition for stick-slip behavior associated with earthquake nucleation (i.e., potentially unstable slip) (e.g., Marone 1998; Scholz 1998).

Of specific interest is whether the response of fracture permeability is linked to the frictional strength and frictional stability or not (Fig. 1c). To explore such a link, fracture permeability may be monitored as an instantaneous step in shear velocity is applied, to implicitly link observations of frictional and permeability evolution. Abundant experimental studies have revealed the values of frictional strength and frictional stability that

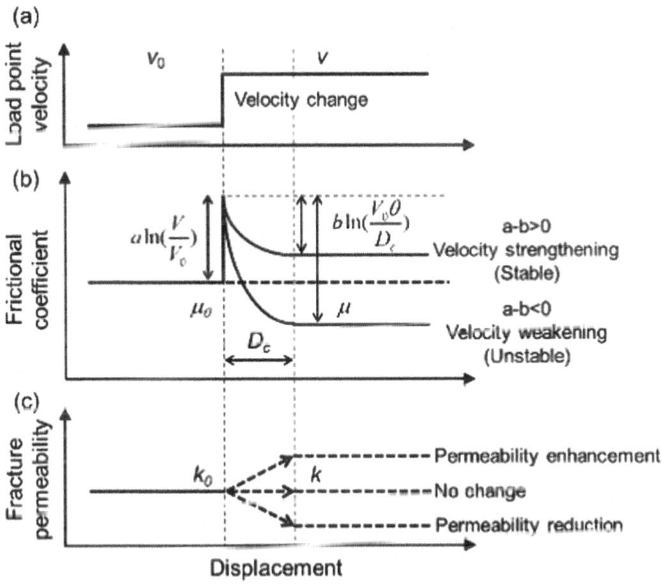

Fig. 1. (a) Step change in the load point velocity during fracture shearing, (b) Idealized rate- and state-friction response to an increase in load point velocity. Two alternative behaviors (i.e., velocity strengthening and velocity weakening) are shown, and (c) Predicted responses of fracture permeability to an increase in load point velocity.

follow a reversal relationship, which is controlled by the mineral compositions. These behaviors may be divided among three mineral groups: tectosilicates, carbonates, and phyllosilicates. In the following we (1) explore such a relationship between the evolution of frictional strength, frictional stability (stable/unstable) and the response of fracture permeability; and (2) study friction-stability-permeability relationships with respect to the mineralogical composition of fractures.

2.1 Experimental Methods

The fracture slip experiments are conducted on L-shaped blocks of Westerly granite (Fig. 2a) that include a pre-cut fracture parallel to the shear direction. The fracture surfaces are pre-roughened with #60 grit (423 μm) silicon carbide to yield a controlled fracture roughness – verified by optical profilometry. Samples are loaded in a single direct-shear configuration within a pressure vessel and biaxial load frame (Fig. 2) to create a true-triaxial stress state (for details of the apparatus, see Samuelson et al. (2009) and Candela et al. (2015)). The normal stress (σ_n) is applied on the fracture plane (45 mm × 50 mm; Fig. 2) via a piston driven by a servo-controlled hydraulic ram and shear stress (τ) is applied by a separate, vertically-oriented, piston driven by a servo-controlled hydraulic ram. The L-shaped block and forcing blocks/platens are unjacketed for measuring rock friction, alone (Fig. 2b), but are jacketed (latex rubber sleeve) for measuring permeability (Fig. 2c).

In the case of the fluid-flow-through experiments, fracture permeability is evaluated based on the cubic law (Witherspoon et al. 1980; Tsang and Witherspoon 1981):

$$k = \frac{e_h^2}{12} \tag{3}$$

where k is the fracture permeability, and e_h is the hydraulic aperture. Using Darcy's law, these parameters are related as:

$$e_h = \left(-\frac{12\eta LQ}{W\Delta P}\right)^{1/3} \tag{4}$$

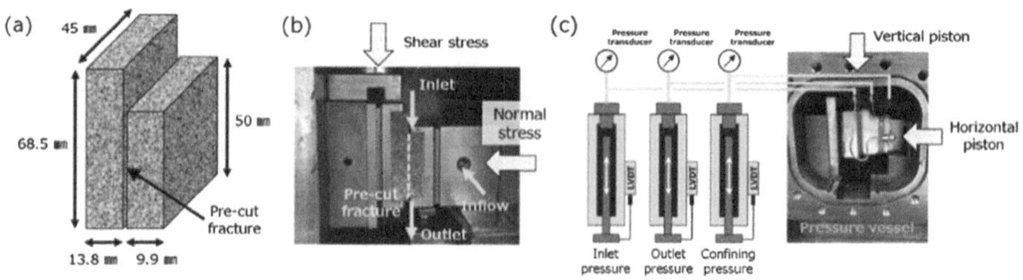

Fig. 2. (a) L-shape block of Westerly granite with a single pre-cut fracture, (b) Experimental configuration for measuring the frictional strength of a rock fracture, and (c) Experimental configuration for measuring the permeability of a rock fracture.

where Q is the flow rate, ΔP is the differential pore pressure ($P_{in} - P_{out}$), η is the fluid viscosity (1.002×10^{-3} Pa · s at 20 °C for distilled water), L is the flow path length (50 mm), and W is the width of the fracture.

To investigate the effect of minerology, we prepare two types of samples (1) intact natural samples and (2) sintered samples. We collect six intact natural samples including five shales (*i.e.,* Green River shale, Opalinus claystone, Marcellus shale, Tournemire shale, and Longmaxi shale) and one tuff (*i.e.,* Newberry tuff). The sintered samples are consolidated from five different minerals: dolomite, calcite, albite, quartz and illite that are common in major fault zones.

All of the samples are characterized by XRD and the weight percentage of each mineral group is illustrated in Fig. 3a. Due to the low stiffness of these samples compared to the Westerly granite, we use a triaxial testing apparatus that independently applies confining pressure and differential (end-to-end) pore pressure while the sample is sheared at a prescribed velocity. This allows the concurrent measurement of the evolution of fracture permeability and friction. The sample coupons are packed within a pair of steel shearing platens (Fig. 3b).

The initial offset of platens is ~8 mm for slip displacement during sliding. The platen-offset gap is filled with filler as a seal. The side and bottom contacts between the sample coupon and the platen surfaces are packed with Teflon to prevent fluid leakage. The assembled platens are packed within a membrane to isolate the sample from the

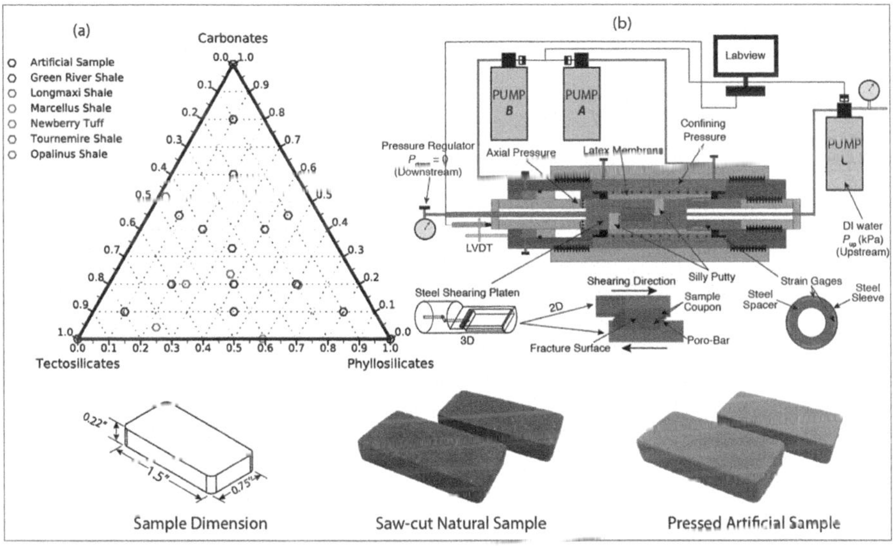

Fig. 3. (a) Ternary diagram of mineral groups (tectosilicate, carbonate, and phyllosilicate) for both natural samples and artificial samples. (b) Experimental setup to measure friction-permeability evolution: Pump A (ISCO 500D) controls the confining pressure (normal stress) applied on the fracture. Pump B (ISCO 500D) controls pressure that provides the source of shear stress applied on the fracture. Pump C (ISCO 500D) injects the fluid at a prescribed flow rate or pressure, allowing the fluid source located at the origin of the fracture and flow along the fractures.

confining fluid. A steel sleeve covers the load cell to prevent the effect of applied confining pressure. We apply a confining stress (normal stress) of 3 MPa and set a constant upstream fluid pressure during axial shear displacement (at constant rate). The minimum flow rate of each pump is 0.001 ml/min and the display resolution of the pump pressure transducer is 1.0 kPa. A load cell with a resolution of 0.3 kPa is used to measure the axial stress. At room temperature, the minimum measurable permeability is 1.0×10^{-16} m^2.

2.2 Results and Analysis

Figure 4 shows the evolution of fracture permeability (k) for the history of shear velocity (v). Fracture permeability decreases with increasing normal stress (from 6 MPa to 12 MPa), which is caused by the closure of the fracture due to the elastic deformations or the damage/crack at the contacting asperities. For these normal stress conditions, fracture permeability also decreases with increasing shear displacement. These reductions in k are due to the change in fracture roughness and the production of wear materials (Wang and Scholz 1994; Faoro et al. 2009; Fang et al. 2017).

In this experiment, we evaluate changes in fracture permeability due to instantaneous steps in shear velocity, and explore the links between frictional stability and permeability change. The responses in k for various combinations of normal stress and velocity step size are shown in Fig. 4. To quantify the permeability change (Δk), we evaluate the fracture permeability both before (k_0) and then after (k) the instantaneous step in shear velocity.

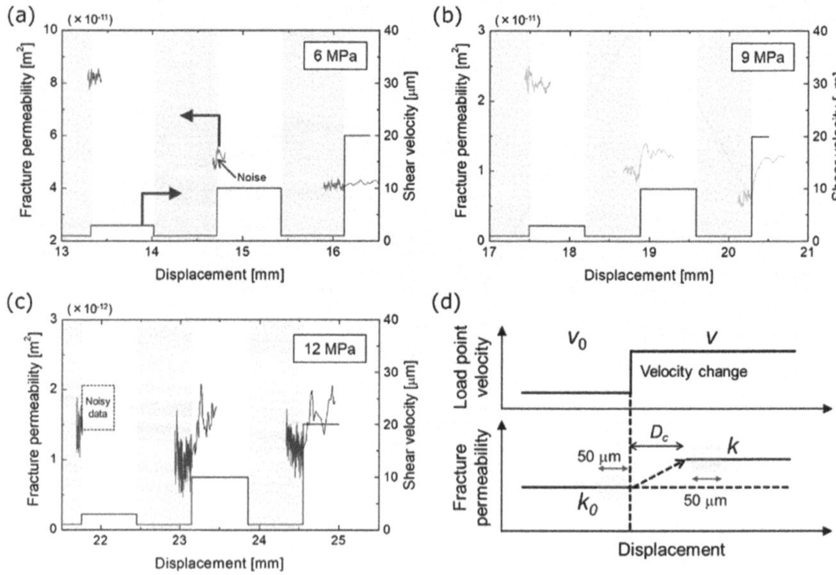

Fig. 4. Experimental results showing fracture permeability and shear velocity versus shear displacement at normal stresses of (a) 6 MPa, (b) 9 MPa, and (c) 12 MPa. In each stage, confining stress is set equivalent to the normal stress (i.e., anisotropic stress condition). Responses in the fracture permeability caused by the shear velocity jump are analyzed.

Both k_0 and k are derived as the ensemble average of fracture permeabilities over the shear-displacement window ± 0.05 mm (see Fig. 4d), therefore, the impacts of permeability fluctuation on the evaluation of k_0 and k are minor. D_c in Fig. 4d is the critical slip distance required to achieve steady-state friction after the velocity step, and is set to 100 μm. The normalized permeability change is then defined as

$$\frac{\Delta k}{k_0} = \frac{k - k_0}{k_0} \tag{5}$$

Because the shear-displacement window for defining $\Delta k/k_0$ is so small (less than ~0.2 mm, see Fig. 4d), the background reduction in k with increasing shear displacement has little influence on the calculation of $\Delta k/k_0$. Considering that the fracture permeabilities after the velocity step are never lower than those before the velocity step (Fig. 4), it is clear that fracture permeability is enhanced due to the velocity step.

The normalized permeability changes ($\Delta k/k_0$) are shown in Fig. 5a as a function of velocity step size ($\ln(v/v_0)$). Note that $\Delta k/k_0$ is not evaluated when the original data include noise/uncertainty or the aforementioned window for ensemble averaging the permeability cannot be placed. The values of $\Delta k/k_0$ are quite small (<0.05) at a normal stress of 6 MPa, and $\Delta k/k_0$ increases with an increase in $\ln(v/v_0)$ at normal stresses of 9 MPa and 12 MPa. $\Delta k/k_0$ ranges from −0.055 to 0.607 in the experiments. On the basis of Fig. 5a, we find that larger values of $\Delta k/k_0$ are recorded when k_0 is small, and that $\Delta k/k_0$ increases with velocity step size. These trends are very similar to those observed in velocity stepping experiments with fine-grained quartz fault gouge, where normalized layer dilation increase with velocity step size (Samuelson et al. 2009).

Based on the experimental results, $\Delta k/k_0$ is shown as a function of the frictional rate parameters $(a - b)$ in Fig. 5b. Figure 5b reveals that almost all points appear exclusively in the second quadrant of the graph, and this region represents both negative values of $(a - b)$ and positive values of $\Delta k/k_0$. This suggests that the permeability enhancement of rock fractures is possibly created by the frictional shearing of velocity-weakening (potentially unstable slip) and may have a connection to the dilation behavior of rock fractures. This is consistent with field observations where

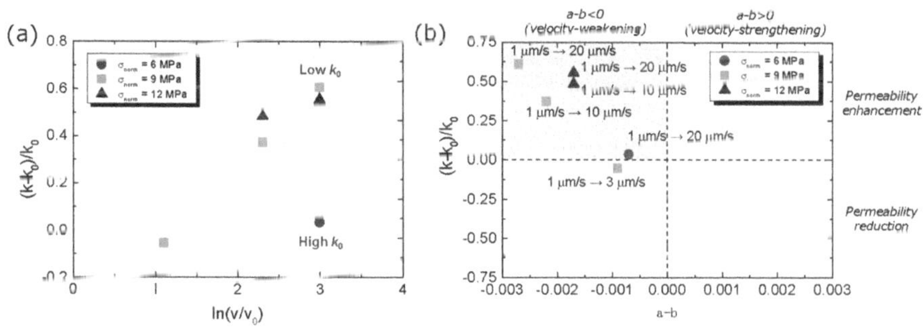

Fig. 5. (a) Relationship between stability parameter (i.e., $a - b$) and velocity step size, (b) Relationship between normalized permeability change (i.e., $(k - k_0)/k_0$) and velocity step size, and (c) Relationship between the instability parameter and normalized permeability change.

fault permeability increases during seismic slip (Guglielmi et al. 2015), and this phenomenon may be related to the mechanisms of permeability enhancement of fracture zones excited by far-field earthquakes (Elkhoury et al. 2006; Xue et al. 2013).

The calculated net fracture permeability monotonically decreases with displacement, consistent with previous observations (Fang et al. 2017). The data fitting method for simulating the permeability change is introduced in Fang et al. (2017). The analyzed frictional parameters and transient permeability change in response to velocity change are shown in Fig. 6 with respect to the selected mineral groups: phyllosilicates, carbonates and tectosilicates. Frictional strength μ and transient permeability change Δk_n and $\Delta k_n/\Delta k_0^i$ decrease with phyllosilicate content but increase with tectosilicate content. In contrast, a reverse trend is observed for frictional stability $(a - b)$. However, the role of carbonate is distinct from the other two mineral groups. The permeability changes decrease with both frictional strength and stability, implying that in the very shallow crust (*i.e.*, low normal stress and low temperature), slip reactivation in calcite-bearing fault gouges is difficult and is unlikely to induce seismic events - and permeability is reduced during the resulting aseismic slip.

As surface contact state, which determines the flow path, is reflected in the frictional strength and stability, we directly correlate the permeability change with friction in Fig. 7. The permeability change Δk has a positive correlation with concurrently

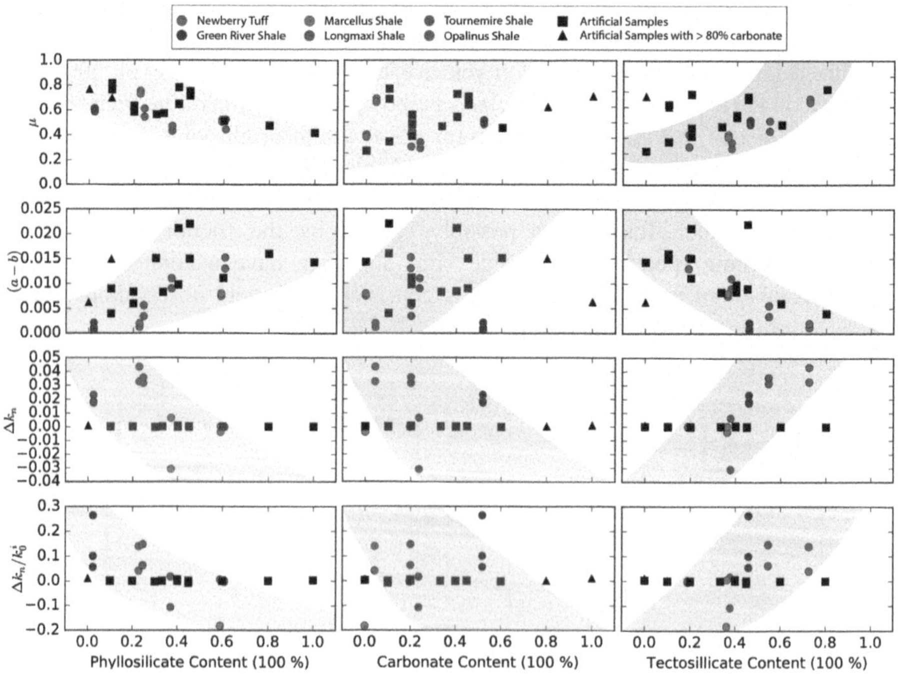

Fig. 6. Effect of mineral composition on friction response (*i.e.*, μ and $(a - b)$) and transient permeability evolution (*i.e.*, Δk_n and $\Delta k_n/\Delta k_0^i$) in response to shear velocity change. (a) Effect of phyllosilicate content; (b) Effect of carbonate content; (c) Effect of tectosilicate content

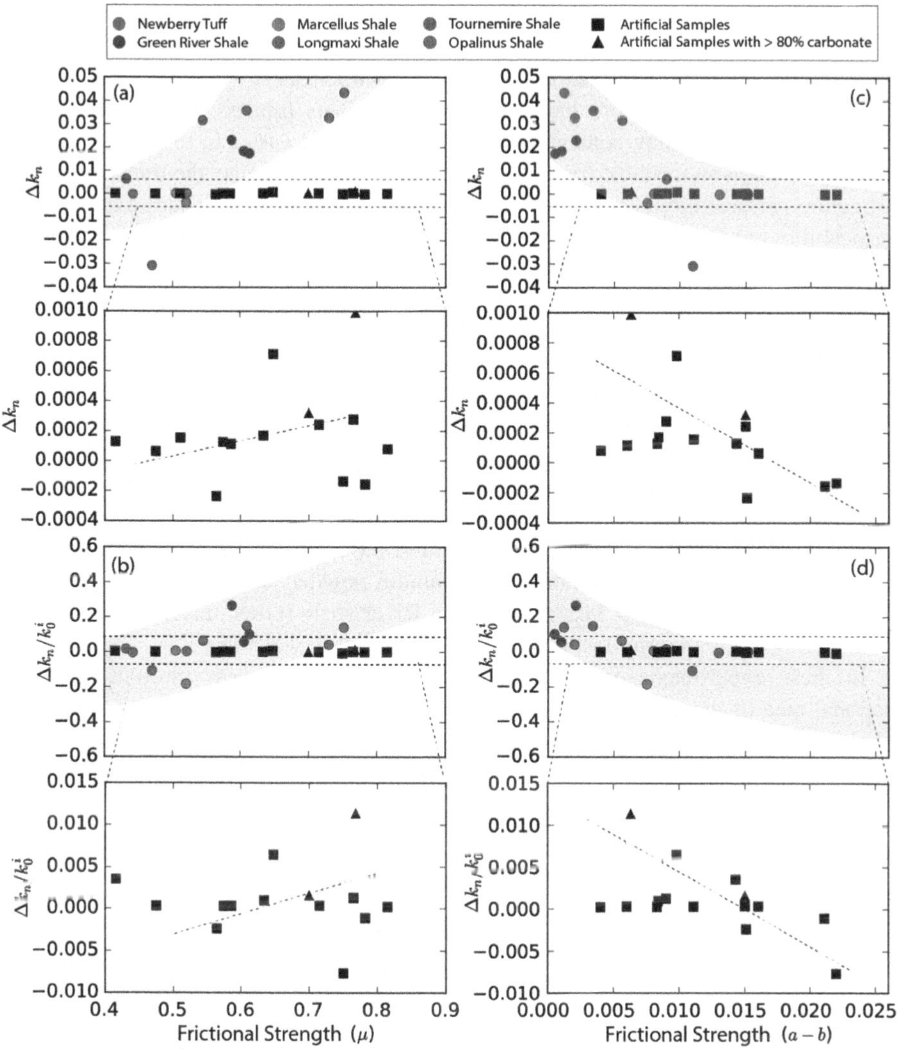

Fig. 7. (a) and (b) Correlation between frictional strength (μ) and permeability evolution (Δk_n and $\Delta k_n / \Delta k_0^i$); (c) and (d) Correlation between frictional stability ($a - b$) and permeability evolution (Δk_n and $\Delta k^i / \Delta k_0^i$).

measured frictional strength μ but a negative correlation with the corresponding frictional stability ($a - b$). This intrinsic linkage of friction and permeability change is directly determined by the asperity contact state and the material properties (*e.g.*, mechanical and swelling) that control the mechanical behaviors of fracture asperities. However, it is worth noting that the magnitude of permeability change in the natural samples is much larger than that of the artificial samples (shown as the solid black symbols in Figs. 6 and 7) – this is due to the very distinct difference in the surface textures.

In summary, with known mineralogical compositions comprising the fracture, the frictional strength and stability of fractures can be estimated. Shear failure is less likely to occur for fractures with higher content of tectosilicates. However, once failure initiates, the fracture is more likely slip unstably. This process is opposite that for fractures with higher clay content - where the fracture is easier to reactivate and will slip stably. When an unstable fracture slides at an accelerating rate, the transient change in fracture permeability can be speculated – those richer in tectosilicates exhibit larger permeability enhancement.

3 Healing and Sealing in the Seismicity-Permeability Cycle

We conducted fluid through-flow experiments with independent application of confining pressure, shear stress and pore pressure (Fig. 8). A double direct shear configuration is accommodated in a triaxial pressure cell. A prismatic rock coupon is sandwiched between half-cylindrical cores. This configuration enables concurrent and continuous measurement of both permeability and friction throughout static and dynamic motion. Experimental variables include surface roughness (ground with #150 grit (rougher) and #600 grit (smoother) aluminum powder) and mineralogy (Westerly granite (WG, tectosilicate > 90%) and Green River shale (GRS, tectosilicate ∼ 46%, carbonate ∼ 52% and phyllosilicate ∼ 2% (Fang et al. 2017))).

In these experiments, permeabilities are calculated for flow through the cross sectional area of the whole sample (cylinder + coupon), i.e.

$$k = \frac{\mu L}{A}\frac{Q}{\Delta P} \tag{6}$$

where μ is fluid viscosity (8.9×10^{-4} Pa · s), L flow path length (23 mm), A is cross sectional area of the sample perpendicular to the flow path (4.71×10^{-4} m^2), Q is flow rate and ΔP is the pressure difference across the sample.

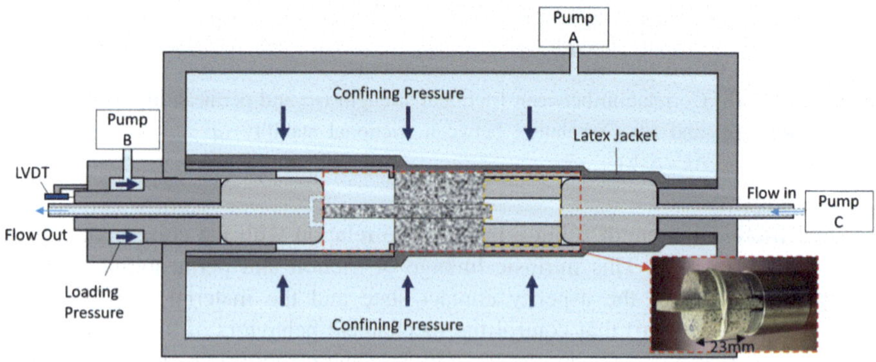

Fig. 8. Experimental configuration. Servo pumps control confining stress (pump A), loading pressure (pump B) and flow pressure difference (pump C). Bottom right picture in (a) shows the Westerly granite sample used in this experiment.

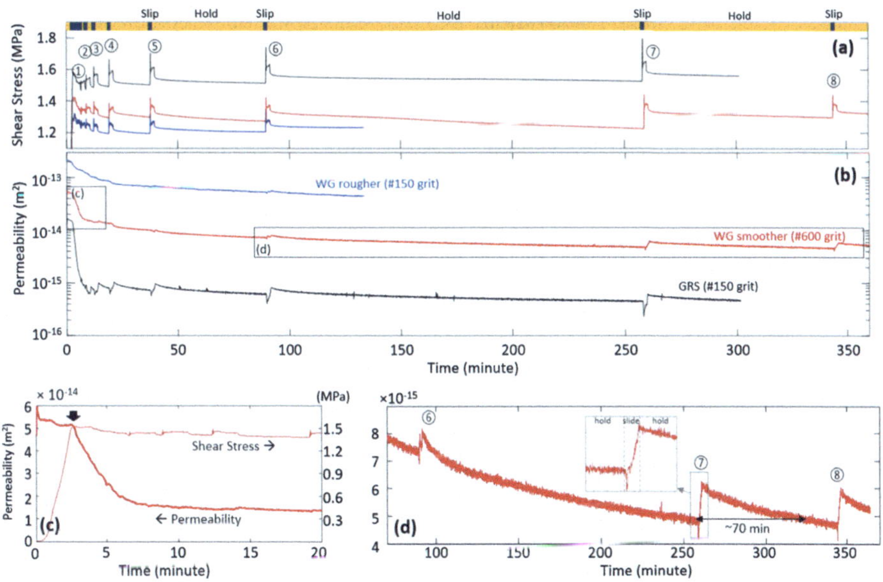

Fig. 9. Friction and permeability response during slide and hold experiments. Each color denotes rock and surface roughness: blue: westerly granite rougher sample (#150 grit), red: WG smoother (#600 grit), and black: Green River shale (#150 grit). (a and b) Shear stress and permeability response during the entire experiment.

Figure 9 shows shear stress and flow rate responses during slide-hold-slide experiments extending for ~6 h. Periods of slip and intervening hold are denoted by the top bar of Fig. 9a with experiments exhibiting typical rate-state frictional response. Applied shear stresses decline during the hold then increase sharply to a peak during reactivation before stabilizing following rate and state friction (Marone 1998). The magnitude of the peak stress increases with the increased duration of the prior hold period, representing time-dependent frictional healing.

Early time response in Fig. 9b shows the decline rate in permeability is anomalously large during the initial shearing-in period for all samples. A zoomed-in view of the initial decline (Fig. 9c) shows that this initial permeability reduction immediately follows the initiation of shear slip (i.e., driven by slip) and the decline rate is significantly reduced after a few millimeters of slip. Normalized permeability reduction during this initial period is greater for weaker rock (GRS) than for stronger/harder rocks (WG) and with smoother rather than rougher surfaces (see Fig. 9b early time). These observations, together with the observed production of comminution products (shale) during the experiment (Fig. 1c) suggest that the initial strong permeability reduction is mainly as a result of comminution/compaction and wear products from the fracture surfaces. This is especially dominant on fresh artificial samples which may not necessarily represent natural faults and fractures.

Further reactivations following incremented durations of static loading (hold periods) show significant permeability enhancements. Figure 9d highlights the permeability responses of the later stage reactivations and inter-slip holds of smoother WG (period identified by rectangles in Fig. 9b). This plot clearly identifies a cyclic repetition of permeability destruction (inter-slip sealing) and then creation (shear permeability enhancement). For instance, permeability continuously decreases during the 10000 s hold between slips ⑥ and ⑦ yielding a permeability reduction of ~37.5% (from $\sim 8 \times 10^{-15}$ m^2 to $\sim 5 \times 10^{-15}$ m^2). In the following slip ⑦, permeability increases by ~25% within 1 mm of shear deformation (<5% of sample length) before permeability again decreases at the conclusion of the slip phase. It takes ~70 min to recover to the original permeability prior to the slip ⑦.

Permeability enhancements resulting from various hold durations are shown in Fig. 10a (absolute increase Δk) and Fig. 10b (normalized increase, $\Delta k/k_0$). The larger ultimate enhancements scale with increasing roughness and hardness, implicating the essential role of shear dilation. The magnitude of absolute permeability enhancement (Fig. 10a) is largest with stiffer/stronger/rougher granite and smallest with weaker/smoother shale. Conversely, the magnitude of normalized permeability enhancement (Fig. 10b) is largest with shale and smallest with rougher granite. This behavior is substantially implied by the cubic law. Although shear aperture increase is largest with a rougher/hard surface, normalized permeability enhancement can be smallest due to the large initial aperture.

The permeability enhancement of westerly granite significantly increases with an extended duration of healing in both absolute (Fig. 10a) and relative magnitudes (Fig. 10b), implying, especially for hard rock, that the magnitude of shear permeability enhancement is dependent on the pre-slip healing/sealing. For the case of the smoother granite (Fig. 10a), no shear permeability enhancement appears after only short periods of healing, ~1% of enhancement induced after 1000 s of healing and ~27% after 10000 s. Permeability enhancement appears approximately log-linear with pre-slip hold duration.

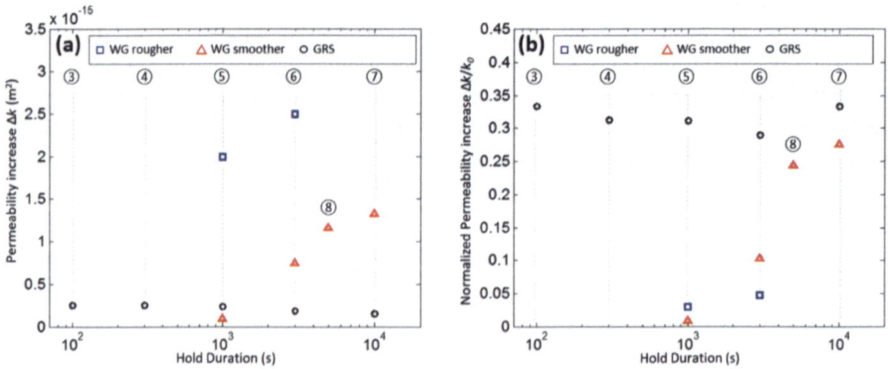

Fig. 10. Sequence of permeability evolution with hold duration. Definition of permeability enhancement Δk and initial permeability k_0 are shown in (a) and (b). The colors and symbols denotes rock and roughness: WG rough (blue square), WG smooth (red triangle), and GRS (black circle).

4 Friction-Stability-Permeability Modelling

Where supercritical CO_2 is injected into aquifers, the subsequent dissolution of CO_2 into the brine may result in substantial reaction and chemical transformation of the minerals comprising the reservoir and caprock, disturb the stress field and alter the permeability. These combined effects may result in weakening of pre-existing faults, transformation of permeability and the triggering of induced-seismicity, via weakening. Pristine and CO_2-altered rocks from the Crystal Geyser fault (Major et al. 2014) define the evolution of Fe-coating (hematite) into altered tribological films (goethite) surrounding the grains with different strength and stability properties. We use this transformed microstructural form to examine ensemble strength and structural stability.

4.1 Numerical Method

In this study, we apply the Distinct Element Method (DEM, Cundall and Strack 1979) to study the shear strength of faults with Fe-coating (Guo and Morgan 2004; Abe and Mair 2009; Rathbun et al. 2013; Ferdowsi et al. 2014). The weakening mechanism at the grain-grain level is simulated by a DEM numerical biaxial shear model (Wang and Elsworth 2016; Wang et al. 2017), in which the weakening effects of trace mounts of talc in a quartz matrix is successfully reproduced. Although available DEM models can simulate the shear strength evolution of fault gouge, few DEM models simulate fault gouge with a microstructural coating, i.e. a frictionally strong skeleton with a frictionally weak coating. We investigate the shear strength evolution of fault gouge materials (quartz) with Fe-coating both before CO_2 alteration (hematite) and after CO_2 alteration (goethite). Specifically, we investigate: (1) the effect of CO_2 alteration of the shear strength of affected faults; (2) the effect of relative coating concentration on the shear strength of affected faults; and (3) the effect of normal stress on the shear strength of CO_2 altered faults. Simulation results show weakening mechanisms of long-term CO_2 alteration on the shear strength of affected faults.

The DEM model (PFC2D, ITASCA Consulting Group) accommodates a scaled representation of the biaxial direct shear test configuration (Mair and Marone 1999), in which the gouge samples are confined and sheared between saw-tooth grooved platens. Specifically, in a full scale biaxial shear experiment, two pieces of synthetic fault gouge samples are sheared in a three-piece apparatus. The schematic of the apparatus is shown in Fig. 11(a). During shear experiments, normal stress is applied to the two side blocks and the center block is driven downwards by a hydraulic piston to apply shear stress. A symmetric half of the biaxial shear experimental configuration is adopted in the DEM model to improve computational efficiency (Fig. 11(b), (d)). The numerical model consists of two shear platens with saw-tooth grooved surfaces. Fault gouge materials are represented as an assembly of circular particles confined between the platens (Fig. 11(c), (e)). The size of the sample (including platens) is approximately 2 cm in length (horizontally) and 4.5 mm in width (vertically). This value may differ slightly for different mineral analog configuration. The skeleton of the confined gouge material is a quartz analog, the Fe-coating materials are a hematite analog (before CO_2 alteration) and a goethite analog (after CO_2 alteration). The radii for the quartz analog in the assembly ranges from approximately 50 to 125 μ in a normal distribution.

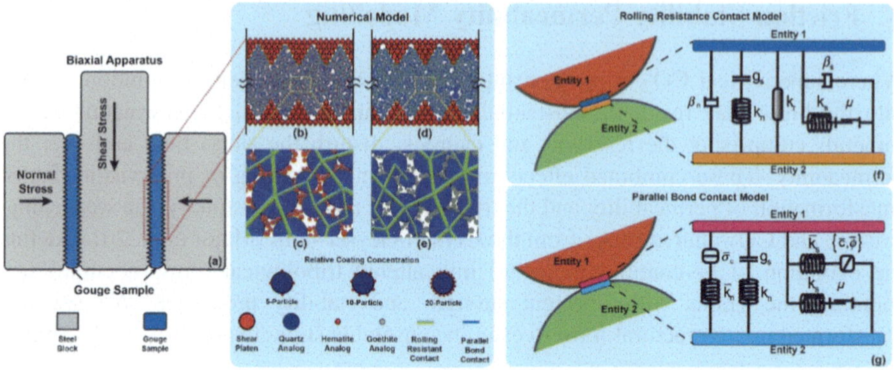

Fig. 11. Schematic of biaxial apparatus (a) (not to scale); DEM model configuration of unaltered gouge sample (b), and its zoomed-in view (c), linear parallel bond model is applied on coating contacts; DEM model configuration of altered gouge sample (d), and its zoomed-in view (e), rolling resistance contact model is installed on coating contacts; relative coating concentration configuration of 5-, 10-, and 20-particles is implemented; (f) layout and mechanical components of rolling resistance linear-elastic contact model; (g) linear parallel bond contact model.

The radii for the Fe-coating range from 5 to 20 μ. Hematite coating particles are attached to the skeleton particles and other coating particles by parallel bonds (Potyondy and Cundall 2004) to simulate cementation; goethite coating is attached to the skeleton particles by parallel bonds, however, a rolling resistance linear-elastic contact model (Iwashita and Oda 1998; Ai et al. 2011; Wensrich and Katterfeld 2012; Jiang et al. 2015) is applied between the goethite coating particles. Additionally, in order to investigate the effect of the amount of coating material on the shear strength evolution of faults, relative concentration of the coating is introduced. The relative concentration is defined as the configuration of an individual skeleton particle with its coating particles (Fig. 11 lower right part). In this study, relative concentrations of 5-particles, 10-particles, and 20-particles are used. Numerical shear experiments are conducted as shown in Fig. 11 until shear displacement reaches 2000 μm.

A rolling resistance linear-elastic contact model is implemented between gouge particle contacts (Fig. 11(f), (g)). The contact model consists of two sets (normal direction and shear direction) of parallel aligned spring and dashpots. A slider is installed in the shear direction to simulate sliding behavior according to a Coulomb friction law. Additionally, a moment resistant component is added to the contact model to counteract rolling. A parallel bond contact model is implemented between quartz and coating analogs. The contact model features bonds in both normal and shear direction to resist normal and shear forces. The bonds will break if pre-set strengths are exceeded. After bonding breakage, contact will evolve by a linear-elastic contact model if the contact is still active.

4.2 Model Configuration

CO_2 altered rocks show significantly reduced fracture roughness, which is plausibly caused by a change in mineralogy and microstructure due to CO_2 alteration. The dissolution of hematitc cementation and precipitation of the low frictional goethite coating may cause significant weakening in the fault zone. Numerical shear experiments are conducted on end-member gouge samples (unaltered/altered by CO_2).

Unaltered samples consist of quartz analog as the skeleton and hematite analog as the cementation. The relative coating concentration of the hematite analog is either 5-particles, 10-particles or 20-particles, respectively for each numerical simulation. The quartz analog skeleton is bonded by a hematite coating. If the bonding is broken due to shear failure, the contact will evolve following a linear-elastic contact model. Normal stress (5, 10, or 15 MPa) is maintained throughout shear experiment. The model configuration before and after shear is shown in Fig. 12(a), (b).

The microscopic imaging analysis shows that after CO_2 alteration, the hematite cementation is dissolved and the goethite coating is found surrounding the skeletal grains (Major et al. 2014). In the numerical model, the altered samples consist of quartz particles as the skeleton with goethite as coating particles. The relative concentration of the goethite coating is either 5-particles, 10-particles, or 20-particles, respectively for each numerical simulation in correspondence with the unaltered samples with hematite cementation. A rolling resistance linear-elastic contact model is installed between goethite coating particles and quartz particles. The normal stress (5, 10, or 15 MPa) is maintained throughout the shear experiment. The model configuration before and after shear is shown in Fig. 12(c), (d).

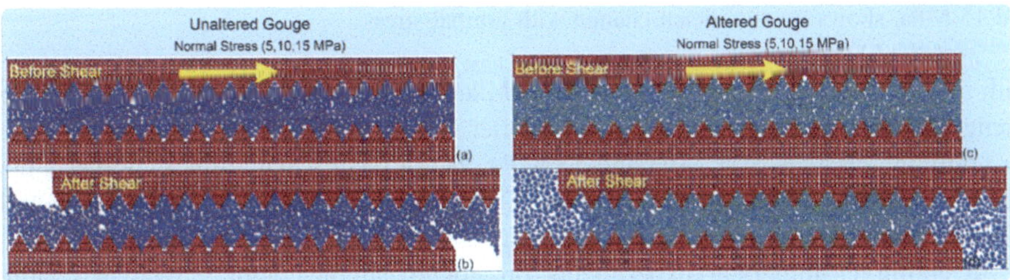

Fig. 12. DEM model configuration with hematite cementation: (a) before shear, and (b) after shear; DEM model configuration with goethite coating: (c) before shear, and (d) after shear.

4.3 Results

The bulk shear strength can be represented by the frictional resistance of the system. The evolution of the coefficient of friction during each test is plotted against shear displacement. The coefficient of friction is calculated by the ratio of the resultant shear stress and confining (normal) stress. Shear strength of the gouge sample before and after CO_2 alteration is hypothesized to be different, i.e. the altered gouge may show lower shear strength due to the dissolution of hematite cementation and precipitation of the low frictional goethite coating. Figure 13(a) (b) (c) show the evolution of friction coefficient of unaltered and altered gouge materials under a confining stress of 10 MPa,

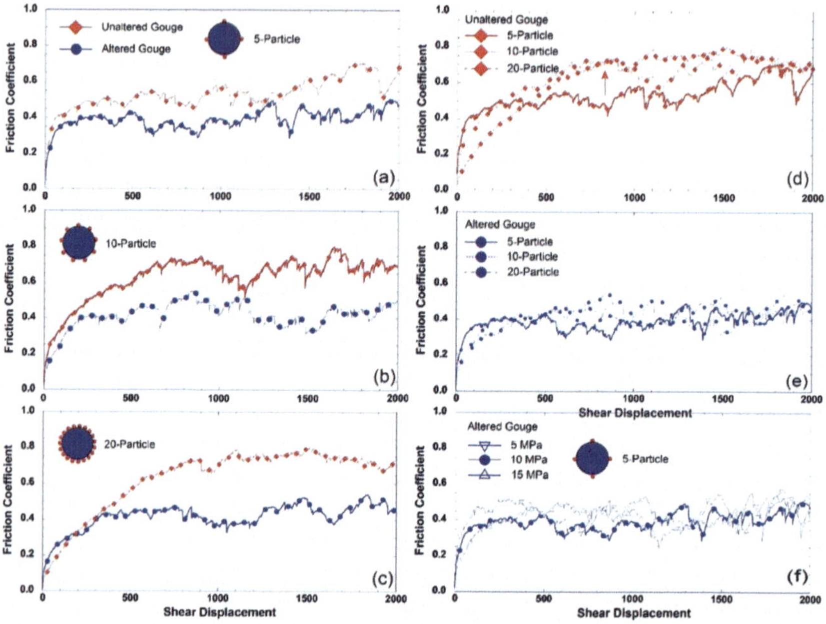

Fig. 13. Friction evolution of unaltered/altered gouge under confining stress of 10 MPa: (a) gouge with relative coating concentration of 5-particles; (b) 10-particles; (c) 20-particles; (d) friction evolution of altered gouge with coating configuration of 5-, 10-, and 20-particles; (e) friction evolution of unaltered with coating configuration of 5-, 10-, and 20-particles; (f) Friction evolution of the altered gouge (5-particle configuration) under normal stress of 5, 10, and 15 MPa, shows no significant change with normal stress.

with relative coating concentration of 5-, 10-, and 20-particles. It is clear that the shear strength is reduced from $\sim$0.60 to $\sim$0.4 after CO_2 alteration. This observation holds for confining stresses of 5 MPa and 15 MPa. These results match the hypothesis that the shear strength of gouge material after CO_2 alteration is lower than the shear strength of the gouge material before CO_2 alteration.

In addition, numerical experiments on gouge samples with a relative coating concentration of 5-particles, 10-particles, and 20-particles are conducted. The evolution of friction coefficient is plotted against shear displacement in Fig. 13(d), (e). It is shown that for unaltered gouge, the shear strength increases (from $\sim$0.65 to $\sim$0.8) with higher relative coating concentration, however, there is no significant change (from $\sim$0.40 to $\sim$0.45) in the shear strength of the altered gouge material with three different relative coating concentrations in this study.

The observations show that unaltered gouge features a higher frictional strength with a higher concentration of hematite coating. This is plausibly due to the increase in the number of parallel bond contacts thus resulting in an increase in shear strength.

For altered gouge material, there is no significant change in shear strength relative to the coating concentration. This is possibly due to the dominant effect of low frictional goethite coating over the quartz skeleton particles. Slip occurs at contacts between coating particles or contacts between coating particles and skeletal particles, at which the shear strength is determined by the frictionally weaker component (coating particles). Therefore, the goethite coating governs the shear strength of the gouge material, even though there is only a trace amount of goethite coating in the gouge (5-particle).

The evolution of shear strength of CO_2 altered gouge material under a confining stress of 5, 10, and 15 MPa is shown in Fig. 13(f). The shear strength of the CO_2 altered gouge materials does not show a significant change with normal stress. This result indicates that the dominant effect of the frictionally weak goethite coating is independent of the change of normal stress in this study. This may further be attributed to the large weakening effect of goethite coating on the shear strength of the faults.

5 Conclusions

Seismicity-permeability relationships can be linked via the dynamic evolution of asperities and contact state of fracture surfaces in which the mineralogical composition defines the mode of seismicity and control the mechanical strength of fracture asperities that eventually affect permeability evolution.

Permeability evolution can be broadly linked to mineralogy and the impact of deformation mode and rheology of individual fractures. This is true both for the evolution of deformation as velocity weakening or strengthening, and also related to the role of healing.

Healing exerts a significant influence on the evolution of permeability, post reactivation. Short inter-seismic times infer compaction after reactivation that evolves to successively greater magnitudes of dilation for longer interseismic periods. This response is inferred through observations of permeability evolution following reactivation.

Microscopic imaging of unaltered vs. CO_2-altered sandstone samples from Crystal Geyser, Utah show that the original hematite cementation is reacted with CO_2, forming goethite precipitated on skeletal grain surfaces as a frictionally weak coating. Mechanical tests show that the altered quartzite samples exhibit a lower fracture toughness than unaltered samples. The DEM model simulates direct-shear experiments on fault gouge samples with hematite/goethite analog as coating. These fault gouge samples reconstruct CO_2 unaltered/altered gouge with idealizations and a hypothesis, i.e. CO_2 altered gouge may have a lower shear strength. Numerical models with relative coating concentrations of 5-, 10-, and 20 particles of unaltered/altered gouge materials under confining stress of 5, 10 and 15 MPa are constructed and sheared. The following conclusions can be drawn from the observations of the numerical simulation results:

1. The shear strength of faults associated with goethite coating are weakened.
2. The weakening effect of goethite coating does not depend on the concentration of the coating particle once the goethite-dominated effect is established.
3. The coating-dominated weakening effect is independent of applied normal stress, implying that coating induced weakening can occur not only at shallow depth but also at greater depths in the Earth crust.

These numerical observations reveal a possible new mechanism of weakening in natural faults, which matches the hypothesis proposed in this study, i.e. trace amounts of low frictional coating can cause significant weakening in shear strength of faults. However, the simulation results are specific to the sample analogs in this study. Laboratory study is needed to validate the numerical results. The modeling results support the argument that the shear strength of faults depends significantly on the mineralogical composition and microstructure, while showing the applicability of DEM modeling in simulating the shear strength evolution in mineralogically complex faults.

Acknowledgements. This work is the result of support provided by DOE Grant DE-FE0023354. This support is gratefully acknowledged.

References

Abe, S., Mair, K.: Effects of gouge fragment shape on fault friction: New 3D modelling results. Geophys. Res. Lett. **36**(23), 2–5 (2009). https://doi.org/10.1029/2009GL040684

Ai, J., Chen, J.F., Rotter, J.M., Ooi, J.Y.: Assessment of rolling resistance models in discrete element simulations. Powder Technol. **206**(3), 269–282 (2011). https://doi.org/10.1016/j.powtec.2010.09.030

Anderson, J.G., Wesnousky, S.G., Stirling, M.W.: Earthquake size as a function of fault slip rate. Bull. Seismol. Soc. Am. **86**, 683–690 (1996)

Anderson, R.N., Zoback, M.D.: Permeability, underpressures, and convection in the oceanic crust near the Costa Rica Rift, eastern equatorial Pacific. J. Geophys. Res. **87**(B4), 2860 (1982)

Bos, B., Spiers, C.J.: Frictional-viscous flow of phyllosilicate-bearing fault rock: microphysical model and implications for crustal strength profiles. J. Geophys. Res. **107**(B2), 2028 (2002). https://doi.org/10.1029/2001JB000301

Candela, T., Brodsky, E.E., Marone, C., Elsworth, D.: Flow rate dictates permeability enhancement during flow pressure oscillations in laboratory experiments. J. Geophys. Res. **120**, 2037–2055 (2015). https://doi.org/10.1002/2014JB011511

Carpenter, B.M., Marone, C., Saffer, D.M.: Frictional behavior of materials in the 3D SAFOD volume. Geophys. Res. Lett. **36**(5), 1–5 (2009). https://doi.org/10.1029/2008GL036660

Collettini, C., Niemeijer, A., Viti, C., Marone, C.: Fault zone fabric and fault weakness. Nature **462**(7275), 907–910 (2009). https://doi.org/10.1038/nature08585

Cundall, P.A., Strack, O.D.L.: A discrete numerical model for granular assemblies. Géotechnique **29**(1), 47–65 (1979). https://doi.org/10.1680/geot.1979.29.1.47

Curtis, J.B.: Fractured shale-gas systems. AAPG Bull. **11**(11), 1921–1938 (2002). https://doi.org/10.1306/61EEDDBE-173E-11D7-8645000102C1865D

Dieterich, J.H.: Modeling of rock friction 1. Experimental results and constitutive equations. J. Geophys. Res. Solid Earth **84**(B5), 2161–2168 (1979). https://doi.org/10.1029/JB084iB05p02161

Elkhoury, J.E., Brodsky, E.E., Agnew, D.C.: Seismic waves increase permeability. Nature **441**, 1135–1138 (2006). https://doi.org/10.1038/nature04798

Ellsworth, W.: Injection-Induced earthquakes. Science **341**(6142), 142–149 (2013). https://doi.org/10.1126/science.1225942

Elsworth, D., Goodman, R.E.: Characterization of rock fissure hydraulic conductivity using idealized wall roughness profiles. Int. J. Rock Mech. Min. Sci. Geomech. Abstr. **23**(3), 233–243 (1986). https://doi.org/10.1016/0148-9062(86)90969-1

Faoro, I., Niemeijer, A., Marone, C., Elsworth, D.: Influence of shear and deviatoric stress on the evolution of permeability in fractured rock. J. Geophys. Res. **114**, B01201 (2009). https://doi.org/10.1029/2007JB005372

Fang, Y., Elsworth, D., Wang, C., Ishibashi, T., Fitts, J.P.: Frictional stability-permeability relationships for fractures in shales. J. Geophy. Res. Solid Earth **122**, 1760–1776 (2017). https://doi.org/10.1002/2016JB013435

Ferdowsi, B., Griffa, M., Guyer, R.A., Johnson, P.A., Marone, C., Carmeliet, J.: Three-dimensional discrete element modeling of triggered slip in sheared granular media. Phys. Rev. E Stat. Nonlinear Soft Matter Phys. **89**(4), 1–12 (2014). https://doi.org/10.1103/physreve.89.042204

Guglielmi, Y., Cappa, F., Avouac, J., Henry, P., Elsworth, D.: Seismicity triggered by fluid injection – induced aseismic slip. Science **348**(6240), 1224–1227 (2015). https://doi.org/10.1126/science.aab0476

Guo, Y., Morgan, J.K.: Influence of normal stress and grain shape on granular friction: results of discrete element simulations. J. Geophys. Res. B Solid Earth **109**(12), 1–16 (2004). https://doi.org/10.1029/2004JB003044

Ikari, M.J., Niemeijer, A.R., Marone, C.: The role of fault zone fabric and lithification state on frictional strength, constitutive behavior, and deformation microstructure. J. Geophys. Res. Solid Earth **116**(8), 1–25 (2011). https://doi.org/10.1029/2011JB008264

Im, K., Elsworth, D., Guglielmi, Y., Mattioli, G.: Geodetic imaging of thermal deformation in geothermal reservoirs - production, depletion and fault reactivation. J. Volcanol. Geoth. Res. **338**, 79–91 (2017). https://doi.org/10.1016/j.jvolgeores.2017.03.021

Ishibashi, T., Asanuma, H., Fang, Y., Wang, C., Elsworth, D.: Exploring the link between permeability and strength evolution during fracture shearing. In: Proceedings of the 50th US Rock Mechanics/Geomechanics Symposium, Houston, Texas (2016)

Iwashita, K., Oda, M.: Rolling resistance at contacts in simulation of shear band development by DEM. J. Eng. Mech. **124**(3), 285–292 (1998). https://doi.org/10.1061/(asce)0733-9399(1998)124:3(285)

Jiang, M., Shen, Z., Wang, J.: A novel three-dimensional contact model for granulates incorporating rolling and twisting resistances. Comput. Geotech. **65**, 147–163 (2015). https://doi.org/10.1016/j.compgeo.2014.12.011

Marone, C.: Laboratory-derived friction laws and their application to seismic faulting. Annu. Rev. Earth Planet. Sci. **26**, 643–696 (1998). https://doi.org/10.1146/annurev.earth.26.1.643

Mair, K., Marone, C.: Friction of simulated fault gouge for a wide range of velocities and normal stresses. J. Geophys. Res. **104**(B12), 28899–28914 (1999)

Majer, E.L., Baria, R., Stark, M., Oates, S., Bommer, J., Smith, B., Asanuma, H.: Induced seismicity associated with Enhanced Geothermal Systems. Geothermics **36**(3), 185–222 (2007). https://doi.org/10.1016/j.geothermics.2007.03.003

Major, J.R., Eichhubl, P., Dewers, T.A., Urquhart, A.S., Olson, J.E., Holder, J.: The effect of CO2-related diagenesis on geomechanical failure parameters: fracture testing of CO2-altered reservoir and seal rocks from a natural analog at Crystal Geyser, Utah. ARMA 14-7463 (2014)

McGarr, A., Simpson, D., Seeber, L.: Case histories of induced and triggered seismicity. Int. Geophys. **81**, 647–661 (2002). https://doi.org/10.1016/S0074-6142(02)80243-1

Moore, D.E., Lockner, D.A.: Frictional strengths of talc-serpentine and talc-quartz mixtures. J. Geophys. Res. Solid Earth **116**(B01403), 1–17 (2011). https://doi.org/10.1029/2010JB007881

Moore, D.E., Rymer, M.J.: Talc-bearing serpentinite and the creeping section of the San Andreas fault. Nature **448**(7155), 795–797 (2007). https://doi.org/10.1038/nature06064

Niemeijer, A., Marone, C., Elsworth, D.: Fabric induced weakness of tectonic faults. Geophys. Res. Lett. **37**(3), 1–5 (2010). https://doi.org/10.1029/2009GL041689

Niemeijer, A.R., Spiers, C.J.: Velocity dependence of strength and healing behaviour in simulated phyllosilicate-bearing fault gouge. Tectonophysics **427**(1–4), 231–253 (2006). https://doi.org/10.1016/j.tecto.2006.03.048

Peng, Z., Gomberg, J.: An integrated perspective of the continuum between earthquakes and slow-slip phenomena. Nat. Geosci. **3**(9), 599–607 (2010). https://doi.org/10.1038/ngeo940

Potyondy, D.O., Cundall, P.A.: A bonded-particle model for rock. Int. J. Rock Mech. Min. Sci. **41**(8 Spec. Iss.), 1329–1364 (2004). https://doi.org/10.1016/j.ijrmms.2004.09.011

Rathbun, A.P., Renard, F., Abe, S.: Numerical investigation of the interplay between wall geometry and friction in granular fault gouge. J. Geophys. Res. Solid Earth **118**(3), 878–896 (2013). https://doi.org/10.1002/jgrb.50106

Ruina, A.: Slip instability and state variable friction law. J. Geophys. Res. (1983). https://doi.org/10.1029/JB088iB12p10359

Samuelson, J., Elsworth, D., Marone, C.: Shear-induced dilatancy of fluid-saturated faults: experiment and theory. J. Geophys. Res. **114**, B12404 (2009). https://doi.org/10.1029/2008JB006273

Schmidt, D.A., Bürgmann, R., Nadeau, R.M., D'Alessio, M.: Distribution of aseismic slip rate on the Hayward fault inferred from seismic and geodetic data. J. Geophys. Res. B Solid Earth **110**(B08406), 1–15 (2005). https://doi.org/10.1029/2004JB003397

Scholz, C.H.: Earthquakes and friction laws. Nature **391**(6662), 37–42 (1998). https://doi.org/10.1038/34097

Tsang, Y.W., Witherspoon, P.A.: Hydromechanical behavior of a deformable rock fracture subject to normal stress. J. Geophys. Res. **86**(B10), 9287–9298 (1981). https://doi.org/10.1029/JB086iB10p09287

Witherspoon, P.A., Wang, J.S.Y., Iwai, K., Gale, J.E.: Validity of cubic law for fluid flow in a deformable rock fracture. Water Resour. Res. **16**(6), 1016–1024 (1980). https://doi.org/10.1029/WR016i006p01016

Walsh, F.R., Zoback, M.D.: Oklahoma's recent earthquakes and saltwater disposal. Sci. Adv. 1–9 (2015). https://doi.org/10.1126/sciadv.1500195

Wang, C., Elsworth, D.: Numerical investigation of the effect of frictionally weak minerals on shears strength of faults. ARMA 16-576 (2016)

Wang, C., Elsworth, D., Fang, Y.: Influence of weakening minerals on ensemble strength and slip stability of faults. J. Geophys. Res. Solid Earth **122**, 7090–7110 (2017). https://doi.org/10.1002/2016JB013687

Wang, W., Scholz, C.: Micromechanics of the velocity and normal stress dependence of rock friction. Pure. appl. Geophys. **143**, 303 (1994). https://doi.org/10.1007/BF00874333

Wensrich, C.M., Katterfeld, A.: Rolling friction as a technique for modelling particle shape in DEM. Powder Technol. **217**, 409–417 (2012). https://doi.org/10.1016/j.powtec.2011.10.057

Xue, L., Li, H.-B., Brodsky, E.E., Xu, Z.-Q., Kano, Y., Wang, H., Mori, J.J., Si, J.-L., Pei, J.-L., Zhang, W., Yang, G., Sun, Z.-M., Huang, Y.: Continuous permeability measurements record healing inside the Wenchuan earthquake fault zone. Science **340**, 1555 (2013). https://doi.org/10.1126/science.1237237

Unsaturated Soils

A Sample Preparation Method Enhancing Shear Strength for Remolded Unsaturated Soil

Yitong Zhao and Fuguo Tong(✉)

China Three Gorges University, Yichang, China
tfg@ctgu.edu.cn

Abstract. Remolded soil is usually substituted for undisturbed soil in shear tests to measure the strength of natural soil. However, the traditional remolding process of unsaturated soil may lead to a wide difference on the strength of soil and therefore fail to represent the strength properties of natural soil. Based on the traditional remolding methods, this paper introduces an experimental method which could enhance the shear strength of remolded soil significantly by readjusting the water distribution and restoring the matric suction between soil layers. Compared to the traditional compacted method, the improved method could effectively raise the cohesion of unsaturated soil, which enables the remolded soil fully to retain its strength and hence to be applied accurately to predict the strength of natural soil.

Keywords: Shear strength · Unsaturated soil · Remolded soil
Matric suction · Cohesion

1 Introduction

The shear strength of soil could be measured through laboratory tests or in-situ tests, while considering the large cost and the inconvenience for the research under different conditions, laboratory tests are more likely to be adopted. For sand and gravel, remolded soil is usually substituted for undisturbed soil since its strength could hold the value basically after the remolding process. The clay remolded by the traditional compacted method, by contrast, with finer granularity and complex structures, presents a vast strength difference from natural soil under non-saturation state (mostly lower than natural soil, according to several previous studies), which forces researchers to use undisturbed or saturated specimen for tests. However, the undisturbed soil is often hard to obtain and the divergence of the sample is also not a negligible issue. The stress-strain characteristics of the saturated soil are apparently unable to determine the properties of real natural unsaturated soil as well. As a result of the absence of reliable values, it is imperative to search for an experimental method which enables remolded unsaturated soil to fully retain its strength and to be applied to predicting the strength of natural soil.

The strength properties of soil are largely dependent on the particle structure, water content and mineral composition. Despite the disturbance of the structure caused by the remolding process, the shear tests' results of remolded sand are still qualified to be the reference for the strength of natural sand or saturated clay respectively, which proved

L. Hu et al. (Eds.): GSIC 2018, *Proceedings of GeoShanghai 2018 International Conference: Multi-physics Processes in Soil Mechanics and Advances in Geotechnical Testing*, pp. 23–29, 2018.
https://doi.org/10.1007/978-981-13-0095-0_2

that the macro-structural changes are not the key reason to the wide difference between remolded soil and natural soil. The relation between water content and shear strength has been investigated by many previous studies. Fredlund et al. [1, 4] had performed a lot of measurements on natural clay and a power law was obtained, that is, the shear strength of clay could be consequently reduced with increasing water content.

But for clay remolded by compacted method, as the water content rises, the shear strength would initially increase and soon dropped after reaching its maximum. Lin and Sun et al. [2, 3] found that the inadequate water-soil interaction was the main factor led to the variation, or to put it another way, the matric suction of natural soil failed to be restored through the remolding process. Expanding upon this idea, the authors aimed at emphasizing the effect of the matric suction on soil strength and put forward an improved remolding method which could recover the suction and thus enhance the shear strength significantly during laboratory tests.

In the present paper, several triaxial-shearing tests are performed on silty clay originated from YiChang, HuBei Province. The experimental results of undisturbed soil, traditional remolded specimens as well as the improved specimens, all in unsaturated conditions, are compared to assess the feasibility of the new sample preparation method.

2 Materials and Methods

To analyse the effects of different remolding methods on soil strength, a series of shear tests were designed to compare the properties of undisturbed soil, compacted specimens and specimens prepared by the improved method.

All soil samples were collected from Three Gorges area, which were identified as silty clay. The sampling depth is between 6 to 8 m. The relative density is 2.76 g/cm^3, the maximum natural saturation is around 66%, the optimum water content is measured as 13.7%.

The undisturbed soil was cut by a thin-walled split mould with the diameter of 39.1 mm and height of 80 mm. A total of 15 specimens were prepared and then divided in five groups. All specimens were dried by drying ovens to the required saturation of 25%, 35%, 45%, 55%, respectively.

The preparation of compacted specimens is strictly following The Standard for Soil Test Method [5]. Disturbed soil was dried, ground and sieved, then mixed with water to make moist paste with saturation of 25%, 35%, 45%, 55%, 65%, 80%, respectively. After the distribution of moisture became even, the soil was compacted in four layers to obtain 6 groups of specimens in line with the dry density of undisturbed soil.

Similar to the previous preparation of traditional remolded soil, the disturbed soil was also dried and ground. To obtain the best compact effect, the dry soil was mixed into wet soil by 50% saturation degree and compacted into specified dimensions. The samples were initially saturated and then dried gradually into designated saturation. After the moisture run through the soil pores fully, 18 columns with varying saturation (35%, 45%, 55%, 65%, 80%) were sheared at a constant rate of 0.5 mm/min under the confining pressure ranged from 100 kPa to 300 kPa, the other 6 columns were also set as alternatives.

The instrument used in shear tests is the SLB-1 stress-strain tri-axial shearing apparatus, which is able to control the shearing tests under equal stress or strain conditions (Fig. 1).

Fig. 1. SLB-1 stress-strain tri-axial shear apparatus.

3 Results and Discussion

3.1 The Comparison of Shear Indexes

It is illustrated that the cohesion of undisturbed soil decreases as the water content rises, which agree with the previous studies on the relationship between soil strength and moisture content (Fig. 2). With the rising water content, the water membrane becomes thicker and impair the cohesion force among soil particles. While friction angle shows a slightly decline as the saturation increases, which may related to the nature of the silty clay used in this work. It can be predicted that the lubrication effect of water would continue to reduce the friction angle of the soil.

For remolded specimens, it can be easily observed that the cohesion of compacted soil decreases not before the saturation degree exceed 45%. This is caused by the traditional remoulding method that directly compact soil paste (dried soil and certain amount of water) in several layers. Shear failure can be easily caused by the weak suction between soil layers and particles when the water content is very low, especially at the interfaces. As the water content goes up, the suction force becomes stronger which contribute to an increasing cohesion and shear strength of the soil, but the strength in this period is still relatively low. After the saturation rises above a certain threshold (45% in this experiment), the bound water membrane becomes thicker, the cohesion of soil could be impaired by the decreasing effective stress, accounting for the subsequent down trend of the curve. However, the cohesion of the sample used new remolding method decreases linearly as the saturation of soil increases, which is consistent with the corresponding relationship of undisturbed soil. Such differences could be explained as follows.

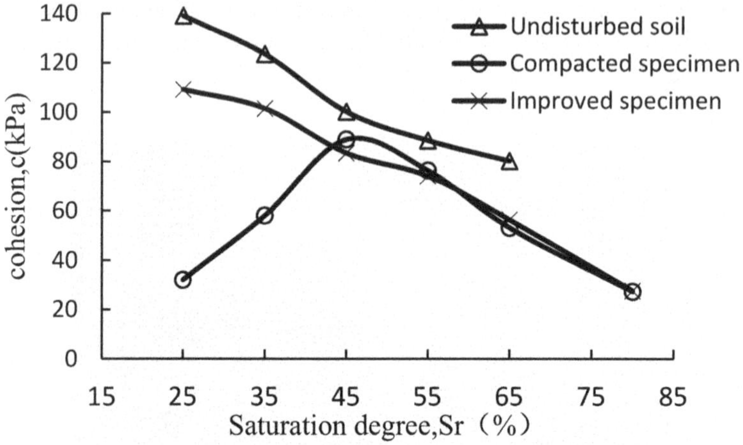

Fig. 2. The cohesion of different specimens under varied saturation degrees.

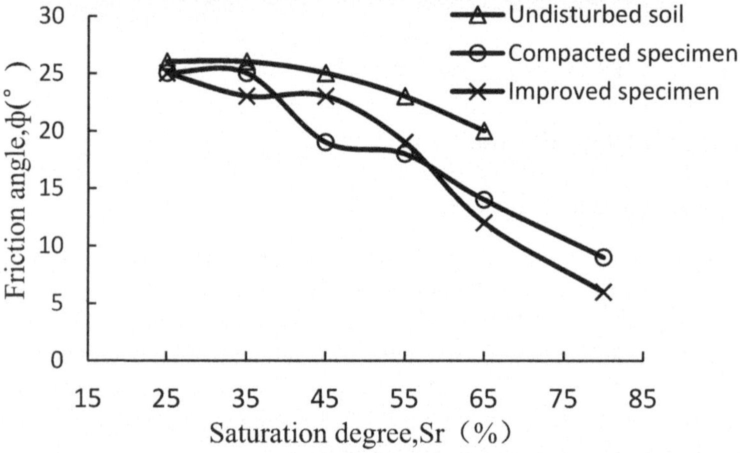

Fig. 3. Trend of friction angle obtained by traditional method and improved method with variation of saturation.

Compared with the traditional sample making, the improved method achieved the best compacting results by making the sample at the optimum moisture content, which insures closer connections between soil particles. Through the process of saturation and slow drying, samples successfully keep its matric suction to the full extent by simulating the way that natural soil loses moisture, which means, a more evenly distribution of water in the columns. Therefore, the cohesion of the samples used improved method shows similar variation with undisturbed soil. The improved remolding method is proved to be able to enhance the shear strength of unsaturated clay effectively from the above analyses.

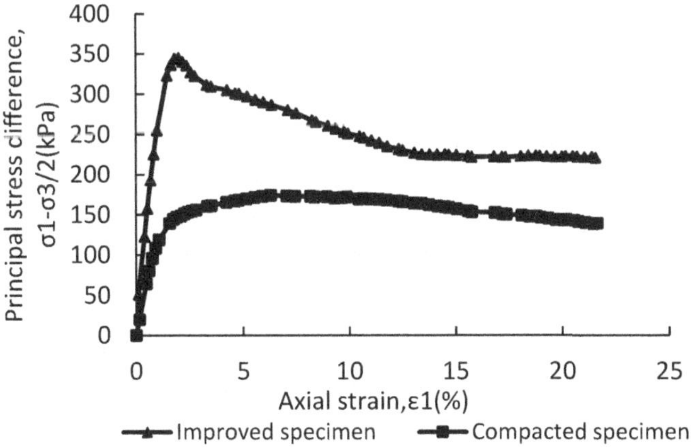

(a) 100 kpa, Sr=25%

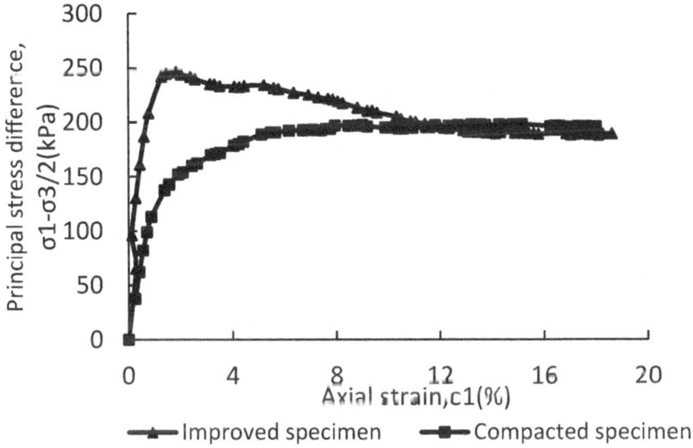

(b) 100kpa, Sr=35%

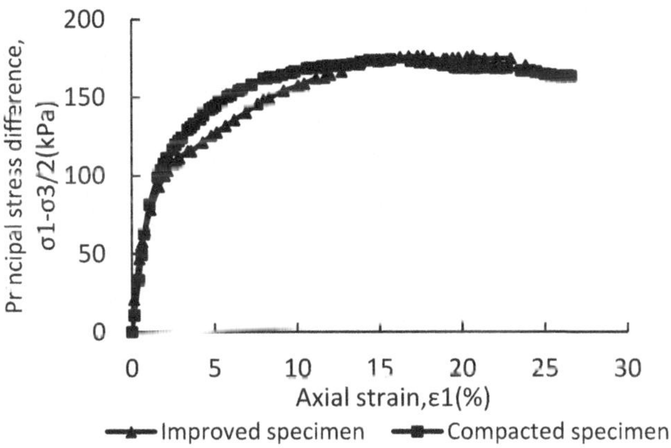

(c) 100kpa, Sr=55%

Fig. 4. The stress-strain curves of two different specimens.

Under both cases the curve of friction angle for different saturations shows basically the same variation, that is, the friction angle decreases with increasing saturation degrees (Fig. 3), which demonstrates that the impacts of matric suction on friction angle turn out to be small. Besides, it can also be observed that the friction angle of remolded soil tends to furthermore sensitive to the varying moisture content, which implies the remolding process caused irreversible damage to the soil structure.

3.2 The Comparison of Shear Characters

The stress-strain curves under 100 kPa are provided in order to analyze how matric suction effect the shear strength and properties of remolded clay (Fig. 4). The variation law under other confining pressures is same, so they won't be covered here.

Figure 4 presented the stress-strain curves for 25%, 35%, 55% saturation degree with two different remolding methods. It can be observed that the peak values of improved specimens are apparently higher than that of traditional specimens. Moreover, as the saturation increases, the stress-strain curve gradually transits from strain softening state to strain hardening state, which is similar to the shear properties of natural structural clay [6, 7]. But for traditional specimens, the stress-strain relation presents properties similar to light sand, that is, stress hardening. On the one hand, the differences in shear properties of the two kinds of specimens demonstrate that matric suction contributes in shear strength differently under different saturations, and also, proved that the shear strength of unsaturated soil could be enhanced by retaining its matric suction. On the other hand, although the peak value of the stress rises with greater matric suction, the stress would then drop rapidly after the maximum value with a relatively low residual strength, which could cause sudden failure especially for cohesive slope. Hence, considering the probability of such brittle failure, it is more secure for engineering projects to evaluate its safety by the improved method developed in this article when conducting soil tests.

4 Summary and Conclusions

From the perspective of recovering the matric suction of soil, this article presents an improved method to remold unsaturated clay through saturation and dehydration process. By comparing a series of shear tests conducted on different specimens, it is found that the cohesion changes of improved specimens under different saturation degrees accord with that of undisturbed soil, moreover, the strength of the two are very close. While the cohesion of specimens prepared by compacted method will increase firstly and then decrease as the saturation grows, the shear strength is also much lower than that of undisturbed soil. In addition, using the improved method could better simulate the shear properties of natural soil and take the sudden failure of dry clay into full consideration when evaluating the stability of realistic slopes. However, the stress-strain relation of compacted specimens is unrepresentative.

References

1. Fredlund, D.G., Morgenstern, N.R., Widger, R.A.: The shear strength of unsaturated soils. Revue Canadienne De Géotechnique **15**(3), 313–321 (1978)
2. Sun, D., Gao, Y.: Water retention behavior of soils with different preparations. Chin. J. Geotech. Eng. **37**, 91–97 (2015)
3. Lin, H., Li, G., Yu, Y.: Influence of matric suction on shear strength behavior of unsaturated soils. Rock Soil Mech. **28**, 1931–1936 (2007)
4. Fredlund, D.G., Rahardjo, H.: Soil Mechanics for Unsaturated Soils. Wiley, New York (1993)
5. GBT 50123-1999: The Standard for Soil Test Method. MOC (Ministry of Construction) (1999)
6. Delage, P., Yu, J.C.: Elasto-plastic behavior of a compacted silt. Int. J. Rock Mech. Mining Sci. Geomech. Abs. **32**, 317–324 (1995)
7. Sun, D., Sheng, D., Sloan, S.W.: Elastoplastic modelling of hydraulic and stress-strain behaviour of unsaturated soils. Mech. Mater. **39**, 212–221 (2007)

Mechanical Behavior of Unsaturated Masado Under Constant-Degree-of-Saturation Condition

X. Y. Qiu[1], X. Xiong[2(✉)], T. Tsunemoto[2], S. Okino[2], and F. Zhang[2]

[1] Shanghai Geotechnical Investigations and Design Institute Co., Shanghai 200093, China
[2] Nagoya Institute of Technology, Nagoya, Japan
101076@tongji.edu.cn

Abstract. Generally speaking, in many cases geomaterials are in unsaturated state. The mechanical properties of unsaturated soil are much more complicated than those of saturated soil. To rationally describe the mechanical properties of unsaturated soil, it is important to find out suitable state variables, such as degree of saturation and skeleton stress that determine the mechanical properties of unsaturated soil. In the existing researches of unsaturated soil, however, element tests under constant-degree-of-saturation condition is not sufficient in relating to the search of constitutive model. In this paper, degree of saturation and the skeleton stress are regarded as the state variables. Oedometer tests and triaxial compression tests under constant-degree-of-saturation condition were conducted to study the mechanical properties of unsaturated Masado (Decomposed granite) at different degrees of saturation. It is found that the degree of saturation and the skeleton stress can be reasonably used as the state variables to describe the mechanical properties of unsaturated soil in proper constitutive model.

Keywords: Unsaturated soil · State variable · Skeleton stress
Degree of saturation · Masado

1 Introduction

Generally speaking, in many cases geomaterials are in unsaturated state. Comparing with saturated soil, the mechanical properties of unsaturated soil are complex, and its experimental and simulation studies are still limited. To rationally describe the mechanical properties of unsaturated soil, the most important thing is to select suitable state variables, which control these mechanical properties. When soil in unsaturated state, it is not only the problem that degree of saturation is less than 1, but also its influence on strength and deformation properties of unsaturated soil. Changes of degree of saturation sometime may even trigger geological disasters, such as landslides or dam breaks, threatening human being and infrastructure. Therefore, it is undoubtedly important to study the mechanical properties of saturated/unsaturated soil, and propose unified constitutive model.

L. Hu et al. (Eds.): GSIC 2018, *Proceedings of GeoShanghai 2018 International Conference: Multi-physics Processes in Soil Mechanics and Advances in Geotechnical Testing*, pp. 30–38, 2018.
https://doi.org/10.1007/978-981-13-0095-0_3

Based on Barcelona Basic Model (Alonso et al. 1990), a number of elasto-plastic constitutive models have been proposed to model the properties of unsaturated soil by different state variables. Some models are proposed in the space of net stress and suction (Cui and Delage 1996; Chiu and Ng 2003), while others are in the space of Bishop's effective stress and suction (Kohgo et al. 1993; Loret and Khalili 2002). In recent years, the effective stress and the degree of saturation as independent variable were used as state variables, such as Zhang and Ikariya (2011), Zhou et al. (2012) and Zhou and Sheng (2015). However, since unsaturated soil experiments require high-precision load control and a lot of time, it is still difficult to verify the rationality of some state variables.

In this paper, the degree of saturation and the skeleton stress are taken as the state variables (Zhang and Ikariya 2011). A kind of decomposed granite, called as Masado with high permeability, was tested with oedometer tests and triaxial compression tests under constant-degree-of-saturation (CDS) condition, in order to investigate its mechanical properties at different degrees of saturation and loading paths. In order to control the degree of saturation, a device called as PVC (Pressure/Volume Controller) made by GDS was utilized and a method to conduct triaxial compression test under CDS condition proposed by Burton et al. (2016) was used in the tests. Based on the test results, an assumption adopted in the model proposed by Zhang and Ikariya (2011) was verified.

2 Brief Description of Constitutive Model

Zhang and Ikariya (2011) proposed an unsaturated soil constitutive model, using skeleton stress and degree of saturation as state variables. In the model, it is assumed that normally consolidated line in unsaturation state (*N.C.L.S.*) is parallel to the normally consolidated line in saturated state (*N.C.L.*) but in a higher position than *N.C.L.*, as shown in Fig. 1, which means that under the same mean stress, unsaturated soil can keep higher void ratio than those of saturated soil. Skeleton stress is a kind of Bishop effective stress, defined as,

$$\sigma''_{ij} = \sigma^t_{ij} - u_a\delta_{ij} + S_r(u_a - u_w)\delta_{ij} = \sigma^n_{ij} + S_r s\delta_{ij} \tag{1}$$

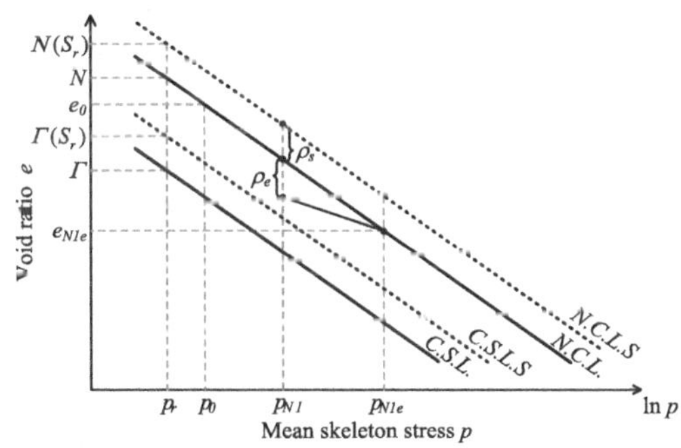

Fig. 1. Moving up of N.C.L. and C.S.L. due to instauration (Zhang and Ikariya 2011).

where σ''_{ij} is skeleton stress tensor, σ^t_{ij} is total stress tensor, σ^n_{ij} is net stress tensor, S_r is degree of saturation, u_a is air pressure, u_w is water pressure and s is suction. The constitutive model is able to describe not only the behavior of unsaturated soil but also saturated soil because the skeleton stress can smoothly shift to effective stress from unsaturated condition to saturated condition. The constitutive model includes nine material parameters, which can be determined by triaxial tests.

3 Sample Material and Test Apparatus

3.1 Test Sample Material

Masado is a typical decomposed granitic that is widely distributed in western Japan. In construction, it is often used to pave the road. Furthermore, compared to silty clay commonly used in unsaturated tests, the permeability of Masado is much larger, which could significantly shorten test time required for unsaturated test. Therefore, Masado was selected for tests under CDS condition.

Figure 2 is the grain grading curve of Masado. Some pre-tests were conducted in according to the JIS A 1205 and JIS A 1202 and the physical material properties of Masado were obtained as shown in Table 1. Moreover, compaction test (JIS A 1210) also conducted to find the optimum moisture content for sample preparation. In this test, only part of Masado with particle diameter less than 2 mm was used.

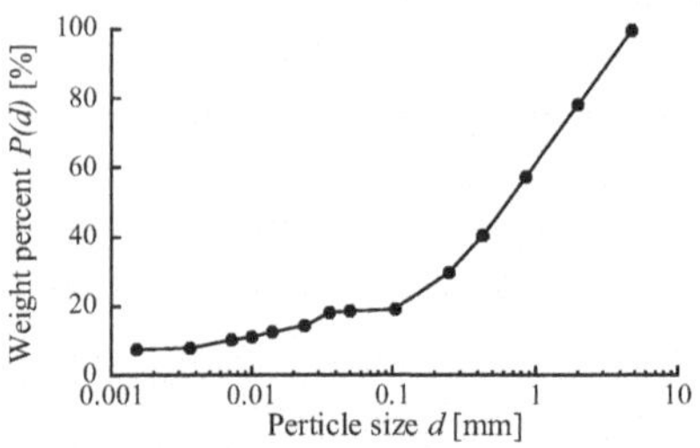

Fig. 2. Grain grading curve of Masado.

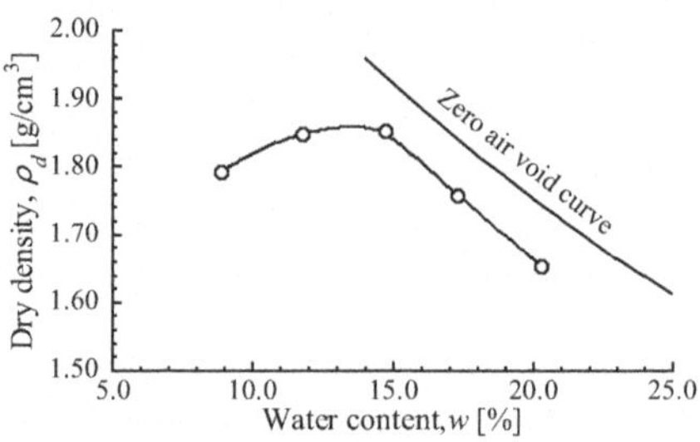

Fig. 3. Compaction curve of Masado.

Table 1. Material properties of Masado

Liquid limit w_L (%)	Plasticity index Ip	Specific gravity G_s	Maximum dry density ρ_d (g/cm^3)	Optimum water content w_{opt}(%)
Non-plastic	Non-plastic	2.66	1.85	13.7

3.2 Oedometer and Triaxial Apparatuses

In this study, the tests were conducted with unsaturated oedometer and triaxial apparatus, which utilized axis-translation method to control suction. Since the apparatuses are sealed during the whole test procedures, the air pressure of the sample could be measured by the pore air piezometer. The PVC and the pore water piezometer are connected to the samples by the ceramic disk of the pedestal. When the air pressure is less than the air entry value (AEV) of the ceramic disk, only the water is allowed to pass through the ceramic disk. Therefore, the pore water pressure of the samples can be measured by the PVC and the pore water piezometer.

For oedometer apparatus, three kinds of pressures are controlled in the unsaturated apparatus by axial pressure, pore air pressure and pore water pressure. Axial pressure and pore air pressure are controlled by the manual regulator or the electric regulator, while pore water pressure is controlled by PVC, which can easily control the suction and degree of saturation in any arbitrary stress path.

For triaxial apparatus, the main feature is that both the porous stone (pore air pressure) and the ceramic disk (pore water pressure) are embedded in the axial cap, and the test time is shortened under double-end-face drainage condition. To obtain the volume change of unsaturated samples, differential manometer is utilized to measure the pressure difference between the burette with the standard water surface and the inner chamber. Unsaturated triaxial apparatus can control four kinds of pressure including axial pressure, confining pressure, pore air pressure and pore water pressure by air pressure.

4 Oedometer Compression Test

4.1 Test Method

In this test, 5 kinds of samples with different initial moisture content were prepared. The tested initial moisture content was determined from the compaction curve of Masado (Fig. 3). Since the optimum moisture content is about 15%, two cases less than 15% (9% and 12%) and two cases larger than 15% (18% and 21%) were selected. The sample for oedometer apparatus was 1.0 cm in height and 6.0 cm in diameter, and was prepared by static compaction method for one layer. The target void ratio was set to 0.65.

For load path, vertical stress was firstly loaded to 20 kPa and then suction was loaded to 50 kPa. After the amount of volume change and the drainage was stable, the vertical stress was loaded continuously to 965 kPa, keeping the degree of saturation being constant. PVC was utilized to keep the degree of saturation constantly during the compression stage by the method proposed by Burton et al. (2016). The pore air pressure was kept constantly and the water content of sample was adjusted by changing the pore water pressure through PVC as following equation,

$$dV_w - S_{r(init.)} dV_v = 0 \tag{2}$$

where dV_w is pore water volume change, $Sr_{(\text{init.})}$ is the initial degree of saturation and $*dV_v$ is the sample volume change.

4.2 Test Results

Figure 4 shows the oedometer test results. Samples with higher initial water content w_0 consolidated under suction $s = 50$ kPa to reach higher degree of saturation. As shown in Fig. 4(a), water content decreased with the reduction of void ratio during compression stage. From the line $wG_s = eS_r$ (dotted line). The degree of saturation was maintained constantly during this stage. The incremental vertical net stress leaded to the increase of suction. In the Fig. 4(b), the suction first rose sharply, followed by a slightly decrease and then increased gradually. The change of suction was different, the higher the initial degree of saturation was, the smaller the suction increment was.

Figure 4(c) and (d) presented the compression curves at different kinds of stress. After the skeleton stress exceeds 100 kPa, the compression curves are nearly straight lines. Under the same skeletons tress, the samples with lower degree of saturation have higher void ratio, which is just the same as the assumption of *N.C.L.* and *N.C.L.S.* (Zhang and Ikariya 2011). In addition, it is also found that the mechanical properties of unsaturated soil can be better represented by the skeleton stress, which also confirmed

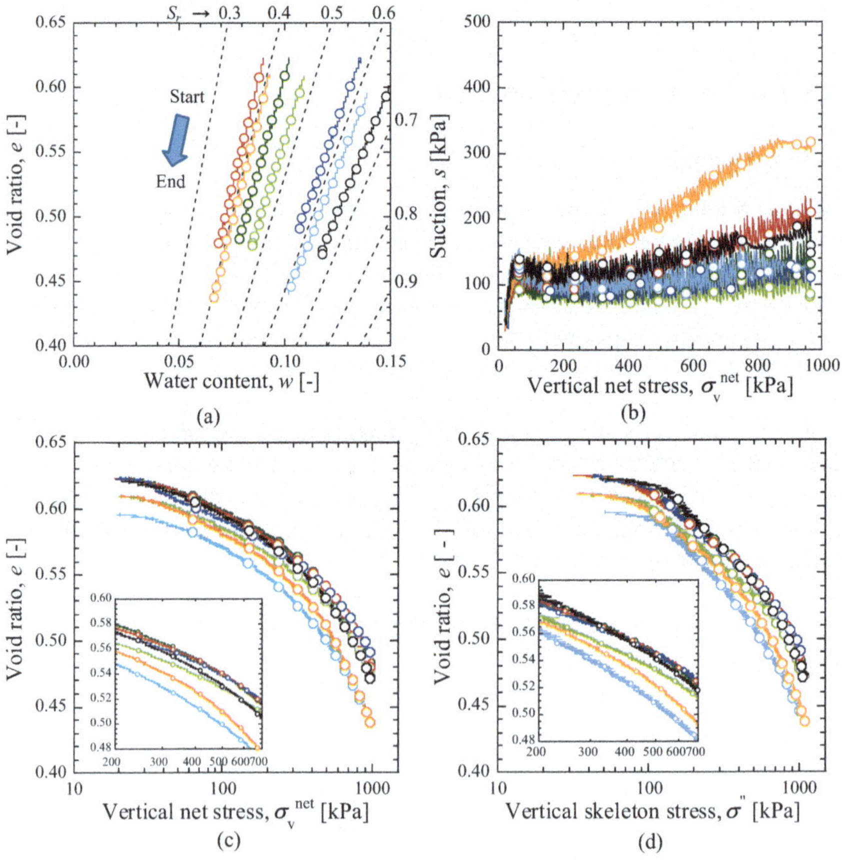

Fig. 4. Results of oedometer tests under CDS condition.

the rationality of utilizing skeleton stress as a state variable in the proposed unsaturated soil constitutive model (Zhang and Ikariya 2011).

5 Triaxial Compression Test

5.1 Test Method

The samples for triaxial apparatus, with the same initial moisture content $w_0 = 15\%$, was 10.0 cm in height and 5.0 cm in diameter. The samples were prepared by static compaction method for three layers and the target void ratio was 0.65.

Figure 5 shows the stress path of unsaturated triaxial compression test under CSD condition. The samples consolidated at different suction and confining stress to reach different degree of saturation. The degree of saturation was controlled to keep constant during shear stage by computer program based on Eq. (2). In all cases, the samples were under drained-vented condition, the shear rate was 0.0025%/min and the maximum deviatoric strain was 15%. The confining stress σ_3^{net} was kept constant during the shear stage.

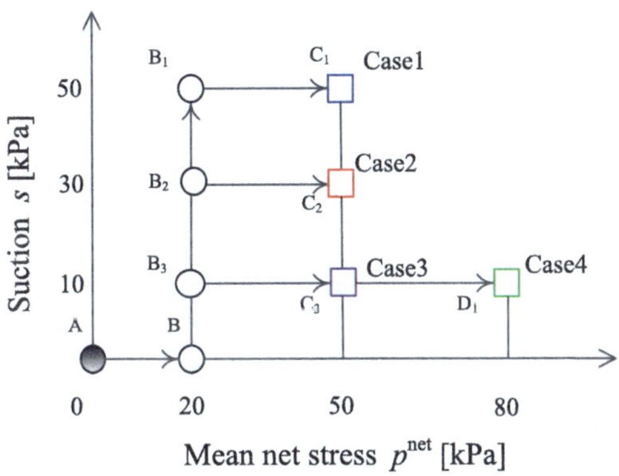

Fig. 5. Stress path of triaxial compression test under CDS condition.

5.2 Test Results

As shown in Fig. 6, the degree of saturation control method (Burton et al. 2016) can keep the degree of saturation constant during the shear stage. In case 1, the suction decreased gradually, while the suctions of Case2, Case3 and Case4 experienced a small reduction firstly and then increased and reach the peak. After the peak, the suction decreased gradually. In the cases with the same confining stress, the initial suction of the sample is large, of which the final suction is also large. In the Case3 and Case4, even the samples confined under the same suction, the increment of suction during shear stage is also different.

As shown in Fig. 7, under the same confining stress, with the increment of initial suction, samples showed hardening and the peak strength increased, while the residual

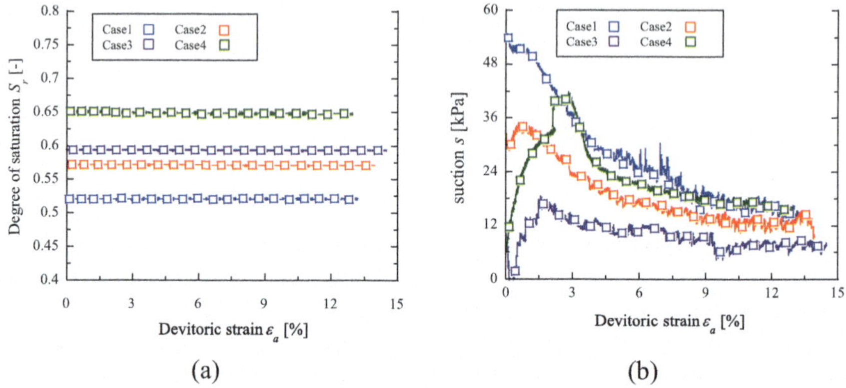

Fig. 6. Degree of saturation and suction in shear stage of triaxial compression test.

strength was almost the same. The volumetric strain of the samples transformed from contraction to dilatancy, and final volumetric strain decreased with the increment of initial suction. The peak strength of Case4 is larger and the volumetric strain transformed from dilatant to contraction with the increment of confining stress.

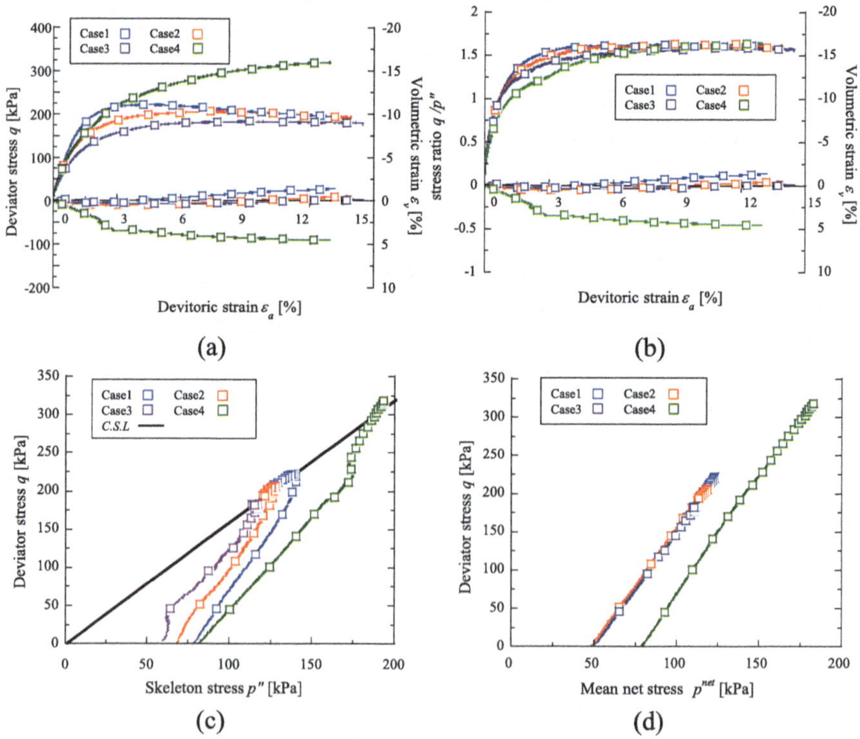

Fig. 7. Results of triaxial compression test under CDS condition.

In addition, it is found that the deviatoric stress ratio q/p'' curves were almost the same at the deviatoric strain about 13% (Fig. 7(b)), which means that with different confining pressure and suction, samples would be sheared to the same critical state lines (Fig. 7(d)). Therefore, by using the degree of saturation and skeleton stress as the state variables, the mechanical behavior of unsaturated soil could be described more reasonably with unified parameters, which verified the rationality of proposed unsaturated soil constitutive model (Zhang and Ikariya 2011). However, since the number of test cases is still not enough, the repeatability of the triaxial compression tests under CDS condition need further confirmation.

6 Conclusion

In this paper, based on the state variables and assumption of proposed unsaturated soil constitutive model (Zhang and Ikariya 2011), a series of oedometer tests and triaxial compression tests under CDS condition were conducted to study the mechanical properties of unsaturated Masado. The main purpose is to verity the rationality of the model by laboratory tests. The main conclusions are made in the following: (1) The loading method under CDS condition (Burton et al. 2016) can successfully keep the degree of saturation approximately constant in oedometer and triaxial compression tests. Therefore, the degree of saturation in this paper could be discussed as a possible state value; (2) According to the tests results, under the same skeleton stress, the samples with lower degree of saturation have higher void ratio, which is just the same as the assumption of *N.C.L.* and *N.C.L.S.* (Zhang and Ikariya 2011); (3) Using the degree of saturation and skeleton stress as the state variables, the mechanical behavior of unsaturated Masado can be described properly by the unified parameters, which also verified the rationality of proposed unsaturated soil constitutive model (Zhang and Ikariya 2011).

References

Alonso, E.E., Gens, A., Josa, A.: Constitutive model for partially saturated soils. Géotechnique **40**(3), 405–430 (1990)

Burton, G.J., Pineda, J.A., Sheng, D., Airey, D.W., Zhang, F.: Exploring one-dimensional compression of compacted clay under constant degree of saturation paths. Géotechnique **66**(5), 435–440 (2016)

Cui, Y.J., Delage, P.: Yielding and plastic behaviour of an unsaturated compacted silt. Géotechnique **46**(2), 291–311 (1996)

Chiu, C.F., Ng, C.W.: A state-dependent elasto plastic model for saturated and unsaturated soils. Géotechnique **53**(9), 809–830 (2003)

Kohgo, Y., Nakano, M., Miyazaki, T.: Theoretical aspects of constitutive modelling for unsaturated soils. Soils Found. **33**(4), 49–63 (1993)

Loret, B., Khalili, N.: An effective stress elastic–plastic model for unsaturated porous media. Mech. Mater. **34**(2), 97–116 (2002)

Zhang, F., Ikariya, T.: A new model for unsaturated soil using skeleton stress and degree of saturation as state variables. Soils Found. **51**(1), 67–81 (2011)

Zhou, A.N., Sheng, D., Sloan, S.W., Gens, A.: Interpretation of unsaturated soil behaviour in the stress-saturation space, I: Volume change and water retention behaviours. Comput. Geotech. **43**, 178–187 (2012)

Zhou, A.N., Sheng, D.: An advanced hydro-mechanical constitutive model for unsaturated soils with different initial densities. Comput. Geotech. **63**, 46–66 (2015)

Compression Behavior of Bentonite Powder/Pellet Mixture

Yang Chen[1,2(✉)], Yu-Jun Cui[2], and Agustin Molinero Guerra[2]

[1] Tongji University, Shanghai, China
103478@tongji.edu.cn
[2] Ecole des Ponts ParisTech, Marne-la-Vallée, France

Abstract. This study deals with the hydro-mechanical behavior of bentonite-based materials used as sealing materials in high-level radioactive waste repositories. Many studies have been done recently on bentonite-based materials, such as the mixture of bentonite and crushed Callovo-Oxfordian (COx) claystone, pre-compacted disks of MX80 bentonite/sand mixture and MX80 bentonite pellets/powder mixture. In this study, the hydro-mechanical behavior of the mixture of MX80 bentonite powder and pellets with a proportion of 20/80 was investigated. Compared to other sealing plugs, this mixture has some operational advantages such as a lower compaction effort and reduction of technological voids between rock and seal. The compression behavior was investigated by performing suction controlled oedometer tests. Results show that the yield stress (σ_v) increases with suction increase; however, this effect is not obvious for suctions ranging from 9 to 138 MPa, suggesting a significant effect of soil microstructure mainly governed by the assembly of pellets.

Keywords: Hydro-mechanical behavior · MX80 pellet/powder mixture
Suction controlled oedometer test

1 Background

1.1 Deep Geological Repository

The main objective in managing and disposing of radioactive (or other) waste is to protect people and environment. This means isolating or diluting the waste so that the rate or concentration of any radionuclides returned to the biosphere is harmless. For disposal, to ensure that no significant environmental releases occur over tens of thousands of years, 'multiple barriers' geological disposal is planned.

1.2 Bentonite-Based Materials

In most cases, bentonite-based materials are used as the sealing materials in this system, because of its properties: high swelling capacity, low permeability and high radionuclide retardation properties. Many studies have been conducted to the performance of buffer/sealing materials such as the experiments in underground research laboratories (URL) in Canada, Switzerland, Belgium and France.

L. Hu et al. (Eds.): GSIC 2018, *Proceedings of GeoShanghai 2018 International Conference: Multi-physics Processes in Soil Mechanics and Advances in Geotechnical Testing*, pp. 39–46, 2018.
https://doi.org/10.1007/978-981-13-0095-0_4

Once the repository is closed and local groundwater conditions are re-established, the water in the host rock formation will move to the repository. The bentonite-based material absorbs water and swells, filling the technical voids as the gaps between the bentonite bricks themselves, between the canister and the bricks, between bricks and the host rock, as well as the fractures in the host rock due to excavation. After that, the subsequent swelling is restrained by the host rock and swelling pressure develops. For ensure the stability of the system, the swelling pressure must be lower than the in situ minor stress: 7 MPa in the Underground Research Laboratory (URL) of Bure site, France (Tang et al. 2011a); 3–4 MPa at Tournemire site, France (Barnichon and Deleruyelle 2009) and 4–5 MPa at Mol site, Belgium (Li et al. 2009).

2 Materials

This work focuses on the HM behavior of a mixture of a powder/pellet MX80 bentonite mixture with a proportion of 20-powder/80-pellets in dry mass. This mixture has been investigated by the institute of radioprotection and nuclear safety (IRSN). The material used in this work (both powder and pellets of bentonite) has been provided by Laviosa Minerals SpA.

2.1 Pellets of MX80 Bentonite

Pellets of MX80 bentonite are produced by compacting powder of MX80 bentonite (as extracted in-situ) in a mould of 7 mm of diameter and 7 mm of height at a water content between 5% and 7%, corresponding to a dry density $\rho_d = 1.998 - 2.120$ Mg/m^3. The pellets have a quasi-cylindrical shape with two half spherical poles on the top and the bottom. Table 1 summarizes their initial properties. The initial suction, s = 132.4 MPa, was measured by a chilled mirror dew point tension-meter (Decagon WP4) and the initial water content, w = 7.25%, was determined after drying the sample (pellet) at 105 °C during 24 h.

Table 1. Initial properties of pellets of bentonite

Initial property	
Initial suction	132.4 MPa
Initial water content	7.25%
Initial dry density	2.12 Mg/m^3
Initial porosity	0.25

2.2 Powder of MX80 Bentonite

Powder of MX80 bentonite was produced by crushing pellets of bentonite. Table 2 shows its initial conditions (Saba et al. 2014). A value of 3.2% of water content was found after drying at 105 °C during 24 h. The initial suction, s = 190.9 MPa, was measured by a chilled mirror dew point tension-meter (Decagon WP4).

Table 2. Initial conditions on MX80 bentonite powder

Initial conditions	
Initial suction	190.9 MPa
Initial water content	3.2%
Unit mass of bentonite particles	2.77 Mg/m^3

2.3 MX80 Pellet/Powder Mixture Preparation

This work aims to study the long term evolution of the HM behavior of the pellet/powder (80/20) MX80 bentonite mixture at a dry density $\rho_d = 1.49\,\mathrm{Mg/m^3}$. It is noted that the global dry density of sample controls the saturated permeability and the swelling pressure (Hoffmann et al. 2007; Saba et al. 2014). The key point of the mixture preparation is to obtain a homogeneous sample. To this purpose, three different protocols were defined in order to choose the one that allowed obtaining a homogeneous mixture. The first protocol consists in filling the cell by packets of three layers of pellets (a layer of pellets is defined as the mass of pellets that fills the base of the sample), and mixing with the corresponding powder (taking into account the proportion 80/20). The second protocol is just like the first one but the cell was filled by packets of one layer of pellets with the corresponding powder. The third protocol consists in filling the cell by layers of pellets and powder separately, which means one single layer of pellets and then the corresponding powder until filling the cell.

In order to decide the protocol which allows obtaining a homogeneous sample, several μ-CT observations were carried out. Three vertical slices corresponding to the three protocols are presented in Fig. 1 (Molinero Guerra et al. 2016). Figure 1(a), corresponding to first protocol, shows a heterogeneous sample because the powder does not fill the voids between the pellets. Figure 1(b), corresponding to the second protocol, presents also a heterogeneous sample because of the voids between pellets and the segregation of powder. Figure 1(c), corresponding to the third protocol, shows a homogeneous sample, though many voids were observed between the pellets. A value of $\rho_d = 1.41\,\mathrm{Mg/m^3}$ and $\rho_d = 1.43\,\mathrm{Mg/m^3}$ were found for the first and

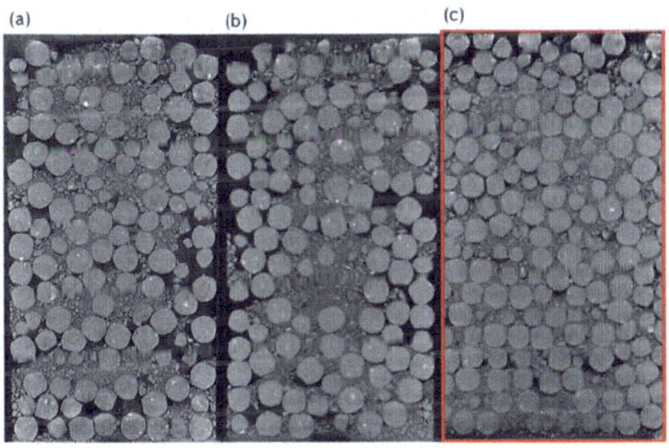

Fig. 1. Microfocus X-ray computed tomography observations of a mixture of powder/pellets of bentonite with a proportion of 20/80 in dry mass obtained by three different protocols (Molinero Guerra et al. 2016)

second protocols respectively, which were smaller than the target dry density ($\rho_d = 1.49\,\mathrm{Mg/m^3}$).

Finally, taking into account the target dry density and homogeneity, the third protocol was chosen in order to prepare the MX80 pellet/powder mixture sample.

3 Experimental Methods

Several oedometer tests under controlled suction conditions were carried out on a mixture of MX80 pellet/powder with a proportion of 80/20 in dry mass prepared at a dry density of $\rho_d = 1.49\,\mathrm{Mg/m^3}$. Both pellets and powder will be at their initial state (s = 138.4 MPa and w = 7.25% for pellets and s = 190.9 MPa and w = 3.2% for powder). In this study, vapour equilibrium technique (VET) was selected for suction control, by using a saturated salt solution. Figure 2 shows the experimental setup of suction controlled oedometer test.

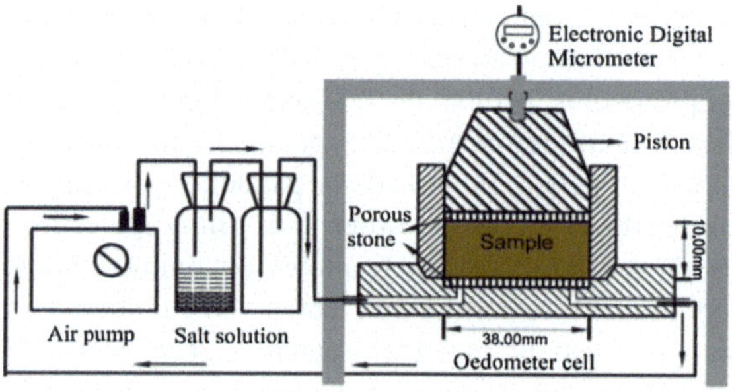

Fig. 2. Experimental setup of suction controlled oedometer test (Wang et al. 2013)

The dimensions of the cell are 50 mm of diameter and 35 mm of height. We can calculate the mass of pellets and mass of powder for each sample.

$$\rho_d = \frac{m_1}{V_1}, V_1 = \frac{\pi D^2 h}{4}, \gamma_d = \rho_d \times g = 1.49\,\mathrm{Mg/m^3} \quad (1)$$

m_1: The total mass of dry pellets and powder
V_1: The volume of cell

$$m_2 = m_{pellet,dry} = 0.8 m_1 \quad (2)$$

$$m_3 = m_{powder,dry} = 0.2 m_1 \quad (3)$$

As it was explained above, the pellets have an initial water content $w_1 = 7.25\%$ and the powder has an initial water content $w_2 = 3.2\%$. Taking into account of the initial water content of both pellet and powder:

$$m_4 = m_{pellets} = (1 + w_1) \times m_2 \tag{4}$$

$$m_5 = m_{powder} = (1 + w_2) \times m_3 \tag{5}$$

After the calculations, we have:

$$m_4 = 87.40\,\text{g} \tag{6}$$

$$m_5 = 21.13\,\text{g} \tag{7}$$

4 Experimental Results

In the test, the displacement Δh was recorded by micrometer every 300 s in the suction controlled and loaded situation. The dimensions of the cell are 50 mm of diameter and 35 mm of height (H_o). When equilibrium was reached, so the evolution of Δh with time is stable, the volumetric strain ε_v can be calculated by this formula:

$$\varepsilon_v = \frac{\Delta h}{H_o} \tag{8}$$

At the end of each loading, the corresponding volumetric strain is obtained. This deformation is linked with the corresponding void radio. This initial void ratio is calculated according to the dry density of MX80 pellet/powder mixture ρ_d and density of solid particle ρ_s. The initial void ratio e_0 was calculated by the formula below:

$$e_0 = \frac{\rho_s}{\rho_d} - 1 \tag{9}$$

The void ratio is calculated at the end of each level of pressure with the following expression

$$\mathrm{e} = (1 + e_0)\varepsilon_v + e_0 \tag{10}$$

The changes in volumetric strain with time during suction imposition for samples equilibrated at 0 MPa, 9 MPa and 138 MPa of suction are presented in Fig. 3. It is noted that higher volumetric strain was observed at 0 MPa of suction. A volumetric strain of 21.1%, 8.9% and −0.5% was found for suction of 0, 9 MPa and 138 MPa, respectively.

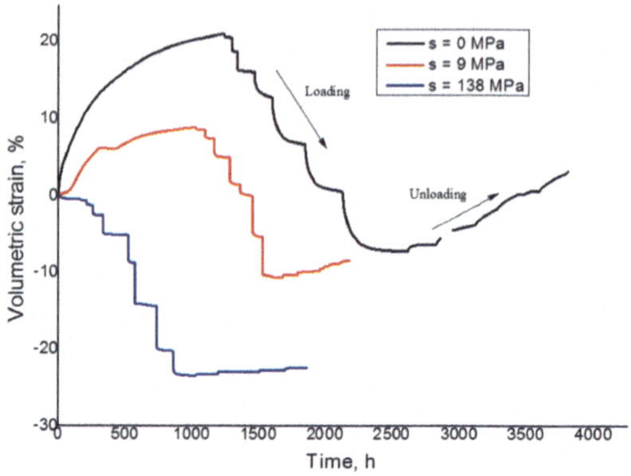

Fig. 3. Evolution of the volumetric strain with time

Once equilibrium arrived at the imposed suction, samples were submitted to controlled-suction compression. The compression curves are presented in Fig. 4 in a diagram giving the changes in void ratio e with respect to vertical net stress $(\sigma_v - u_a)$ in which u_a is the air pressure, equal to the atmospheric pressure. Given the significant concerns about the validity of effective stress in unsaturated soils, it was preferred to use the independent variables approach involving the vertical net stress $(\sigma_v - u_a)$ and suction $(s = u_w - u_a)$, (Coleman 1962).

Figure 5 shows the results of Wang et al. 2013, the compression curves in a diagram giving the changes in void ratio e with respect to vertical net stress $(\sigma_v - u_a)$.

Initial void radios were very different because of the significant dependence of initial swelling with respect to the suction imposed, when the suction increased.

In the Figs. 4 and 5, the compression curves are characterized by an initial linear branch with a low compressibility (pseudo-elastic domain) followed by a second branch with a higher compressibility (plastic domain). In the diagram of Wang, there is a slight upward curvature at higher stresses, and each sample exhibited a slightly

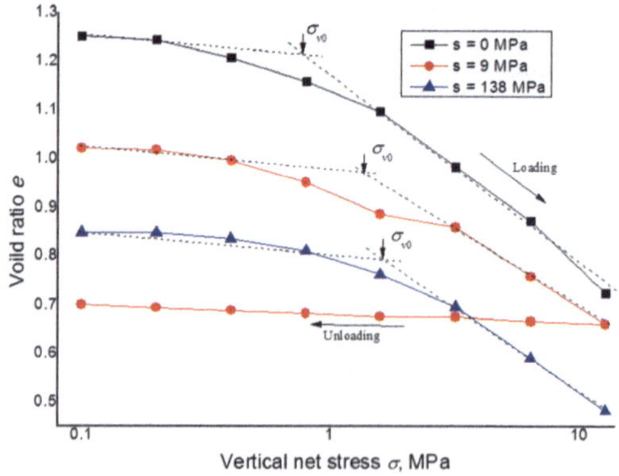

Fig. 4. Void radio of samples versus vertical net stress for 0 MPa, 9 MPa and 138 MPa

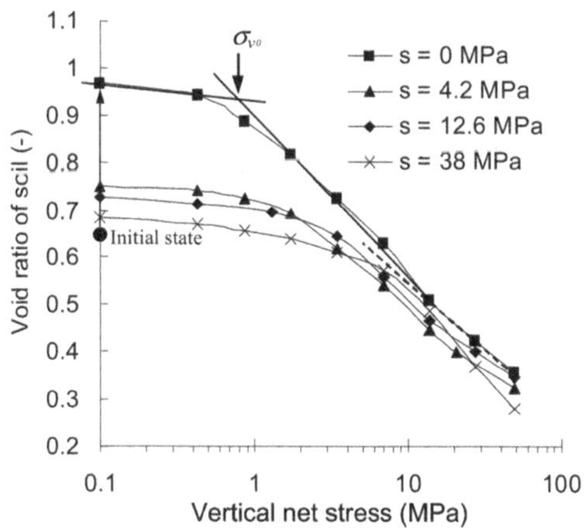

Fig. 5. Void radio of soil versus vertical net stress for different suctions (Wang et al. 2013)

S-shaped compression curve. As suggested by other authors, the points at high stresses were not used for determining the compression coefficient (C_c).

The yield stress (σ_v) delimitates the pseudo-elastic zone and the plastic one. It is noted both in the Figs. 4 and 5 when the suction increased, the yield stress (σ_v) increases too, though entire suction 9 MPa and suction 138 MPa, the increase is not too evident.

References

Barnichon, J.D., Deleruyclle, F.: Sealing experiments at the Tournemire URL. In: Towards Convergence of Technical Nuclear Safety Practices in Europe, EUROSAFE (2009)

Coleman, J.D.: Stress-Strain relations for partially saturated soils. Géotechnique **12**(4), 348–350 (1962)

Hoffmann, C., Alonso, E.E., Romero, E.: Hydro-mechanical behaviour of bentonite pellet mixtures. Phys. Chem. Earth **32**(8–14), 832–849 (2007)

Li, X.L., Bastiaens, W., Van Marcke, P., et al.: Design and development of the large scale in-situ PRACLAY heater test and horizontal HLW disposal gallery seal test in Belgian URL hades. In: International Symposium on Unsaturated Soil Mechanics and Deep Geological Nuclear Waste Disposal, Shanghai, China, pp 24–28 (2009)

Molinero Guerra, A., et al.: In-depth characterisation of a mixture composed of powder/pellets MX80 bentonite. App. Clay Sci. (2016). http://linkinghub.elsevier.com/retrieve/pii/S0169131716304513

Saba, S., Romero, E., et al.: Hydro-mechanical behavior of bentonite-sand mixture used as sealing materials in radioactive waste disposal galleries. Université de Paris Est (2014)

Tang, C.S., Tang, A.M., Cui, Y.J., Delage, P., Barnichon, J.D., Shi, B.: A study of the hydro-mechanical behavior of compacted crushed argillite. Eng. Geol. **118**(3–4), 93–103 (2011a)

Wang, Q., Minh, A., et al.: The effects of technological voids on the hydro-mechanical behavior of compacted bentonite – sand mixture. Soils Found. **53**(2), 232–245 (2013)

Comparison of Wetting Front Advancing Method and Instantaneous Profile Method for the Hydraulic Conductivity Measurements

Xu Li(✉), Yang Wu, Jiankun Liu, and Lihong Chen

Key Laboratory of Urban Underground Engineering of Ministry of Education, Beijing Jiaotong University, Beijing 100044, China
ceXuLi2012@163.com

Abstract. Wetting front advancing method is of lower cost and faster experimental progress for measuring the unsaturated hydraulic conductivities. In this paper, a new method is proposed to identify the wetting front, in which the monitored data is used to infer wetting front. Data analysis of three column tests reported by earlier scholar proves that the new method is applicable. Wetting front advancing method can be used to analyze wetting column test data and offer high quality results of unsaturated hydraulic conductivities. The limitation of a clear wetting front that can be observed is not necessary. The remove of this limitation will greatly expand the applicability of the wetting front advancing method.

Keywords: Wetting front advancing method · Soil column test
Instantaneous profile method · Unsaturated hydraulic conductivities

1 Introduction

Wetting front advancing method (Li et al. 2009) can be used to measure the unsaturated hydraulic conductivities of soil. It is of lower cost and faster experimental progress comparing the instantaneous profile method. Both the capillary rising process (Lu and Likos 2004) and infiltration process can be used in analyzed by the wetting front advancing method. However, the accuracy of these two methods is still lack of a thorough comparison.

In this paper, the data of soil column test reported by (Cui et al. 2008; Ng and Leung 2012) are analyzed by both wetting front advancing method and instantaneous profile method. The accuracy of two methods and the key parameters involved in these methods are also discussed.

2 Soil Column Test and the Determination of Wetting Front Advancing Velocity

Soil column test is often used to study the water infiltration or capillary process in single layer or multilayer soils (Li et al. 2009). Li et al. (2009) initially reported that a clear wetting front that be observed is required. However, it is not right. In fact, wetting

L. Hu et al. (Eds.): GSIC 2018, *Proceedings of GeoShanghai 2018 International Conference: Multi-physics Processes in Soil Mechanics and Advances in Geotechnical Testing*, pp. 47–53, 2018.
https://doi.org/10.1007/978-981-13-0095-0_5

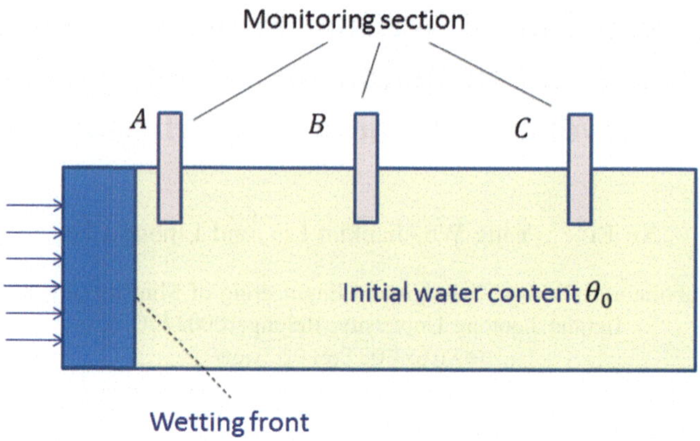

Fig. 1. Soil column test

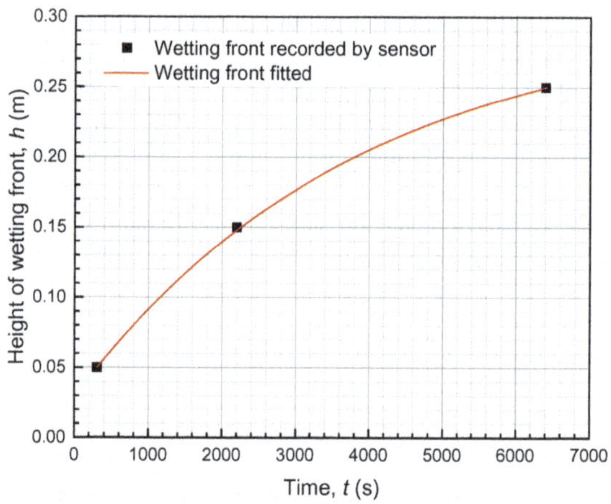

Fig. 2. Fitted wetting front advancing curve

front can be extracted from the monitored data in the cases where no clear wetting front (WF) is observed.

As shown in Fig. 1, there are three monitoring sections A, B, and C. If a water content θ_c is used as an indicator of wetting front, θ_c may be monitored in sections A, B and C at time of t_A, t_B and t_C, respectively. The distances between the infiltration boundary and sections A, B, and C are h_A, h_B and h_C. The relation between distance and time is plot in Fig. 2 and fitted by exponential function. Further, the wetting front velocity can be extracted as,

$$v(t) = \frac{\partial h}{\partial t} \tag{1}$$

With using the assumption that the wetted zone is advancing uniformly, the hydraulic conductivity can be calculated as

$$k = \frac{(\theta_2 + \theta_1) - 2\theta_0}{2(\psi_1 - \psi_2 + a\gamma_w v dt)} \gamma_w v^2 dt \quad (2)$$

where θ_1, θ_2 are the volumetric water content monitored at time t_1, t_2; ψ_1, ψ_2 are the suction at time t_1, t_2; θ_0 is the initial water content; γ_w is the bulky weight of water; a is the sine of the flow direction, as 1 for downward flow, −1 for upward flow, and 0 for horizontal flow.

3 Verification of the Fitted Wetting Front Advancing Velocity

The data from Li et al. (2009) are used to verify the fitting wetting front advancing velocity. Four monitoring sections are used Li et al. (2009). The water content records at these four sections are illustrated in Fig. 3. Giving a specific critical volumetric water content θ_c, the time when wetting front arrives the monitored sections can be extracted from the water content record, as t_A, t_B, t_C and t_D shown in Fig. 3. θ_c is supposed to be larger than the initial water content θ_0 and smaller than the maximum water content in the primary rising stage, as the available range for θ_c illustrated in Fig. 3.

The capillary rising column test of sandy silt reported in Li et al. (2009) is used as an example. For that test, the relation between the locations of monitoring sections h and the wetting front arriving time t is plotted in Fig. 4a. Three θ_c values as 0.06, 0.1 and 0.15 are used to investigate the influence of the specific θ_c on the wetting front

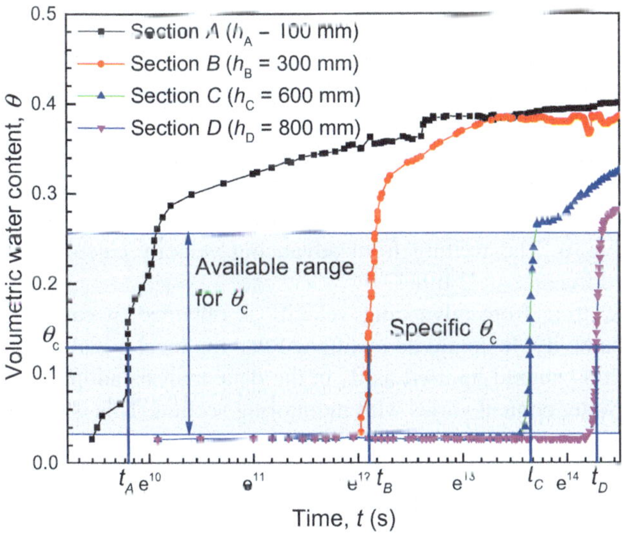

Fig. 3. Identify the time when wetting front arrives the monitoring section based on the water content records in the capillary rising column test of sandy silt

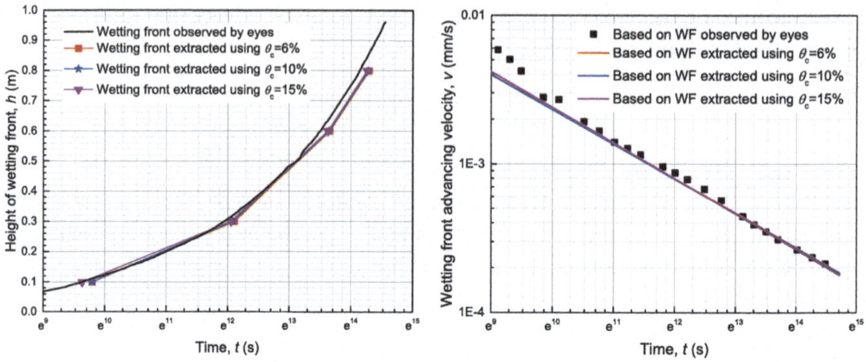

(a) Wetting front advancing curve (b) Wetting front advancing velocity

Fig. 4. Wetting front advancing process determined by various critical volumetric water content

advancing curve. It can be seen from Fig. 4a that these curves are very close. Further, the wetting front advancing velocity extracted from these curves are nearly identical, as shown in Fig. 4b. That's to say, the error from a specific θ_c is negligible if θ_c is picking from the reasonable range.

From Fig. 4, it can be concluded that a clear wetting front is not a necessary condition for the wetting front advancing method and the remove of this limitation will greatly expand the applicability of the wetting front advancing method.

4 Data Analysis for the Column Test Reported by Cui et al. (2008)

Cui et al. (2008) reported a valuable flux-controlled column test. The test data also can be analyzed by wetting front advancing method to obtain hydraulic conductivity.

Cui et al. (2008) did not record the observed wetting front curve, but reported the suction profiles recorded at different time and soil water characteristic curve (Fig. 7 and Eq. 3 in Cui et al. 2008). Based on the suction profiles and soil water characteristic curve, the water content profiles at different time can be obtained, as plotted in Fig. 5.

Using θ_c as 0.16, 0.17, and 0.18, the wetting front advancing process obtained is illustrated in Fig. 6. The wetting front advancing velocity calculated by these three values have a difference of 2 times.

After the wetting front advancing velocity is obtained, it can used to solve the hydraulic conductivity. It should be mentioned that the local initial water content at the monitoring section should be used as θ_0 in the data analysis adopting WFAM in case that the initial water content varies with monitoring section. This is an upward capillary test, the sine of the flow direction, a will be used as −1.

For the data at section A (h_A = 50 mm), the hydraulic conductivities calculated by wetting front advancing method (WFAM) is compared with those calculated by instantaneous profile method (IPM), which is conducted and reported by Cui et al. (2008). As shown in Fig. 7, the results from WFM are more reasonable than those from IPM. WFM shows great advantage in this data analysis.

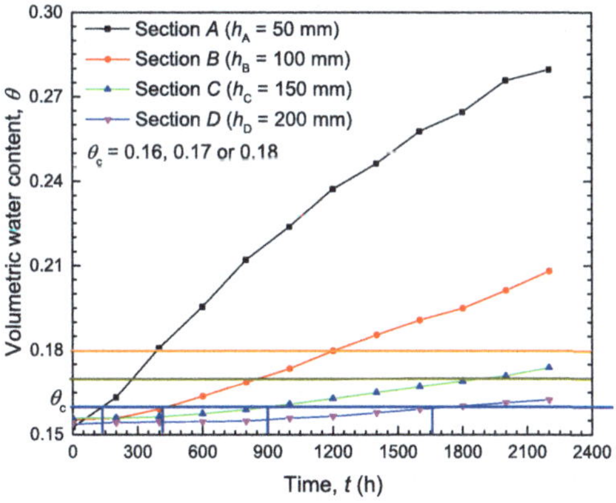

Fig. 5. Water content profiles at different time

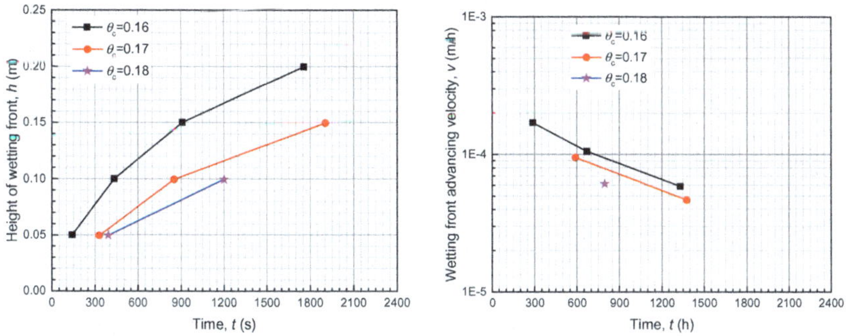

(a) wetting front advancing process (b) wetting front advancing velocity

Fig. 6. The wetting front advancing velocity in the water injection process

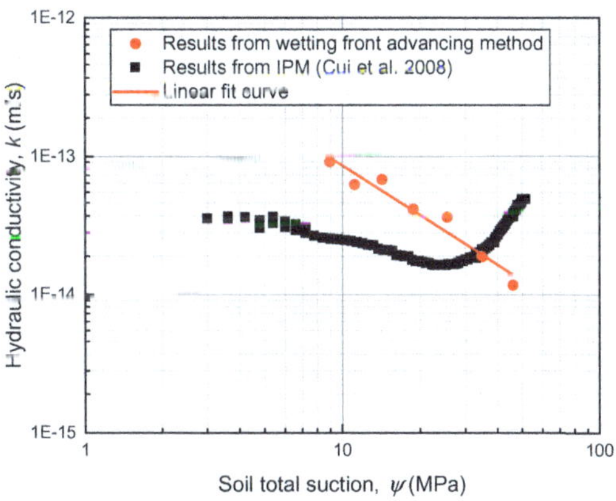

Fig. 7. Hydraulic conductivities at section A (h_A = 50 mm)

5 Data Analysis for the Column Test Reported by Ng and Leung (2012)

Ng and Leung (2012) reported another valuable data of column test where both the suction profiles and water content profiles at various time are reported (as those shown in Figs. 5 and 6 in Ng and Leung 2012). Based on these data of wetting test under zero vertical stress (SC0 test in Ng and Leung 2012), the wetting front advancing velocity is calculated and plotted in Fig. 8.

Similarly to the process used in Sect. 4, the wetting front advancing velocity is used to solve the hydraulic conductivity. The local initial water content at monitoring section is also used as θ_0 in the WFAM. This is a downward infiltration test, the sine of the flow direction, a will be used as 1.

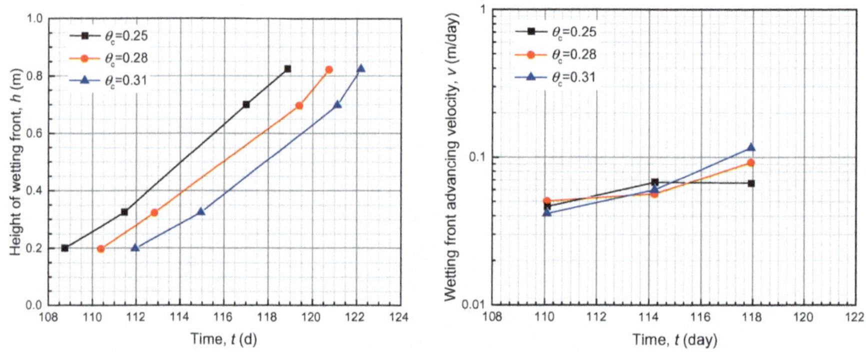

(a) wetting front advancing process (b) wetting front advancing velocity

Fig. 8. Wetting front advancing velocity analysis in infiltration process

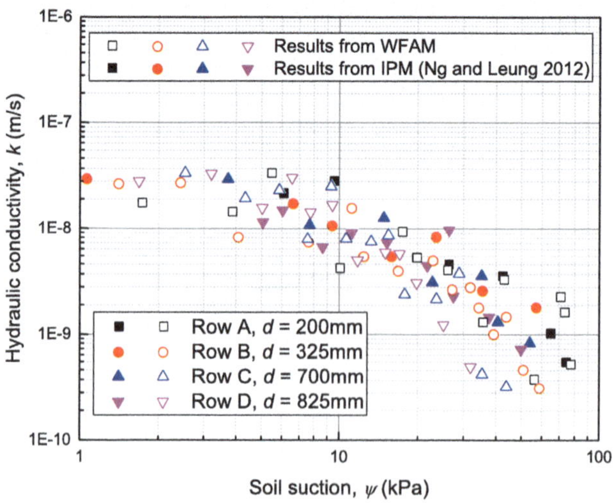

Fig. 9. Wetting front advancing velocity analysis in the infiltration process

The hydraulic conductivities calculated by using the wetting front advancing method are compared with those reported by Ng and Leung (2012) in Fig. 9. It is clear that the results from WFAM and IPM are consistent for this test.

6 Summary

As proved by the data analysis, wetting front advancing method is applicable to wetting column test and can offer high quality results for unsaturated hydraulic conductivities. The limitation of a clear wetting front that can be observed is not necessary and the wetting front can be inferred from the data monitored by sensor. The remove of this limitation will greatly expand the applicability of the wetting front advancing method.

However, wetting front advancing method is still not applicable to the drying process of soil (Fujimaki and Inoue 2003).

Acknowlegements. This research is supported by the National Natural Science Foundation of China (Nos. 51479001, 41471052).

References

Cui, Y., Tang, A.M., Loiseau, C., Delage, P.: Determining the unsaturated hydraulic conductivity of a compacted sand–bentonite mixture under constant-volume and free-swell conditions. Phys. Chem. Earth Parts A/B/C **33**, S462–S471 (2008)

Fujimaki, H., Inoue, M.: A flux-controlled steady-state evaporation method for determining unsaturated hydraulic conductivity at low matric pressure head values. Soil Sci. **168**, 385–395 (2003)

Li, X., Zhang, L.M., Fredlund, D.G.: Wetting front advancing column test for measuring unsaturated hydraulic conductivity. Can. Geotech. J. **46**, 1431–1445 (2009)

Lu, N., Likos, W.J.: Rate of capillary rise in soil. J. Geotech. Geoenviron. Eng. **130**, 646–650 (2004)

Ng, C., Leung, A.K.: Measurements of drying and wetting permeability functions using a new stress-controllable soil column. J. Geotech. Geoenviron. Eng. **138**, 58–68 (2012)

A Testing Device for Shear Strength of Sliding-Zone Soil and Its Application

Hao Li(✉) and Xiaojun Li

Xi'an University of Science and Technology, Xi'an 710054, ShaanXi, China
857841207@qq.com

Abstract. The shear strength of sliding-zone soil is the focus of slope stability evaluation. The shear strength of undisturbed sliding-zone soil is hard to attain by traditional testing method owning to its softy and high water content. A new testing method and device is put forward in this paper, in which the shear test of sliding-zone soil can be conducted in the circular knife, which reduces the disturbance and improves the accuracy. Meanwhile, the shear strength of soil under different water content can be obtained. In combination with the microscopic test technique, the microstructures of the shear surface of the sliding zone can also be observed. Results show that: (1) the method and device for determining the strength of the sliding-zone soil are effective and can be used to study the strength of the sliding-zone soil. (2) Due to the shear stress, minerals in the middle zone break away from their original position and they are squeezed into smaller ones. (3) It's proved that the landslides are related to the watercourse formed initially in the sliding zone.

Keywords: Shear strength · Sliding-zone soil · Water content

1 Introduction

The shear strength parameters of the slip zone soil are the basis for the quantitative evaluation of the landslide stability, the calculation of the landslide thrust and the remediation project. If the angle of internal friction varies from 1 to 2°, the thrust of the landslide may increase exponentially [1, 2]. Therefore, the accurate of soil strength parameters is of great significance [3]. A large number of studies have shown that the activities of the landslide depends on the changes in the stress state and intensity of sliding-zone soil, and these changes ultimately depend on the type, microstructure, material composition and formation mechanism of the slip zone and so on [4, 5].

Many of scholars have been studying the change of soil under the conditions of changing stress from the perspective of combining the macro performance with microcosmic changes. Among them, Liao et al. [6], Li et al. [7] studied the natural and saturated samples strength characteristics in sliding zone under different stress state and different water content using the ring shear method. Li [8], Li [9] believe that the results of three-axis cutting method is satisfying, but the sliding surface is difficult to determine while the operation is more difficult; remoded soil can gain the remain intensity by multiple straight shear, but the structure of the undisturbed soil can not be taken into account. The in-situ experiment of straight shear is more realistic. But owning to the

L. Hu et al. (Eds.): GSIC 2018, *Proceedings of GeoShanghai 2018 International Conference: Multi-physics Processes in Soil Mechanics and Advances in Geotechnical Testing*, pp. 54–61, 2018.
https://doi.org/10.1007/978-981-13-0095-0_6

limitation of the pit excavation and sample making, it is difficult to popularize. Surface strength measured by natural sliding surface shear test is the most ideal, but the sampling is more difficult. Xing et al. [10, 11] found that there were not direct relationships between the internal friction angle and the water content, The internal friction angle of the same soil was essentially a stable value.

Li et al. [12] suggested that the study of the mineral composition of sliding soil should be strengthened, and that the combination of multidisciplinary theory and method should be studied for micro and macro research. Zhang et al. [13] found that the sliding zone soil strength decreased rapidly with the water reaction. Shi et al. [14] studied the material composition and microstructure characteristics of sliding soil by using the microscopic identification and X-ray diffraction detection. Liu et al. [15] found that the microscopic morphology of the soil particles on the shear plane has obvious fractal characteristics by using scanning electron microscopy. Jiang et al. [16] quantitatively analyzed the inner and outer microstructures of Zhuhai seabed soft soil shear zone and found that there were significant differences in the grain arrangement between the shear zone and its edge.

These studies have contributed to the understanding of the macro-microscopic properties of sliding-zone soil. However, in the existing test method, there are some problems such as the difficulty of the indoor test sampling and the inability to observe the microcosmic change during the sliding process, especially when the water content is increased. Aiming at the above problems, a test method is proposed in which the disturbance of the soil under high water content can be avoided and the changes of different mineral forms and composition could be observed in the shearing process.

Taking into account the previous studies, some scholars believe that the water content on the remodeling of the internal friction angle of loess has little effect. In this article, from the perspective of water content impact on the cohesion, the Xi'an - Tongchuan Expressway along the Jingyang a landslide area Q2 reshaping loess is selected. Using the independently invented shear instrument, the change of microstructure and mineral shape in sliding zone can be observed. The microstructural characteristics of the shear zone are analyzed from the perspective of deformation and evolution of the particles in order to establish the relationship between the microscopic change and the macroscopic manifestation in the shear zone. It has certain theoretical and engineering practice for the slope stability analysis and prevention.

2 Test Equipment and Sample Preparation Steps

2.1 Shear Test Equipment, Sample Preparation and Test Procedures

Shear test equipment

The schematic of the strip-map shear device in the shear test equipment (Fig. 1) is shown in Fig. 2. Strip type pure shear device consists of left and right fixture and a pair of loading fixture in the middle part. The size of those are demonstrated in Fig. 2 (A-A′, B-B′) and Fig. 2 (C-C′). During the test, the position of the left and right pairs can not be changed, and the loading fixture move relatively vertically under the vertical external loads. At the same time ring knife matched with the pure shear device is

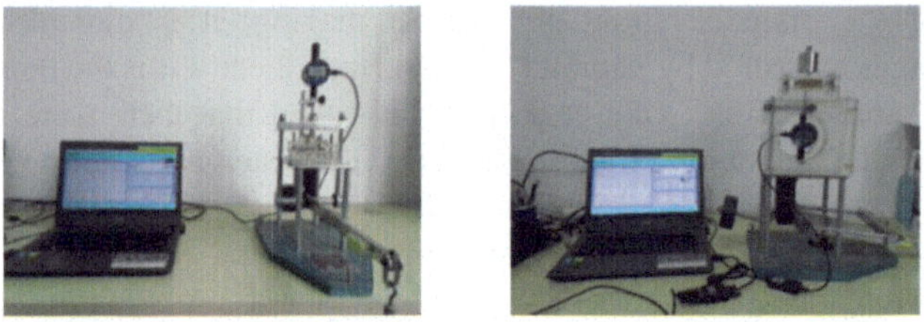

(a) loading by consolidation apparatus (b) loading by weight

Fig. 1. Experimental apparatus

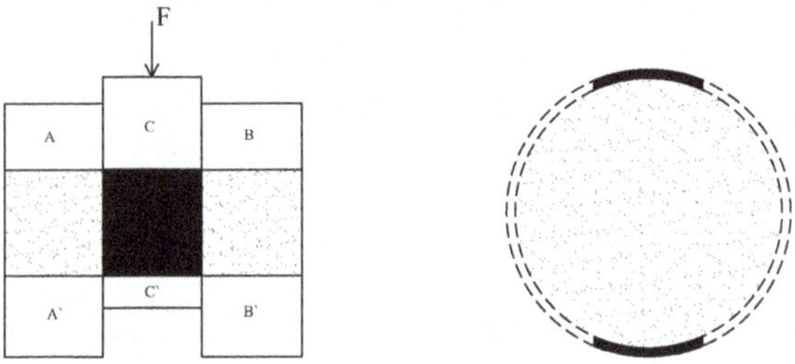

(a) Shear device (b) The four ring knife

A-A '—Left fixture; B-B' — Right fixture; C-C '—Loading fixture

Fig. 2. Strip-map shear device design schematic

designed. The ring knife consists of four arc-shaped part with 20 mm in height, 61.8 mm in diameter and 2 mm in wall thickness. In the preparation of soft original sliding zone soil, the four ring knife are stitched and tape is used to wrap around the ring knife to fix ring knife. The soft sample is prepared by sample cutting, and the adhesive tape is cut with a blade after the belt shearing device is installed. The large ring knife at both sides are removed while, two small ring knife are still remain, as shown in Fig. 2. The device prevents the sample from being pushed out of the ring knife and causes the sample to be disturbed.

For simplicity, the consolidation device has been modified serving as the loading device of this test as shown in Fig. 1(a), but when the water content is too high, the weight is directly loaded as shown in Fig. 1(b). The vertical displacement of the loading chuck is collected by means of an electronic digital micrometer, and the displacement is stored in the computer synchronously with time.

This device can not only be used to study the impact of water on the shear strength of soil. In combination with the microscopic test technique, the microstructures of the shear surface of the sliding zone can also be observed.

Pure shear test sample preparation

The loess selected for this experiment is a landslide and loess (Q_2) in Jingyang. The soil samples were light grayish yellow and mainly consist of silty clay with high water content. The soil is homogeneous and some areas have white salt. The natural water content was measured to be 9% and the dry density was 1.45 g/cm^3. Plastic limit water content is 17.26% and liquid limit water content is 26.3%.

In order to verify that if the equipment can be used to determine the strength of soft sliding zone soil under high water content, We dry soil samples naturally. After grinding, soil pass over 2 mm sieve. The dry density is 1.45 g/cm^3 and water content is 17%, 20%, 23%, and 26%, respectively. Four groups of remolded loess were prepared by pressing method. There are 3 samples in each group. The sample is cylindrical with a height of 20 mm and an inner diameter of 61.8 mm.

2.2 SEM Test Equipment, Sample Preparation and Test Procedures

SEM sample preparation

The soil sample with dry density of 1.45 g/cm^3 and 9% water content is selected. Developed shear equipment is used for research. As a preliminary attempt, this test only carried out a microscopic observation of the shear plane after load. A force of 38.5 N is applied to the sample, and the soil samples are cut and the two shear bands appeared. The configured epoxy resin solution is dropped into the shear slit and dropped once every hour until the epoxy solution was no longer infiltrated to cure it. The sample was allowed to stand for 24 h and placed in a thermostat and dried at a temperature of 60 °C. The soil samples are taken out of the ring knife after drying cut along the vertical direction of the shear band. A preliminary grinding and polishing of the section after the cutting shear band is carried out and a spray treatment is performed on the surface of the shear joint (Fig. 3).

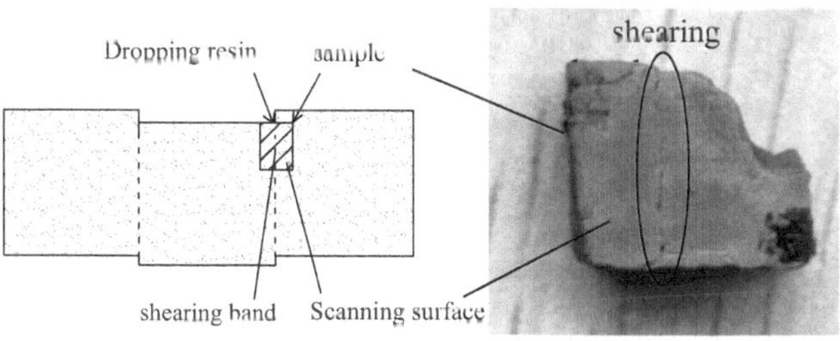

Fig. 3. Sample preparation for SEM

Electron microscopy scanning test

In this study, JSM-6390A scanning electron microscopy was used. The prepared sample is fixed with a conductive tape on a sample table and then sample table is placed in the scanner. After vacuumization, scanning start.

The mineral morphology of the edge of the shear zone and the mineral composition were magnified by 500 times to obtain the microstructure image. At the same time, the energy spectra of the seven elements of AL, Ca, Fe, K, Mg, Na and Si were selected for the post-processing.

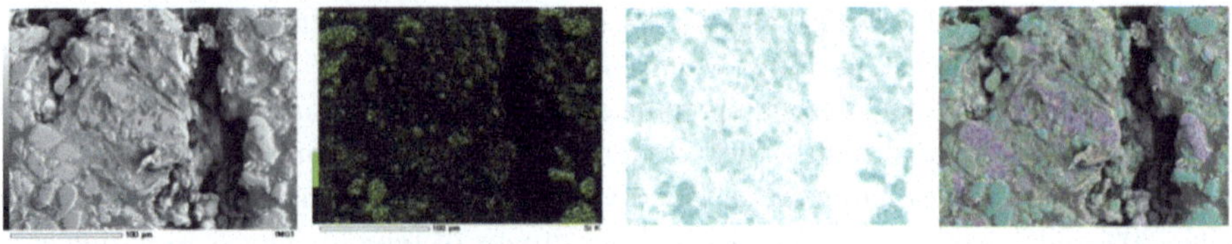

(a) base map (b) Si element (c) Give color (d) color film

Fig. 4. Microscopic image processing

The scanned energy spectrum image, such as the energy spectrum scanning image of Fig. 4(b) Si element, is endowed with the emerald color by using the Photoshop software, as shown in Fig. 4(c). Similarly, the other six elements are given different colors, and then the processed seven elements of the energy spectrum scan images superimposed on the base map. Color scanned films are shown in Fig. 4(d). The colored film is spliced to obtain the mineral morphology of the shear zone [17].

3 Experimental Results Analysis

3.1 Analysis of Shear Test Results

The shear test was used to complete the pure shear test of Jingyang Q_2 remolded loess in a dry density of 1.45 g/cm^3, four groups of water, and a total of 12 samples. There were three parallel samples in each group. The shear strength of the remolded loess is presented in Fig. 5.

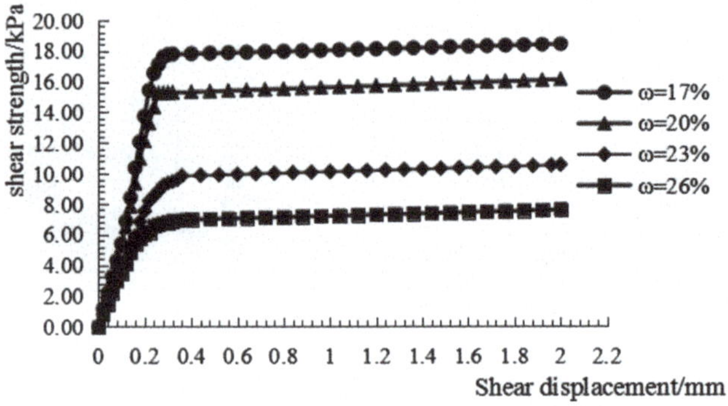

Fig. 5. The relational curves of shear strength with shear displacement of remolded loess

The results show the shear strength of remolded loess at the dry density of 1.45 g/cm^3 decreases with the increase of water content. At the same time, the reliability of the test equipment was verified.

3.2 Multi-scale Structural Analysis of Shear Zone

The chromatographic image of the processed scanning electron microscope is spliced as shown in Fig. 6(d).

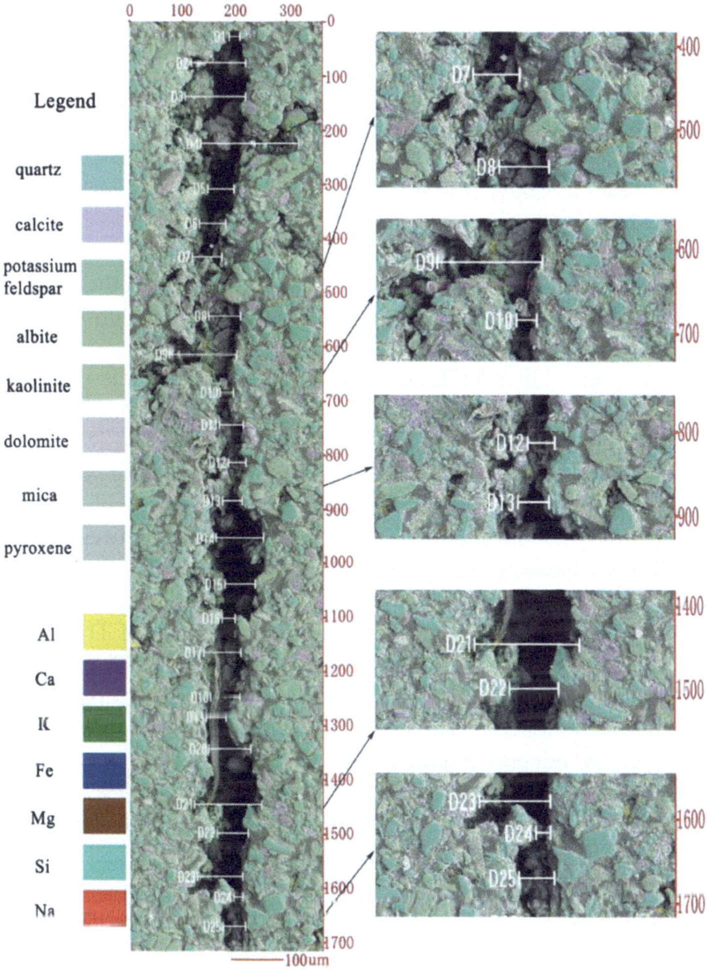

Fig. 6. Microstructure morphology of minerals in shearing band

The maximum slit width D4 is 183.46 μm, the shortest seam width D24 is 17.32 μm, and the average seam width is 65.56 μm, as shown in Table 1, using the computer. From Fig. 6 we can see Jingyang reshaping loess in the quartz, feldspar, carbonate mineral particles are relatively small. The minerals in the middle of the shear slit are separated from the original position by the shear stress, and the friction is increasing in the seam, so that the mineral particles are squeezed, rubbed, cut and become small particles of particles. At the same time the soil particles in the seam

Table 1. Statistics on seam widths of shearing band

Numbering	D1	D2	D3	D4	D5
Seam width/μm	19.90	102.33	113.18	183.46	49.354
Numbering	D6	D7	D8	D9	D10
Seam width/μm	49.35	55.56	59.69	126.87	24.55
Numbering	D11	D12	D13	D14	D15
Seam width/μm	45.22	32.56	37.21	88.89	56.59
Numbering	D16	D17	D18	D19	D20
Seam width/μm	24.55	69.77	55.30	36.96	78.30
Numbering	D21	D22	D23	D24	D25
Seam width/μm	126.36	58.65	85.01	17.31	42.12

move, friction makes small particles which located in two sides of the shear surface rub off. The particles in the seam were run off due to epoxy resin solution falling into the shear cracks in the process. The resulting shear seam is wider.

When epoxy resin solution infiltrated, the small mineral particles in cracks washed and blocked in the middle of D7 and D8 to form mineral aggregates. The same is true for small particles around D10.

There are many obvious mineral aggregates formed in the shear seam between D12 and D13 by shear action. The right side of the shear surface can be clearly observed the quartz mineral formed by shear action on the fracture surface between D12 and D13, but because the corresponding half of the other quartz mineral in the shear seam by rubbing, Cut, and the scouring of epoxy resin, it can not find the corresponding broken quartz. A long strip of filamentous objects located in between D17 and D21, the analysis is that a fiber embed into the grinding of the sample. The formation of the shortest shear gap located in D24, the mineral fracture and broken phenomenon appeared between D24 and D26. The particles can clearly be observed in the scanning plane. The degree of mineral fragmentation was serious. The entire seam showed dog-like staggered, wide width of the gap, no regular.

Soil under the shear action, the length of the shear zone is different, the internal small particles may be washed away by the epoxy resin solution, also proved that the slide in the sliding process, the belt will form a good water channel. When the rain is encountered, the landslide is prone to secondary slippage.

4 Conclusion and Suggestion

The pure shear test was carried out for Jingyang high water content remodeling loess. Through the microstructure test of soil shear joint, the following conclusions are obtained:

(1) The effectiveness of independently developed test equipment was verified and it can be used under high water content. The shear strength of remolded loess decreased with the increase of water content under the condition of dry density of 1.45 g/cm^3.

(2) Minerals in the middle of the shear fractures, due to shear stress, breaks away from the original position and the friction resistance increase. Consequently, mineral particles are squeezed, rubbed, cut to smaller mineral particles. At the same time, because of the movement of the soil particles, the tiny mineral particles in the two sides of the shear fractures are rubbed off, which makes the fracture wider.
(3) Through the measurement of the fracture width, it is proved that the second landslide is prone to slip due to the loose channel formed in the first sliding process.

References

1. Chen, C.-c., Zhang, J.-m.: Research on the determination of shear strength of sliding-zone soil. J. China Foreign Highway **32**(06), 65–70 (2012)
2. Sun, Z., Li, Z.-j., Xu, Y.-b., et al.: Parameters analysis and determination of shear strength for slip soil of Pangduo landslide. Water Res. Hydropower Northeast China **10,** 4–5+14+71 (2010)
3. Pan, H.: Method for determining strength parameters of slip soil. West-China Explor. Eng. **08**, 37–38 (2007)
4. Liu, X.-l., Deng, J.-h., Li, G.-t.: Shear strength properties of slip soils of land slides. An over view. Rock Soil Mech. **25**(11), 1849–1854 (2004)
5. Wang, J.: Experimental Study on Engineering Properties of Slip Soils of Landslides along G 212 Highway in Long-nan. Lan zhou University, Lan zhou (2006)
6. Liao, J.-m., Wu, Y.-p., Ou, G.-z., et al.: Shear strength the behavior of slide zone soil in ring shear tests. J. Wuhan Univ. Technol. **35**(10), 92–95 (2013)
7. Li, L.-y., Dang, J.-q., Wang, W.: Research on ring-shear test of loess shear resistance strength in Jingyang. J. Xi'an Univ. Technol. **31**(01), 113–119 (2015)
8. Li, T.-d.: Determination method of shear strength of sliding-zone soil. Mount. Res. **2**(01), 25–30 (1984)
9. Li, X.-g.: Determination of shear strength parameters of sliding-zone soil. West-China Explor. Eng. **02**, 37–38 (2010)
10. Xing, X.-l., Li, T.-l., Ju, K.-l., et al.: Research on the strength parameters of unsaturated loess. J. Eng. Geol. **23**(2), 252–259 (2015)
11. Xing, X.-l., Li, T.-l., Li, P., et al.: Variation regularities of loess shear strength with the moisture content. Hydrogeol. Eng. Geol. **41**(3), 53–59+97 (2014)
12. Li, X., Liang, S.-y.: Progresses in sliding zone soil of landslides. Adv. Earth Sci. **05**, 484–491 (2010)
13. Zhang, Y.-c., Yang, G.-h., Hu, H.-y., et al.: A test research on slip soil of some landslide. In: The Third National Geotechnical and Engineering Academic Conference, pp. 231–236 (2009)
14. Shi, G., Bao, X.-b., Lian, C.: Study on shear strength of slip soil of Yangjiao landslides group. Yangtze River **42**(22), 54–56 (2011)
15. Liu, D., Chen, X.-p.: Microscopic observation and analysis of ring shear surface of slip zone soil. Chin. J. Rock Mech. Eng. **32**(9), 1827–1834 (2013)
16. Jiang, M.-j., Peng, L.-c., Zhu, H.-h., et al.: Microscopic investigation on shear band of marine clay in Zhuhai, China. Rock Soil Mech. **31**(7), 2017–2029 (2010)
17. Li, X.-j., Wang, X.-h., Xie, X.-t., et al.: A mineral component identification method based on mineral standard color database. China: CN104700097A, 10 June 2015

Consolidation Model for Gassy Soils Considering the Compression Coefficient of Air-Water Mixture

Fan-Ge Zhang[1], Kan Liu[1,2](✉), Xiao-Jun Zhu[2], and Min Yang[1]

[1] Department of Geotechnical Engineering, Tongji University, Shanghai, China
tjliukan@gmail.com
[2] Shanghai Shen Yuan Geotechnical Engineering Co., Ltd., Shanghai, China

Abstract. The gassy soil or quasi-saturated soil which containing entrapped gas bubbles is a special type of unsaturated soils. Gassy soils could exist in many situations such as offshore geotechnical engineering and water infiltration under heavy rains. For this type of soil, the deformations due to the air compression cannot be ignored, which is usually described as the compressibility of the air-water mixture. This paper aims to calculate the compression coefficient of air-water mixture and propose the consolidation equation of gassy soils. Based on the conceptual model of small bubbles, deformation characteristics of entrapped gas bubbles, Bolye's law of ideal gas and surface tension effects, the compression coefficient of air-water mixture in gassy soils is derived. The presence of gas could influence the process of soil consolidation, and the entrapped gas bubbles are considered to be drained off along with pore water in the process of consolidation. The deformation of the soil during consolidation could be divided into two parts: one part is the compression deformation due to draining of air-water mixture, and the other part is the compression of the air-water mixture. The derived compression coefficient equation of air-water mixture is used to establish the consolidation equation of gassy soils. Calculation results using our model are compared with the experimental data from literatures, and the results show that the proposed model could properly reflect the consolidation behavior of gassy soils.

Keywords: Gassy soils · Air-water mixture · Compression coefficient Consolidation equation

1 Introduction

In many practical projects, soils are unsaturated. Researchers have already done a great deal of theoretical and experimental research on the unsaturated soils. All unsaturated soil may be separated into three major categories [1]: (1) unsaturated soil with connected pore air and connected pore water; (2) unsaturated soil with connected pore air and disconnected pore water; (3) unsaturated soil with disconnected pore air and connected pore water in the skeleton. The third type of soil is usually called gassy soil. Faybishenko [2] has proposed a concept of quasi-saturation to describe the porous materials with entrapped gas bubbles.

L. Hu et al. (Eds.): GSIC 2018, *Proceedings of GeoShanghai 2018 International Conference: Multi-physics Processes in Soil Mechanics and Advances in Geotechnical Testing*, pp. 62–70, 2018.
https://doi.org/10.1007/978-981-13-0095-0_7

In many cases, the pore air and water are assumed to be a compressible mixture so as to simplify the analysis, and then the consolidation behavior can be analyzed [3–5]. Due to the biochemical process, soils in the coastal area mostly contain gas bubbles. Some researchers carried out a thorough research on gassy soils with the reference of offshore geotechnical projects. Nageswaran [6] and Thomas [7] have studied the consolidation characteristic of coastal gassy soils from the perspectives of sample preparation, consolidation test, theoretical analysis and numerical simulation. Wheeler [1] has explored the existing form and size of pore gas bubbles. And on this basis, a conceptual model of gassy soil was proposed to predict the consolidation properties. The size of entrapped gas bubble has a direct impact on soil mechanical properties. The research on soils containing either large or small bubbles has been carried out much further by Pietruszczak [8]. There are many other researchers who have done relevant studies on quasi-saturated soils [9].

This paper focus on the third type of unsaturated soil which is mentioned above. The gas bubbles are small enough so that disconnected pore air and connected pore water are existed in the skeleton of the soils. It is assumed that the air-water mixture of the gassy soil is compressible. On the basis of this assumption, an equation is derived to calculate the compression coefficient of air-water mixture. And then the consolidation equation of gassy soils is developed based on compression coefficient. Finally the calculation results are compared with the experimental data from literatures.

2 The Compression Coefficient of Air-Water Mixture

The existence of gas bubbles would significantly increase the compressibility of the air-water mixture. Fredlund [5] pointed out that as low as one percent of air content could have a significant impact on the compressibility. Thus, the compressibility of the air water mixture due to existence of gas bubbles cannot be neglected.

The general definition for the volume compressibility C is:

$$C = -\frac{1}{V}\frac{dV}{du} \tag{1}$$

in which u is the pore fluid pressure or the reference pressure; V is the total volume

For the air-water mixture, the total volume V is the sum of the pore water volume V_w and the gas bubbles volume V_g, it can be expressed as:

$$V = V_w + V_g \tag{2}$$

Thus the compressibility of the air water mixture can be calculated using the following equation:

$$C_{gw} = -\frac{1}{V_{g0} + V_{w0}}\frac{d(V_g + V_w)}{du} \tag{3}$$

in which u is a reference pressure for the mixture.

For simplification of analysis, some assumptions are made that the shapes of all the entrapped gas bubbles are uniform spheres, all the gas bubbles are of a size, the pressures in the bubbles are the same, which is the pore air pressure u_g.

The entrapped gas bubbles in the air-water mixture are regarded as the ideal gas, following Bolye's law on the condition of constant temperature, so the relation between the gas bubbles volume and the pore air pressure is as follows:

$$V_{g0}(u_{g0} + P_a) = V_g(u_g + P_a) \tag{4}$$

where the subscript $_0$ represents the initial state; *Pa* is the atmospheric pressure, equal to 101.3 kPa.

Considering the impact of the interface tension of air and water, the relation of the pore air pressure u_g and the pore water pressure u_w can be expressed as

$$u_g = u_w + 2q/r \tag{5}$$

where q is the surface tension coefficient; r is the radius of curvature of the gas bubble. The value of q decreases with increasing air pressure, however, Schuurman [10] reported that q value variation can be neglected and q is equal to 7.4×10^{-3} N/m. For small bubbles, r is equal to bubble radius.

According to Henry's laws, the volume of gas dissolved in water can be calculated as follows:

$$V_{da} = HV_w \tag{6}$$

where H is the coefficient of solubility. The value of H decreases with increasing air pressure, but the changes could be neglected and H is equal to 0.02 at 20 °C.

Based on the above, as the pore water pressure u_w changes, the gas volume V_g changes accordingly, so the pore water pressure u_w is used as a reference pressure. Because of the incompressibility of pure water without air, the pore water volume V_w can be considered to maintain invariable, then Eq. (3) becomes:

$$C_{gw} = -\frac{1}{V_{g0} + V_{w0}} \frac{dV_g}{du_w} \tag{7}$$

Based on the Bolye's law, the following equation is obtained:

$$(V_g + V_{da})(u_g + P_a) = (V_{g0} + V_{da0})(u_{g0} + P_a) \tag{8}$$

The relation between the pore water pressure and the pore air pressure is:

$$u_{g0} = u_{w0} + 2q/r_0, \; u_g = u_w + 2q/r \tag{9}$$

Assuming the shapes of all the entrapped gas bubbles are uniform spheres and the number of gas bubbles in the air-water mixture to be N, then:

$$V_{g0} = N\frac{4}{3}\pi r_0^3,\ V_g = N\frac{4}{3}\pi r^3 \tag{10}$$

This results in:

$$\frac{r}{r_0} = (\frac{V_g}{V_{g0}})^{\frac{1}{3}} \tag{11}$$

Substituting Eqs. (9) and (11) into Eq. (8), following relation is obtained:

$$\frac{dV_g}{du_w} = \left[-\frac{V_{g0}+V_{da0}}{(V_g+V_{da})^2}(u_{w0}+\frac{2q}{r_0}+P_a)+\frac{2q}{3r_0}\frac{1}{V_g}(\frac{V_{g0}}{V_g})^{\frac{1}{3}}\right]^{-1} \tag{12}$$

which inserted in Eq. (7) gives:

$$C_{gw} = \frac{1}{V_{g0}+V_{w0}}\left[\frac{V_{g0}+V_{da0}}{(V_g+V_{da})^2}(u_{w0}+\frac{2q}{r_0}+P_a)-\frac{2q}{3r_0}\frac{1}{V_g}(\frac{V_{g0}}{V_g})^{\frac{1}{3}}\right]^{-1} \tag{13}$$

For those cases when the water saturation is close to 100% or equal to 100%, only gas dissolved in water is left. The calculated C_{gw} is still significantly affected by the pore water pressure u_w. The reason is that the dissolved air also obey Bolye's law in the equation. However, Dorsey [11] had measured the compressibility of saturated water with dissolved air at different temperatures and pressures, and found that this compressibility has little changes at different temperatures and pressures. Researchers such as Fredlund had argued that this compressibility value can be taken as 4.58×10^{-7} kPa^{-1} from Dorsey's tests.

This article suggests that it is more reasonable to deal with the dissolved gas by referring to Dorsey's test results. We will take into account the dissolved gas and water as a whole whose compressibility C_w is equal to 4.58×10^{-7} kPa^{-1}. S_O is the initial saturation degree. Therefore, the compressibility of the air-water mixture C_{gw} can be modified into:

$$C_{gw} = \frac{1}{V_{g0}+V_{w0}}\left[\frac{V_{g0}}{V_g^2}(u_{w0}+\frac{2q}{r_0}+P_a)-\frac{2q}{3r_0}\frac{1}{V_g}(\frac{V_{g0}}{V_g})^{\frac{1}{3}}\right]^{-1}+S_0C_w \tag{14}$$

3 Consolidation Model for Gassy Soils

Through the analysis of compressibility of the air-water mixture above, the deformation of the gassy soils during consolidation could be divided into two parts: one part is the compression deformation due to draining of air-water mixture, and the other part is the compression of the air-water mixture. So the volume conservation equation of gassy soils in the process of consolidation can be expressed as:

$$dQ = dV_v + dV_{gw} \tag{15}$$

where dQ is the total volume of air-water mixture outflowing from a soil element; dV_v is the volume of the air-water mixture due to pore pressure dissipation; dV_{gw} is the volume of the compression of the air-water mixture.

Based on the analysis above, the constitutive equation of one-dimensional consolidation for gassy soils can be written as:

$$\frac{k}{\gamma_{gw}}\frac{\partial^2 u_w}{\partial z^2}dxdydzdt = \frac{1}{D}\frac{\partial u_w}{\partial t}dxdydzdt + \frac{n}{D_{gw}}\frac{\partial u_w}{\partial t}dxdydzdt \tag{16}$$

where k is the permeability coefficient of air-water mixture; γ_{gw} is the unit weight of air-water mixture; n is the porosity of gassy soil; D is the compression modulus of the soil skeleton; D_{gw} is the compression modulus of the air-water mixture.

The unit weight of air-water mixture γ_{gw} can be obtained by definition:

$$\gamma_{gw} = S\gamma_w \tag{17}$$

where S is the saturation of gassy soil at any moment; γ_w is the unit weight of water.

Equation (16) can be written as the standard form of one-dimensional consolidation equation:

$$C_v\frac{\partial^2 u_w}{\partial z^2} = \frac{\partial u_w}{\partial t} \tag{18}$$

where C_v is the consolidation coefficient. Considering the compressibility of air-water mixture, C_v can be written as:

$$C_v = \frac{k/(S\gamma_w)}{\frac{a}{1+e} + \frac{e}{1+e}C_{gw}} \tag{19}$$

The initial and boundary condition in the one-way drainage test can be expressed as:

$$\begin{array}{lll} t = 0, & 0 \le z \le H, & u_w = \sigma; \\ 0 < t < \infty, & z = 0, & u_w = 0; \\ 0 < t < \infty, & z = H, & \frac{\partial u_w}{\partial z} = 0. \end{array}$$

in which, z is the depth of the soil sample, σ is the total stress applied, H is the maximum water discharging distance.

When C_v is constant, the theoretical solution of Eq. (18) can be derived:

$$u_w = \sigma \sum_{m-0}^{\infty} \frac{2}{M} \sin \frac{Mz}{H} \exp(-M^2 T_v) \tag{20}$$

in which,

$$M = \frac{1}{2}(2m+1)\pi, T_v = \frac{C_v}{H^2} t \tag{21}$$

The average degree of consolidation can be defined using pore pressure:

$$U = 1 - \sum_{m=0}^{\infty} \frac{2}{M^2} \exp(-M^2 T_v) \tag{22}$$

In fact, Eq. (18) is a partial differential equation with varied coefficient. Because C_v is concerned with the pore water pressure u_w, it is necessary to dividing time steps into much smaller intervals and the consolidation coefficient is considered as a constant at each interval during each time step. Substituting the consolidation coefficient into Eq. (20), the pore water pressure u_w can be obtained. And then C_v in the next interval can be calculated by u_w. After obtaining the pore water pressure and the average degree of consolidation, the deformation of sample can be calculated:

$$s(t) = U(t) \cdot s_f \tag{23}$$

where $s(t)$ is deformation at time t, $U(t)$ is the average degree of consolidation at time t, s_f is the final displacement which can be computed with the volume compressibility.

Specifically, at fully saturated state, the effect of pore air is neglected, which means C_{gw} is equal to 0, the derived solution will degenerate into Terzaghi's consolidation equation, and the calculated results are the same.

4 Model Validation

For validating the consolidation model, the calculation results are compared with the experimental data from Nageswaran [6]. In the literature, a technique was presented to make remoulded gassy soils with different gas contents. Using the remoulded samples, tests under different initial degrees of saturation and different consolidation pressures were carried out. The initial thickness of each sample is 40 mm and diameter 100 mm. Some of the test results are used to compare with the calculation results. The initial degrees of saturation S are 80%, 90% and 95% while the consolidation pressure levels include 50 kPa, 100 kPa and 200 kPa. Multi-stage loading was adopted. The loading interval is about 11 h, corresponding data are extracted from the calculation results. The comparison of calculation results and experimental data is shown as Figs. 1, 2 and 3.

In general, the model simulations compare well with the experimental results. As shown, the total settlement increases with the increase in gas content. However the time

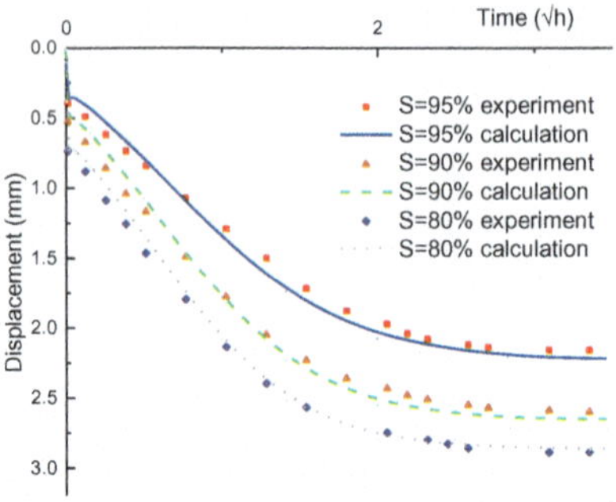

Fig. 1. The displacement-time curves under the consolidation stress of 50 kPa.

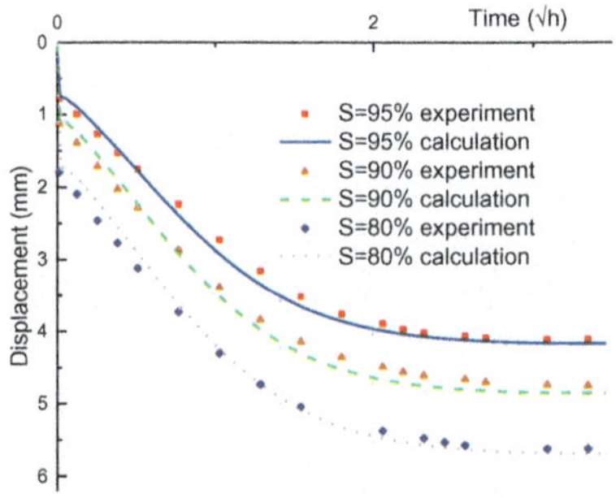

Fig. 2. The displacement-time curves under the consolidation stress of 100 kPa.

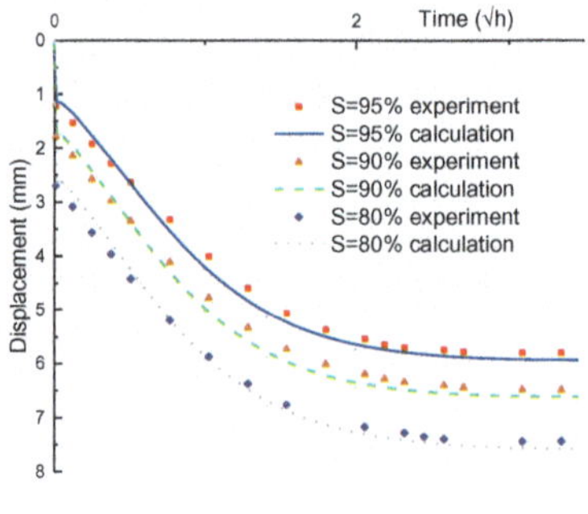

Fig. 3. The displacement-time curves under the consolidation stress of 200 kPa.

taken to achieve basic stable settlement from these curves for different gassy soils shows no significant variation. In the first 3–4 h, the displacement increases rapidly. And then the increase is slowed and the displacement gradually flatten within about 11 h. Significantly, there is an obvious instantaneous deformation at the initial loading. This is due to the compressibility of the air-water mixture in the gassy soils at the beginning of loading. The instantaneous deformation increases with the gas content and consolidation stress. Therefore, this directly caused the different total settlement of the samples.

5 Conclusion

In this study, based on some simplified assumptions, the compression coefficient of air-water mixture in gassy soil is derived. And then, the consolidation equation of gassy soils is established. Compared with experimental results presented in the literature, the following conclusions are obtained:

(1) The proposed one-dimensional consolidation equation of gassy soil could properly reflect the consolidation behavior of the gassy soil even though some simplified assumptions are made. The calculation results compare well with the experimental results.
(2) Immediate displacement change occurs in a gassy soil as the air-water mixture is compressed when the pressure is applied. And the subsequent consolidation rate decreases as the gas bubbles expand.
(3) Under the same consolidation pressure, the lower the saturation, the larger the displacement. Therefore, it is reasonable to consider the compressibility of the pore air-water mixture during the consolidation computation of gassy soil.

The article just discusses the preliminary research achievement about the compressibility of gassy soil. Based on this research, follow-up studies will be taken.

Acknowledgements. The authors acknowledge the support received from Dr. Jianfeng Xue and related organizations.

Funding. This work is supported by the National Natural Science Foundation of China under Grant No. 41572258.

References

1. Wheeler, S.J.: The undrained shear strength of soils containing large gas bubbles. Géotechnique **38**(3), 399–413 (1988)
2. Faybishenko, B.A.: Hydraulic behavior of quasi-saturated soils in the presence of entrapped air: laboratory experiments. Water Resour. Res. **31**(10), 2421–2435 (1995)
3. Wei, H., Zhan, L., Chen, Y.: Compressibility and consolidation of unsaturated soils with high degree of saturation and its application. Chin. J. Geotech. Eng. **28**(2), 264–269 (2006)
4. Xu, H.: Studies on the consolidation theory for unsaturated soft soils containing biogas [Ph.D. thesis]. Zhejiang University, Hangzhou (2011)
5. Fredlund, D.G., Rahardjo, H., Fredlund, M.D.: Unsaturated Soil Mechanics in Engineering Practice. Wiley, Hoboken (2012)
6. Nageswaran S.: Effect of gas bubbles on the sea bed behavior [Ph.D. thesis]. The University of Oxford, Oxford (1983)
7. Thomas S. D: The consolidation behaviour of gassy soil [Ph.D. thesis]. The University of Oxford, Oxford (1987)
8. Pietruszczak, S., Pande, G.N.: Constitutive relations for partially saturated soils containing gas inclusions. J. Geotech. Eng. **122**(1), 50–59 (1996)

9. Sills, G.C., Wheeler, S.J., Thomas, S.D., et al.: Behaviour of offshore soils containing gas bubbles. Géotechnique **41**(2), 227–241 (1991)
10. Schuurman, I.E.: The compressibility of an air/water mixture and a theoretical relation between the air and water pressures. Geotechnique **16**(4), 269–281 (1966)
11. Dorsey, N.E.: Properties of Ordinary Water-Substance. Reinhold, New York (1940)

Thermal Water Retention Characteristics of Compacted Bentonite

Haiquan Sun(✉), David Mašín, and Jan Najser

Faculty of Science, Institute of Hydrogeology, Engineering Geology and Applied Geophysics, Charles University, Albertov 6, 128 43 Prague, Czech Republic
haiquan.sun@natur.cuni.cz

Abstract. Compacted bentonites are popularly being considered as buffer or backfill material in high level nuclear waste repositories around the world. These bentonites may undergo various conditions including hydration from around geo-environmental water and heating by the radiation of the used fuel during its period of operation. Water retention properties of as-compacted Czech bentonite B75 with three initial dry densities (1.27 g/cm^3, 1.60 g/cm^3 and 1.90 g/cm^3) and bentonite powders were investigated within temperature 20–80 °C at unconfined conditions. Vapor equilibrium method was used to control constant relative humidity. The influence of temperature on water retention properties was analyzed and discussed. Results show that the temperature decreased water retention capacity for all cases. The water retention capacity is lower at high temperature especially at lower suction. The temperature has more significant effect on drying path than wetting path. Furthermore, the volume swelling decreased with the increased temperature upon saturation. The hysteretic behavior decreased with the increase of temperature for all studied materials.

Keywords: Bentonite · Temperature · Water retention capacity

1 Introduction

Compacted bentonites are usually considered as engineered barrier material or backfill material in high level nuclear waste disposal. These compacted bentonites are used to fill the gap between canister and host rock, or as backfill material to seal the nuclear waste disposal tunnel galleries. Under repository conditions, the compacted bentonite will be hydrated by absorbing underground water from the host rock. Because the manufactured bentonite pellets are usually of cylinder shape, unavoidable voids between bentonite pellets. The voids between the compacted bentonite pellets will be filled by the swelling volume of compacted bentonite when subjected to hydration. During this period, the bentonite swells freely, as it is in unconfined conditions. Once they fill the gaps, surrounding host rock will constraint the bentonite volume increase, will be at constant volume conditions. The swelling pressure develop among host rock, bentonite and canister.

Water retention capacity of compacted bentonite is one of the important characteristics of hydro-mechanical behavior. During the operation of the repository,

L. Hu et al. (Eds.): GSIC 2018, *Proceedings of GeoShanghai 2018 International Conference: Multi-physics Processes in Soil Mechanics and Advances in Geotechnical Testing*, pp. 71–78, 2018.
https://doi.org/10.1007/978-981-13-0095-0_8

radiogenic heat released from the waste container invade the bentonite. For the safety of the repository, it is necessary to study the influence of temperature on water retention properties of compacted bentonite.

It is mostly accepted that the temperature can change the liquid–gas interfacial tension and thus, in turn, it affects water retention curves [1]. The increased temperature can decrease the swelling pressure [2]. Numerical modelling of the behavior of unsaturated soil under non-isothermal conditions has been subject of many studies. For example, Yao and Zhou [3] studied the change of yield surface under heating, Zhou et al. [4] and François and Laloui [5] developed models, which were based on the existing elastoplastic models, extended from isothermal condition to non-isothermal condition. Mašín [6] developed a THM double structure hypoplastic model for expansive soils.

In this paper, several tests will be described which were performed to investigate the influence of temperature on water retention curves of compacted bentonite and bentonite powders under unconfined conditions. The controlled temperature was 20 °C and 80 °C respectively. The initial dry density of compacted bentonite was 1.27 g/cm^3, 1.6 g/cm^3 and 1.9 g/cm^3 respectively and other bentonite was also studied in powder state. The results will be compared and discussed.

2 Materials

The commercial Czech B75 bentonite extracted from the Cerny vrch deposit (north-western region of the Czech Republic), was used in this study. It is a calcium magnesium bentonite with a montmorillonite content around 60% and initial water content about 10%. The plastic limit, liquid limit and specific density of solid particles were 65%, 229%, and 2.87, respectively. These compacted bentonites and powders has the same mineral compositions. For more detailed information about the Czech bentonite B75, see Sun et al. [7].

3 Methods

Vapour equilibrium method described in Delage et al. [8] was applied to control the suction. The relative humidity in the closed desiccator was controlled by different saturated salt solutions. The total suction has a unique relationship with relative humidity described by the following equation:

$$S_t = (RT\rho_w/w) * \ln(1/RH) \tag{1}$$

Where S_t is the total suction (kPa); R is the molar gas constant, which equals to 8.314462 J/(mol K); T is the absolute temperature (K); ρ_w is the density of water; w is the molecular mass of water vapour, which equals to 18.016 g/mol; RH is relative humidity of the system which is defined as the ratio of partial pressure of vapour over saturation vapour pressure.

The relative humidity and total suction of each saturated salt solution under 20 °C and 80 °C applied in this paper are listed in Table 1. The prepared samples of each different initial dry density were dried in the oven at 110 °C for over 24 h. Then, the samples were cut into small irregular pieces weighing between 0.8 and 1.5 g. These small pieces were placed on the ceramic plate above the saturated salt solutions in each desiccator for wetting test. For 80 °C, all the desiccators were put in the temperature controlled environment. The samples could freely expand during wetting. It usually took about 2 months to reach equilibrium. More detailed about the volume measurement procedures are given in Sun et al. [12]. Thus, the void ratio and degree of saturation can be obtained by knowing the sample volume and water content.

In water retention curve measurements half of the samples was tested directly under wetting path starting from the oven-dried state, while the other half was equilibrated at 3.29 MPa at 20 °C using vapour equilibrium method and tested under drying path at 20 °C and 80 °C. The detailed experimental procedures are shown in Fig. 1.

Table 1. Relative humidity and suction imposed by saturated saline solution for vapour equilibrium methods. (adopted from Greenspan [9], OIML [10], Tang [11]).

Salt solutions	Relative humidity (%)		Suction (MPa)	
	20 °C	80 °C	20 °C	80 °C
LiCl·H2O	11.31	10.51	290.05	361.17
CH3COOK	13.11		270.39	
MgCl2·6H2O	33.07	26.05	147.26	215.65
K2CO3	43.16	35.42	111.82	166.39
NaBr	59.1	51.4	69.99	106.70
NaCl	75.47	76.29	37.45	43.39
KCl	83.1	78.9	21.47	37.99
K2SO4	97.2	95.5	3.78	7.38

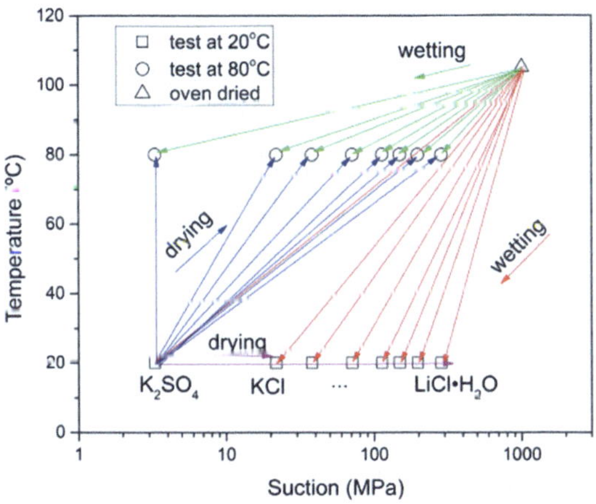

Fig. 1. Wetting and drying paths under 20 °C and 80 °C.

4 Results and Discussion

4.1 Water Retention Curves at 20 °C

Figure 2(a) shows the water retention curves of samples at three initial dry densities (wetting and drying path). It is clear that the initial dry density had little influence on the water content, which increases almost linearly with logarithm of suction. Much more significant was the effect of suction on void ratio (Fig. 2(b)). As most water is in the micro-pores at high suctions, these results suggest that micro-pores are only little influenced by the compaction pressures as if their volume would be affected, also their water retention capacity would change. Figure 2(c) shows the relationship between degree of saturation and suction. It is affected by the dry density substantially, but not due to the effect of water content, which is more or less the same at the three dry densities for the given suction, but due to the effect of different void ratios.

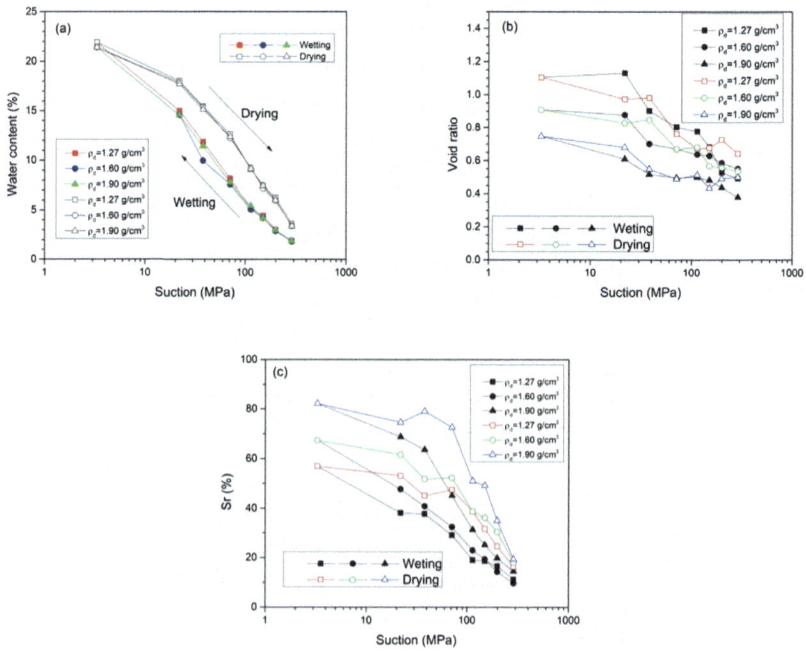

Fig. 2. Water content (a), void ratio (b) and degree of saturation (c) versus suction for water retention curve measurements under wetting and drying path at three different dry densities at 20 °C.

4.2 Temperature Effects on WRC

The water retention curves at unconfined conditions, following a wetting-drying path, are determined for two dry densities of 1.27 g/cm^3 and 1.90 g/cm^3 and temperature of 20 °C and 80 °C. The wetting path starts from the oven dried samples (suction nearly 1000 MPa [13]) successive to suction of 3.29 MPa, and then, a drying path continuing

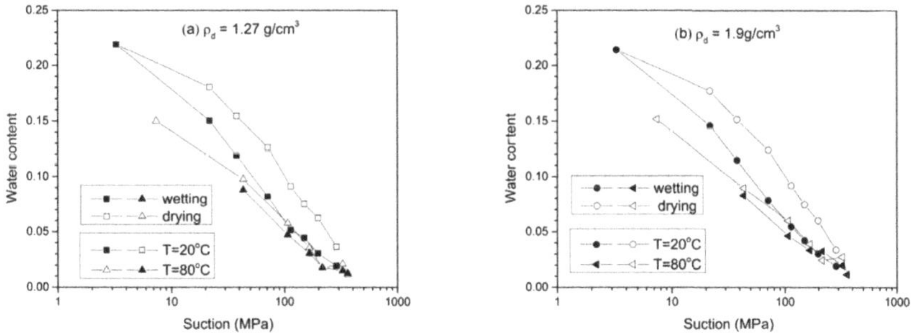

Fig. 3. Water retention curve of compacted bentonite with initial dry density of 1.27 g/cm^3 (a) and 1.90 g/cm^3 (b) under 20 °C and 80 °C.

to 286.7 MPa. The wetting and drying curves of compacted bentonite were obtained in terms of equilibrium water content and suction in Fig. 3.

The water retention capacity for the same dry density is lower for the higher temperature, especially for the lowest suction. It can be seen that the equilibrium water content decreases as the temperature increases for both low and high dry density samples. The water retention capacity decreases less with the increase of total suction. The lower retention capacity at higher temperature is clear in the wetting after drying path for both dry densities (Fig. 3). It is more obvious when the wetting and drying paths are separated at each dry density as shown in Fig. 4. As for wetting test (Fig. 4a, c), the temperature seems to have a minor influence on the water retention capacity compared with drying path (Fig. 4b, d). In the low suction range, the water retention capacity decreases more with increase of temperature. On the contrary, it decreases less with increasing temperature in the high suction range.

The water retention curves of bentonite powders at 20 °C and 80 °C are shown in Fig. 5. It is obvious that increase of temperature decreases the water retention capacity like compacted samples. As mentioned before (Fig. 2), the equilibrium water content versus suction is practically independent of initial dry density. So, the highest initial dry density of 1.90 g/cm^3 was chosen to compare with bentonite powders as shown in Fig. 6. It can be seen that both compacted bentonite and bentonite powder follows the same trend with only a little effect of density on water retention curves at 20 °C. However, water retention properties of bentonite powder seem more sensitive to increase of temperature, the water retention capacity decrease more when compared to compacted bentonite at 80 °C.

It can be seen that the temperature not only affects the water retention capacity but also influences the hysteretic behavior. The hysteretic loop was narrower with the increase of temperature (Figs. 3, 5 and 6). It means that the hysteretic behavior will be less obvious at higher temperature.

Figure 7 shows the change of void ratio during wetting under unconfined conditions for low and high initial dry densities. For higher suction (100 MPa in this paper), the volume changes are small and less depended on temperature (Fig. 7a). Upon wetting, volume swelling is higher at 20 °C for both dry densities.

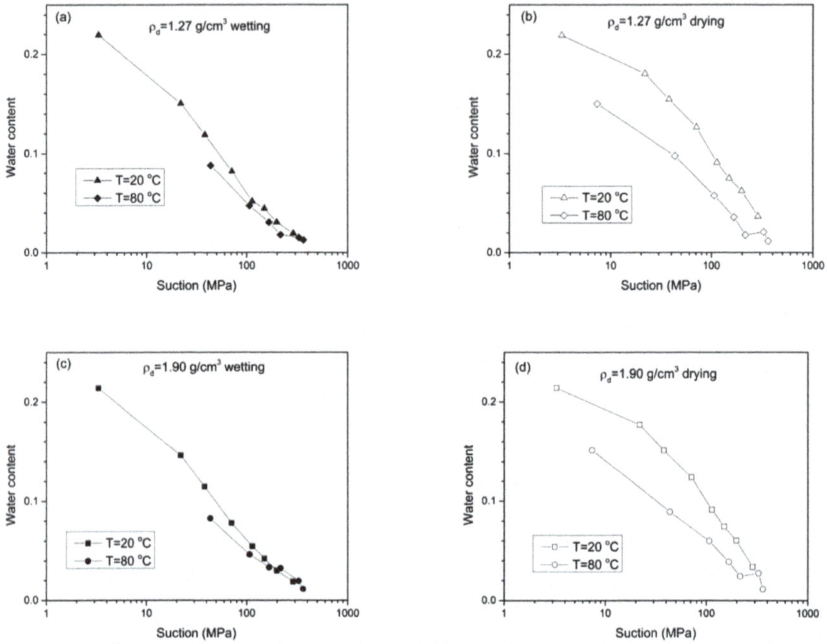

Fig. 4. Wetting (a) and drying (b) paths of initial dry densities of 1.27 g/cm^3 and wetting (c) and drying (d) paths of 1.90 g/cm^3 samples under 20 °C and 80 °C.

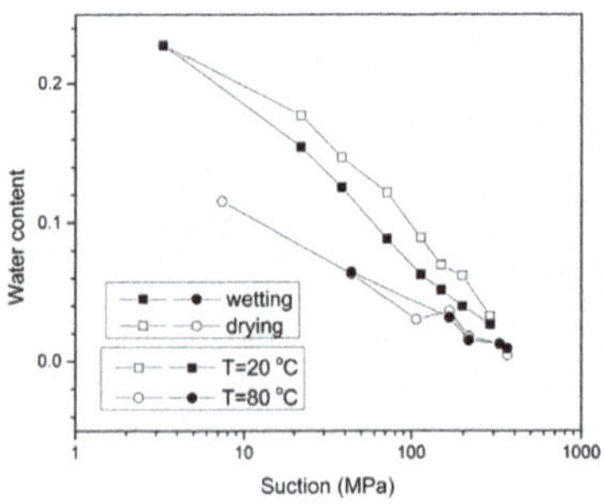

Fig. 5. Drying and wetting path for bentonite powders under 20 °C and 80 °C.

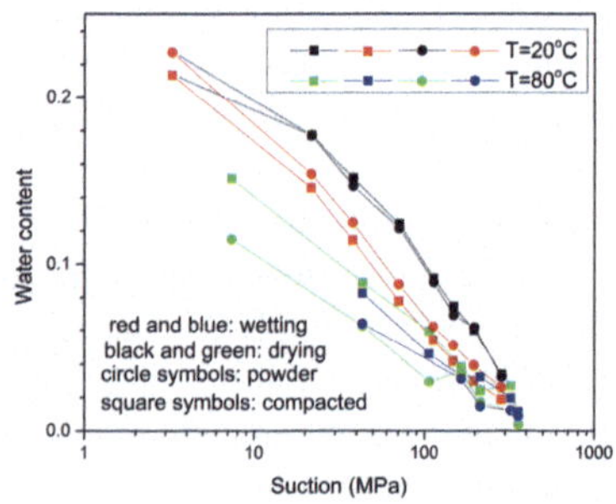

Fig. 6. Comparison of thermal WRC between compacted bentonite and powders.

The variation of pore water properties can affect the water retention capacity of compacted bentonite. While the temperature may have an influence on the water surface tension, the liquid-gas interfacial tension, contact angle between water and soil, these factors affect water retention capacity. For Czech bentonite B75, capillary forces dominated the behavior of compacted bentonite in the low suction range (<20 MPa) [6].

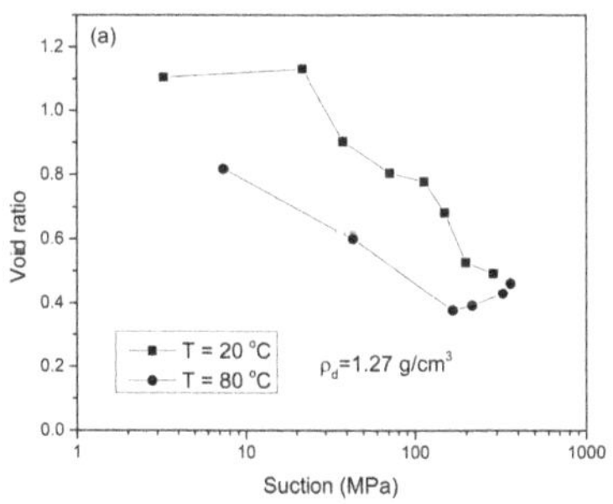

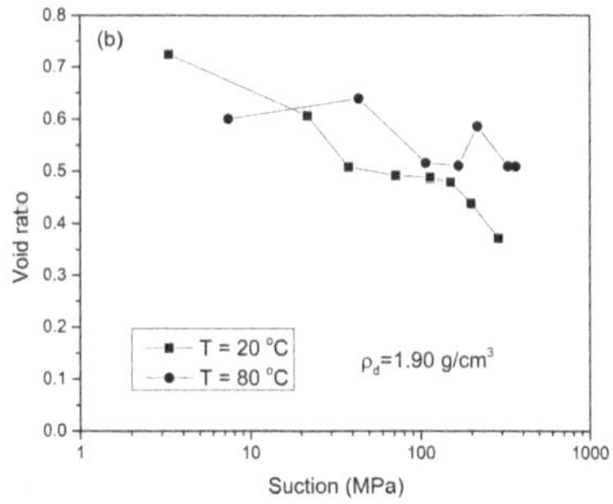

Fig. 7. Suction versus void ratio at wetting path with 20 °C and 80 °C for initial dry density of 1.27 g/cm^3 (a) and 1.90 g/cm^3 (b).

The decrease of surface tension caused by increasing temperatures resulted in low water retention capacity. This phenomenon was observed in Figs. 3, 5 and 6. However, the capillary model may be invalid in high suction range. The physical-chemical properties of intra-aggregate water and the alteration of microstructure by temperature dominate the water retention capacity. The increased temperature can drive the intra-aggregate water into inter-aggregate pores, which would explain the lower water retention capacity at higher temperature at a certain suction [14]. The results of Czech bentonite B75 are consistent with this theory and shows that the temperature has a great influence on the water retention capacity.

5 Conclusions

The water content vs. logarithm of suction relationships show nearly linear trend independent of the initial dry density. The water retention capacity decreased with increasing temperature for both compacted bentonite and bentonite powder. The water retention capacity is lower at high temperature especially at lower suctions. The temperature decreases more water retention capacity of bentonite powder than compacted bentonite. The hysteretic behavior of compacted bentonite is less obvious (hysteretic curve is narrower) with temperature increase for both compacted samples and powders. Also, the volume swelling is higher at lower temperature. The temperature has a more significant effect on drying path than wetting path for compacted Czech bentonite B75.

Acknowledgements. Financial support by the research grant GACR 15-05935S of the Czech Science Foundation is greatly appreciated. This project receives funding from the Euratom research and training programme 2014–2018 under grant agreement No 745942. The first author acknowledges support by the grant No. 846216 of the Charles University Grant Agency. The authors are grateful to the Centre of Experimental Geotechnics of Czech Technical University (prof. J. Pacovský, Dr. J. Svoboda, Dr. R. Vašíček) for providing access to their thermal facilities.

References

1. Grant, S.A., Salehzadeh, A.: Calculation of temperature effects on wetting coefficients of porous solids and their capillary pressure functions. Water Resour. Res. **32**(2), 261–270 (1996)
2. Romero, E., Gens, A., Lloret, A.: Suction effects on a compacted clay under non-isothermal conditions. Géotechnique **53**(1), 65–81 (2003)
3. Yao, Y.P., Zhou, A.N.: Non-isothermal unified hardening model: a thermo-elasto-plastic model for clays. Geotechnique **63**(15), 1328 (2013)
4. Zhou, A.N., Sheng, D., Li, J.: Modelling water retention and volume change behaviours of unsaturated soils in non-isothermal conditions. Comput. Geotech. **55**, 1–13 (2014)
5. François, B., Laloui, L.: ACMEG-TS: a constitutive model for unsaturated soils under non-isothermal conditions. Inte. J. Numer. Anal. Meth. Geomech. **32**(16), 1955–1988 (2008)
6. Mašín, D.: Coupled thermo hydro mechanical double-structure model for expansive soils. J. Eng. Mech. **143**(9), 04017067 (2017)
7. Sun, H., Mašín, D., Boháč, J.: Experimental characterization of retention properties and microstructure of the Czech bentonite B75. In: Proceedings of the 19th International Conference on Soil Mechanics and Geotechnical Engineering, Seoul, pp. 1249–1252 (2017)
8. Delage, P., Howat, M.D., Cui, Y.J.: The relationship between suction and swelling properties in a heavily compacted unsaturated clay. Eng. Geol. **50**, 31–48 (1998)
9. Greenspan, L.: Humidity fixed points of binary saturated aqueous solutions. J. Res. Natl. Bur. Stan. **81**(1), 89–96 (1977)
10. The International Organization of Legal Metrology (OIML): The scale of relative humidity (RH) of air certified against saturated salt solutions. OIMLR 121, France (1996)
11. Tang, A.M., Cui, Y.J.: Controlling suction by the vapour equilibrium technique at different temperatures and its application in determining the water retention properties of MX80 clay. Can. Geotech. J. **42**(1), 287–296 (2005)
12. Sun, H., Mašín, D., Najser, J., Neděla, V.: Bentonite micro-structure evolution in wetting-drying cycles studied by ESEM, MIP and WRC measurements, Géotechnique, (under view) (2018)
13. Nelson, J.D., et al.: Foundation Engineering for Expansive Soils (2015). ISBN: 978-0-470-58152-0
14. Villar, M.V., Lloret, A.: Influence of temperature on the hydro-mechanical behaviour of a compacted bentonite. Appl. Clay Sci. **26**(1), 337–350 (2004)

A Constitutive Model for Unsaturated Soils Using Degree of Capillary Saturation and Effective Interparticle Stress as Constitutive Variables

Annan Zhou(✉), Shengshen Wu, and Jie Li

Royal Melbourne Institute of Technology (RMIT), Melbourne 3000, Australia
annan.zhou@rmit.edu.au

Abstract. This paper discusses the role of the degree of capillary saturation in modelling the coupled hydro-mechanical behaviour of unsaturated soils and proposes a new constitutive model for unsaturated soils by using the degree of capillary saturation and the effective inter-particle stress. In the model, the shear strength, yield stress and deformation behaviour of unsaturated soils are governed directly by the above two constitutive variables. The model is then validated against a variety of experimental data in the literature, and the results show that a reasonable agreement can be obtained using this new constitutive model.

Keywords: Unsaturated soil · Degree of capillary saturation
Constitutive model

1 Introduction

The objective of this study is to establish a constitutive model for unsaturated soils by acknowledging that (i) pore water consists of capillary water and adsorbed water, and (ii) they contribute very differently to the constitutive behaviour of unsaturated soils. Because the capillary water exists among soil particles and the pressure of capillary water affects the contact stress among soil particles. Therefore, the stress carried by the capillary water is classified as an inter-particle stress. Compared with the capillary water, the contribution from the adsorbed water to the shear strength and deformation of a soil is very limited [1–3]. In principle, this is because adsorbed water wraps the surface of each soil particle. Thus, the stress carried by the adsorbed water is reasonably treated as an intra-particle stress.

One of the most widely-used constitutive variables for unsaturated soil modelling is the Bishop effective stress [4]. The formulation of σ'_{ij} can be written as follows

$$\sigma'_{ij} = \sigma_{ij} + \chi s \delta_{ij} = \bar{\sigma}_{ij} - u_a + \chi(u_a - u_w)\delta_{ij} \tag{1}$$

where σ_{ij} is the net stress with $\sigma_{ij} = \bar{\sigma}_{ij} - u_a$, $\bar{\sigma}_{ij}$ is the total stress, u_a is the pore air pressure, χ is the effective stress parameter, s is the suction ($s = u_a - u_w$), u_w is the pore water pressure and δ_{ij} is the Kronecker delta. In this study, the degree of capillary

L. Hu et al. (Eds.): GSIC 2018, *Proceedings of GeoShanghai 2018 International Conference: Multi-physics Processes in Soil Mechanics and Advances in Geotechnical Testing*, pp. 79–86, 2018.
https://doi.org/10.1007/978-981-13-0095-0_9

saturation (S') is selected as the basic constitutive variable to highlight that only the capillary water affects the strength and deformation of unsaturated soils. Specifically, on one aspect, the degree of capillary saturation is used for the effective stress parameter, i.e., $\chi = S'$, and the effective stress when $\chi = S'$ is referred to as the effective inter-particle stress (σ'_{ij}) to emphasize that the intra-particle stress associated with the adsorbed water pressure has been ruled out. On the other aspect, the slope of the NCLs is also a function of the degree of capillary saturation, i.e., $\lambda = \lambda(S')$ to underline that the mechanical state of an unsaturated soil is related to the capillary water only. It is important to note that, although the constitutive relationship is initially established in the space of $\left\{\sigma'_{ij}, S'\right\}$, it can be generalised in the space of primary variables $\{\sigma_{ij}, s, S\}$ that is in accordance with variables adopted for finite element methods [5–7].

2 Constitutive Equations

2.1 Hydraulic Equations

Sharing the theoretical concept delivered by Or and Tuller [8] and practical method by Khlosi et al. [9], a simple equation of water retention curve was proposed to consider capillarity and adsorption separately [1], which can be written as follows:

$$S = S' + S'' \tag{2}$$

where S is the degree of saturation; S' and S'' are the capillary component and adsorptive component of the degree of saturation respectively. Following Khlosi et al. [9], the two-parameter equation proposed by Kosugi [10] to quantify the capillary component (i.e., capillary water retention curve, CWRC) was employed here:

$$S' = (1 - S'')C(s), \text{ and } C(s) = \frac{1}{2}\operatorname{erfc}\left((\sqrt{2}\zeta)^{-1}\ln(s/s_{\mathrm{m}})\right) \tag{3}$$

where erfc() the complementary error function, s_{m} the suction that corresponds to the median pore radius (r_{m}), and ζ^2 the variance of the log-transformed pore radius. s_{m} varies from s_{mR} corresponding to the main drying branch to s_{mA} corresponding to the main wetting branch. Zhou [11] stated that the contact angle (θ) for the main drying branch and the main wetting branch are equal to the receding contact angle (θ_{R}) and advancing contact angle (θ_{A}), respectively. Also, only on the main drying/wetting branches, the contact angle (θ) is independent of the change of the suction. For the scanning processes, the contact angle is approaching to θ_{R} for drying and to θ_{A} for wetting. Based on the concept introduced by Zhou [11], the term $C(s)$ can be revised to consider hysteretic behaviour due to the variation of contact angle caused by the suction change as follows

$$C(s) = \frac{1}{2}\operatorname{erfc}\left((\sqrt{2}\zeta)^{-1}\ln[(s\cos\theta_{\mathrm{R}})/(s_{\mathrm{mR}}\cos\theta)]\right) \tag{4}$$

where θ_R and θ_A are the receding and advancing contact angles respectively. For simplicity, θ_R is usually assumed to be 0 and θ_A can be calibrated by the main wetting branch. s_{mR} stands for the suction that corresponds to the median pore radius (r_m) in the main drying process, which can be calibrated by the main drying branch. The variation of the contact angle due to suction has been provided by Zhou [11].

$$d\theta = -\frac{\beta ds}{s\tan\theta} \text{ and } \beta = \begin{cases} [(\cos\theta - \cos\theta_R)/(\cos\theta_A - \cos\theta_R)]^b & ds \geq 0 \\ [(\cos\theta_A - \cos\theta)/(\cos\theta_A - \cos\theta_R)]^b & ds < 0 \end{cases} \quad (5)$$

where b is a parameter to adjust the rate of contact angle change due to suction change, which can be calibrated by scanning wetting or drying tests.

In addition to the contact angle hysteresis, the mechanical loading changes soil's pore distribution and further affects its water retention behaviour [12–17]. The mechanical compression due to net stress increase decreases the median pore radius (r_m) as well as the variance (ζ^2) [18]. This phenomenon is termed as the mechanical shift of the water retention curve. The following equation is proposed

$$r_m = r_{m0}(1 - \varepsilon_{v\sigma})^a \text{ and } \zeta^2 = \zeta_0^2(1 - \varepsilon_{v\sigma})^a \quad (6)$$

where r_{m0} and ζ_0^2 is the median pore radius and the variance at the reference state (i.e., $\varepsilon_{v\sigma} = 0$), $\varepsilon_{v\sigma}$ is the volumetric strain due to the mechanical loading, a is a parameter to consider the mechanical effects on pore size distribution. Per the Young-Laplace equation, for a given contact angle, suction is in inverse proportion to pore radius. Therefore, term $C(s)$ can be further upgraded to the following expression to consider the mechanical shift.

$$C(s) = \frac{1}{2}\operatorname{erfc}\left(\left[\sqrt{2}\zeta_0(1 - \varepsilon_{v\sigma})^{a/2}\right]^{-1}\ln[(s(1 - \varepsilon_{v\sigma})^a)/(s_{mR0}\cos\theta)]\right) \quad (7)$$

where s_{mR0} is the suction that corresponds to the median pore radius at the reference state ($\varepsilon_{v\sigma} = 0$) and the receding contact angle ($\theta = \theta_R$, ζ_0^2) the variance of the log-transformed pore radius at the reference state ($\varepsilon_{v\sigma} = 0$). Both can be easily calibrated by test results obtained from the drying branch of conventional water retention experiments with a constant net stress. The volumetric strain due to hydraulic loading (ε_{vs}) has been considered when we calibrate s_{mR0} and ζ_0^2. This is the reason why only the volumetric strain due to mechanical loading ($\varepsilon_{v\sigma}$) is involved in Eq. (7).

The adsorptive component S'' (i.e., adsorbed water retention curve, AWRC) can be described by the following equation [1]:

$$S'' = (1 - P_{cc})\Theta_a/\Theta_s \quad (8)$$

where Θ_s is the volumetric water content at the fully saturated state, Θ_a is the maximum volumetric water content due to adsorption ignoring capillary condensation due to the mutual influence of adjacent adsorptive water films, and P_{cc} stands for the possibility of capillary condensation ($0 \leq P_{cc} \leq 1$). The simplest equation meets the requirement of P_{cc} can be written as: $P_{cc} = S'$. Specifically, Θ_a can be described by the equation proposed by Campbell and Shiozawa [19], i.e.,

$$\Theta_a = \Theta_s A(s), \text{ and } A(s) = \alpha(1 - \ln s / \ln s_d) \tag{9}$$

where α is the parameter that is related to the maximum degree of saturation due to adsorption (without considering capillary condensation) when the suction is equal to 1 kPa. s_d is the suction at oven dryness. Experimental results have shown that oven dryness generally corresponds to a finite suction of 10^6 kPa. Therefore, the adsorbed water retention curve can be specified as

$$S'' = A(s)(1 - S') \tag{10}$$

Combining Eqs. (2), (3) and (10) yields the following closed-form equations for WRC, CWRC and AWRC:

$$\begin{cases} \text{WRC:} & S = S' + S'' = \frac{C(s)+A(s)-2C(s)A(s)}{1-C(s)A(s)} \\ \text{CWRC:} & S' = \frac{C(s)-C(s)A(s)}{1-C(s)A(s)}, \text{ and AWRC: } S'' = \frac{A(s)-C(s)A(s)}{1-C(s)A(s)} \end{cases} \tag{11}$$

2.2 Mechanical Equations

Realising the inter-particle water (or water bridges or menisci) is composed of the capillary water only while the adsorptive water forms the water film wrapping particles, the effective inter-particle stress equation for the shear strength of unsaturated soil was suggested by Zhou et al. [1], highlighting only the capillary water contributes to the shear strength under a given suction.

$$\sigma'_{ij} = \sigma_{ij} + S's\delta_{ij} \tag{12}$$

where σ'_{ij} is the effective inter-particle stress, and S' can be determined by Eq. (11). In the space of deviator stress (q) and mean effective inter-particle stress (p'), the shear strength of unsaturated soil can be written as [20, 21]

$$q = Mp'. \tag{13}$$

The experimental validation for Eqs. (12) and (13) on predicting unsaturated soil strength can be found in [1].

Realising that the degree of capillary saturation contributes much more to unsaturated soil's mechanical behaviour than the degree of adsorptive saturation, the NCLs of unsaturated soils are proposed as below, by adopting the degree of capillary saturation (S') other than the degree of saturation (S) as the key variable.

$$v = N - \lambda \ln p', \tag{14}$$

where p' is the mean effective inter-particle stress defined in Eq. (12), v the specific volume, N the intercept of the NCL with the v-axis when $\ln p' = 0$, and λ the elasto-plastic compression index representing the slope of the NCL, which is assumed as a function of the degree of capillary saturation, i.e. $\lambda = \lambda(S')$.

$$\lambda = \lambda_0 - (\lambda_0 - \lambda_d)(1 - S')^k, \tag{15}$$

where λ_0 is the elastoplastic compression index for the fully saturated soil, λ_d is the elastoplastic compression index for when the soil contains zero capillary water (i.e. oven dryness), and k is a coupling parameter which can be determined by a drying test. For the elastic response, the following equation is employed.

$$dv = -\kappa dp'/p', \tag{16}$$

where κ is the elastic compression index representing the slope of the unloading and reloading line (URL). For κ, an equation like λ is employed.

$$\kappa = \kappa_0 - (\kappa_0 - \kappa_d)(1 - S')^k, \tag{17}$$

where κ_0 is the elastoplastic compression index for the fully saturated soil, and κ_d is the elastoplastic compression index for when the soil contains zero capillary water (i.e., oven dried). For the clayey soil, the capillary water can only be fully removed by very high suction and the compressibility of the clayey soil at a very high suction is far less than its compressibility at the fully saturated state. Therefore, for simplicity, in this paper, we assume $\lambda_d = \kappa_d = 0$ to simplify Eqs. (15) and (17) as

$$\kappa = \kappa_0 - \kappa_0(1 - S')^k, \text{ and } \lambda = \lambda_0 - \lambda_0(1 - S')^k. \tag{18}$$

The yield surface (or the loading collapse surface) function in the isotropic stress states

$$f_{LC} = (p')^{1-(1-S')^k} - p'_c = 0. \tag{19}$$

The Modified Cam-clay model is employed here to extend the isotropic yield surface to a triaxial stress state [22–24]

$$f = p' + \frac{q^2}{M^2 p'} - (p'_c)^\delta = 0, \text{ where } \delta = \frac{1}{1 - (1 - S')^k}. \tag{20}$$

3 Validations

Li et al. [25] presented a series of suction controlled triaxial tests on the compacted Zaoyang clay. The model parameters for predictions are listed in Table 1.

Table 1. Model parameters for the compacted unsaturated Zaoyang clay

Mechanical parameters (4)				Hydraulic parameters (6)					
λ_0	κ_0	M	k	s_{mR0}	ζ_0	α	θ_A	a	b
0.11	0.02	0.95	5.0	400 kPa	3.0	0.45	70°[a]	13	0.5[a]

[a]Note: θ_A (for wetting) and b (for scanning) are not necessary because only constant suction tests are involved in this section.

Figures 1 and 2 present a series of unsaturated triaxial compression tests on the compacted Zaoyang clay. The suction is set to be 100 kPa and 200 kPa in Figs. 1 and 2, respectively. For each figure (i.e., for each suction level), the stress-strain relationship (q vs ε_1), volume change (v vs ε_1) and the change of the degree of saturation (S vs ε_1) in constant-suction triaxial compression are presented in subfigure a, b and c, respectively. In addition, for each figure (i.e., for each suction level), three different

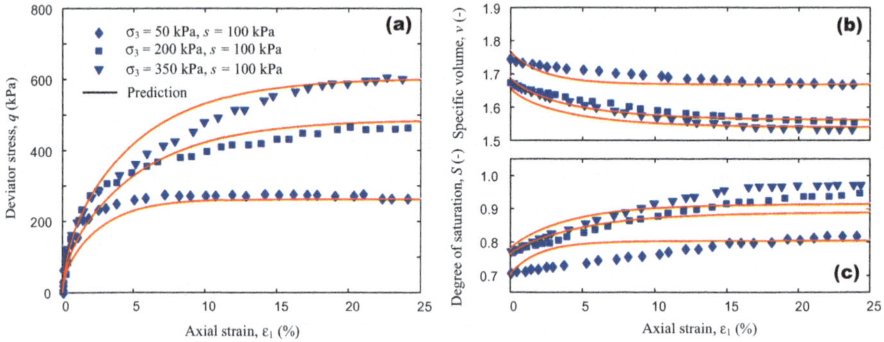

Fig. 1. Stress-strain relationship, volume change and saturation change for the unsaturated Zaoyang clay (s = 100 kPa) in triaxial compressions and model simulations.

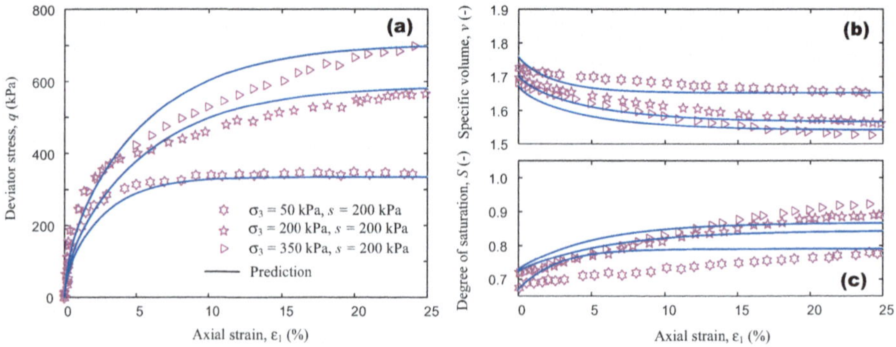

Fig. 2. Stress-strain relationship, volume change and saturation change for the unsaturated Zaoyang clay (s = 200 kPa) in triaxial compressions and model simulations.

confining stresses (net stresses = 50, 200 and 350 kPa) are involved in the constant suction triaxial compressions. The different symbols are used to present the observed test results and solid curves are employed to stand for model predictions. As shown in Figs. 1 and 2, the model predictions quantitatively capture the test data reasonably well for the compacted samples with various suctions (100 and 200 kPa) and various confining stresses (50, 200 and 350 kPa).

4 Conclusions

A new constitutive model for unsaturated soils using the degree of capillary saturation and effective inter-particle stress is proposed in this paper, where the shear strength, yield stress and deformation behaviour of unsaturated soils are governed by two constitutive variables. The proposed constitutive model can capture the observed mechanical and hydraulic behaviours with a limited number of parameters. The capacity of the model has been validated against a variety of experimental data in the literature.

References

1. Zhou, A., Huang, R.-Q., Sheng, D.: Capillary water retention curve and shear strength of unsaturated soils. Can. Geotech. J. **53**(6), 974–987 (2016)
2. Lu, N., Godt, J.W., Wu, D.T.: A closed-form equation for effective stress in unsaturated soil. Water Resour. Res. **46**(5), W05515 (2010)
3. Konrad, J.-M., Lebeau, M.: Capillary-based effective stress formulation for predicting shear strength of unsaturated soils. Can. Geotech. J. **52**(12), 2067–2076 (2015)
4. Bishop, A.W.: The principle of effective stress. Teknisk Ukeblad **106**(39), 859–863 (1959)
5. Sheng, D., Sloan, S.W., Gens, A., Smith, D.W.: Finite element formulation and algorithms for unsaturated soils. Part I: theory. Int. J. Numer. Anal. Meth. Geomech. **27**(9), 745–765 (2003)
6. Zhou, A.N., Zhang, Y.: Explicit integration scheme for non-isothermal elastoplastic model with convex and nonconvex subloading surfaces. Comput. Mech. **55**, 943–961 (2015)
7. Zhang, Y., Zhou, A.: Explicit integration of a porosity-dependent hydro-mechanical model for unsaturated soils. Int. J. Numer. Anal. Meth. Geomech. **40**(17), 2353–2382 (2016)
8. Or, D., Tuller, M.: Liquid retention and interfacial area in variably saturated porous media: Upscaling from single-pore to sample-scale model. Water Resour. Res. **35**(12), 3591–3605 (1999)
9. Khlosi, M., Cornelis, W.M., Douaik, A., van Genuchten, M.T., Gabriels, D.: Performance evaluation of models that describe the soil water retention curve between saturation and oven dryness. Vadose Zone J. **7**(1), 87–96 (2008)
10. Kosugi, K.: Lognormal distribution model for unsaturated soil hydraulic properties. Water Resour. Res. **32**(9), 2697–2703 (1996)
11. Zhou, A.N.: A contact angle-dependent hysteresis model for soil–water retention behaviour. Comput. Geotech. **49**, 36–42 (2013)
12. Sheng, D., Zhou, A.N.: Coupling hydraulic with mechanical models for unsaturated soils. Can. Geotech. J. **48**(5), 826–840 (2011)

13. Zhou, A.N., Sheng, D.: An advanced hydro-mechanical constitutive model for unsaturated soils with different initial densities. Comput. Geotech. **63**, 46–66 (2015)
14. Zhou, A.N., Sheng, D., Sloan, S.W., Gens, A.: Interpretation of unsaturated soil behaviour in the stress-saturation space, I: volume change and water retention behaviours. Comput. Geotech. **43**, 178–187 (2012)
15. Zhou, A.N., Sheng, D., Sloan, S.W., Gens, A.: Interpretation of unsaturated soil behaviour in the stress-saturation space, II: constitutive relationships and validations. Comput. Geotech. **43**, 111–123 (2012)
16. Zhou, A.N., Sheng, D., Carter, J.P.: Modelling the effect of initial density on soil-water characteristic curves. Geotechnique **62**(8), 669–680 (2012)
17. Zhou, A.N., Sheng, D., Li, J.: Modelling water retention and volume change behaviours of unsaturated soils in non-isothermal conditions. Comput. Geotech. **55**, 1–13 (2014)
18. Oualmakran, M., Mercatoris, B.C.N., François, B.: Pore-size distribution of a compacted silty soil after compaction, saturation, and loading. Can. Geotech. J. **53**(12), 1902–1909 (2016)
19. Campbell, G.S., Shiozawa, S.: Prediction of hydraulic properties of soils using particle-size distribution and bulk density data. In: van Genuchten, M.T., Leij, F.J., Lund, L.J. (eds.) Indirect Methods for Estimating the Hydraulic Properties of Unsaturated Soils, pp. 317–328. Univ. of California, Riverside (1992)
20. Yao, Y., Hu, J., Zhou, A., Luo, T., Wang, N.: Unified strength criterion for soils, gravels, rocks, and concretes. Acta Geotech. **10**(6), 1–11 (2015)
21. Sheng, D., Zhou, A.N., Fredlund, D.G.: Shear strength criteria for unsaturated soils. Geotech. Geol. Eng. **29**(2), 145–159 (2011)
22. Yao, Y.P., Hou, W., Zhou, A.N.: UH model: three-dimensional unified hardening model for overconsolidated clays. Geotechnique **59**(5), 451–469 (2009)
23. Yao, Y.P., Kong, L.M., Zhou, A.N., Yin, J.H.: Time-dependent unified hardening model: three-dimensional elastoviscoplastic constitutive model for clays. J. Eng. Mech. **141**(6), 04014162 (2015)
24. Yao, Y.P., Sun, D.A., Matsuoka, H.: A unified constitutive model for both clay and sand with hardening parameter independent on stress path. Comput. Geotech. **35**(2), 210–222 (2008)
25. Li, J., Yin, Z.-Y., Cui, Y., Hicher, P.-Y.: Work input analysis for soils with double porosity and application to the hydromechanical modeling of unsaturated expansive clays. Can. Geotech. J. **54**(2), 173–187 (2016)

Soil Water Retention Surface Determination Using a New Triaxial Testing System

Lin Li[1(✉)], Xiong Zhang[2], and Peng Li[3]

[1] University of Alaska Anchorage, Anchorage, AK 99508, USA
lli10@alaska.edu
[2] Missouri University of Science and Technology, Rolla, MO 65409, USA
zhangxi@mst.edu
[3] Chang'an University, Xi'an 710064, Shaanxi, China
Lipeng2013@chd.edu.cn

Abstract. The soil water retention curve has been widely used to characterize the relationship between the water content and the soil water potential. The conventional pressure plate method for the soil water retention curve determination is very time consuming. Also, soil volume change during testing is not available. Recently, a new triaxial testing system has been developed for unsaturated soil characterization. In this study, this new system was utilized to determine the soil water retention surface through a series of constant water content consolidation tests on unsaturated soils with different moisture contents. Results from these tests indicate that the new triaxial system is a time-efficent option for soil water retention surface determination.

Keywords: Soil water retention surface · Triaxial test
High-suction tensiometer

1 Introduction

The soil water retention curve is the relationship between the water content and the soil water potential. In the past, the soil water retention curve was commonly used to estimate in situ soil suctions by measuring the unsaturated soil water content. Several models (Williams et al. 1983; Hutson and Cass 1987; Rossi and Nimmo 1994; Assouline et al. 1998) were proposed to predict the soil water retention curves for unsaturated soils based upon their physical properties. A review of the existing soil water retention curve models was presented by Sillers and Fredlund (2001). In the past, based on principle of axis translation technique, the pressure plate equipment has been commonly used for the soil water retention curve determination (i.e. ASTM C1699 2009; Richards 1941; and Péron et al. 2007). However, for the pressure plate test, due to the low permeability of unsaturated soils, several weeks are required for the soil water retention curve determination of one soil. Also, during testing, soil volume change, which is required for calculation of soil void ratio or degree of saturation, cannot be measured. Many research efforts (Lourenço et al. 2011; Padilla et al. 2005) have been dedicated to rapid and accurate determination of soil water retention curve. In Padilla et al. (2005), the Fredlund SWCC device was adopted soil water retention

L. Hu et al. (Eds.): GSIC 2018, *Proceedings of GeoShanghai 2018 International Conference: Multi-physics Processes in Soil Mechanics and Advances in Geotechnical Testing*, pp. 87–94, 2018.
https://doi.org/10.1007/978-981-13-0095-0_10

curve determination. Vertical load was applied to the tested sample which guaranteed the contact between the sample and oedometer wall. As a result, volume change due to suction change could be determined simply by monitoring the axial displacement of the top plate. In Lourenço et al. (2011), a high-suction tensiometer was used to measure soil suction during drying process from which the soil water retention curve was extracted. However, the soil volume content cannot be measured during testing.

In Li and Zhang (2015), a new triaxial testing system was developed based on modifications on a conventional triaxial test apparatus which is for saturated soils. In the new triaxial testing system, high-suction tensiometers were equipped to measure soil suction during triaxial testing which is different from the suction-controlled triaxial test in which the axis-translation technique is adopted to control soil suction (Bishop and Donald 1961). To install the tensiometers, two holes are drilled in the base of the triaxial cell as shown in Fig. 1a. In this study, the high-suction tensiometers developed in Li and Zhang (2014) were used for soil suction measurement. The tensiometers were saturated and saturated in the triaxial chamber as shown in Fig. 1b. In this triaxial testing system, the photogrammetry-based method developed in Zhang et al. (2015) is adopted for soil volume change measurement during testing. A series of constant water content tests were performed on unsaturated soils with different moisture contents. A soil water retention surface, proposed by Salager et al. (2010), was then extracted to characterize soil water retention behavior with consideration of the effect of both suction and soil deformation.

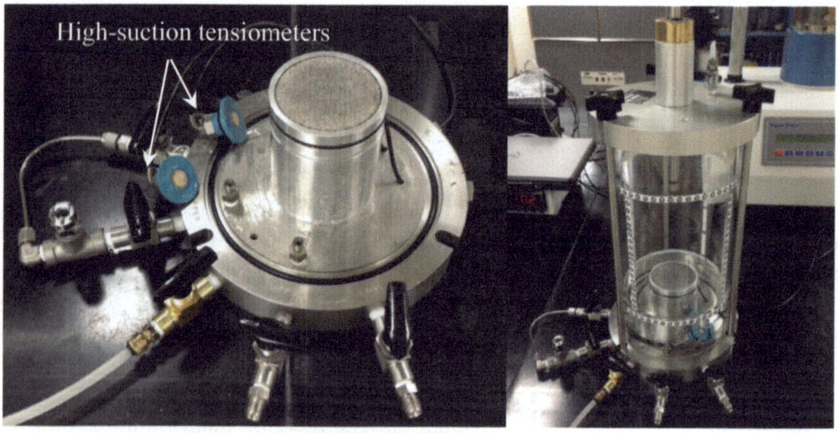

(a) Modification on cell pedestal; (b) High-suction tensiometer saturation

Fig. 1. Modification for soil suction measurement

2 Constant Water Content Test on Unsaturated Soils

2.1 Sample Preparation

Locally available Fairbanks silt was used to fabricate several unsaturated soil specimens (71 mm in diameter and 142 mm in height) using the static compaction method (Ladd 1978) as shown in Fig. 2a. The properties of the used soil are summarized in Table 1. The soil cylinders were compacted in 10 layers. The surface of each soil layer was scarified to ensure good contact between adjacent soil layers as shown in Fig. 2b. Before testing, the soil specimen was completely sealed in the plastic cup for two days to reach matric suction equilibrium as shown in Fig. 2c. After suction equilibrium, the constant water content consolidation tests were conducted.

(a) (b) (c)

Fig. 2. Specimen preparation using the static compaction method

Table 1. Soil properties of the used soil mixture

Soil properties	Value
Maximum dry density	1.836 g/cm^3
Optimum moisture content	15%
Specific gravity	2.7
Plastic limit	18.2
Liquid limit	19.7
Plastic index	1.5

2.2 Undrained Consolidation Tests

After saturation and calibration, tensiometers can then be used for matric suction measurement. Before sample installation, two holes were cut on the membrane, which would be used to cover the soil specimen during testing. The specimen was mounted on the pedestal of the triaxial cell and carefully sealed with O-rings. A small suction (−5 kPa) was applied to the inside of the soil specimen after this. Vacuum grease was smeared on the back of the grommet, which was used to hold the tensiometer. By carefully stretching the membrane, the tensiometer can be placed on the sample surface through two holes on membrane. Two saturated tensiometers were placed at the middle of the specimen (see Fig. 3). The use of vacuum grease ensured a good seal between the grommet and the membrane, and a negative pressure was applied to the inside of the soil specimen held the sensor in place during tensiometer installation. Some measurement targets were posted on the membrane surface after this with the help of vacuum grease. A series of images around the specimen were then captured to determine the volume of the soil specimen before any load was applied. The triaxial cell was installed and filled up with water after this as shown in Fig. 3.

Fig. 3. System setup for the constant water content consolidation test

A confining pressure of 50 kPa was then applied to hold the tensiometer in place, and the applied suction inside of the soil specimen was released simultaneously. The reading of the tensiometers required a certain time to reach equilibrium. After the equilibrium, isotropic load was applied in steps to a maximum net confining load of 600 kPa for the water retention surface determination. Usually, approximately 20 min, dependent on soil suction, were required to ensure the stabilization of the high-suction tensiometer readings. Once suction stabilized, the volume of the soil specimen was measured by using the photogrammetry-based method. Then, isotropic compression test was performed. Confining pressure was gently increased (i.e. loading) or decreased (i.e. unloading) to a target value and followed by another suction stabilization period. The time required for new suction equilibrium also depends on soil suction level. After suction equilibrium, the volume of the soil specimen was measured again.

2.3 Experimental Results

During isotropic compression test, matric suctions of the test specimen under different net confining stresses was monitored via high-suction tensiometers. A typical matric suction variation for the soil specimen with water content of 11.84% due to isotropic load was presented in Fig. 4. During isotropic loading, there was an immediate drop of suction associated with the increase of isotropic load, followed by a small increase in suction with time and finally reached equilibrium. The entire isotropic loading process lasted for approximately 5 h.

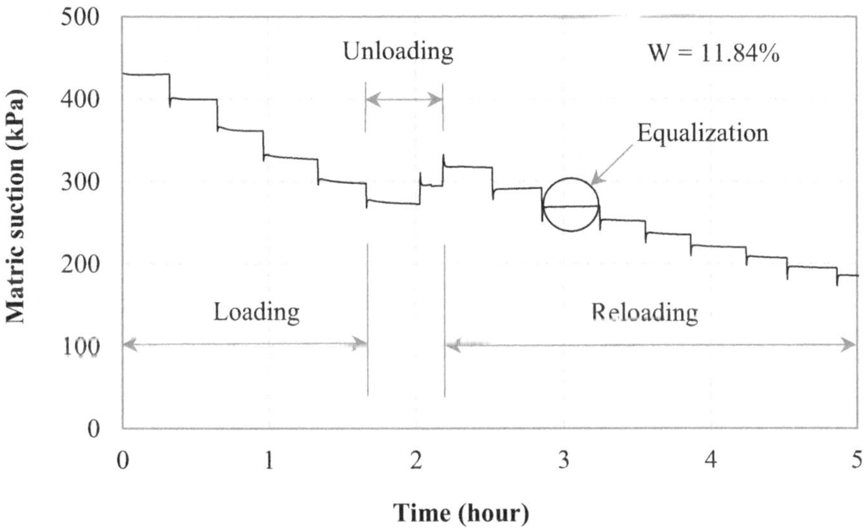

Fig. 4. A typical high-suction tensiometer response during isotropic loading

The representative soil suctions recorded by the tensiometers after each equalization stage were plotted in Fig. 5a. Generally, soil suction decreased with increasing isotropic load. Also, soils with low water contents had higher initial suction and experienced the larger reduction in suctions after the same loading procedure. When there is an unloading at the mean net stress of 300 kPa, there were irrecoverable suction changes due to the soil plasticity. The soil volume change was measured based on the measurement targets on soil surface. Details on the soil volume measurement method using the photogrammetry-based method can be found in Zhang et al. (2015) and Li et al. (2015). During isotropic loading, with increasing confining load, the soil volume,

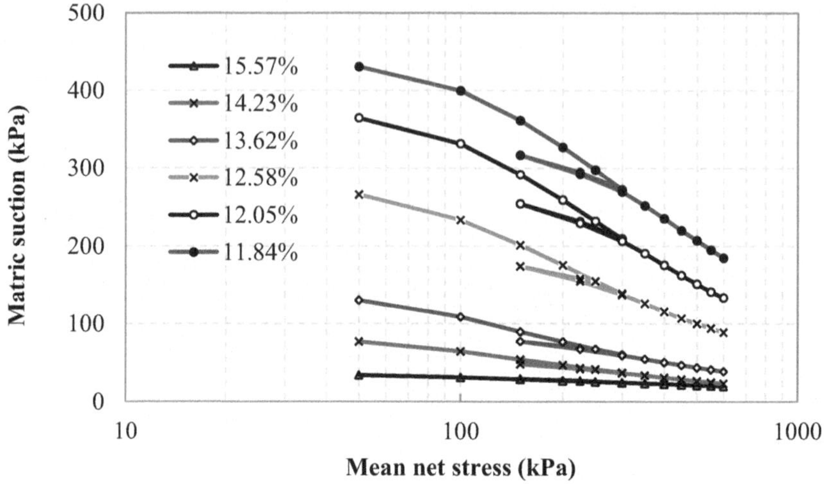

(a) Soil suction variation during isotropic loading

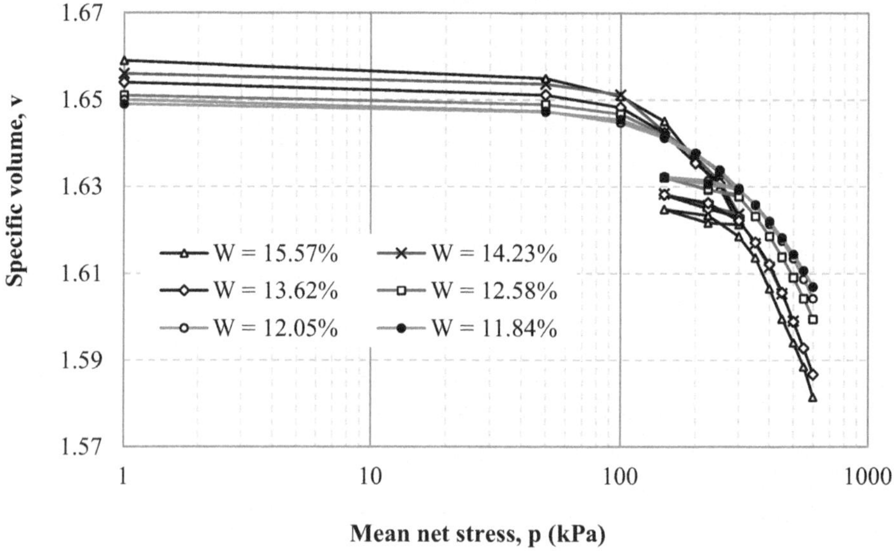

(b) Soil volume change during isotropic loading

Fig. 5. Soil response during isotropic loading under undrained condition

continuously decreased as shown in Fig. 5b. The soil specimens with the highest water content experienced the highest volumetric reduction when subjected to the same loading procedure. This was because the moisture lubricated the soil particles and led to smaller modulus during the isotropic loading. The plastic soil deformation was also identified when there was an unloading as shown in Fig. 5b.

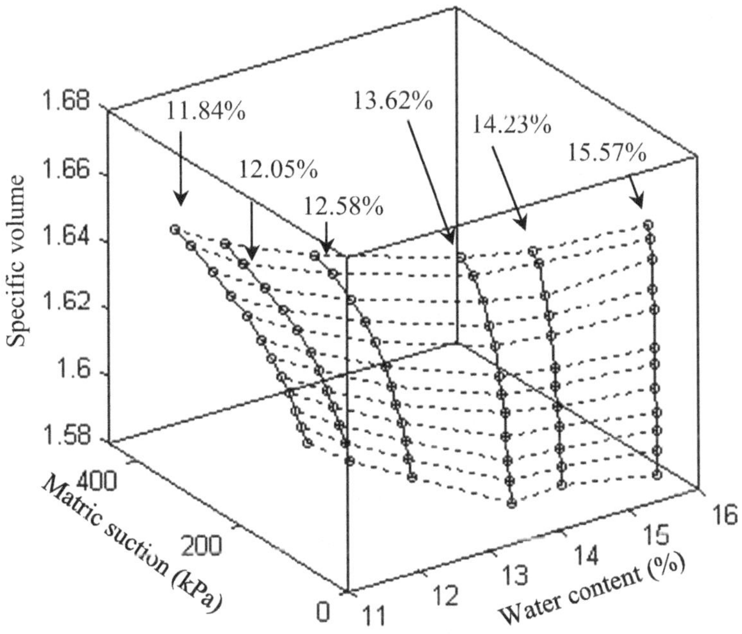

Fig. 6. Soil water retention surface

With the soil volume, suction, and water content, the soil water retention curves for six soil specimens were extracted and plotted in Fig. 6 in which the unloading and reloading was not included. With six soil water retention curves, a soil water retention surface could then be created as shown in Fig. 6 in which the significant influence of soil specific volume on the soil water retention capability was clearly shown. With this surface, the soil suction can be predicted with given specific volume and water content.

3 Conclusions

In this study, to determine the water retention surface of an unsaturated soil, a new triaxial testing system was adopted for the constant water content triaxial tests. During triaxial testing, soil suction and volume changes were measured using high-suction tensiometers and the photogrammetry-based method, respectively. Through a series of isotropic consolidation tests on the unsaturated soil, the soil water retention surface was determined. Compared with the conventional pressure plate method, the soil volume changes during testing could be recorded which makes the soil water retention curve more accurate and representative.

References

Assouline, S., Tessier, D., Bruand, A.: A conceptual model of the soil water retention curve. Water Resour. Res. **34**(2), 223–231 (1998)

ASTM: C 1699: Standard Test Method for Moisture Retention Curves of Porous Building Materials Using Pressure Plates. Annual Books of ASTM Standards, Philadelphia, PA (2009)

Hutson, J.L., Cass, A.: A retentivity function for use in soil–water simulation models. J. Soil Sci. **38**(1), 105–113 (1987)

Li, L., Zhang, X.: A new triaxial testing system for unsaturated soil characterization. Geotech. Test. J. **38**(6), 823–839 (2015)

Li, L., Zhang, X.: Development of a new high-suction tensiometer. In: Soil Behavior and Geomechanics, pp. 416–425 (2014)

Li, L., Zhang, X., Chen, G., Lytton, R.: Measuring unsaturated soil deformations during triaxial testing using a photogrammetry-based method. Can. Geotech. J. **53**(3), 472–489 (2015)

Lourenço, S.D.N., Gallipoli, D., Toll, D.G., Augarde, C.E., Evans, F.D.: A new procedure for the determination of soil-water retention curves by continuous drying using high-suction tensiometers. Can. Geotech. J. **48**(2), 327–335 (2011)

Padilla, J.M., Perera, Y.Y., Houston, W.N., Fredlund, D.G.: A new soil–water characteristic curve device. In: Tarantino et al. (eds.) Proceedings of the Advanced Experimental Unsaturated Soil Mechanics, EXPERUS 2005. Balkema Publishers, pp. 15–22 (2005)

Péron, H., Hueckel, T., Laloui, L.: An improved volume measurement for determining soil water retention curves. Geotech. Test. J. **30**(1), 1 (2007)

Richards, L.A.: A pressure-membrane extraction apparatus for soil solution. Soil Sci. **51**(5), 377–386 (1941)

Rossi, C., Nimmo, J.R.: Modeling of soil water retention from saturation to oven dryness. Water Resour. Res. **30**(3), 701–708 (1994)

Williams, J., Prebble, R.E., Williams, W.T., Hignett, C.T.: The influence of texture, structure and clay mineralogy on the soil moisture characteristic. Aust. J. Soil Res. **21**, 15–32 (1983)

Zhang, X., Li, L., Chen, G., Lytton, R.L.: A photogrammetry-based method to measure total and local volume changes of unsaturated soils during triaxial testing. Acta Geotech. **10**(1), 55–82 (2015)

Sillers, W.S., Fredlund, D.G.: Statistical assessment of soil-water characteristic curve models for geotechnical engineering. Can. Geotech. J. **38**(6), 1297–1313 (2001)

Bishop, A.W., Donald, I.B.: The experimental study of partly saturated soil in the triaxial apparatus. In: Proceedings of the 5th International Conference on Soil Mechanics and Foundation Engineering, Paris 1, pp. 13–21 (1961)

Salager, S., El Youssoufi, M.S., Saix, C.: Definition and experimental determination of a soil-water retention surface. Can. Geotech. J. **47**(6), 609–622 (2010)

Ladd, R.S.: Preparing test specimens using under compaction. Geotech. Test. J. **1**(1), 16–23 (1978)

Comparative Study on the Stress-Dilatancy of Xinjiang Loess Under Saturated and Unsaturated Conditions

Yue-feng Zhou(✉), Bi-wei Gong, Jun Tong, and Cong-an Li

Key Laboratory of Geotechnical Mechanics and Engineering of Ministry of Water Resources, Changjiang River Scientific Research Institute, Wuhan 430010, China
zhou.yuefeng@163.com

Abstract. The amount and scale of engineering construction in Xinjiang develop greatly in recent years. A channel slope in north Xinjiang is an infilled project, which is compacted using local loess material. Being different from the structural feature of natural loess, the infilled material needs to be sufficiently compacted. Study on the stress state, deforming feature and failing pattern of the material is important to reveal the slope stability issues. In this study, saturated and unsaturated triaxial shear tests were performed to study the stress-strain relationship and the mobilized shear strength, and to further investigate the dilatancy of the material. Saturated loess and unsaturated loess at three different suction levels were analyzed respectively. The research results are as below: Firstly, the mobilized shear strengths at peak state and at residual state increase with confining stress and suction level. Secondly, the soil dilatancy is affected by matric suction and stress level. The mobilized angle of dilation at peak state increases significantly with matric suction, but decrease with confining stress.

Keywords: Reconstituted loess · Matric suction · Dilatancy

1 Introduction

China's BELT & ROAD initiative is a significant measure for regional coordinated development. The BELT, i.e. the Silk Road Economic Belt, goes through loess regions including Xinjiang Province. The amount and scale of engineering construction in Xinjiang develop greatly in recent years. A channel slope in north Xinjiang is an infilled project, which is compacted using local loess material. Being different from the structural feature of natural loess, the infilled material needs to be sufficiently compacted.

Soil dilatancy is affected by degree of compaction, stress level and matric suction etc. in the macro scale. Above factors appear as particle arrangement and inter-particle pressure etc. in the micro scale. Under different degrees of saturation, the contribution of matric suction on inter-particle pressure is different, demonstrating dilative or contractive behavior. Previous studies (Alonso et al. 1990; Bruchon et al. 2013; Wheeler and Sivakumar 1995; Ng and Chiu 2003) have shown that the shear strength of unsaturated soil nonlinearly varies with the matric suction. For unsaturated sand,

L. Hu et al. (Eds.): GSIC 2018, *Proceedings of GeoShanghai 2018 International Conference: Multi-physics Processes in Soil Mechanics and Advances in Geotechnical Testing*, pp. 95–101, 2018.
https://doi.org/10.1007/978-981-13-0095-0_11

Donald (1956) introduced that the shear strength increased to a peak value and then reduced with the increase of matric suction, i.e. Donald Effect. Dai et al. (2014) reported that particle material demonstrated lower dilatancy at a certain wetting state than at saturated and dry states. Being different with the structural feature of undisturbed loess, reconstituted loess needs to be sufficiently compacted in engineering construction. Whereas, the evolution of soil dilatancy with matric suction is still not clear. On the one hand, the plastic flow rule for unsaturated soil is related to its dilatancy, which is currently based on theoretical deduction from saturated soil. On the other hand, the mobilized unsaturated strength is affected by water, which can provide theoretical direction for engineering design.

For natural soils, mineral composition, structural feature and particle size distribution can inevitably bring complicated influence on soil dilatancy. In above several aspects, reconstituted loess is a typical unsaturated soil without special features. This article studies the stress-dilatancy of unsaturated soil at different stages of SWCC.

2 Introduction

2.1 Test Material

Table 1 summarizes the basic physical properties of the studied loess, which were determined in accordance with BS1377 (British Standard Institution 1990). Figure 1 shows the particle size distribution of the loess, which mainly consisted of silt-sized particles.

Table 1. Physical properties of the loess

Optimum water content/%	Max. dry density 10^3 kg/m^3	Specific gravity	Liquid limit %	Plastic limit %	Coefficient of permeability/cm/s
13	1.820	2.71	23	13.3	6×10^{-5}

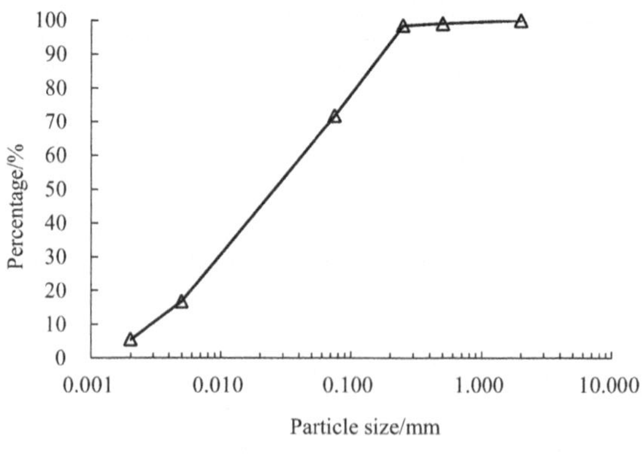

Fig. 1. The particle size distribution curve of the loess

According to Inspection and assessment specification for constructional quality of hydraulic and hydroelectric engineering, the quality control of the channel construction should be strictly inspected based on the design degree of compaction. In this project, the degree of compaction should not be smaller than the design value, i.e. Rc = 0.95. To examine soil behavior at a normal compacted state, a typical degree of compaction is considered in laboratory tests, i.e. Rc = 0.96.

HITACHI SU8010 Field Emission Scanning Electron Microscopy was adopted to examine the micro structure of the soil at Rc = 0.96, the image of which is shown at the scale of 500 in Fig. 2. The soil skeleton demonstrates a relative low clarity, whose surface was adhered by pieces of debris. The intergranual pores are filled up by minerals, indicating lower porosity at micro scale.

Fig. 2. SEM image for the soil sample at Rc = 0.96

2.2 Test Program

A Fredlund SWCC apparatus was employed to investigate the soil-water characteristics of the reconstituted loess at the designed degree of compaction, i.e. Rc = 0.96 (Table 2). Using a GDS triaxial testing apparatus, two consolidated drained triaxial tests were performed on saturated specimens, which represent a special matric suction level of 0 kPa. A GDS double-cell triaxial testing apparatus for unsaturated soil is adopted to perform unsaturated triaxial tests at constant suctions of 30 kPa and 120 kPa, with accurate measurement of volumetric strain for dilatancy analysis.

Table 2. Testing results of the loess

Test no.	SWCC96	T96-1	T96-2	T96-3	T96-1	T96-2	T96-3
Net mean stress $(\sigma_3 - u_a)$/kPa	/	40	40	40	80	80	80
Matric suction s/kPa	0–1200	0	30	120	0	30	120

3 Tests Results

3.1 SWCC Curve

After initial saturation, the specimen was subjected to drying by increasing the matric suction, which is achieved by increasing the air pressure while maintaining a small axial load. The back pressure was increased by 5–400 kPa in each step, which was maintained for the equalization of matric suction for 2 to 7 days because of the low coefficient of permeability of the unsaturated soil and HAECD. Using the measured final water content, The soil water characteristic curve of loess can be deduced from the drying path (Fig. 2). An air-entry value of 8 kPa and a residual volumetric water content is roughly 11% are expected for the studied loess at the controlled degree of compaction (Fig. 3).

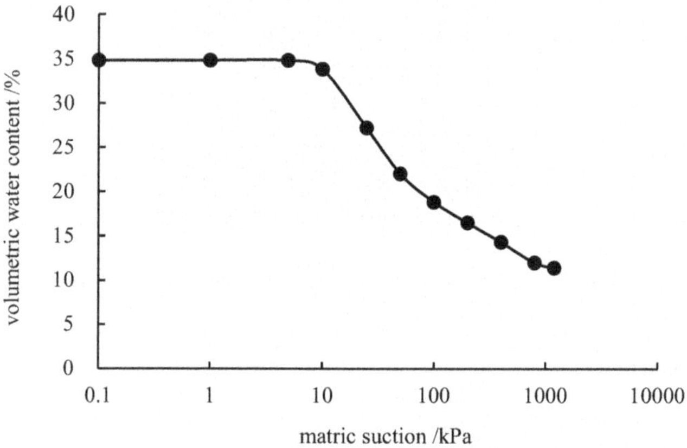

Fig. 3. The soil-water characteristic curve of the loess

3.2 Stress and Deformation at Different Suctions

Figure 4 summarizes the results of the constant suction tests at the confining stress of 40 kPa. Three levels of matric suction were investigated, i.e. 0 kPa, 30 kPa and 120 kPa. When subjected to drained compression, the reconstituted specimens required about 2–3% axial strain to attain the peak strength (Fig. 4(a)). The deviator stress then showed a remarkable reduction with further compression. The mobilized strengths both at peak state and at residual state increased with suction level. For example, the mobilized peak strength for the specimen at a suction of 0 kPa was 118 kPa, and enhanced to 338 at 120 kPa suction.

In Fig. 4(b), the volumetric strain rose slightly when the axial strain is below 1% for the specimen at a suction of 0 kPa. It then decreased continuously, suggesting a dilative deformation in the compressive shearing process (Fig. 4(b)). The specimens at suctions of 30 kPa and 120 kPa demonstrated dilative behavior from the very beginning of shearing. The soil dilatancy remarkably increased with the suction level.

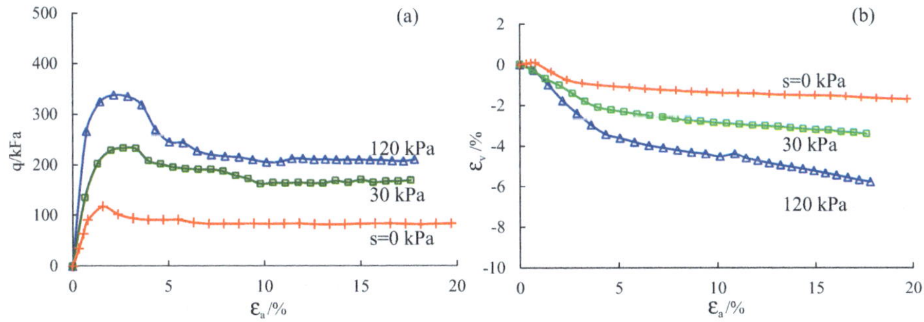

Fig. 4. The test results at a confining stress of 40 kPa and different suction levels

The final volumetric strain did not reach constant value in the three tests, suggesting that the dilative behavior persists over the test processes.

Figure 5 summarizes the results of the constant suction tests at the confining stress of 80 kPa, which demonstrate essentially similar stress and deformation trends with the test results at the confining stress of 40 kPa. The deviator stress for the specimen at a suction of 0 kPa rose to a peak at an axial strain of approximately 2%, and then exhibited a slight strength reduction with limited strain-softening behavior. In Figs. 4 and 5, it can be seen that the mobilized strength increases together with the confining stress, whereas the dilatancy demonstrates an opposite trend.

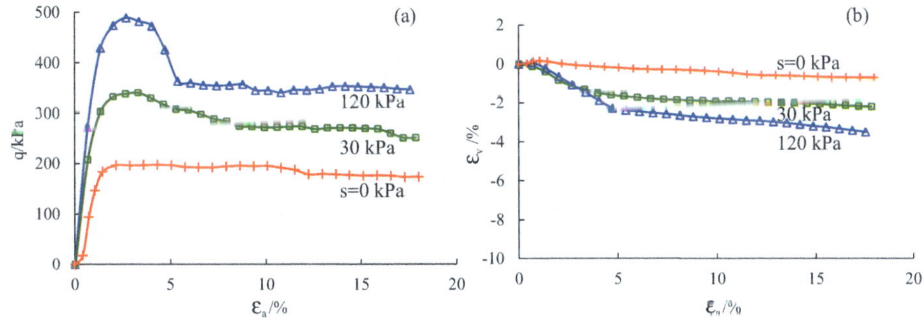

Fig. 5. The test results at a confining stress of 80 kPa and different suction levels

4 Discussion

4.1 Mobilized Peak Strength

The peak states obtained at three different suction levels are shown in Fig. 6, together with a strength envelope for the same soil at saturated state in a companion article (Zhou et al. 2017). Based on limited test results, the intercepts of strength lines show values of 0 kPa, 56 kPa and 92 kPa for the soil at suction levels of 0 kPa, 30 kPa and 120 kPa, implying cohesions of 0 kPa, 28 kPa, 47 kPa respectively. The mobilized

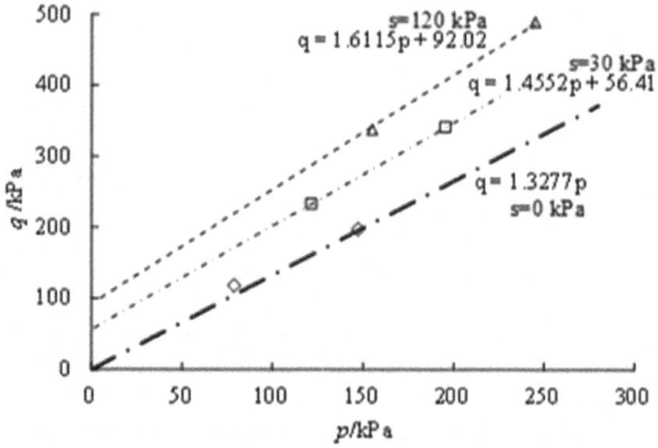

Fig. 6. The mobilized peak strength at different suction levels

internal friction angles at peak state are 33°, 36°, 39° at suction levels of 0 kPa, 30 kPa and 120 kPa, respectively.

4.2 Mobilized Peak Strength

The angle of dilation ψ depicts the relationship between volumetric strain and deviatoric strain at plastic state. It essentially reflects the flow rules in constitutive modeling. The angle of dilation is initially defined in slip surface analysis of slope under plane strain condition. It can be expressed as a general three-dimensional form below:

$$\sin(\psi) = -3\delta\varepsilon_v / (6\delta\varepsilon_q - \delta\varepsilon_v) \quad (1)$$

The mobilized angle of dilation at peak state is summarized in Fig. 7. Under the suction level of 0–120 kPa, the value varies from 13° to 22° at a confining stress of 40 kPa, and from 4° to 16° at a confining stress of 80 kPa. Figure 7 shows that the angle of dilation increases significantly with the matric suction, but decrease with the confining stress.

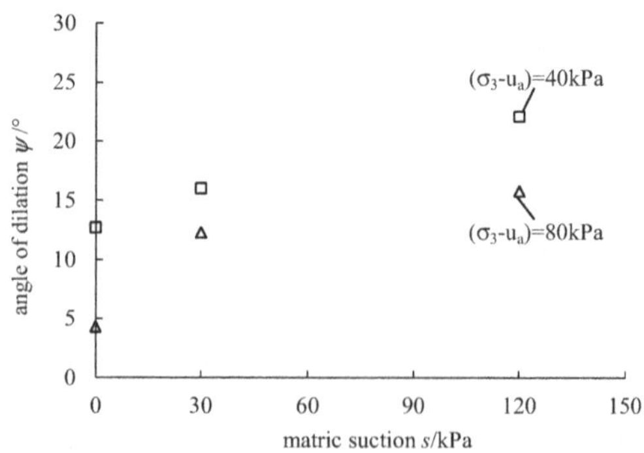

Fig. 7. The variation in the angle of dilation with matric suction at different stress levels

5 Conclusions

In this study, two sets of constant suction tests were performed to examine the stress-dilatancy of Xinjiang loess. The following preliminary conclusions were drawn from the test results.

(1) Both at peak state and at residual state, the mobilized shear strength increases with the confining stress and the suction level.
(2) The soil dilatancy is affected by the matric suction and the stress level. The mobilized angle of dilation at peak state increases significantly with the matric suction, but decreases with the confining stress.

Acknowledgements. The authors acknowledge the financial support of National Natural Science Foundation Project of China (Grant No. 515090181) and the Scientific Activity Project for Returned Oversea Scholars of MHRSS (Grant No. CKSD2016310/YT).

References

Alonso, E.E., Gens, A., Josa, A.: A constitutive model for partially saturated soils. Geotechnique **40**(3), 405–430 (1990)
Bruchon, J.-F., Pereira, J.-M., Vandamme, M., et al.: Full 3D investigation and characterisation of capillary collapse of a loose unsaturated sand using X-ray CT. Granul. Matter **15**(6), 783–800 (2013)
Donald, I.: Shear strength measurements in unsaturated non-cohesive soils with negative pore pressures. In: Proceedings of the 2nd Australia–New Zealand Conference on Soil Mechanics and Foundation Engineering. Technical Publications Ltd., Wellington (1956)
Dai, B.B., Yang, J., Zhou, C.Y.: A preliminary investigation on effect of particle size on mechanical behavior of granular materials. Rock Soil Mech. **35**(7), 1878–1884 (2014)
Ng, C.W.W., Chiu, C.F.: Laboratory study of loose saturated and unsaturated decomposed granitic soil. J. Geotech. Geoenviron. Eng. **129**(6), 550–559 (2003)
Wheeler, S.J., Sivakumar, V.: An elasto-plastic critical state framework for unsaturated soil. Géotechnique **45**(1), 35–53 (1995)
Zhou, Y.F., Cheng, Z.L., Gong, B.W., et al.: The dilatancy and instability of an infilled material in a channel slope in Xinjiang. Chin. J. Rock Mech. Eng. (2017, accepted). In Chinese

Finite Element Modeling of the Bearing Capacity for Transmission Tower Foundations on Expansive Soil

Zheng Su[1,2], Xilin Lü[1,2(✉)], Jiangu Qian[1,2], and Daokun Qi[3]

[1] Key Laboratory of Geotechnical and Underground Engineering of Ministry of Education, Tongji University, Shanghai 200092, China
xilinlu@tongji.edu.cn
[2] Department of Geotechnical Engineering, Tongji University, Shanghai 200092, China
[3] State Grid Henan Economic Research Institute, Zhengzhou 450052, China

Abstract. To satisfy the increasing power demands, lots of transmission lines are constructed in the middle-west of China. Expansive soil widely distributes in these areas, and the safety of foundation for transmission line tower needs to be evaluated. Based on consolidated drained triaxial tests, the cohesion and internal friction angle of unsaturated expansive soil were obtained. The compressive and uplift bearing capacities of the plate foundation for transmission line tower foundation were analyzed by 3D finite element modeling. The failure mode of the plate foundation under compressive load shows a local shear mode. For the uplift bearing capacity, the failure mode shows an annular apophysis mode around the plate foundation at the surface. The influence of the saturation of soil on the bearing capacity is more obvious than the uplift bearing capacity of a plate foundation.

Keywords: Expansive soil · Shear strength · Transmission line tower
Bearing capacity · Numerical simulation

1 Introduction

With the rapid development of economy, more and more transmission lines are being constructed in the middle-west of China, i.e. Henan province, where natural resources and electric power are lacking. It is urgent to construct transmission lines in these areas to satisfy the power demands of people's living and industrial production. The expansive soil distributed area usually affects the safety of the foundation for transmission lines. It is important to study the mechanical behavior of expansive soil and its influence on the bearing capacity of foundation.

Mechanical behaviors of unsaturated soil are usually described by the two independent stress state variables [1–5], especially the theory based on the two stress state variables $(\sigma - u_a)$ and $(u_a - u_w)$ [6, 7]. The shear strength of expansive soil has been widely studied in the past decades. Based on wetting-drying cycle tests of Nanyang unsaturated expansive soil, Miao et al. [8] investigated the soil-water characteristic curve and shear strength. A hyperbolic equation of suction stress and suction strength

L. Hu et al. (Eds.): GSIC 2018, *Proceedings of GeoShanghai 2018 International Conference: Multi-physics Processes in Soil Mechanics and Advances in Geotechnical Testing*, pp. 102–110, 2018.
https://doi.org/10.1007/978-981-13-0095-0_12

was presented to describe the influence of suction. Sheng et al. [9] compared three kinds of shear strength equations to experimental data, and the empirical and phenomenological equations were shown be applicable to all kinds of unsaturated soil. Based on the in-suit plate loading tests on unsaturated expansive soil, Xu [10] analyzed the bearing capacity characteristic of unsaturated expansive soil. The prediction agrees well with the measured results of in-situ test of Ningxia and Handan expansive soils. Pacheco et al. [11] discussed the distinctions between shallow and deep failure modes of transmission tower in tension load, and proper modification was applied to the Grenoble models to apply it to inhomogeneity soil caused by compacted backfill.

This paper studied the compressive and uplift bearing of power transmission tower on expansive soil. The Mohr-Coulomb criterion was adopted to describe the shear strength of unsaturated expansive soil. Based on consolidated triaxial tests, the cohesion and friction angle of unsaturated expansive soil with varying saturation degree was characterized. The compressive and uplift bearing behaviors of a plate foundation on expansive soil of different saturation degree were analyzed by 3D FEM, and the influence of saturation degree on bearing capacity was discussed.

2 Shear Strength of the Expansive Soil

2.1 Strength Criteria

Due to the complexity strength criteria by Bishop and Fredlund for unsaturated soil, the conventional Mohr-Coulomb was adopted to describe the undrained shear strength of expansive soil. The shear strength of unsaturated expansive soil strength consists of 4 components in total stress for the complicacy; those parts related to c', φ', suction force, and expansion force. The strength caused by the expansion force and the suction can be included in the cohesion

$$c_{total} = c' + \tau_s + \tau_{ps} \tag{1}$$

where τ_s is the strength induced by suction, and τ_{ps} is the strength of expansion.

The friction angle can similarly be converted to an equivalent friction angle; the unsaturated expansive soil strength criterion can be simplified to

$$\tau_f = c_{total} + \sigma * tan\varphi_{total} \tag{2}$$

The above formulas can be taken as an equivalent Mohr-Coulomb criterion, and strength parameters could be obtained by triaxial tests.

2.2 Triaxial Tests on Expansive Soil

The expansive soil samples were taken from Nanyang, Henan Provence. The soil was crushed firstly, and then sieved through 2 mm sieve. The samples were prepared according to the designed initial water content. The conventional GDS stress path triaxial apparatus was carried out to obtain the influence of initial water content on shear strength. The testing sample is outlined in Table 1. The prepared sample is shown

Table 1. Triaxial tests on expansive soil

Initial water content (%)	Initial saturation (%)	Confining pressure (kPa)
7.65	25	50, 100, 200
13.78	45	
19.90	65	
26.03	85	

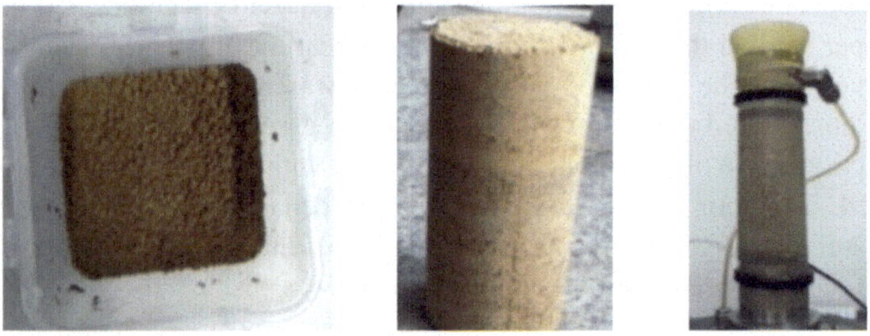

(a) Soil sample (b) Test sample (c) Triaxial test

Fig. 1. Preparation of the testing sample

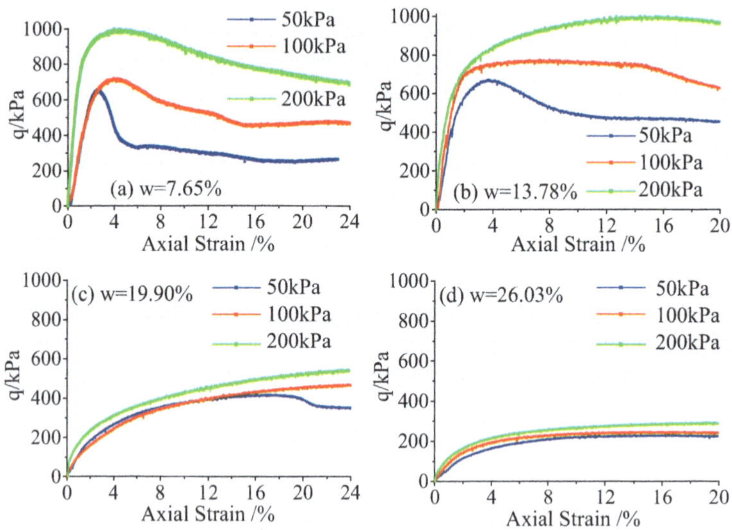

Fig. 2. Triaxial test results under different initial water content

in Fig. 1. The drainage valve at sample top was turned on and the one at bottom was turned off to exhaust the air in sample but keep the water.

According to the test results with different initial water contents, the stress-strain curve turns out to exhibit significant softening at a lower initial water content (w = 7.65%). As is shown in Fig. 2, the deviatoric stress reduces dramatically after it approaches the peak, and this behavior is more obvious under a lower confining pressure condition. As the water content increases, the stress-strain curve exhibits a significant hardening behavior, except for at a lower confining pressure, i.e. 50 kPa.

According to the Mohr circles and strength envelopes, the shear strength of the tested sample with different initial water content is shown in Fig. 3, and the obtained cohesion and friction angle are shown in Fig. 4 and Table 2. With the increase in the initial water content, the cohesion increases slightly and then decreases, and the friction angle decreases following a s-type curve. The reason why cohesion varies is liquid bridge.

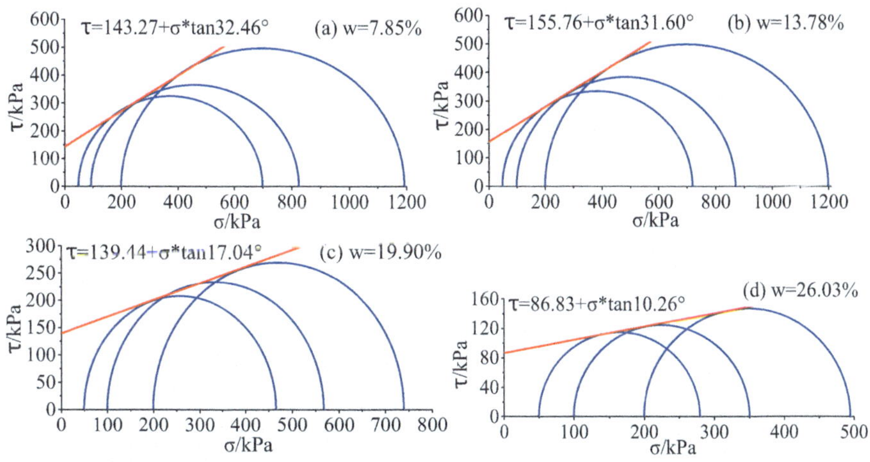

Fig. 3. Mohr circles and strength envelopes with different initial water content

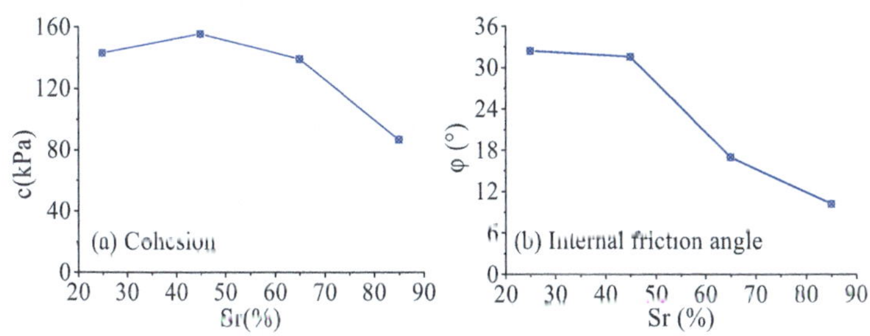

Fig. 4. Strength parameters of expansive soil in different initial saturation

Table 2. Strength parameters of expansive soil of different initial water content

Water content w (%)	Saturation Sr (%)	Cohesion c (kPa)	Friction angle φ (°)
7.65	25	143.27	32.46
13.78	45	155.76	31.60
19.90	65	139.44	17.04
26.03	85	86.83	10.26

The liquid bridge is zero when the soil is dry and fully saturated, and it increases initially and then decreases with water content.

3 Numerical Modeling of the Bearing Capacity

3.1 Numerical Simulation Procedure

The loading behavior of the transmission tower foundation was simulated by finite element simulation. The compressive and uplift bearing capacities of a plate foundation were studied, and the adopted meshes are shown in Fig. 5. The C3D8R element was chosen for the soil and foundation. The initial saturation degree of soil was assumed uniformly distributed. The friction coefficient between the soil and foundation was 0.32. The strength criterion and parameters with different water content and different saturation are obtained as shown in Table 2. Besides, the unit weight of expansive soil is γ = 18.5 kN/m^3, elastic modulus E = 60 MPa, Poisson's ratio ν = 0.26, and the unit weight γ = 23.0 kN/m^3, elastic modulus E = 200 GPa, Poisson's ratio ν = 0.33 of the foundation.

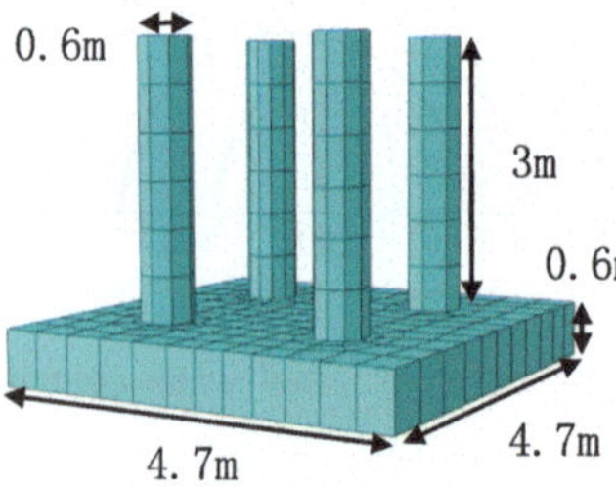

(a) Foundation model

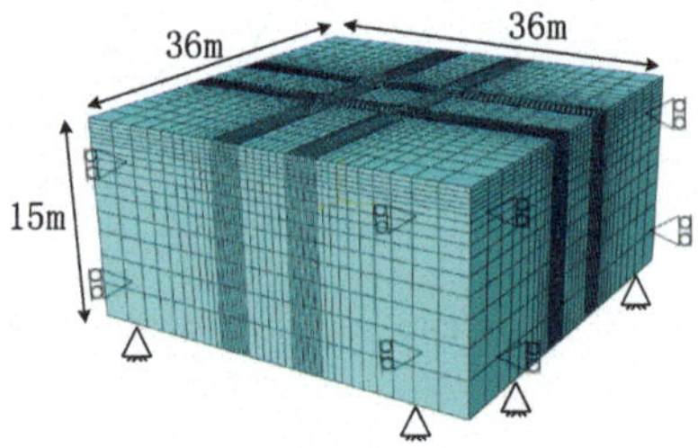

(b) Soil model

Fig. 5. Plate foundation model and the mesh

3.2 Bearing Capacity

The bearing capacity of a tower foundation on expansive soil can be obtained by applying a vertical load on the top of the foundation. The load-displacement curve of the tower foundation can be obtained according to the results. The obtained stress, displacement and plastic strain distributions are shown in Fig. 6. According to the results, the soil around the plate foundation rises to be an annular apophysis; the soil beneath the foundation goes down with the downward movement of foundation as shown in Fig. 6-(b). The Mises stress and equivalent plastic strain zone of the plate foundation is arc-shaped beneath the foundation. It can be concluded that the failure mode of the foundation tends to be a local shear failure mode.

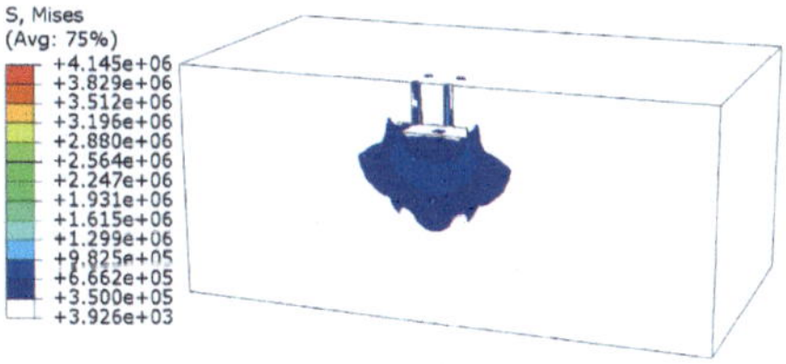

(a) Stress contour of soil

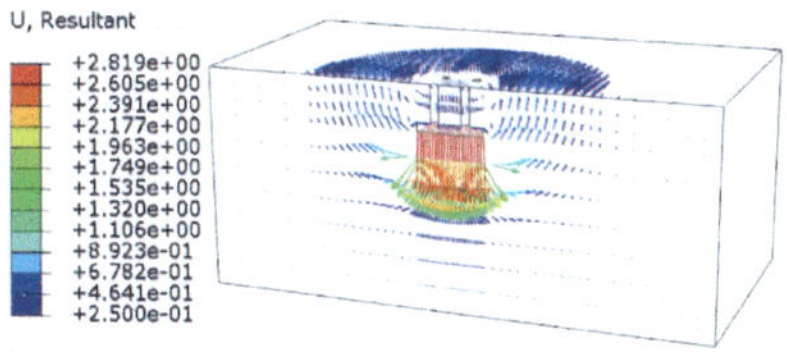

(b) Displacement vector of soil

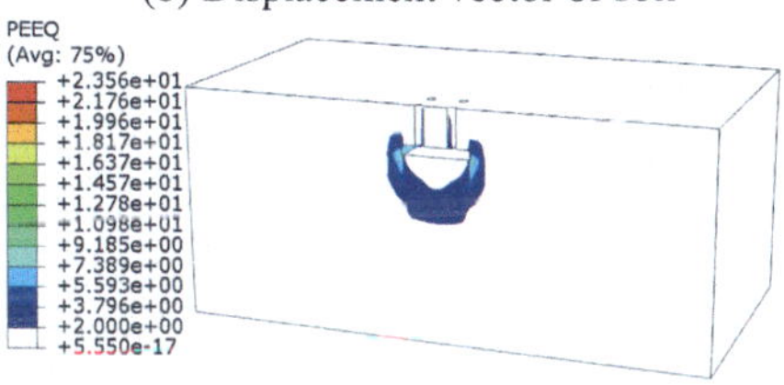

(c) Equivalent plastic strain contour of soil

Fig. 6. Compressive bearing capacity of the foundation (Sr = 85%)

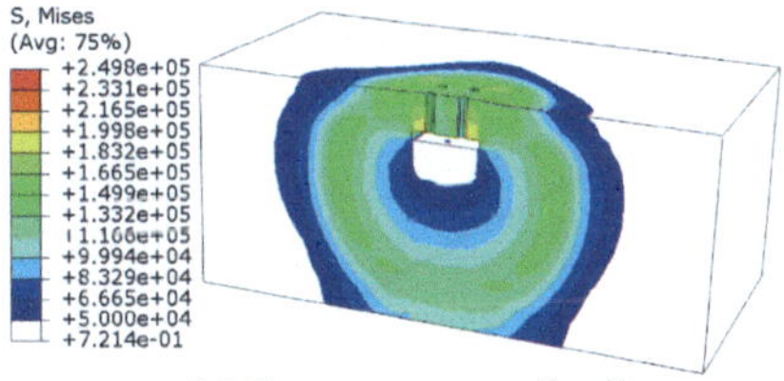

(a) Stress contour of soil

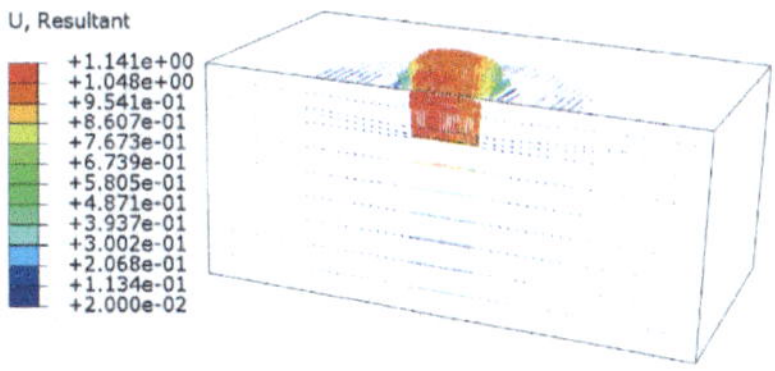

(b) Displacement vector of soil

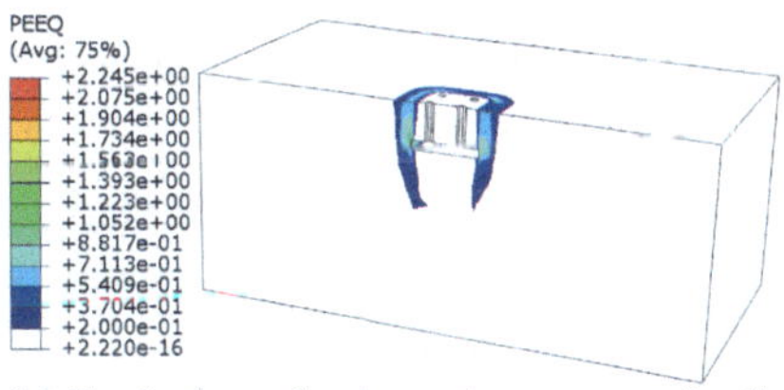

(c) Equivalent plastic strain contour of soil

Fig. 7. Uplift bearing capacity of the foundation (Sr = 85%)

3.3 Uplift Bearing Capacity

Since the transmission tower inevitably suffers lateral loads caused by line weight and wind, an uplift force would be applied on the tower foundation. The uplift bearing capacity of the tower foundation plays an important role in the safety of a transmission tower. A vertical uplift load was applied on the top of the foundation to simulate the vertical uplift force. As a result, the limit uplift bearing capacity of the tower foundation buried in the expansive soil can be obtained. The results of the simulations of the foundation are shown in Fig. 7. The region of stress distribution of plate foundation concentrates around the foundation and tends to be a vertical annular apophasis. There is an annular apophysis around the plate foundation, even though it is not comparable to that in compressive condition. The equivalent plastic strain zone of the foundation concentrates around the foundation and is not connected.

3.4 Variation with the Degree of Saturation

With the increase in displacement, the force on the foundation increases linearly at first, and then it increases slowly or even decreases, as is shown in Fig. 8. There exists an obvious inflection point on the load-displacement curve of the plate foundation. The

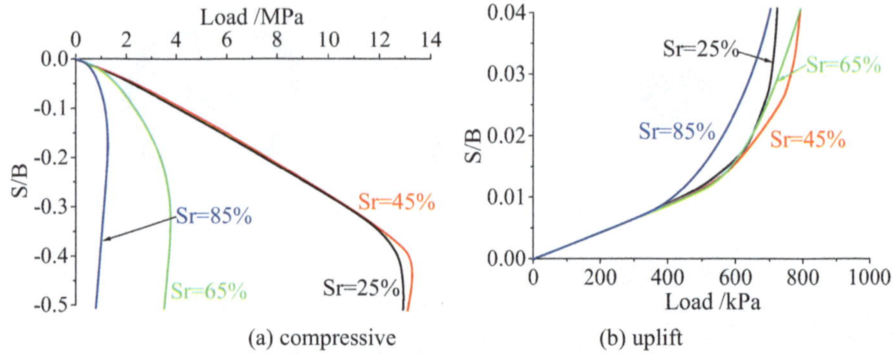

Fig. 8. Load-displacement curves

Table 3. Ultimate bearing capacity of footings

Saturation (%)	25	45	65	85
Ultimate bearing capacity (MPa)	12.86	13.26	3.77	1.25
Ultimate uplift bearing capacity (kPa)	698.12	755.65	581.51	498.96

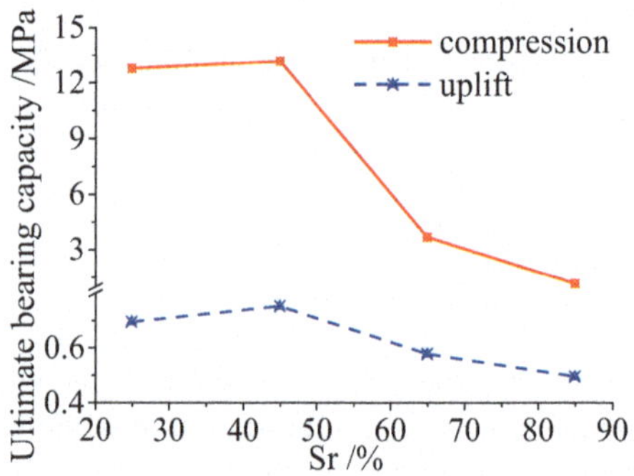

Fig. 9. Relationship between ultimate bearing capacity and the saturation of footing foundation

ultimate bearing capacities of the foundation with different saturation are shown in Table 3, and the relationship curves are shown in Fig. 8. The uplift and compressive bearing capacity of the foundation increases slightly when the saturation increases from 25% to 45%, and then it decreases dramatically with saturation. Besides, the limit bearing capacity of plate foundation reduces significantly with the saturation. As is shown in Table 3, the limit bearing capacity of plate foundation decreases nearly 90.28% when the saturation increases from 25% to 85%, and the limit uplift bearing capacity decreases to be 60%–70% of the initial value.

It can be found from Fig. 9 that the limit compressive bearing capacity decreases dramatically with saturation, and the influence of saturation degree of soil on the compressive bearing capacity is obvious than uplift bearing capacity.

4 Conclusion

The mechanical behavior of unsaturated expansive soil was studied by triaxial tests, and the equivalent strength parameters of Mohr-Coulomb criterion were obtained. Based on the obtained parameters, the compressive and uplift bearing capacities of the plate foundation of tower foundation were simulated. With the increase in the initial water content of expansive soil, the cohesion increases slightly and then decreases significantly, and the friction angle decreases with the increase in the initial water content by following a S-type curve. The friction angle decreases at a slow rate when the water content is low or high, while it changes significantly in the medium interval of water content. Results show that the equivalent plastic strain zone in compressive bearing capacity condition is arc-shaped, and the failure mode of the plate foundation is local shear mode. For the uplift bearing capacity, the region of stress distribution concentrates around the foundation, and the displacement of the soil around the foundation concentrates on the top of the foundation. There is an annular apophysis developed around the plate foundation on the surface. The difference between the compressive and uplift bearing capacities is significant when the saturation is low, and the difference is small when the degree of saturation becomes high.

Acknowledgments. The work was financially supported by the State Grid Corporation of China (through Grant No. 5217L0160001) and the Fundamental Research Funds for the Central Universities, these support are gratefully acknowledged.

References

1. Fredlund, D.G. (ed.): Engineering Approach to Soil Continua. Transportation and Geotechnical Group, University of Saskatchewan (1974)
2. Barden, L., Madedor, A.O., Sides, G.R.: Volume change characteristics of unsaturated clay. ASCE J. Soil Mech. Found. Div. **95**, 33–51 (1969)
3. Burland, J.B. (ed.): Some Aspects of the Mechanical Behavior of Partly Saturated Soils. Butterworth and Company (Australia) Ltd., Sydney (1965)
4. Bishop, A.W., Blight, G.E.: Some aspects of effective stress in saturated and unsaturated soils. Geotechnique **13**, 177–197 (1963)
5. Coleman, J.D.: Stress/strain relations for partly saturated soil. Geotechnique **12**(4), 348–350 (1962)
6. Fredlund, D.G., Rahardjo, H. (eds.): Soil Mechanics for Unsaturated Soils. Wiley, New York (1993)
7. Fredlund, D.G., Morgenstern, N.R., Widger, R.A.: The shear strength of unsaturated soils. Can. Geotech. J. **15**(3), 313–321 (1978)
8. Miao, L., Liu, S., Lai, Y.: Research of soil–water characteristics and shear strength features of Nanyang expansive soil. Eng. Geol. **65**, 261–267 (2002)

9. Sheng, D., Zhou, A., Fredlund, D.G.: Shear strength criteria for unsaturated soils. Geotech. Geol. Eng. **29**, 145–159 (2011)
10. Xu, Y.: Bearing capacity of unsaturated expansive soils. Geotech. Geol. Eng. **22**, 611–625 (2004)
11. Pacheco, M.P., Danziger, F.A.B., Pereira, P.C.: Design of shallow foundations under tensile loading for transmission line towers: an overview. Eng. Geol. **101**(3–4), 226–235 (2008)

Climate Change Impacts on Reactive Soil Movements

Xi Sun, Jie Li(✉), Annan Zhou, and Hamayon Tokhi

School of Engineering, RMIT University, Melbourne, VIC 3001, Australia
jie.li@rmit.edu.au

Abstract. Climate affects the extent of soil moisture variation likely to occur through time and hence the amount of foundation movement likely to occur at a reactive soil site. The Thornthwaite Moisture Index (TMI) as a useful climate parameter has been widely employed in Australia to estimate the depth of design soil suction change (H_s) which is needed for the determination of characteristic ground movement (y_s). Precipitation and temperature are the primary weather parameters required for the TMI computation. By applying the projected rainfall reduction and temperature increase in 2030, 2050 and 2070 in the TMI calculation, the effects of climate changes on expansive soil movements and site classification can be quantified by the use of the predicted TMI. In this study, TMI values of various areas of the State of Victoria were calculated by using climate projections generated from 23 climate models. These predicted TMI indices were then used to delineate TMI isopleth lines on the map of the State of Victoria to visualize and compare climate conditions in 2030, 2050 and 2070. These maps show an overall significant growth of drying for the State of Victoria, where the most noticeable increase of aridity is expected to occur in 2070. The general decrease in TMI value across various areas of Victoria indicates a prominent reduction in average soil moisture availability. This implies that residential slabs are likely to experience greater ground movements due to the greater depth of H_s, which in turn may result in a higher incidence of slab edge heave and an increase in the occurrence of distortion of residential buildings built on expansive soils.

Keywords: Thornthwaite moisture index · Expansive soil · Climate change

1 Introduction

There is now broad agreement that the earth's climate is changing as a result of human activities, particularly emission of greenhouse gases (GHGs), are very likely to be the dominant cause. A common tool used for projecting climate change is a climate model which is a mathematical representation of the Earth's climate system. Current climate models predict that the steady increase of global average annual carbon dioxide (CO_2) will, directly and indirectly, impact the performance of infrastructures. It was projected that temperature would rise across the Australian continent with the increase of approximately 1 °C and 3 °C in 2030 and 2070 respectively and this pattern would vary little seasonally. Different from the temperature pattern which is always increasing, the trend of precipitation change exhibit both increase and decrease [1].

L. Hu et al. (Eds.): GSIC 2018, *Proceedings of GeoShanghai 2018 International Conference: Multi-physics Processes in Soil Mechanics and Advances in Geotechnical Testing*, pp. 111–119, 2018.
https://doi.org/10.1007/978-981-13-0095-0_13

The use of the climate index, TMI, as a predictor of the depth of design soil suction changes is now widely accepted by geotechnical engineers and practitioners in Australia. Over the last 20 years or so, many researchers have published studies including TMI maps for various parts of Australia. The first TMI contour map of the continent of Australia was produced by Aitchison and Richards [2]. Based on this map, Smith [3] created a TMI contour map for the State of Victoria by using the same TMI values that were derived by Aitchison and Richards [2]. This map has been endorsed by the Australian Residential Slabs and Footings Standard AS2870 [4] and used with minor modification as guidance for design engineers to infer the depth of design soil suction change, H_s, which is required for estimation of the characteristic surface movement, y_s. In this study, three TMI based climate zoning maps of the State of Victoria for 2030, 2050 and 2070 are produced.

Climate affects the extent of soil moisture variation likely to occur through time [5–7] and hence the amount of foundation movement likely to occur on moisture reactive foundation soils. The extent of variation in the soil moisture profile between the extremes of wet and dry is greater in a more arid climate. This study assesses the effects of climate variations (the projected changes in temperature and precipitation for 2030, 2050 and 2070) on the depth of design soil suction changes (H_s) that can be estimated by using the TMI.

2 Overview of Thornthwaite Moisture Index

The Thornthwaite Moisture Index (TMI) was originally derived from precipitation minus evaporation (P-E) Index by Thornthwaite [8] for the purpose of climate classification. In Australia, this climate parameter has been widely used in geotechnical engineering and other disciplines to evaluate the soil moisture changes in an active zone, predict the depth of seasonal moisture changes and classify climatic zones. TMI is mainly a function of rainfall and potential evapotranspiration (*PET*). A negative TMI indicates an arid climate with a deficit of precipitation relative to *PET* and generally low moisture in the soil. A positive TMI indicates a humid climate with a surplus of precipitation and generally high soil moisture. Zero TMI index means that, over the long term, under average conditions, the inflow from precipitation just equals the loss of soil moisture through evapotranspiration that would occur if the notional reference vegetative cover was actually present.

TMI computation equation was originally pioneered by C. W. Thornthwaite in 1948. It was quickly accepted by the world scientific community and has been widely used by many researchers across a variety of disciplines with success over the last several decades. The original TMI equation is computationally intensive due to the need to determine moisture surplus and deficit required for performing water-balance approach. This equation was revised by Thornthwaite and Mather [9] and further modified by Mather [10], leading to the following simplified TMI equation (Eq. 1) which eliminates the need to carry out the water-balance approach and requires only the annual precipitation (P) and the adjusted potential evapotranspiration (*PET*).

$$TMI = 100\left(\frac{P}{PET} - 1\right) \quad (1)$$

TMI indices can vary greatly if different PET estimation methods are employed. A relatively simple temperature-based PET equation (Eq. 2) pioneered by Thornthwaite [8] has been widely used by many researchers in Australia. A review and comparison of different assumptions and methods for calculating TMI can be found in Sun *et al.* [11].

$$PET_i = e_i\left(\frac{D_i N_i}{30}\right) \quad (2)$$

where D_i is the day length correction factor for the month i; N_i is the number of days in the month i; and e_i is the non-adjusted potential evapotranspiration (cm) for the month i given as:

$$e_i = 1.6\left(\frac{10t_i}{H_y}\right)^a \quad (3)$$

where t_i is the mean monthly temperature in °C and calculated as the average of t_{max} and t_{min}. The heat index for month i is determined as follows:

$$h_i = (0.2t_i)^{1.514} \quad (4)$$

The annual heat index, H_y is simply calculated by summing the 12 monthly heat index values. The power term a in Eq. 3, is given as:

$$a = 6.75 \times 10^{-7}H_y^3 - 7.71 \times 10^{-5}H_y^2 + 0.017921H_y + 0.49239 \quad \text{and} \quad 0 < a < 4.25 \quad (5)$$

3 TMI Determination for Victoria

Climate model which is a mathematical representation of the Earth's climate system has been commonly utilized for climate change projections. A total of 23 climate models from the Coupled Model Inter-comparison Project 3 (CMIP3) database have been adopted to project spatially dependent future climate for the continent of Australia [1]. 40 emission scenarios have been developed under Special Report on Emissions Scenarios (SRES) [12]. A1FI (fossil fuel intensive) may be considered as the extreme high greenhouse gas emissions (GHGs) scenario, A1B (balanced) is the medium, and A1T (predominantly non-fossil fuel) is the low emission scenario. Probability distributions have been employed to present the uncertainties caused by the differences among results of climate models and 50th percentile or the median gives the best estimate.

Climate projections from 23 models provide by CSIRO, BOM and DCCEE [12] are used to determine precipitation and temperature needed for the calculation of TMI for

49 stations for the State of Victoria in 2030, 2050 and 2070. Mid-range emission scenario (A1B scenario) is adopted in 2030 since there are marked variations among results of climate models due to near-term changes in climate, while extreme climate change scenario (A1FI scenario) is employed for 2050 and 2070 projections as climate changes are more dependent on the greenhouse gas emissions scenario.

The methodology used to predicting and quantifying climate condition by the use of TMI for various localities of the State of Victoria in 2030 (A1B scenario), 2050 (A1FI scenario) and 2070 (A1FI scenario) are described as follows:

- Collection of climate parameters in 1990

Weather parameters (i.e. monthly precipitation and monthly temperature) from various weather stations across the State of Victoria in 1990 were purchased from the Bureau of Meteorology (BOM). It should be pointed out that although there are 204 accessible weather stations across the State of Victoria, only 49 of them have the fully recorded historical climatological data required for this study.

- Projection of climate trend

The projected seasonal precipitation and temperature trend maps for the State of Victoria in 2030 (A1B scenario), 2050 (A1FI scenario) and 2070 (A1FI scenario), based on the best estimate (i.e. 50th percentile), was obtained from *'Climate Change In Australia'* website [13]. All the available weather stations of the State of Victoria were plotted on the projected seasonal trend maps by using their coordinates (i.e. latitude and longitude) so that the threshold of seasonal variation trend for each weather station can be obtained in accordance with the given color scale bar (Fig. 1). Seasonal average variation trend (i.e. the average of the maximum and minimum threshold of the variation trend) was adopted in the calculation.

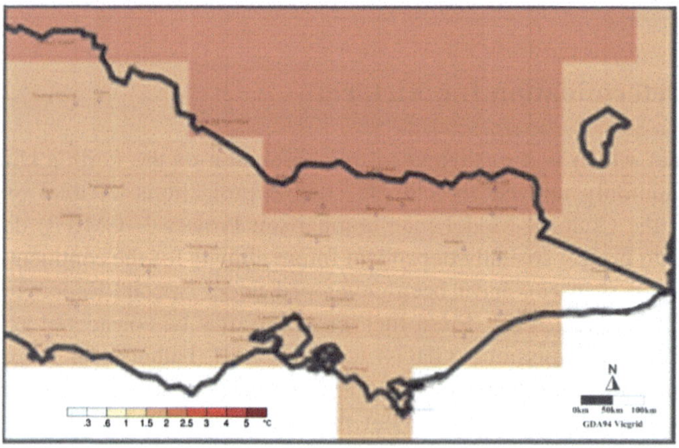

Fig. 1. The projected summer temperature map for the State of Victoria in 2050 under A1FI emission scenario.

- Determination of TMI in 2030, 2050 and 2070

The precipitation and temperature data in 2030, 2050 and 2070 were obtained using the seasonal average precipitation and temperature trend that is projected based on weather condition in 1990. These predicted weather parameters were then used for TMI calculation. The Thornthwaite *PET* model (i.e., Eq. 2) and the simplified Thornthwaite equation (i.e., Eq. 1) were used for TMI computation.

4 The Delineation of TMI Isopleth Map for Victoria

A few different versions of TMI based contour map of the State of Victoria have been published over the last ten years. Lopes and Osman [14] created TMI maps for the State of Victoria using the climatic data from the period of 1948 to 2007. Leao and Osman [15] calculated TMI using the 1913–2012 data. More recently, Li and Sun [16] produced three climate zoning maps of the State of Victoria from 1954 to 2013. In this study, The calculated TMI indices were employed for the delineation of isolines on the map of the State of Victoria for 2030, 2050 and 2070 by the use of Surfer® 11 to evaluate TMI changes. Climate zones of Victoria are classified in accordance with Table 1.

Table 1. Relationship between TMI, H_s and climatic zone [4].

Climatic zone	Description	TMI	H_s (m)
1	Alpine/wet coastal	>+10	1.5
2	Wet temperate	−5 to +10	1.8
3	Temperate	−15 to −5	2.3
4	Dry temperate	−25 to −15	3.0
5	Semi-arid	−40 to −25	4.0
6	Arid	<−40	>4.0

TMI-based contour maps (Figs. 2, 3 and 4) clearly indicate that the State of Victoria will become warmer and drier, and a significant increase in clay soil movement is expected for the State. It seems the most noticeable climate change will occur in 2070 due to the significant TMI variation compared to TMI map for 2030 and 2050. Generally speaking, the arid zone (zone 6, TMI < −40) were mainly distributed in the north-west region of the State while the wettest zone (i.e. zone 1, TMI > 10) spread out along the southern and eastern coastal areas. The substantially negative TMI values present in the far north-west area can be attributed to the long period and high frequency of drought, whereas large positive TMI values occur in southern regions are the result of the moist coastal climate. The overall decrease in TMI values across various locations of Victoria represents a prominent increase in aridity and hence a significant reduction in average soil moisture availability, which could potentially result in an increase in the severity of cracking of residential footings.

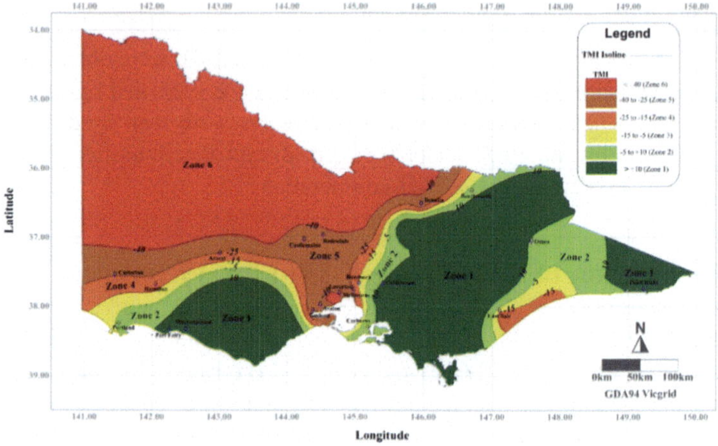

Fig. 2. TMI contour map for the State of Victoria in 2030.

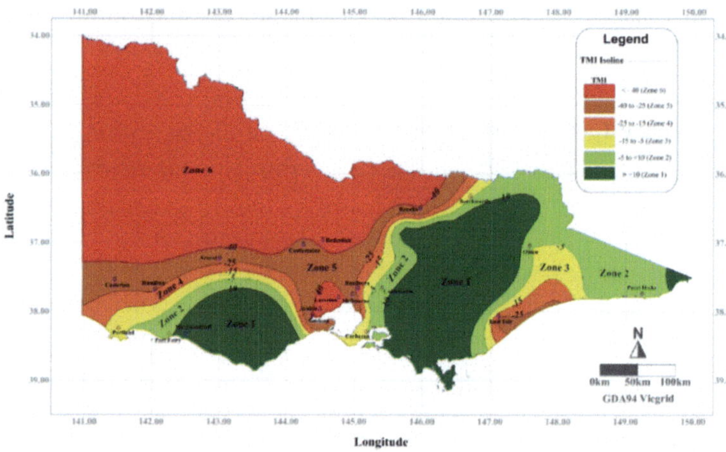

Fig. 3. TMI contour map for the State of Victoria in 2050.

Comparing Figs. 2, 3 and 4 reveal that climate Zones 6 have spread into the metropolitan area of Melbourne, taking up a large portion of previous Zone 4 and Zone 5. It is worth noting that soil in south-western suburbs of Melbourne including Laverton, and Avalon is expected to experience the growth of desiccation than elsewhere. This implies the greater depth of design soil suction change and the higher incidence of excessive ground surface movements. In December 2011, the Housing Industry Association (HIA) estimated that more than 1000 new houses in the western suburbs of Melbourne were damaged due to soil heave. These new houses have all been built in drier weather between 2003 and 2010 and have been subjected to larger ground differential movements induced by abnormal moisture changes after the

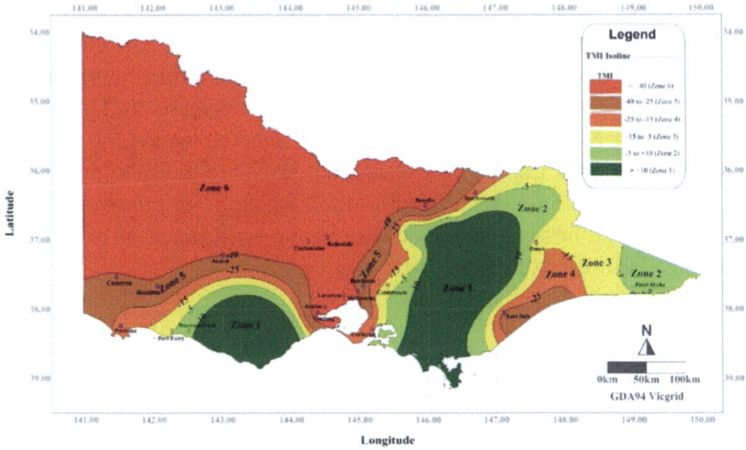

Fig. 4. TMI contour map for the State of Victoria in 2070.

construction of gardens/lawns and watering system around buildings and the breaking of the long drought in 2011 [17, 18].

The changes in TMI isopleth in Figs. 2, 3 and 4 depict:

- Climate Zone 6 (Arid climate) at north-west is expected to suffer the severe aridity and expand further towards south and east to include Castlemaine, Redesdale, Benalla, Geelong and Melbourne CBD and its inner south-west suburbs (i.e., Avalon, Laverton), leading to a marked reduction of Zone 5.
- Climate Zone 5 (Semi-arid climate) has a significant contraction in the central of the State and moves to south and east to include Casterton, Hamilton, Ararat, and Bundoora. The prevailing dry temperate climate (i.e., Zone 4) in East Sale area is replaced by Zone 5.
- Climate Zone 4 (Dry temperate climate) moves to east and south-west to include Portland, Cerberus and Beechworth which are previously predominated by Zone 2.
- Climate Zone 3 (Temperate climate) move towards to south and expands further to eastern coastal areas to include Coldstream and Omeo, resulting in a significant contraction of Zone 2.
- Climate Zone 2 (Wet temperate climate) moves to south coastal in Port Fairy and Warrnambool areas, and east coastal regions to include Point Hicks.
- Climate Zone 1 (Wet coastal climate) retract to the south coast, and a notable contraction takes place in the eastern part of the State.

5 Conclusions

By the use of the projected rainfall reduction and temperature increase from 23 climate models [13], the effects of changes in climate (temperature and precipitation) on TMI and H_s can be evaluated. In this paper, a method for predicting TMI has been presented.

The predicted TMI values were employed to produce three TMI isopleth maps for the State of Victoria in 2030, 2050 and 2070. These maps have shown that Victorian climate will become drier in terms of annual TMI, where the most noticeable increase of aridity is expected to occur in 2070. The general decrease in TMI value across various areas of the State of Victoria indicates a marked reduction in average soil moisture availability. This implies that residential slabs are likely to experience larger ground movements due to the greater depth of H_s, which in turn may result in a higher incidence of slab edge heave and an increase in the occurrence of distortion of residential buildings built on expansive soils.

References

1. CSIRO and BOM: Climate change in Australia. Technical report, 140 p (2007)
2. Aitchison, G.D., Richards, B.G.: A broad-scale study of moisture conditions in pavement subgrades throughout Australia: Parts 2 & 3. In: Moisture Equilibria & Moisture Changes in Soils Beneath Covered Areas. Butterworths, Sydney (1965)
3. Smith, R.: Estimating soil movements in new areas. In: Seminar-Extending the Code Beyond Residential Slabs and Footings. The Institution of Engineers, Australia (1993)
4. AS2870: Australian Standard: Residential Slabs and Footings. Standards Australia (2011)
5. Sun, X., Li, J., Zhou, A.N.: Assessment of the impact of climate change on expansive soil movements and site classification. Aust. Geomech. J. **52**(3), 39–50 (2017)
6. Yao, Y.P., Zhou, A.N.: Non-isothermal unified hardening model: a thermo-elastoplastic model for clays. Geotechnique **63**, 1328–1345 (2013)
7. Sun, D., Wenjing, S., Yan, W., Li, J.: Hydro-mechanical behaviours of highly compacted sand-bentonite mixture. J. Rock Mech. Geotech. Eng. **1**(2), 79–85 (2010)
8. Thornthwaite, C.W.: An approach toward a rational classification of climate. Geogr. Rev. **38** (1), 55–94 (1948)
9. Thornthwaite, C.W., Mather, J.R.: The water balance, vol. 8, no. 1, pp. 1–104. Laboratory of Technology, Publications in climatology (1955)
10. Mather, J.R.: Climatology: Fundamentals and Applications, pp. 113–131. McGraw Hill, New York (1974)
11. Sun, X., Li, J., Zhou, A.N.: Evaluation and comparison of methods for calculating Thornthwaite Moisture Index. Aust. Geomech. J. **52**(2), 61–75 (2017)
12. IPCC SRES (Special Report on Emissions Scenarios): Special Report on Emissions Scenarios: A Special Report of Working Group III of the Intergovernmental Panel on Climate Change. 599 p. Cambridge University Press, Cambridge (2000)
13. CSIRO, BOM and DCCEE: Climate change in Australia. http://www.climatechange inaustralia.gov.au/. Accessed 10 Jan 2014
14. Lopes, D., Osman, N.Y.: Changes of Thornthwaite's total moisture indices in Victoria from 1948–2007 and the effect on seasonal foundation movement. Aust. Geomech. J. **45**(1), 37–48 (2010)
15. Leao, S., Osman-Schlegel, N.Y.: TMI for urban resilience: measuring and mapping long-term climate change on soil moisture. In: 7th Australasian Housing Researcher's Conference, Fremantle, WA, Australia (2013)

16. Li, J., Sun, X.: Evaluation of changes of Thornthwaite Moisture Index in Victoria. Aust. Geomech. J. **50**(3), 39–49 (2015)
17. Li, J., Cameron, D.A., Ren, G.: Case study and back analysis of a residential building damaged by expansive soils. Comput. Geotech. **56**, 89–99 (2014)
18. Li, J., Zou, J., Bayetto, P., Barker, N.: Shrink-swell index database for Melbourne. Aust. Geomech. J. **51**(3), 61–76 (2016)

Role of Matric Suction on Shear Strength of Unsaturated Compacted Soil at Low Confining Stress

Ali Murtaza Rasool[1,2(✉)] and Jiro Kuwano[1]

[1] Saitama University, Saitama 338-8750, Japan
ali_eng@hotmail.com
[2] National Engineering Services Pakistan (Pvt.) Ltd., Lahore 54580, Pakistan

Abstract. In order to deal with some typical geotechnical problems, such as stability of slopes and bearing capacity of foundations shear strength is considered as a fundamental soil property. The influence of matric suction on shear strength has been recognized and various types of laboratory element tests with suction control or measurement has been performed by many researchers. This study aimed at understanding the role of matric suction on the transformation of shear strength and failure mechanism in unsaturated soil. The method of study was via a series of advanced laboratory element tests of two specific types; (1) constant water content (CW) tests, and (2) fully undrained (FU) tests. The tests were performed using the axis translation technique to measure matric suction. The results indicate that the change in water content has an influence on initial matirc suction. The relationship between matric suction and shear strength is non-leaner, and change in matric suction affects the mechanical behavior of unsaturated soil.

Keywords: Unsaturated soil · Shear strength · Water content
Matric suction

1 Introduction

In order to deal with some typical geotechnical problems, such as stability of slopes and bearing capacity of foundations, shear strength is considered as a fundamental soil property. The shear strength of soil is known to depend on many factors, like the range of applied stresses and drainage characteristics of the soil. In the case of unsaturated soil, the influence of matric suction has been recognized and various types of tests with suction control or measurement had been performed by many researchers [3, 4, 6, 7]. Consolidated drained (CD), constant water content (CW) and undrained (UD) triaxial tests are some of the unsaturated tests with suction control and measurement. Many laboratory tests (like CD and CW) on unsaturated soil have been conducted under a constant air pressure, however, the drained conditions for water and air cannot always be attained in engineering problems [2]. For example, the air pressure in river embankment increases during the seepage process and may vary during soil compaction. The air pressure changes in such cases, therefore, it is important to investigate the air-water-soil interaction and a thorough understanding of the mechanical behavior

L. Hu et al. (Eds.): GSIC 2018, *Proceedings of GeoShanghai 2018 International Conference: Multi-physics Processes in Soil Mechanics and Advances in Geotechnical Testing*, pp. 120–128, 2018.
https://doi.org/10.1007/978-981-13-0095-0_14

of unsaturated soils under low confining stresses is important for predicting the stability and deformation of the surface layer in natural slope or embankment [1].

For this reason, we have conducted a series of triaxial tests on silty soil under undrained conditions for the water known as constant water content (CW) test and undrained conditions for both water and air known as fully undrained (FU) test. In FU tests, after applying axis translation technique, during the shear process neither pore water nor pore air was allowed to flow out from the soil and both the pore water and pore air pressure were accurately measured. Whereas in CW tests, during the shear process the specimen was loaded maintaining the pore air pressure constant and drained and pore water pressure in an undrained mode. This research focuses on the surface layer of slopes or embankments which are normally under a state of low confining stress, therefore, the soil specimens in this study are tested at low confining stress.

The main goal of this work is to investigate the role of matric suction on shear strength behavior of unsaturated silty soil at low confining stress under different drainage conditions. The change in matric suction due to water content is investigated and the effect of matric suction on volume change, saturation ratio and effective stress is also studied and discussed.

2 Experimental Setup and Methodology

In this part we will discuss the experimental setup, physical properties of soil and technique to prepare specimen required to perform the test.

2.1 Experimental Setup

Figure 1 gives a schematic diagram of a modern unsaturated triaxial testing device for cylindrical specimens 50 mm in diameter and 100 mm in height. In order to separate the routes for the measurement and the control of the pore air pressure and the pore water pressure, a membrane filter and a PTFE (polytetrafluoroethylene, known as Teflon) sheet were used. The PTFE sheet was placed on the top of the specimen to cut off the flow of water and to measure the air pressure. The membrane filter was installed in the lower pedestal to cut off the flow of air and to measure the water pressure. The thin membrane filter with pores of 0.45 mm has air entry value of 420 kPa. For fully undrained tests, no pore water was allowed to flow out in both the isotropic consolidation and monotonic shearing process. The salient feature of this triaxial apparatus is that both pore air and water pressures can be measured separately. Pore air pressure transducer was installed in the top cap and connected to air regulator in order to give continuously supply of the air throughout the test. In addition to this, a solenoid valve was also installed in the air drainage line inside the top cap to control drained/undrained pore air. The change in volume of the specimen during the triaxial tests was measured as the change in the water level in the inner cell by LCDPT (Low capacity differential pressure transducer). The monotonic loading was applied under strain-controlled conditions. The vertical load of the soil specimen was measured by the inner load cell. The cell pressure and pore water pressure was measured by cell and pore water pressure transducers. All instruments were connected with dynamic strain amplifiers which were

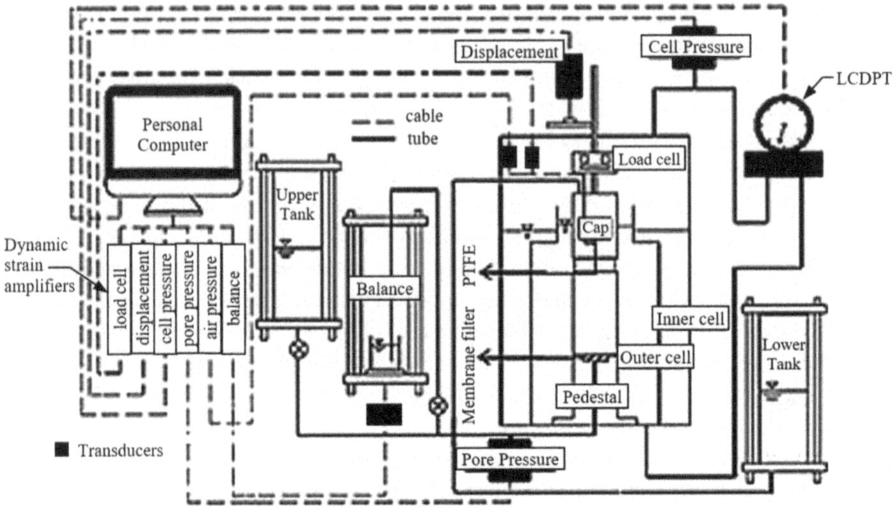

Fig. 1. Schematic diagram of triaxial test apparatus

then connected to A/D board and finally to PC. The dynamic strain amplifiers have a voltage range of 0–10 V and PC has a software for controlling instruments.

2.2 Physical Properties of Soil

The silty soil known as "DL clay" in Japan was tested in this study. DL clay is homogenous and easy to obtain. It is larger in grain size than average clay and is composed of 90% silt and 10% clay. The soil particle density is 2.635 g/cm^3 and the liquid limit is non-plastic. The reason of using DL clay is that it has lower initial suction than kaolin clay under the same degree of saturation.

2.3 Sample Preparation

The soil specimens used in this research were prepared by the static compaction method. The purpose of using static compaction as opposed to dynamic compaction is to obtain a more homogeneous specimen in terms of density [9]. Prior to performing the compaction, the dry DL clay was mixed well with water to make up a water content of 10%, 15%, 20% and 25%. After mixing with the water the specimens were statically compacted in a layer of 20 mm (each) in a special apparatus with a hydraulic jack in order to obtain homogeneity. The specimens were compacted to achieve a degree of compaction of about 80%. The pre-consolidation pressure on the soil at the time of sample preparation was more than during test process, therefore the soil used in termed as over consolidated soil. According to Nishimura unsaturated soils near the ground surface and artificially compacted soils are commonly over-consolidated due to change in the environment [8].

3 Experimental Results

In this part, we will discuss the measurement of initial suction at various water content and test results of constant water content tests and fully undrained test and finally, we will compare and discuss the results of these two test series.

3.1 Initial Suction

The previous researches show that initial suction influences the behavior of unsaturated soil. However, the suction depends on pore-water distribution and degree of saturation. Figure 2 shows the variation of the monitored values of initial suction with time of statically compacted specimens with a water content of 10%, 15%, 20% and 25% when they were set in the triaxial apparatus. Due to different values of water content, the initial suction varies and we can see that the higher the water content less is the initial suction and less is the time required for stabilizing initial suction and vice versa. The less time in stabilizing initial suction is due to use of a thin membrane filter having air entry value of 420 kPa.

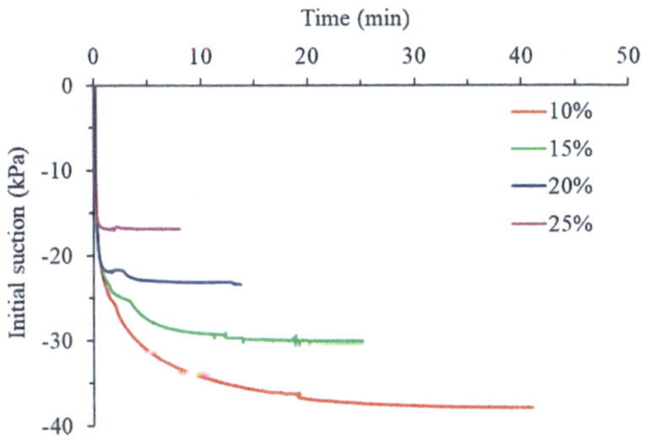

Fig. 2. Measurement of initial suction

3.2 Constant Water Content Test

In constant water content test after measurement of the initial suction, the matric suction was applied to soil specimen using the axis translation technique [4]. The technique basically translated the reference origin of pore water pressure from standard atmospheric condition to the final air pressure in the chamber. During the shear process, the pore air pressure was drained and controlled whereas the pore water pressure was measured. Figure 3 depicts the volumetric strain and stress-strain relations under different matric suction levels for CW tests. Due to high past pressure and a high degree of compaction, the soil behavior was typical of over-consolidated soil hence the volume increased with increase in axial strain. The effect of over-consolidation can also be seen in a stress-strain curve in form of first a peak and then a post peak failure response. The stress-strain relationship of specimen CW10 showed more brittle pattern than other specimens, this tendency seems to influenced by high matric suction that increases the stiffness of soil, this behavior was also observed by Kato et al. [5], however, the brittleness decrease with a decrease in matric suction. The shear strength of soil also changed with matric suction and the maximum value of shear strength was observed by specimen with 15% water content. The peak shear strength value of all specimens were ranged between 0–2.75%, up to this value there was no much difference in volumetric strain, however, afterward, some change was observed which was due to post peak response of specimens. The pore water pressure was undrained and measured during the shearing process so slight increase in pore water pressure was also observed.

Figure 4(a) shows the saturation ratio versus axial strain during shearing, a slight decrease in saturation ratio was observed in all specimens, the decrease in saturation ratio was due to the dilative behavior of soil. Figure 4(b) illustrates the stress path in terms of deviator stress and the mean effective stress. The stress paths are sloped at 3.0, as the soil used in this study was over-consolidated, therefore, the stress path first crossed critical state line reach the maximum point and ultimately reverse its direction and fall on the critical state line. It can be seen that the final points of the stress paths all meet at the critical state line with a gradient of 1.1 passes through the origin.

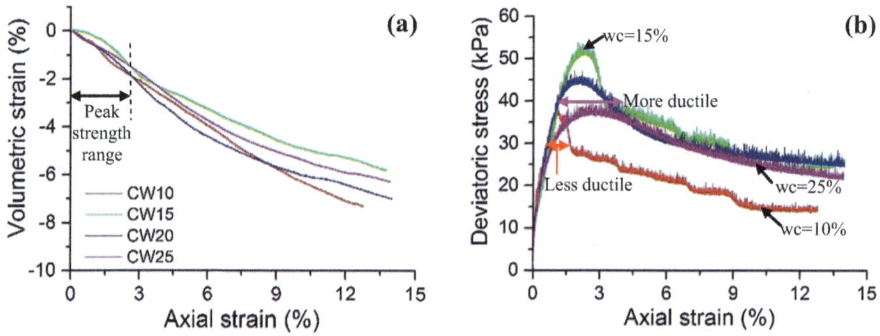

Fig. 3. Constant water content test results: (a) volumetric strain (b) stress-strain relation

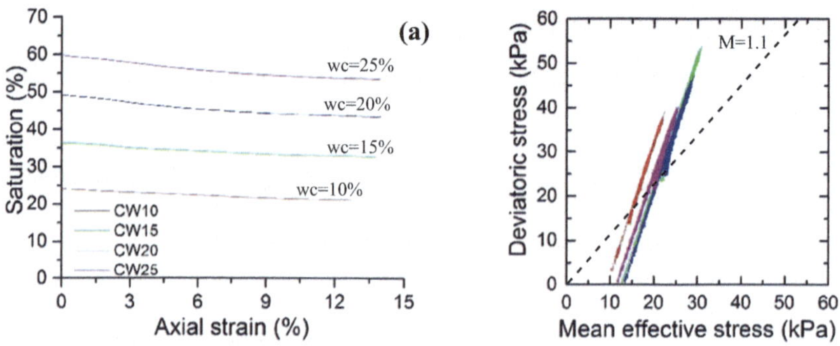

Fig. 4. Constant water content test results: (a) saturation ratio (b) stress path

3.3 Fully Undrained Tests

Most of the laboratory tests on unsaturated soils have been conducted under constant air pressure. However, the drained conditions for water and air cannot always be attained in practice [2]. In the present study, we have conducted triaxial tests on silty clay under fully undrained (FU) conditions. After measuring the initial suction and applying axis translation technique (as explained earlier), the specimens were sheared in undrained conditions for both air and water i.e. both pore air and pore water pressure were measured. Figure 5 illustrates the volumetric strain and stress-strain relationship for various matric suction levels of FU tests. Due to a high degree of compaction, the soil showed dilative behavior and volume increase with an increase in axial strain. The deviatoric stress also increase with axial strain and value of deviatoric stress varied with matric suction, the effect of high degree of compaction can also be seen in form of peak and post peak failure response. The peak deviatoric stress was obtained by specimen prepared with 15% water content. However, for same matric suction values the specimens tested in fully undrained conditions showed little bit more increase in deviatoric stress than specimen tested in constant water content conditions due to increase in pore air pressure during shearing. Unlike CW tests the peak values of deviatoric stress were ranged between 0–4.5% and residual strength of specimens was not meeting at a point. Due to the dilative behavior of soil, a slight decrease in saturation ratio was also observed during shearing.

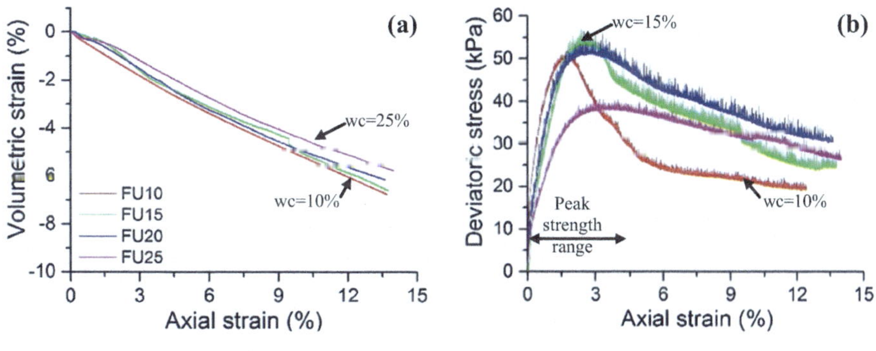

Fig. 5. Fully undrained test results: (a) volumetric strain (b) stress-strain relationship

Figure 6(a) shows the change in pore air pressure during the shear process. It can be seen that pore air pressure increase with an increase in monotonic loading then decrease and finally become constant. The increase in pore air pressure was due to compression of air inside the specimen, however, the air dissolved as the soil skeleton start broken. This result in a decrease in pore air pressure and eventually it stabilized and become constant. Figure 6(b) indicates the stress path in terms of deviator stress and the mean effective stress. Likewise, CW tests stress paths are sloped at 3.0, the stress path first crossed the critical state line reach the maximum point and ultimately reverse its direction and fall on the critical state line. The stress ratio at failure is about 1.1 which is same as CW tests, however, in the case of FU tests, the gradient is intercepted at 4 kPa. The same value of

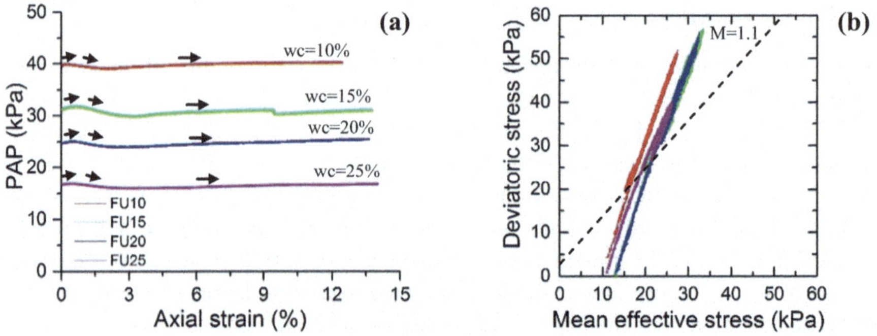

Fig. 6. Fully undrained test results: (a) pore air pressure (b) stress path

the gradient along CSL indicate that the behavior of unsaturated soils can be described well through the use of mean effective stress, the difference in intercept is due to different drainage conditions of tests.

3.4 Discussion

The results from constant water content tests and fully undrained tests are compared and influence of matric suction on mean effective stress and deviatoric stress is plotted in Figs. 7 and 8. It can be seen that the values of both mean effective stress and deviatoric stress are a little bit high for FU tests as compared to CW because of effect of undrained pore air pressure in fully undrained conditions. Figure 7 illustrates that the relation between mean effective stress and matric suction is non-linear. The values increased with increasing matric suction for suction range up to 32 kPa which corresponds to 15% water content but decreased with increasing matric suction for the suction range greater than 32 kPa. The same non-linear behavior was observed in case of deviatoric stress with increasing matric suction. It can be seen that for the soil tested under same confining stress, the maximum deviatoric stress is obtained at matric suction of 32 kPa, after which it drops. This shows that the shear strength behavior of

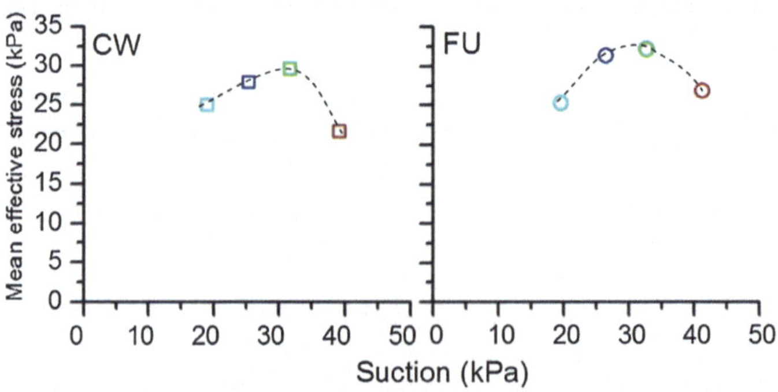

Fig. 7. Relationship between mean effective stress and suction at peak for CW & FU tests

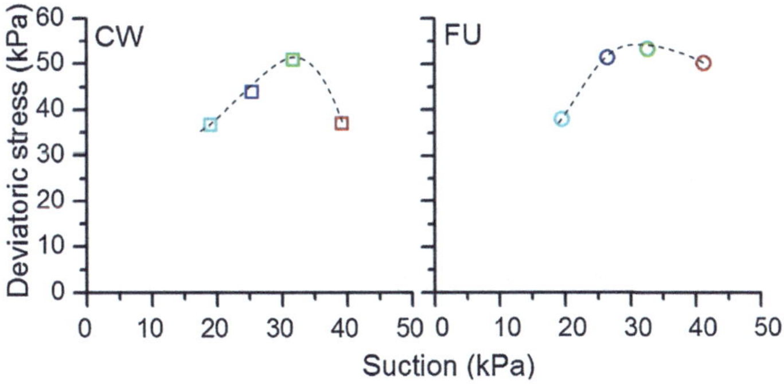

Fig. 8. Relationship between deviatoric stress and suction at peak for CW & FU tests

unsaturated soils can be described well through the mean effective stress. The behavior of deviatoric stress against matric suction is same as mean effective stress, however, the values are different. Finally, it can be said that it the mean effective stress that controls the shear strength of unsaturated soils.

4 Conclusion

In this paper, the role of matric suction on the shear strength of unsaturated silty soil was studied under constant water content and fully undrained conditions. The test results showed that matric suction decrease with increase in water content. During shearing, no much change in volumetric strain was observed, the saturation ratio decrease because of dilative behavior. In case of fully undrained tests, pore air pressure also changed during shearing because of change in air voids. The intercept of stress path changed because of post peak shear behavior. Peak shear strength and effective stress were observed at the same matric stress suction which showed that the mean effective stress controls the shear strength behavior of unsaturated soils.

References

1. Byeong, S.K., Seong, W.P., Takeshita, Y., Kato, S.: Effect of suction stress on critical state of compacted silty soils under low confining pressure. Int. J. Geomech. **16**(6), 1–11 (2016)
2. Fusao, O., Takeshi, K., Hirokata, S., Young, S.K., Norisuke, N., Sayuri, K.. Experimental study on behavior of unsaturated compacted silt under triaxial compression. Soil Found. **50** (1), 27–44 (2010)
3. Fredlund, D.G., Rahardjo, H.: Soil Mechanics for Unsaturated Soils. Wiley, New York (1993)
4. Hilf, J.W.: An investigation of pore water pressure in compacted cohesive soils. U.S. Dept. Interior Bureau Reclamation Technical Memo 654 (1956)
5. Kato, S., Kawai, K., Yoshimura, Y., Sunden, W.: Effect of suction on unconfined compressive strength and undrained shear strength of a compacted silty clay. In: Proceedings of the 3rd International Conference on Unsaturated Soils, Brazil, pp. 513–518 (2002)

6. Meen, W.G., Yong, M.W.: Failure of soil under water infiltration. Eng. Geol. **181**, 124–141 (2014)
7. Melinda, F., Rahardjo, H., Han, K.K., Leong, E.C.: Shear strength of compacted soil under infiltration conditions. J. Geotech. Geoenvironmental Eng. **130**(8), 807–817 (2004)
8. Nishimura, T., Fredlund, D.G., Gan, J.K.M., Hirabayashi, Y.: Total stress ratio and shear strength parameters for an unsaturated compacted soil. In: Proceedings of 11th Asian Regional Conference on Soil Mechanics and Geotechnical Engineering, Korea, pp. 125–129 (1999)
9. Rasool, A.M., Kuwano, J., Tachibana, S.: Behavior of compacted unsaturated soil in isotropic compression, cyclic and monotonic shear loading sequence in undrained condition. In: Proceedings of 6th International Symposium on Deformation Characteristics of Geomaterials, Argentina, pp. 267–274 (2015)

Thermodynamic Response of Crystalline Swelling and Double-Layer Swelling of Compacted Bentonite

Yonggui Chen(✉) and Lina Liu

Key Laboratory of Geotechnical and Underground Engineering of Ministry of Education, Department of Geotechnical Engineering, Tongji University, Shanghai 200092, People's Republic of China
cyg@tongji.edu.cn

Abstract. Strong thermodynamic disequilibrium existed in HLW repository site often deteriorates the hydration swelling capacity of compacted bentonite, including the crystalline swelling and the double-layer swelling. In order to investigate the temperature effect on the crystalline swelling capacity and the double-layer swelling capacity, four swelling pressure tests and four swelling strain tests were conducted on compacted Gaomiaozi (GMZ) bentonite specimens saturated with de-ionized water under the temperature of 20 °C to 80 °C. Main observations show that the crystalline swelling capacity and the double-layer swelling capacity generally decrease with the temperature increases; whereas the crystalline swelling pressure and the double-layer swelling strain present a slight increase tendency under higher temperatures, especially the temperature 60 °C and 80 °C.

Keywords: Compacted bentonite · Crystalline swelling
Double-layer swelling · Swelling pressure · Swelling strain

1 Introduction

The increase in environmental radioactive waste as a consequence of nuclear industrial development is an issue that the society must face. The presence of hazardous radionuclide is detrimental to a variety of living species, especially human beings and animals. Whatever progressive technique is developed in the future for recycling and reusing spent fuel, there will always be some inevitable residue that requires final disposal.

As a suitable artificial barrier material, bentonite is often used in multiple barrier systems of high-level radioactive waste repository (Saiyouri et al. 2000; Lemaire et al. 2004; Komine 2004; Autor 2006; Ye et al. 2013) because of its high sorption capacity for retardation of radionuclide, hydration swelling capacity for relieving the surrounding rock stress, low hydraulic conductivity against fluid transport and radionuclide migration, good self-sealing potential for fractures healing, and adequate rheological deformability for permanent isolation of radioactive waste, etc. (Pusch 1983; Dixon 2000; Komine 2004; Lloret and Villar 2007; Castellanos et al. 2008; Siddiqua et al. 2011; Zhu et al. 2013). In reality, strong thermodynamic disequilibrium

L. Hu et al. (Eds.): GSIC 2018, *Proceedings of GeoShanghai 2018 International Conference: Multi-physics Processes in Soil Mechanics and Advances in Geotechnical Testing*, pp. 129–138, 2018.
https://doi.org/10.1007/978-981-13-0095-0_15

caused by the nuclear decay heat of 50–210 °C from waste canister often existed in HLW repository site (Garcíagarcía et al. 2006). These good buffer performances, especially the hydration swelling capacity of compacted bentonite, may be severely threatened. Hence, it is necessary to investigate its thermodynamic response to carry out the hydration swelling capacity evaluation of artificial barrier material.

Based on the hydration swelling capacity of compacted bentonite, major contributions have been conducted. Norrish (1954) thought that crystalline swelling and double electric layer swelling controlled mesoscopic swelling process. For the adsorption process of montmorillonite to water molecules, Bird (1984) found that the polar water molecules entered into the crystal structure of smectite clays and adsorbed on its crystal layers, resulting in the interlayer spacing expansion of montmorillonite following with the crystal layers increasing from one layer to four layers. On the basis of the findings of Norrish (1954) and Bird (1984), Suzuki et al. (2005) graphically expressed the relationship between microscopic and mesoscopic swelling, thought that stage I-the adsorbed water occupied the interlayer space between smectite layers of a quasicrystal under moderate humidity conditions, while the micro-pores among quasicrystals remained dry; stage II-the aggregate swelling could no longer be recognized even if the quasicrystals swelled and filled the micropores after contacting with the solution; stage III-the aggregates continued to swell as quasicrystals swelled further after filling micropores; and from stages II to III, thick quasicrystals might split into several thinner quasicrystals with fewer layers while maintaining a constant interlayer distance. Laird (2006) thought that six separate processes controlled the swelling of smectites in aqueous systems, that is, crystalline swelling, double-layer swelling, formation and breakup of quasicrystals, cation demixing, co-volume swelling and Brownian swelling; and also showed that basic mechanism and forces were different.

Contributions to the temperature effect on the hydration swelling capacity are also being done (Pusch et al. 1990; Lingnau et al. 1996; Xie et al. 2007; Ye et al. 2013). Heating effect on the swelling pressure of montmorillonite depends on the predominant cation in the exchange complex; generally speaking, the swelling pressure increases in sodium montmorillonite while it decreases in calcium montmorillonite (Pusch et al. 1990; Lingnau et al. 1996); the reason invoked is that the net effect of two mechanisms —lattice contraction due to dehydration of the interlamellar space and increase of the osmotic pressure at stack contacts—is different for the two types of clay, namely Lattice contraction predominates in Ca bentonite while the increase in osmotic pressure predominates in Na bentonite (Pusch et al. 1990). In addition to the effect of the predominant cation in the exchange complex, these factors of overconsolidation ratio, plasticity, expansibility and water content etc. (Villar and Lloret 2004) also affect the hydration swelling capacity of compacted bentonite. Highly overconsolidated clay exhibit a volume expansion when they are heated (Baldi et al. 1988; Sultan et al. 2002; Romero et al. 2005). The volume change induced by temperature increases with the plasticity of clay (Demars and Charles 1982).

These research findings focus more on the temperature effect on the hydration process, as well as the interaction of temperature with some factors. However, Strong thermodynamic disequilibrium effect on the hydration swelling capacity, including the crystalline swelling and the double-layer swelling, and even the effect mechanism of compacted bentonite is very rare. Hence, based on the observation of the hydration swelling capacity of compacted bentonite under different temperatures, this paper mainly investigates the temperature effect on the crystalline swelling capacity and the double-layer swelling capacity. Four swelling pressure tests and four swelling strain tests were conducted on compacted GMZ bentonite specimens saturated with de-ionized water under the temperature of 20 °C to 80 °C.

2 Materials and Methods

2.1 Choice of Raw Materials

GMZ bentonite was taken from Gaomiaozi in the Inner Mongolia Autonomous Region, 300 km northwest from Beijing, China. It is a light gray powder sodium bentonite with a montmorillonite fraction of about 75.4% in mass, with small quantities of Quartz (around 11.7%), Cristobalite (around 7.3%), Feldspar (around 4.3%), Kaolinite (around 0.8%) and Calcite (around 0.5%). The total cation exchange capacity is 62.59–82.06 meq/100 g, with 29.66–38.48 meq/100 g for the major exchangeable cations Na, 19.73–23.18 meq/100 g for Ca, 8.74–13.40 meq/100 g for Mg, and 0.47–1.01 meq/100 g for K (Ye et al. 2014).

The value obtained for the external specific surface area is 33.90 m^2/g and the total specific surface area obtained is about 570 m^2/g (Ye et al. 2010, 2013). The bulk chemical components of the specimen obtained using the X-ray fluorescence spectrometry (XRF) include SiO_2, Al_2O_3, Fe_2O_3, Na_2O, CaO, K_2O, MgO, FeO, TiO_2, P_2O_5, MnO and some lanthanides, such as La, Ce, Nd, and so on (Chen et al. 2012).

2.2 Laboratory Test Methods

GMZ bentonite power with an initial water content of 11.92% was first compacted in a stainless steel ring mold to obtain the cylindrical specimen of 50 mm diameter and 10 mm height of a target dry density of 1.7 g/cm^3. In this process, the specimen was confined in the ring preventing it from swelling deformation laterally. After that, the specimen with the metallic confining ring was immediately installed into the testing cell for the swelling capacity test under the constant volume condition (Fig. 1) and the constant vertical stress condition (Fig. 2).

To measure the temperature effect on the swelling pressure, the de-ionized water was infiltrated into the processed specimens. These specimens had been completely saturated and the difference in suction potential between the bentonite and the hydration solution approached zero. The swelling pressure test was considered as completed. The kinetics development of the swelling pressure was recorded by the data logger every two minutes.

What's more, to measure the temperature effect on the swelling strain, compacted bentonite specimens were infiltrated into the de-ionized water from the bottom under constant vertical stress 0.1 MPa. The one-dimensional vertical deformation was recorded by the dial indicator. Consequently, vertical swelling deformation occurred when absorbing the de-ionized water. The test was considered to be completed when the vertical deformation recorded was stable for 72 h at least.

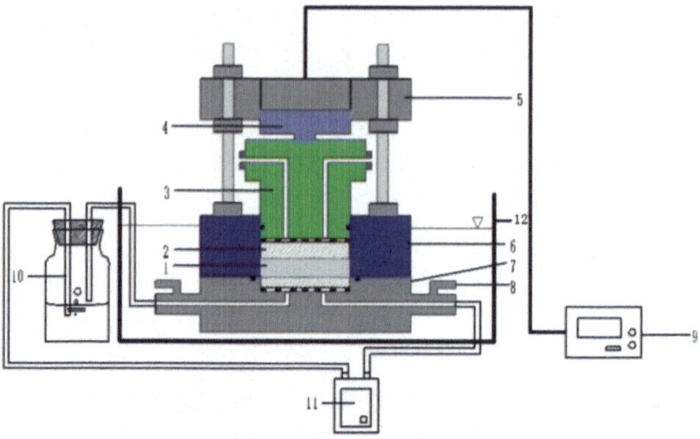

Fig. 1 Multifunction setup for swelling pressure and hydraulic conductivity test (1: Sample; 2: Porous stone; 3: Piston; 4: Load sensor; 5: Top cover; 6: Ring; 7: Basement; 8: Valve; 9: Data logger; 10: Reagent bottle; 11: Peristaltic pump; 12: Digital thermostat water bath pot)

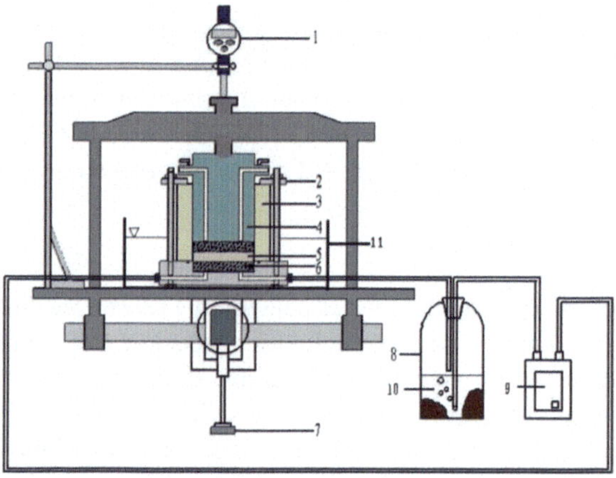

Fig. 2 Schematic temperature-controlled one-dimensional swelling deformation apparatus (1: Dial indicator; 2: Flange; 3: Ring; 4: Piston; 5: Specimen; 6: Porous stone; 7: Weights; 8: Reagent bottle; 9: Peristaltic pump; 10: De-ionized water; 11: Digital thermostat water bath pot)

3 Results and Discussion

The evolution process curve of the swelling pressure and the swelling strain over time under different temperatures was shown as Fig. 3. The evolution process curve of the swelling pressure is mainly divided into three stages for these specimens saturated with de-ionized water at different temperatures based on its inflection point, that is, stage I (the crystalline swelling), stage II (the breakup and formation of quasicrystal), and stage III (the double-layer swelling) (Fig. 3a) (Suzuki et al. 2005; Laird 2006). The evolution process curve of the swelling strain is divided into three stages based on Sridharan et al. (2004) research results, that is, initial swell, primary swell and

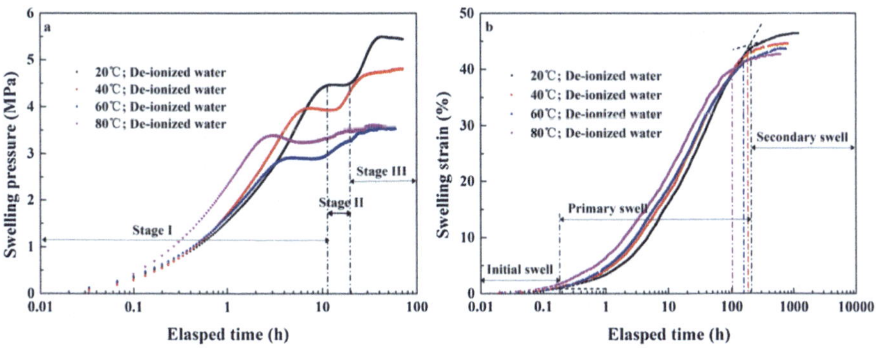

Fig. 3. The evolution process curve of the swelling pressure and the swelling strain over time under different temperatures

secondary swell (Fig. 3b) (Sridharan et al. 2004; Rao and Thyagaraj 2007); primary swell and secondary swell are controlled by the crystalline swelling and the double-layer swelling, respectively.

For the evolutionary process curve of the swelling pressure, the inflection points are conducted to distinguish the crystalline swelling and double-layer swelling. What's more, for the evolutionary process curve of the swelling strain, Sridharan et al. (2004) research results are cited in the paper, that is, the tangential intersection of the swelling strain curve of the approximately linear growth is used. In order to obtain the specific data, the evolution process curves of the swelling pressure and the swelling strain of the approximately linear growth are fitted.

3.1 Temperatures Effect on the Crystalline Swelling

The adsorbed water occupies the interlayer space between smectite layers and makes bound water membrane thickness thicker, resulting in the thickness of quasicrystals increasing; thicker quasicrystals split into several thinner quasicrystals with fewer layers and fill the micropores after contacting with aqueous systems. Crystalline swelling is a process whereby 0 to 4 discrete layers of water molecules are intercalated between individual 2:1 layers within a smectite quasicrystal (Laird 2006). The distance between the bentonite layers can gradually equal to 9.6Å, 12.6Å, 15.6Å, 18.6Å and 21.6Å, which corresponds to 0, 1, 2, 3 and 4 moisture layers, respectively (Saiyouri et al. 2000) (Fig. 4).

Test results show that the crystalline swelling pressure and the crystalline swelling strain generally decrease with the temperature increases (Fig. 5). 4.47 MPa, 3.98 MPa, 2.91 MPa and 3.21 MPa of the crystalline swelling pressure, and 43.91%, 42.61%, 41.20% and 40.02% of the crystalline swelling strain are given corresponding to the temperature 20 °C, 40 °C, 60 °C and 80 °C, respectively.

Crystalline swelling is composed of two parts, that is, the direct adsorption of water molecules and the indirect adsorption of water molecules, which are controlled by hydrogen chemical bonds and exchangeable cations presented on the surface of the

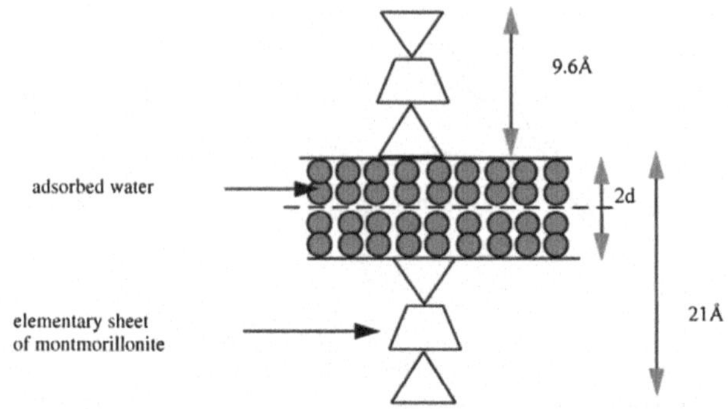

Fig. 4. Schematic distance between two layers of smectite (Saiyouri et al. 2000)

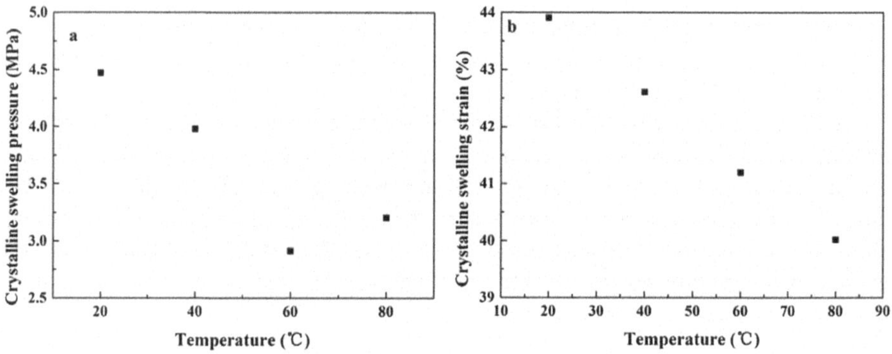

Fig. 5. Crystalline swelling pressure and crystalline swelling strain under different temperatures

crystal layer, respectively (Savage et al. 1992; Dougherty 1998; Laine and Karttunen 2010; Esrafili, 2013; Mafy et al. 2015). As one of the intermolecular forces, hydrogen chemical bonds are susceptible to temperature (Olphen 1965; Mafy et al. 2015) because that high temperature often causes more intense molecular thermal motion. There is no doubt that the increase in energy will weaken hydrogen chemical bonds interaction, so that the crystalline swelling capacity is weakened compared to the temperature 20 °C.

Ignoring the general decrease tendency presented by the crystalline swelling pressure, it shows a slight increase when the temperature is 60 °C to 80 °C. This paper attributed it to the thermal expansion effect of the crystal, which was caused by high temperatures. The microscopic mechanism of the thermal expansion effect can be interpreted as: the spacing of atoms increases with temperature because of an asymmetric potential energy curve between atoms. A balance between strong forces of attraction and repulsion controlled crystalline swelling, which was more easily modeled by both the potential energy of attraction and repulsion. The extent of crystalline swelling was determined by a balance between potential energy of repulsion (G_{rp}), attraction (G_{at}) and mechanical resistance (G_{rs}), namely

$$G_{rp} = G_{at} + G_{rs} \tag{1}$$

where G_{rs} was defined as positive when mechanical resistance contributed to the potential energy of attraction and negative when it contributed to the potential energy of repulsion (Laird 1996).

The thermal expansion effect makes mechanical resistance contribute to the potential energy of repulsion, that is, mechanical resistance (G_{rs}) is defined as negative. Therefore, an unbalance between potential energy of attraction and repulsion is produced, which promotes the crystalline swelling capacity.

3.2 Temperatures Effect on the Double-Layer Swelling

Double-layer swelling occurs between quasicrystals against within quasicrystals of crystalline swelling (Laird 2006). The interlayer cations of smectite would stoichiometrically balance the negative surface charge arising from isomorphous substitution (Marcelja and Quirk 1992; Muurinen and Lehikoinen 1999; Mata et al. 2005). The adsorption layer is formed because that the interlayer cations were absorbed and were agglomerated on the surface of the clay particles. Meanwhile, due to the thermal motion of moleculars, the interlayer cations would be free from the surface of clay particles which further forms the diffusion layer (Tripathy et al. 2004). Thus, a total double-layer structure composed of an adsorption layer and a diffusion layer has eventually formed.

The double-layer swelling pressure and the double-layer swelling strain generally decrease with the temperature increases (Fig. 6). 0.96 MPa, 0.84 MPa, 0.61 MPa and 0.18 MPa of the double-layer swelling pressure, and 2.5%, 2.1%, 2.5% and 2.78% of the double-layer swelling strain are given corresponding to the temperature 20 °C, 40 °C, 60 °C and 80 °C, respectively.

It was not difficult to imagine that high temperatures would promote the movement of anions and cations in the adsorption layer and the diffusion layer, and thereby cause the electric potential and even the double-layer repulsion to decrease. However, the double-layer swelling strain shows a slight increase under higher temperatures,

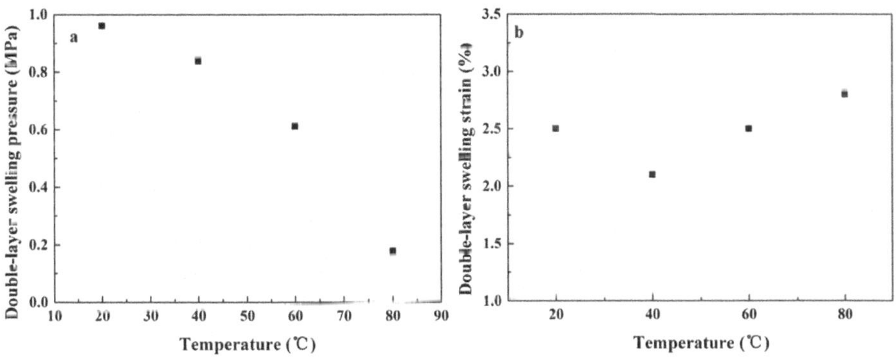

Fig. 6. Double-layer swelling pressure and double-layer swelling strain under different temperatures

especially the temperature 60 °C and 80 °C. It may be related to the thermal expansion effect. According to the relationship between the saturated vapor pressure and temperature (Fredlund and Rahardjo 1993), that is

$$P_v = B \times \exp\left(-\frac{F}{R_v T}\right) \tag{2}$$

where P_v was defined as the saturated vapor pressure; B and F were the constant, namely 2.315×10^8 kPa and 2489.3 kJ/kg, respectively; R_v was defined as the water vapor content, namely 0.4615 kJ/(kg·K); T was defined as the temperature (K).

The saturated vapor pressure was calculated as 2.236 kPa, 7.655 kPa, 21.529 kPa and 53.856 kPa in turn when the temperature is 20 °C to 80 °C. Compared to the temperature 20 °C, the saturated vapor pressure increases about 223.82%, 810.70% and 2178.17% corresponding to the temperature 40 °C, 60 °C and 80 °C, respectively. What's more, the strength of compacted bentonite would decrease because of its hydration process. Therefore, the reduction in strength provided an opportunity for airflow to slowly penetrate bentonite particles.

4 Conclusions

In this paper, eight swelling tests were conducted on compacted GMZ bentonite specimens with an initial dry density of 1.70 g/cm^3. Temperature 20°C to 80°C was conducted to investigate the temperature effect on the crystalline swelling capacity and the double-layer swelling capacity. Main conclusions are shown as follows:

(1) Compared to the temperature 20 °C, the crystalline swelling capacity and the double-layer swelling capacity generally decrease with the temperature increases. The crystalline swelling pressure and the double-layer swelling strain show a slight increase under higher temperatures, especially the temperature 60 °C and 80 °C, which may be related to the thermal expansion effect.

(2) The crystalline swelling process and the double-layer swelling process of the compacted bentonite are controlled by their initial stress state.

References

Autor, S.: The study of Spanish clays for their use as sealing materials in nuclear waste repositories: 20 years of progress. J. Iberian Geol. **32**(1), 15–36 (2006)

Baldi, G., Hueckel, T., Pellegrini, R.: Thermal volume changes of the mineral–water system in low-porosity clay soils. Can. Geotech. J. **25**(4), 807–825 (1988)

Bird, P.: Hydration-phase diagrams and friction of montmorillonite under laboratory and geologic conditions, with implications for shale compaction, slope stability, and strength of fault gouge. Tectonophysics **107**(3), 235–260 (1984)

Castellanos, E., Villar, M.V., Romero, E., Lloret, A., Gens, A.: Chemical impact on the hydro-mechanical behaviour of high-density FEBEX bentonite. Phys. Chem. Earth Parts A/B/C **33**, S516–S526 (2008)

Chen, Y.G., He, Y., Ye, W.M., Lin, C.H., Zhang, X.F., Ye, B.: Removal of chromium(III) from aqueous solutions by adsorption on bentonite from Gaomiaozi. China. Environ. Earth Sci. **67**(5), 1261–1268 (2012)

Demars, K.R., Charles, R.D.: Soil volume changes induced by temperature cycling. Can. Geotech. J. **19**(2), 188–194 (1982)

Dixon, D.A.: Porewater salinity and the development of swelling pressure in bentonite-based buffer and backfill materials. Posiva Report (2000)

Dougherty, R.C.: Temperature and pressure dependence of hydrogen bond strength: A perturbation molecular orbital approach. J. Chem. Phys. **109**(17), 7372–7378 (1998)

Esrafili, M.D.: A theoretical investigation of the characteristics of hydrogen/halogen bonding interactions in dibromo-nitroaniline. J. Mol. Model. **19**(3), 1417 (2013)

Fredlund, D.G., Rahardjo, H.: Soil Mechanics for Unsaturated Soils. Jons Wley & Sons, New York (1993)

Garcíagarcía, S., Jonsson, M., Wold, S.: Temperature effect on the stability of bentonite colloids in water. J. Colloid Interface Sci. **298**(2), 694–705 (2006)

Komine, H.: Simplified evaluation on hydraulic conductivities of sand - bentonite mixture backfill. Appl. Clay Sci. **26**(1–4), 13–19 (2004)

Laird, D.A.: Model for crystalline swelling of 2:1 phyllosilicates. Clays Clay Miner. **44**(4), 553–559 (1996)

Laird, D.A.: Influence of layer charge on swelling of smectites. Appl. Clay Sci. **34**(1–4), 74–87 (2006)

Laine, H., Karttunen, P.: Long-Term Stability of Bentonite: A Literature Review. POSIVA, Working Report 2010-53, Finland (2010)

Lemaire, T., Moyne, C., Stemmelen, D.: Imbibition test in a clay powder (MX-80 bentonite). Appl. Clay Sci. **26**(1–4), 235–248 (2004)

Lingnau, B.E., Graham, J., Yarechewski, D., Tanaka, N., Gray, M.N.: Effects of temperature on strength and compressibility of sand-bentonite buffer. Eng. Geol. **41**(1–4), 103–115 (1996)

Lloret, A., Villar, M.V.: Advances on the knowledge of the therm-hydro-mechanical behaviour of heavily compacted "FEBEX" bentonite. Phys. Chem. Earth Parts A/B/C **32**(8–14), 701–715 (2007)

Mafy, N.N., Afrin, T., Rahman, M.M., Mollah, M.Y.A., Susan, M.A.B.H.: Effect of temperature perturbation on hydrogen bonding in aqueous solutions of different urea concentrations. RSC Adv. **5**(73), 59263–59272 (2015)

Marcelja, S., Quirk, J.P.: Salt penetration into electrical double layers. Langmuir **8**(11), 2778–2780 (1992)

Mata, C., Guimarães, L.N., Ledesma, A., Gens, A., Olivella, S.: A hydro-geochemical analysis of the saturation process with salt water of a bentonite crushed granite rock mixture in an engineered nuclear barrier. Eng. Geol. **81**(3), 227–245 (2005)

Muurinen, A., Lehikoinen, J.: Porewater chemistry in compacted bentonite. Eng. Geol. **54**(1), 207–214 (1999)

Norrish, K.: The swelling of montmorillonite. Discuss. Faraday Soc. **18**, 120–134 (1954)

Olphen, H.V.: Thermodynamics of interlayer adsorption of water in clays. I—Sodium vermiculite. J. Colloid Sci. **20**(8), 822–837 (1965)

Pusch, R.: Use of clays as buffers in radioactive repositions. KBS Report, pp. 46–83 (1983)

Pusch, R., Karnland, O., Hokmark, H.: GMM - a general microstructural model for qualitative and quantitative studies of smectite clays (1990)

Rao, S.M., Thyagaraj, T.: Role of direction of salt migration on the swelling behaviour of compacted clays. Appl. Clay Sci. **38**(1), 113–129 (2007)

Romero, E., Villar, M.V., Lloret, A.: Thermo-hydro-mechanical behaviour of two heavily overconsolidated clays. Eng. Geol. **81**(3), 255–268 (2005)

Saiyouri, N., Hicher, P.Y., Tessier, D.: Microstructural approach and transfer water modeling in highly compacted unsaturated swelling clays. Mech. Cohesive-frictional Mater. **5**(1), 41–60 (2000)

Savage, D., Bateman, K., Hill, P., Hughes, C., Milodowski, A., Pearce, J., Rae, E., Rochelle, C.: Rate and mechanism of the reaction of silicates with cement pore fluids. Appl. Clay Sci. **7**, 33–45 (1992)

Siddiqua, S.S., Blatz, J.B., Siemens, G.S.: Evaluation of the impact of pore fluid chemistry on the hydromechanical behaviour of clay-based sealing materials. Can. Geotech. J. **48**(2), 199–213 (2011)

Sridharan, A., Gurtug, Y.: Swelling behaviour of compacted fine-grained soils. Eng. Geol. **72**(1), 9–18 (2004)

Sultan, N., Delage, P., Cui, Y.J.: Temperature effects on the volume change behaviour of boom clay. Eng. Geol. **64**(2), 135–145 (2002)

Suzuki, S., Prayongphan, S., Ichikawa, Y., Chae, B.G.: In situ observations of the swelling of bentonite aggregates in NaCl solution. Appl. Clay Sci. **29**(2), 89–98 (2005)

Tripathy, S., Sridharan, A., Schanz, T.: Swelling pressures of compacted bentonites from diffuse double layer theory. Can. Geotech. J. **41**(3), 437–450 (2004)

Villar, M.V., Lloret, A.: Influence of temperature on the hydro-mechanical behaviour of a compacted bentonite. Appl. Clay Sci. **26**(1–4), 337–350 (2004)

Xie, M.L., Wang, W.Q., De Jonge, J., Kolditz, O.: Numerical modelling of swelling pressure in unsaturated expansive elasto-plastic porous media. Transp. Porous Media **66**(66), 311–339 (2007)

Ye, W.M., Chen, Y.G., Chen, B., Wang, Q., Wang, J.: Advances on the knowledge of the buffer/backfill properties of heavily-compacted GMZ bentonite. Eng. Geol. **116**(1), 12–20 (2010)

Ye, W.M., Borrell, N.C., Zhu, J.Y., Chen, B., Chen, Y.G.: Advances on the investigation of the hydraulic behavior of compacted GMZ bentonite. Eng. Geol. **169**(6), 41–49 (2014)

Ye, W.M., Wan, M., Chen, B., Chen, Y.G., Cui, Y.J., Wang, J.: Temperature effects on the swelling pressure and saturated hydraulic conductivity of the compacted GMZ01 bentonite. Environ. Earth Sci. **68**(1), 281–288 (2013)

Zhu, C.M., Ye, W.M., Chen, Y.G., Chen, B., Cui, Y.J.: Influence of salt solutions on the swelling pressure and hydraulic conductivity of compacted GMZ01 bentonite. Eng. Geol. **166**(10), 74–80 (2013)

Evaluation of the Auto-Correlation Distance of Unsaturated Soils

Xiaohui Tan[1,2(✉)], Muzi Hu[1], C. Hsein Juang[2], Peng Li[3], and Mengfen Shen[2]

[1] Hefei University of Technology, Hefei, Anhui 230009, China
tanxh@hfut.edu.cn
[2] Clemson University, Clemson, SC 29634, USA
[3] Rongzhi College of Chongqing Technical and Business University, Chongqing 401320, China

Abstract. The value of a soil parameter may vary greatly in space from one point to another due to different depositional and post-depositional conditions. The phenomenon is known as soil variability which is characterized by auto-correlation of soils. The extent of auto-correlation is generally expressed by its auto-correlation distance. This paper examines the auto-correlation distance properly for unsaturated soils through laboratory tests focusing on three types of soil parameters, including the physical, deformation, and hydraulic parameters. Test results show that the variability of the physical parameters (water content, bulk density, void ratio, specific gravity, free swelling ratio, liquid limit, plastic limit, and plasticity index) is very small; the variability of the deformation parameters (shrinkage limit, maximum linear shrinkage ratio, coefficient of contraction, volume shrinkage ratio and volume expansion ratio) is moderate; while the variability of the hydraulic parameters (soil-water characteristic curve fitting parameters, and the saturated hydraulic conductivity) is relatively large. Generally, the horizontal auto-correlation distance is greater than the vertical auto-correlation distance.

Keywords: Spatial variability · Auto-correlation distance
Correlation function · Unsaturated soil · Soil parameter

1 Introduction

Due to different depositional and post-depositional conditions, the values of a soil parameter are correlated but may vary greatly in space from one point to another even in the same soil layers. This correlation and variability can be expressed by random field theory which was pioneered by Vanmarcke [1]. The probabilistic modeling of soil profiles by random field theory is the basis for many geotechnical reliability analyses.

In the random field theory, a soil parameter can be described statistically by the mean, the coefficient of variation (COV) and the auto-correlation distance (ζ). The auto-correlation distance defines the minimum distance within which the values of a soil parameter are significant correlated. The means and the COVs of soil parameters can be calculated by probabilistic statistics [2, 3]. However, the auto-correlation

L. Hu et al. (Eds.): GSIC 2018, *Proceedings of GeoShanghai 2018 International Conference: Multi-physics Processes in Soil Mechanics and Advances in Geotechnical Testing*, pp. 139–146, 2018.
https://doi.org/10.1007/978-981-13-0095-0_16

distance needs more complex tests and calculations [4], which makes it relatively hard to obtained

Many methods have been proposed to determine the auto-correlation distances of soils [2, 5–7]. The values of auto-correlation distances obtained by different scholars vary greatly due to different test methods and conditions. On the other hand, most published values of auto-correlation distances were obtained using shear strength properties [8], site investigation data [9] (e.g., cone penetration test) or physical properties of soils [10]. Much less attention has been paid to the hydraulic parameters of unsaturated soils due to the time-consuming testing required for unsaturated soil parameters [11, 12]. Therefore, it is very important to eliminate the influence of different test conditions on the auto-correlation distances of different soil parameters by using the same soil. Meanwhile, it is also necessary to test and evaluate the auto-correlation distances of hydraulic parameters of unsaturated soils because many constructions are built in unsaturated soils.

Based on the above review, a series of laboratory tests were carried out and three types of soil parameters (the physical, deformation, and hydraulic parameters) were obtained for the unsaturated soil in Hefei, China. The means, the COVs, and the auto-correlation distances of these parameters were calculated and assessed.

2 Experimental Investigation

A construction site in Hefei, China was selected as research area. This construction site consists mainly of thick clayey layer. For the evaluation of horizontal auto-correlation distances of soil parameters, we collected undisturbed soil columns in 10 boreholes. These boreholes were located on the same cross section and the distance between each two adjacent boreholes was 10 m (Fig. 1) according to the literature review [10, 11]. In each borehole, the test soil was sampled at 10 m below the ground level. All these soil columns were unsaturated clays. Each soil column's height was 60.0 cm and the diameter was 7.2 cm. On the other hand, continuous soil columns were sampled along one borehole for the evaluation of vertical auto-correlation distances of soil parameters. The total thickness of the clayey layer in this borehole was 10.2 m and the continuous soil sample was divided into 12 segments of equal length. So there were all together 21 undisturbed soil columns. These undisturbed soil columns were then cut into small pieces to carry out laboratory tests.

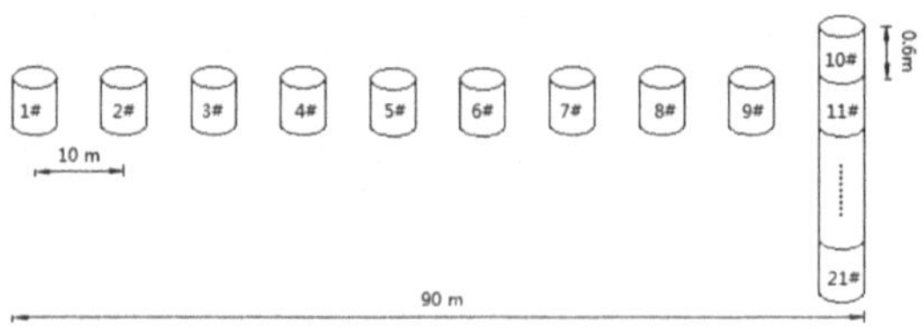

Fig. 1. Locations of soil samples

Because the laboratory tests for unsaturated soil is very time consuming, these undisturbed soil columns were wrapped very carefully to protect them from losing water before being tested. The detailed procedure for wrapping and preserving these soil columns is referred to Li (2017) [13].

For assessing the auto-correlation distance of soil parameters, many laboratory tests were carried out. These tests can be classified as physical parameter tests, deformation parameter tests, and hydraulic parameter tests. These tests are carried out according to the Standard for soil test method in China (GB/T 50123-1999) [14].

The physical parameter tests include the testing of soil's water content (w_0), bulk density (ρ_0), void ratio (e_0), specific gravity (G_s), free swelling ratio (δ_{ef}), liquid limit (w_L), plastic limit (w_p), and plasticity index (I_p). These parameters can be readily calculated based on their definitions and the equations listed in the Standard [14].

The deformation parameter tests consist of the testing of shrinkage limit (w_n), maximum linear shrinkage ratio (δ_m), coefficient of contraction (λ_n), volume shrinkage ratio (δ_v), and volume expansion ratio (η_v). The first four parameters (w_n, δ_m, λ_n, and δ_v) could be measured by shrinkage test [15]. Through the shrinkage test, the relationship between linear shrinkage ratio (δ) and water content (w) could be obtained, and parameters w_n, δ_m, and λ_n could be calculated by fitting the $\delta \sim w$ relationship using Weibull model [16]. Parameter η_v (volume expansion ratio) could be measured by humidification test. Both volume shrinkage ratio (δ_v) and volume expansion ratio (η_v) could be calculated by dividing the change of soil sample's volume to the original volume before testing.

The hydraulic parameter tests consisted of the testing of soil-water characteristic curve (SWCC) and unsaturated hydraulic conductivity curve (UHC). The SWCC can reflect the relationship between the suction and the water content of unsaturated soil, whereas the UHC can describe the relationship between the suction and the hydraulic conductivity of unsaturated soil. Many models have been proposed to describe the SWCC and UHC, among which the van Genuchten models for the SWCC and UHC are widely used in geotechnical engineering [17, 18], and they are expressed as follows:

$$w = w_r + \frac{w_s - w_r}{[1+(\psi/a)^n]^m},\ k = ks\frac{\{1-(\psi/a)^{n-1}[1+(\psi/a)^n]^{-m}\}^2}{[1+(\psi/a)^n]^{m/2}} \tag{1}$$

where ψ is the suction; w_s and w_r are the saturated and residual water content, respectively; a, n and m are curve fitting parameters and $m = 1 - 1/n$ is assumed in van Genuchten model (Parameter a corresponds to the suction at the inflection point on the curve; parameter n is related to the pore size distribution of the soil and it corresponds to the slope of the curve at the inflection point); and k_s is the saturated hydraulic conductivity. Therefore, the hydraulic parameters are a, n, w_s, w_r, and k_s.

For obtaining the SWCC in both lower suction and higher suction range, the osmotic method and the filter paper method were used together for suction measurements [19]. Variable head permeability test was used to test the saturated hydraulic conductivity of soil samples. For ensuring the test accuracy, all tests were repeatedly conducted in three times, and the averages of the three tests were used for the following analysis.

3 Calculation of Auto-Correlation Distance

There are several methods for the calculation of auto-correlation distance (ζ), such as average distance method, statistical simulation method, recursive space method, curve limit method, and correlation function method, among which the correlation function method (CFM) is the most widely used in geotechnical engineering. The CFM is to calculate the auto-correlation distance by fitting the empirical sample auto-correlation function using a theoretical correlation function. The empirical sample correlation function $R(h)$ can be estimated using the following equations [20]:

$$R(h) = \frac{1}{s^2(n-j-1)} \sum_{i=1}^{n-j} x(z_i)x(z_{i+j}) \tag{2}$$

where $h = j\Delta z$ is the distance of two points with j intervals and Δz is the sampling distance; n is number of samples; s^2 is the sample variance; $x(z_i)$ and $x(z_{i+j})$ are measured values of soil parameter at locations of z_i and z_{i+j}, respectively.

By fitting the relationship between $R(h)$ and h using a specific theoretical correlation function, the fitting parameter of correlation function and the auto-correlation distance can be obtained. Five popular correlation functions and the corresponding calculation equations for auto-correlation distance are listed in Table 1 [4, 8].

Table 1. Auto-correlation models and corresponding auto-correlation distances

Model	Correlation function	Auto-correlation distance (ζ)
Single exponential (SNX)	$R(h) = \exp(-bh)$	$2/b$
Squared exponential (SQX)	$R(h) = \exp[-(bh)^2]$	$\pi^{0.5}/b$
Cosine exponential (CSX)	$R(h) = \exp(-bh)\cos(bh)$	$1/b$
Second-order Markov (SMK)	$R(h) = (1 + bh)\exp(-bh)$	$4/b$
Triangular (BIN)	$R(h) = 1 - bh$ for $h \leq 1/b$ $R(h) = 0$ for $h \geq 1/b$	$1/b$

Generally, no auto-correlation model is preferable over others based on physical motivations [21]. Therefore, the empirical sample correlation data is fitted by each of the five models listed in Table 1 using least-squares regression procedures, and then the best correlation function model is selected by comparative assessment of the goodness-of-fit (e.g., the determination coefficient) of the five models.

4 Results and Discussions

As described in Sect. 2, the physical parameters, the deformation parameters of η_v and δ_v, and the hydraulic parameters of k_s can be calculated by simple formulas, while the deformation parameters of w_n, δ_m, and λ_n and the hydraulic parameters of a, n, w_s, and

w_r need to be calculated through curve fitting. The fitted $\delta \sim w$ curves (from shrinkage tests) and $w \sim \psi$ curves (from SWCC tests) for the 21 sets of data are shown in Fig. 2 (a) and (b), respectively. It can be easily found that the fitting parameters of these curves vary apparently.

According to the descriptions in Sect. 3, the five correlation function models listed in Table 1 are adopted for fitting the empirical sample auto-correlation function. The cosine exponential model (CSX) is found to be superior to other models because the result by CSX yields the maximum determination coefficients.

The statistics (range, mean, COV, and auto-correlation distance) of the physical, deformation, and hydraulic parameters are listed in Table 2. Because the values of residual water content (w_r) are very small and they have little influence on the mechanical and hydraulic properties of unsaturated soils, the statistics of w_r are not listed in Table 2.

Table 2. Statistics of soil parameters.

Parameters		Range	Mean	COV	Auto-correlation distance (m)	
					Horizontal	Vertical
Physical parameters	Water content w_0 (%)	19.6–28.5	23.8	0.09	8.93	0.69
	Void ratio e_0	0.55–0.73	0.63	0.05	6.04	0.82
	Density ρ_0 (g/cm^3)	1.96–2.09	2.03	0.02	6.57	0.64
	Specific gravity G_s	2.60–2.76	2.69	0.01	4.47	0.63
	Plastic limit w_P (%)	18.1–29.9	25.6	0.12	5.20	0.93
	Liquid limit w_L (%)	36.7–65.9	54.3	0.15	7.02	1.06
	Plasticity index I_P	18.5–38.3	28.7	0.21	7.72	1.20
	Free swelling ratio δ_{ef} (%)	12.5–71.3	44.9	0.32	11.53	1.49
Deformation parameters	Shrinkage limit w_n (%)	7.2–14.2	10.7	0.16	9.64	1.29
	Maximum linear shrinkage δ_m (%)	2.2–9.3	6.4	0.30	5.67	0.80
	Coefficient of contraction λ_n	0.21–0.56	0.45	0.19	3.99	0.72
	Volume shrinkage ratio δ_v (%)	9.0–9.2	18.6	0.29	4.78	1.14
	Volume expansion ratio η_v (%)	2.3–19.7	10.1	0.52	4.80	0.80
Hydraulic parameters	Parameter$a$ (kPa)	1.5–44.3	13.3	0.84	4.12	0.88
	Parameter n	1.07–1.14	1.10	0.02	16.07	1.22
	Saturated water content w_s (%)	24.0–38.9	31.2	0.13	9.14	1.14
	Saturated hydraulic conductivity k_s (m/s)	(0.22–3.65) $\times 10^{-8}$	1.19×10^{-8}	0.84	4.19	0.93

4.1 Statistics of Soil Parameters

As can be seen from the values listed in Table 2, the range of COVs of the physical parameters is between 0.01 and 0.32, and the mean of these COVs is 0.12. Furthermore, the COVs of the three basic physical parameters (w_0, e_0, and G_s) are less than 0.1, which means that the variability of these physical parameters is very small.

The range of COVs of the deformation parameters is between 0.19 and 0.52 and the mean of these COVs is 0.29. This value is higher than the mean of COV of physical parameters.

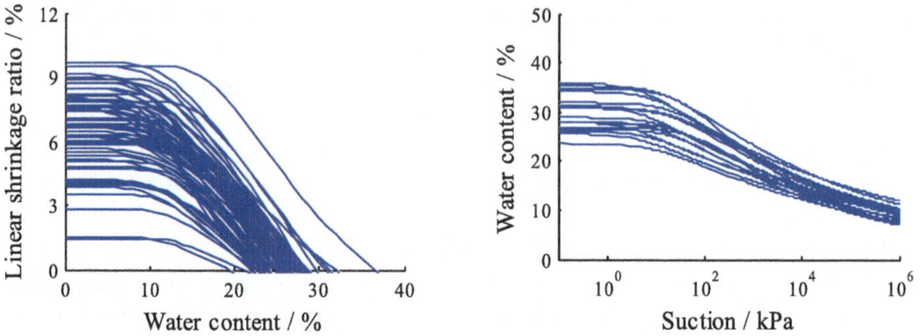

Fig. 2. (a) Shrinkage curves of tested soils. (b) Soil-water characteristic curves of tested soils.

The range of COVs of the hydraulic parameters is between 0.02 and 0.84, and the mean of these COVs is 0.46. The high mean value of COV of hydraulic parameters is due to the high variability of parameter a and k_s. The variability of parameter n is very small, which can be seen from the approximately parallel linear segments of the SWCCs in Fig. 2.

As a whole, the variability of the physical parameters is very small; the variability of the deformation parameters is moderate; while the variability of the hydraulic parameters is relatively large.

4.2 Auto-Correlation Distances of Soil Parameters

From Table 2, we can find that the average values of the horizontal auto-correlation distances for the physical, deformation and hydraulic parameters are 7.2 m, 5.8 m, and 8.4 m, respectively. Similarly, the average values of the vertical auto-correlation distances for the physical, deformation and hydraulic parameters are 0.9 m, 1.0 m, and 1.0 m, respectively. By comparing the values of the horizontal and vertical auto-correlation distances, it is found that the horizontal auto-correlation distances are larger than the vertical auto-correlation distances for all the three types of parameters. This conclusion is consistent with the existing studies [2, 22].

From the values of the auto-correlation distances listed in Table 2, we can find that there are no apparent differences for the physical, deformation and hydraulic parameters of the unsaturated clay. This can be further proved by Wilcoxon one-way rank analysis method [13]. Gao (1996) [23] also addressed that the auto-correlation distance was an inherent attribute of soils. This conclusion is very important for the selection of the auto-correlation distances of unsaturated soil. For example, the measurements of the auto-correlation distances of hydraulic parameters are usually complicated and time-consuming. But according to this conclusion, the auto-correlation distances of hydraulic parameters can be approximately evaluated by the physical parameters which are relatively easy to measure. The founding provides insights in evaluation of the auto-correlation distance of hydraulic parameters under the situation of limited hydraulic tests.

5 Conclusions

To evaluate the auto-correlation distances of unsaturated soils, a series of laboratory tests were carried out in this study to obtain three types of soil parameters. Test results showed that the variability of the physical parameters is very small, the variability of the deformation parameters is moderate, and the variability of the hydraulic parameters is relatively large. The auto-correlation distance is an inherent attribute, and there is no apparent difference among the auto-correlation distances of the physical, deformation, and hydraulic parameters of the unsaturated soil.

Acknowledgements. The financial support from the National Natural Science Foundation, China (41572282, 41372281) and the Natural Science Foundation of Anhui, China (JZ2015AKZR0045) are greatly acknowledged.

References

1. Vanmarcke, E.H.: Probabilistic modeling of soil profiles. J. Geotech. Div. **103**(11), 1227–1247 (1997)
2. Phoon, K.K., Kulhawy, F.H.: Characterization of geotechnical variability. Can. Geotech. J. **36**(4), 612–624 (1999)
3. Akbas, S.O., Kulhawy, F.H.: Characterization and estimation of geotechnical variability in ankara clay: a case history. Geotech. Geol. Eng. **28**(5), 619–631 (2010)
4. Onyejekwe, S., Kang, X., Ge, L.: Evaluation of the scale of fluctuation of geotechnical parameters by autocorrelation function and semivariogram function. Eng. Geol. **214**, 43–49 (2016)
5. Spry, M.J., Kulhawy, F.H., Grigoriu, M.D.: Reliability-based foundation design for transmission line structures: volume 1, geotechnical site characterization strategy: final report (1988)
6. Li, X.Y., Xie, K.H., Yu, Y.: Research of the characteristics of correlation distance on soil properties indexes. Chin. J. Geotech. Eng. **36**(8), 91–95 (2003)
7. Yan, S.W., Zhu, H.X., Liu, R., Sun, W.H.: Study on methods for estimating correlation distance of soil layers. Rock Soil Mech. **28**(8), 1581–1586 (2008)
8. Phoon, K.K., Quek, S.T., An, P.: Identification of statistically homogeneous soil layers using modified bartlett statistics. J. Geotech. Geoenvironmental Eng. **129**(7), 649–659 (2003)
9. Haldar, S., Babu, G.L.S.: Design of laterally loaded piles in clays based on cone penetration test data: a reliability-based approach. Géotechnique **59**(7), 593–607 (2009)
10. Firouzianbandpey, S., Griffithsd, V., Ibsenl, B., Andersenl, V.: Spatial correlation length of normalized cone data in sand: case study. Can. Geotech. J. **51**(8), 844–857 (2014)
11. Moradi, F., Moosavi, A.A., Moghaddam, B.K.: Spatial variability of water retention parameters and saturated hydraulic conductivity in a calcareous Inceptisol (Khuzestan province of Iran) under sugarcane cropping. Archives of Agronomy and Soil Science (2016)
12. Gupta, N., Rudra, R.P., Parkin, G.: Analysis of spatial variability of hydraulic conductivity at field scale. Can. Biosyst. Eng. **48**(1), 55 (2006)
13. Li, P.: Assessment of spatial variability of unsaturated clay in Hefei. Master Dissertation, Hefei University of Technology (2014)
14. Standard for soil test methodGB/T 50123-1999

15. Li, P., Tan, X.H., Wang, X., Xin, Z.Y., Wang, X.E.: Experimental study on volume deformation characteristics of expansive soils during shrinkage and swelling process. Ind. Constr. **45**(9), 93–98 (2015)
16. Wang, X.E., Tan, X.H., Xin, Z.Y., Xu, Q., Xie, Y.: Experimental study of shrinkage properties of expansive soil. Chin. J. Geotech. Eng. **37**(s2), 107–114 (2015)
17. Sillers, W.S., Fredlund, D.G.: Statistical assessment of soil-water characteristic curve models for geotechnical engineering. Can. Geotech. J. **38**(6), 1297–1313 (2001)
18. Genuchten, M.T.V.: A closed-form equation for predicting the hydraulic conductivity of unsaturated soils. Soil Sci. Soc. Am. J. **44**(5), 892–898 (1980)
19. Shen, M.F., Tan, X.H., Xin, Z.Y., Xie, Y., Xu, Q.: Laboratory research of soil water characteristic curve by osmotic method and filter paper method. Electron. J. Geotech. Eng. **18**, 5421–5434 (2013)
20. Fenton, G.A.: Random field modeling of CPT data. J. Geotech. Geoenvironmental Eng. **126** (12), 486–498 (1999)
21. Spry, M.J., Kulhawy, F.H., Grigoriu, M.D.: Reliability-based foundation design for transmission line structures: geotechnical site characterisation strategy. Report EL-5507(1). Palo Alto: Electric Power Research Institute (1988)
22. Cho, S.E., Park, H.C.: Effect of spatial variability of cross-correlated soil properties on bearing capacity of strip footing. Int. J. Numer. Anal. Meth. Geomech. **34**(1), 1–26 (2010)
23. Gao, D.Z.: Safety index and its application in geotechnical engineering design. Geotech. Invest. Surv. **1**, 1–6 (1996)

Relationship of Collapse Potential and Swell Pressure with Suction of Unsaturated Expansive Soil

Saloni P. Pandya(✉), Narendra Sarswat, and Ajanta Sachan

Indian Institute of Technology Gandhinagar, Gandhinagar, Gujarat, India
pandya_saloni@iitgn.ac.in

Abstract. Expansive soils in unsaturated state cause problems to several civil engineering structures due to its transient volume change behavior. Present study is focused on evaluation of swelling and collapsible characteristics of unsaturated compacted expansive soil specimens and their relationship with suction. A series of volume change swell pressure tests and double oedometer tests were conducted at varying degree of saturation to evaluate swelling pressure and collapse potential of Bharuch expansive soil respectively. Suction measurements of soil were conducted by employing WP4C Dew point Potentiameter at different water content. Presence of higher negative pore pressure induced higher swelling pressure, which was observed to vary exponentially with total suction. Results indicated significant influence of soil suction on soil fabric arrangement, which substantially affected collapse behaviour of expansive soil. Collapse potential of Bharuch expansive soil was observed to decrease with increase in water content due to reduction in suction of soil indicating metastable state of soil at lower degree of saturation.

Keywords: Total suction · Collapse potential · Swell pressure

1 Introduction

Unsaturated soils comprising of three phase system (air, water & soil) and possessing water content less than complete saturation ($S_r < 100\%$) are encountered in numerous geotechnical engineering practices especially in arid and semi arid regions. Zone above natural hydraulic datum (water table) consisting of unsaturated soil is commonly referred as vadose zone; and most of the man-made structures such as roads, embankments, airfields, earth dams, tunnels, natural slope linings, covers of waste containment are preferred to be constructed in this zone of the soil layer to ease out overall construction process. Soil layer above hydraulic datum possess negative pore pressure, which is commonly known as soil suction. Presence of soil suction influences the overall mechanical response of soil in unsaturated state [1]. Expansive soils in their unsaturated state are encountered widely in several parts of the world. Heavy financial losses occur every year due to presence of such soils [1]. Major problem with the expansive soil is due to its shrinkage-swelling characteristics leading to severe damages to geotechnical structures. Geometric arrangement (fabric) within soil skeleton would influence the

L. Hu et al. (Eds.): GSIC 2018, *Proceedings of GeoShanghai 2018 International Conference: Multi-physics Processes in Soil Mechanics and Advances in Geotechnical Testing*, pp. 147–155, 2018.
https://doi.org/10.1007/978-981-13-0095-0_17

volume change behavior of soil [2]. Transient change in the suction of soil with change in degree of saturation subjected to environmental changes would cause the alteration in soil fabric. Urbanization leading to sudden upsurge in water level within soil strata due to improper drainage conditions could cause swelling of soil mass followed by hydro-consolidation (collapse behaviour) of expansive soil under given loading conditions [3]. Volume change within the soil mass i.e. swelling and collapse would depend upon negative pore pressure present within the soil mass. The present research is focused on the influence of varying suction pressures on collapse potential and swelling characteristics of expansive soil.

2 Material Properties and Specimen Preparation

Soil selected for the present analysis was collected from Bharuch (Gujarat, India). Sample was collected from 0.5 m depth to ensure absence of organic matter. Grain size distribution of soil exhibited 6% sand, 35% silt and 59% clay. Atterberg's limits of the soil were attained; liquid limit (LL) of 69%, plasticity limit (PL) of 30%, shrinkage limit (SL) of 10%. Specific gravity of soil was obtained to be 2.74. The collected soil was classified as CH soil (cohesive soil with high compressibility) as per the IS (Indian standard) classification system. DFSI (Differential free swell Index) of collected soil was obtained to be 104%, which indicated the soil to be highly expansive in nature. X-ray diffraction analysis revealed that soil possess Montmorillonite, Halloysite and Quartz minerals. Maximum dry density (MDD) and optimum moisture content (OMC) of Bharuch expansive soil were obtained to be 1.46 g/cc and 24% respectively employing standard proctor test. Specimens S1, S2, S3, S4 and S5 of the Bharuch expansive soil were prepared at dry density (ρ_d) of 1.46 g/cc & varying water content (w) and corresponding degree of saturation (S_r) as listed in Table 1. Moist tamping method [4] was used to prepare specimens for obtaining total suction, collapse potential and swell pressure of Bharuch expansive soil.

Table 1. List of unsaturated soil specimens of Bharuch expansive soil

Specimen No.	w (%)	As-compacted water content	S_r (%)
S1	17	7% dry of OMC	53
S2	21	3% dry of OMC	66
S3	24	OMC	75
S4	27	3% wet of OMC	84
S5	31	7% wet of OMC	97

3 Experimental Methodology

3.1 Total Suction Measurements Using Dew Point Potentiameter

Dew point Potentiameter was used to evaluate total suction (ψ) of the soil. Dew point Potentiameter works on principle of Chilled Mirror Hygrometer technique [5, 6]. WP4C Dew point potentiameter developed by Decagon devices was used in the present study (Fig. 1). The technique determined the total suction of soil specimen by evaluating relative humidity (p/p_0) employing Kelvin's equation (Eq. 1) and displaying total suction (ψ) as an output.

$$\psi = \frac{RT}{M}\ln\frac{p}{p_o} \tag{1}$$

Where, R is the gas constant, T is the Kelvin temperature, M is the molecular mass of water, p is the vapour pressure of air and p_0 is the saturation vapour pressure at sample temperature. The Dew point Potentiameter method could cover large range of suction (0 kPa to 300 MPa) and was reported to be the most quick and accurate method to measure total suction [7–9]. The main components of the Dew Point Potentiameter comprises of mirror, photo detector cell, temperature sensor (thermopile) and Fan. Specimen filling half of the PVC cup, provided by the manufacturer was placed in the block. The fan in the block could speed up the equilibration process. A beam of light was directed onto the mirror and reflected into the photo detector cell. When condensation would occur, the photo detector cell would detect the variance in reflectance. The temperature, at which condensation appeared on the mirror, could be detected by the thermocouple attached to the mirror. The temperature sensor measured the temperature of the specimen. At equilibrium, water potential (Total suction) was equivalent to the relative humidity of the air in the chamber (p/p_0). In the current study, specimens were prepared at 1.46 g/cc and varying water contents: 17%, 21%, 24%, 27% and 31%. Soil-water mixture was filled and compacted to obtain required bulk density up to half height of the PVC cups to avoid contamination inside dew point potentiameter. Total suction was evaluated at 20 °C. Dew point potentiameter's block temperature (T_b) was set at 20 °C and temperature equilibrium was acquired in 20–25 min. It was ensured that sample temperature (T_s) was lesser than block temperature (T_b) to avoid condensation inside the block by placing as-compacted specimen in biological incubator with temperature control. It took 30–45 min for as-compacted specimen to reach temperature less than block temperature (20 °C). This could be ensured by placing the specimen in the block and checking the variation of Ts-Tb. If the value of Ts-Tb < 0, the specimen was allowed for determination of Total suction. In continuous mode of measurements, five consecutive readings of total suction for each specimen were recorded and averaged to obtain total suction of the specimen at that particular water content and dry density. Each readings took around half an hour for measurement. Entire test for each as-compacted specimen was completed in the period of three and half hours.

3.2 Collapse Potential Determination Using Double Oedometer Setup

Collapse potential suggests the extent of collapse of soil skeleton subjected to sudden inundation or vertical stress augmentation. In the present study, collapse potential (CP) of Bharuch expansive soil was determined by employing double-oedometer test [2]. In 1989, the double-oedometer test was introduced as one of the most accurate and less time consuming technique to obtain collapse potential (CP) of soils [9]. The mechanism and methodology of double-oedometer test were introduced and explained

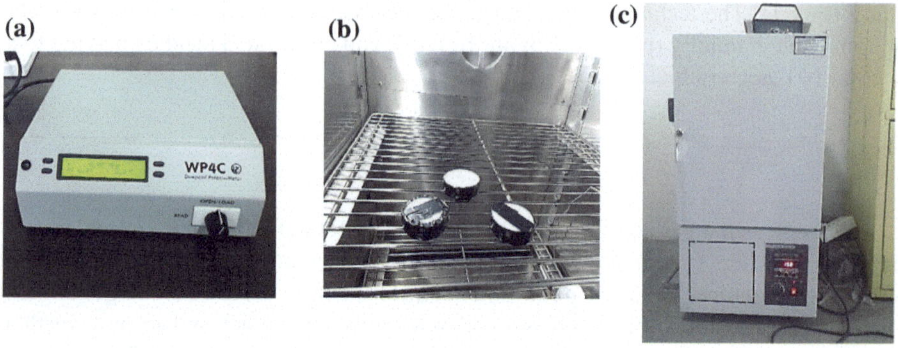

Fig. 1. Total suction measurement of Bharuch expansive soil. (a) WP4C Dew Point Potentiameter, (b) Specimens inside biological incubator, (c) Biological incubator.

in detail by [9]. This technique was further used by various researchers such as [10]. The same mechanism was adopted in the current study. In this procedure, two identically compacted specimens were placed in the consolidation mould of two different 1D consolidation setups. One of them was loaded under as-compacted conditions (unsaturated) and the other one was initially saturated for 24 h under seating load of 5 kPa. Both the specimens were further loaded sequentially; 10 kPa, 20 kPa, 50 kPa, 100 kPa, 200 kPa, 400 kPa and 800 kPa. Unsaturated state of the specimen was preserved by placing a thin plastic sheet between porous stones and damp filter paper as shown in Fig. 2.

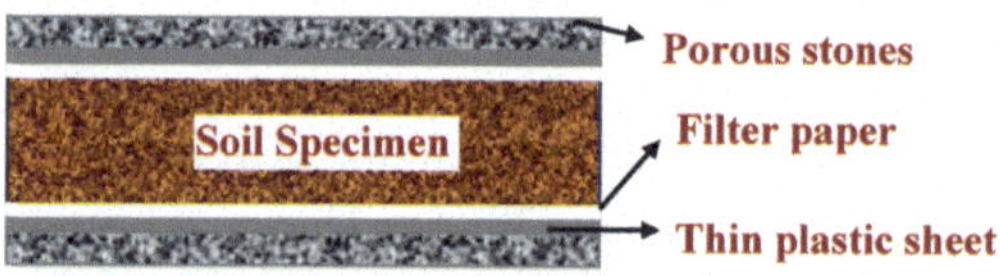

Fig. 2. Schematic diagram depicting as-compacted specimen in double oedometer test

Porous stone acted as a rigid plate for the as-compacted (unsaturated) specimen. Air/water outlets of the oedometer mould were sealed using insulation (electrical) tape to prevent evaporation of water from the unsaturated specimen. The entire setup was covered with plastic shrink wrap. Vertical stresses were incremented on both specimens (as-compacted & saturated) after each specimen reached equilibrium under applied vertical stress. Equilibrium time of 24 h under each load was considered in the present analysis. Equilibrium void ratio (after 24 h) indicating volumetric strain for each vertical stress increment was evaluated for both specimens. Collapse potential of soil is expressed as the difference between volumetric strains at a particular vertical stress normalized by initial void ratio of the specimen. It was determined using expression given by [10] represented in Eq. 2.

$$CP = \frac{e_i - e_f}{1 + e_0} \quad (2)$$

Where, e_i = equilibrium void ratio from 1D consolidation oedometer test of as-compacted specimen, e_f = equilibrium void ratio from 1D consolidation oedometer test of saturated specimen (inundated specimen), e_0 = initial void ratio of the specimen, CP = collapse potential.

3.3 Swell Pressure Determination Using Volume Change Method

Swelling pressure (P_s) can be described as the pressure exerted by the expansive soil, when change in the volume of soil has been restricted. Swelling pressure (P_s) of the Bharuch soil was determined employing volume change method using conventional consolidation set-up. In this test, the as-compacted specimen prepared in the oedometer ring was placed in the consolidation set up and water was allowed to penetrate the soil specimen. Under seating load of 5 kPa, specimen was allowed to swell until the volume change became constant. Subsequently, the specimen was loaded sequentially from 10 kPa, 20 kPa, 50 kPa, 100 kPa, 200 kPa, 400 kPa and 800 kPa. The pressure increment, at which soil specimen reached its initial void ratio (e_0), was determined as swelling pressure of soil from e vs log σ_v'. Swelling pressure of expansive soil was determined at varying water content ranging from 17% to 31% at 1.46 g/cc dry density.

4 Results and Discussion

4.1 Relationship of Swell Pressure with Total Suction

Total suction (water potential) indicates the tendency of water to move from one place to other due to capillarity, surface tension and osmosis. Water potential is the quantification of the potential energy between water within the soil mass and the free water. It varies with the change in concentration of solutes or due to variation in adhesive/cohesive forces binding the water to solid surfaces termed as matric potential (matric suction). As depicted from Fig. 3, the total suction of the Bharuch soil was found to be decreasing with the increase in water content. As degree of saturation would increase, the proportion of the bulk water within the particles would increase in unsaturated soil [4]. This would cause gradual disappearance of the air-water interface and dilute the salt concentration indicating reduction in total suction of soil. Variation of swell pressure with degree of saturation and total suction are represented in Fig. 4(a) and (b) respectively. Swelling pressure of Bharuch expansive soil was observed to vary exponentially with total suction, as shown in Fig. 4(b). Swell pressure of Bharuch soil was observed to decrease with increment in the degree of saturation. However the specimen possessing higher suction was observed to possess higher swelling capacity. Suction would govern the degree of development of diffused double layer [11]. Both matric and osmotic component of the total suction would control inter-layer spacing between structural units of Montmorillonite swelling mineral [12]. Larger values of total suction indicated smaller inter-layer spacing yielding higher negative charge on

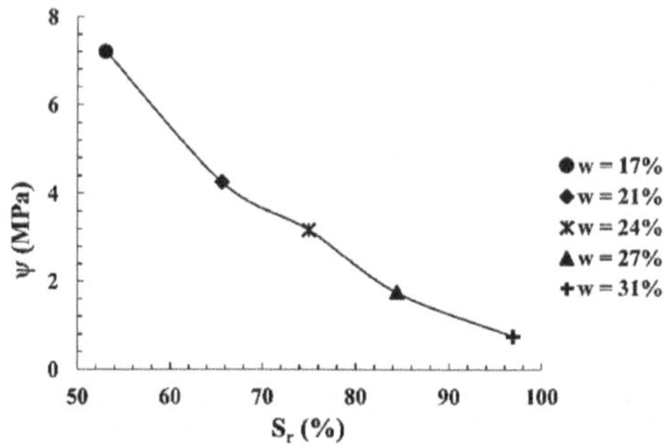

Fig. 3. Variation of total suction with degree of saturation of Bharuch expansive soil

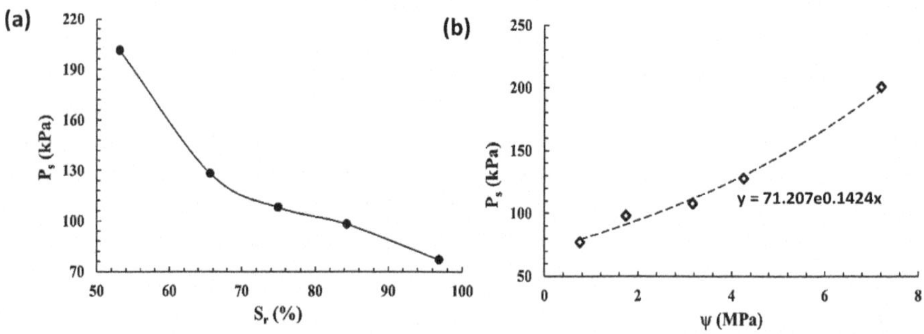

Fig. 4. Variation in Swell pressure of Bharuch expansive soil. (a) Degree of saturation vs Swell pressure, (b) Matric suction vs Swell pressure

the particle, which would led to higher affinity of water for equilibrium. This would induce higher swelling pressure of expansive soil containing higher matric suction. Increase in degree of saturation reduced the affinity for water, which led to lower values of swell pressure for the specimens possessing lesser suction values.

4.2 Relationship of Collapse Potential with Total Suction

Collapse potential is the quantitative measurement of soil-skeleton collapse due to sudden inundation or vertical stress augmentation. Higher values of collapse potential indicate higher susceptibility to collapse. Soil skeleton collapse of any soil is associated with localized deformations. Variation of collapse potential of Bharuch expansive soil with increase in vertical stress at different water content is shown in Fig. 5. Collapse characteristics of expansive soil were observed to increase with the increase in vertical stress. The observation was found to be analogous for all specimens. Higher collapse potential of Bharuch soil was observed at lower degree of saturation. Collapse potential at 400 kPa and 800 kPa vertical stresses at varying degree of saturation are listed in Table 2. The specimens possessing higher suction exhibited higher collapse as compared to specimens with lower suction values. Collapse potential for specimen S1 was

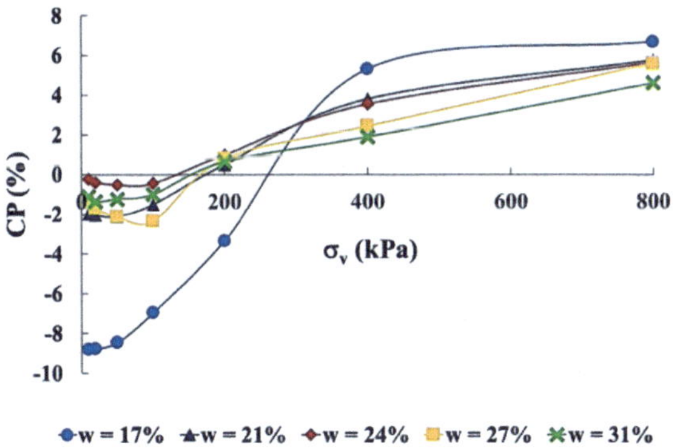

Fig. 5. Collapse potential of Bharuch expansive soil with varying water content at different vertical stress

Table 2. Collapse potential of Bharuch expansive soil with varying degree of saturation and as-compacted total suction

w (%)	S_r (%)	ψ (MPa)	CP(%) @ σ_v = 800 kPa	CP(%) @ σ_v = 400 kPa
17	53	7.20	6.7	5.3
21	66	4.25	5.8	3.8
24	75	3.17	5.6	3.5
27	84	1.75	5.6	2.5
31	97	0.76	4.6	1.9

observed to be 2.15 times and 1.45 times higher as compared to specimen S5 at 400 kPa and 800 kPa vertical pressures respectively. With upsurge in water content, collapse potential of specimen was observed to decrease.

Matric suction component of the total suction governs the strength behavior of soil in unsaturated state. The osmotic component of the total suction is negligible as compared to matric component and hence can be neglected [5]. Total suction directly can be taken equal to matric suction assuming negligible contribution of osmotic suction in the mechanical behaviour of soil. Soil particles under unsaturated condition are held together by capillary force arising due to capillarity and surface tension phenomena (matric suction).

It can be referred as additional inter-particle resistance [4] due to presence of menisci or air-water interface. This inter-particle resistive force would tend to bring particles towards each other leading to improvement in the strength and stiffness of soil. This might lead to closer and random orientation of particles joined by menisci. With sudden inundation (saturation), flattening of the menisci would take place. This might result in progressive reduction of inter-particle resistive force causing rapid collapse of the unstable soil skeleton. Specimens possessing higher suction would be bonded with higher intensity as additional inter-particle resistive force would be large. Hence, it would cause larger magnitude of collapse within soil skeleton on saturation. However, at higher water content (low suction), menisci would be nearly flattened resulting into uniform arrangement of particles, which might lead to lower amount of collapse on saturation. Due to application of vertical stress, the sliding and rotation amongst soil particles would occur within the soil mass.

This might lead to straining of the menisci, which would eventually weaken the resistive force leading to the collapse of the soil structure. Hence, the specimen possessing lower suction would undergo higher collapse indicating friable and unstable nature of soil structure at lower degrees of saturation.

4.3 Relationship of Swelling Pressure with Collapse Potential

From the current research, it could be identified that swelling pressure and collapse potential would be governed by the negative pore water pressure (suction) of unsaturated expansive soil at given dry density and water content. Thus, it could be justified that the suction could alter the geometric arrangement of particles, which would further control the volumetric deformation (Swell and collapse) of soil. Polynomial variation was observed between swelling pressure and collapse potential of the Bharuch expansive soil (Fig. 6a and b). The suction behavior of unsaturated soil was reported to be influenced by its particle size distribution [1]. In expansive soils, the amount of clay content is found to be higher. The smaller the particle size, the higher would be the matric suction component of the soil suction. Hence, it could be understood that the intrinsic connection between swelling pressure and collapse potential would be the particle size for the expansive soil. Thus, it could be attributed that larger the amount of smaller particles (clay) present in the soil mass, larger would be the value of suction. Hence, the higher values of swelling pressure and collapse potential would be obtained. Analogous observations were found regarding relationship of swelling and collapse behavior from the present study.

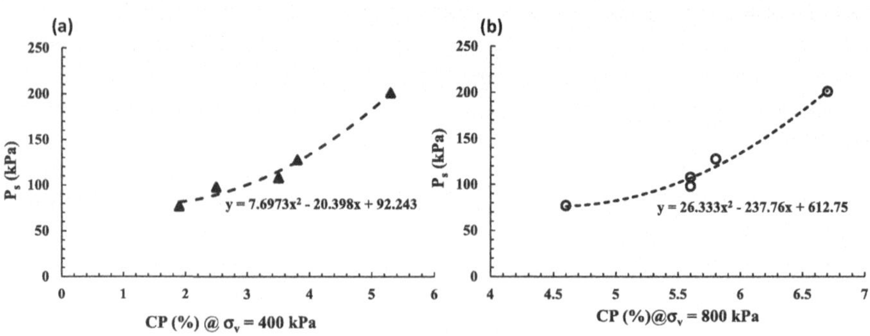

Fig. 6. Variation of swelling pressure with collapse potential of Bharuch expansive soil. (a) CP @ σ_v = 400 kPa, (b) CP @ σ_v = 800 kPa

5 Conclusion

Key observations from the current study are as follows:

- Total suction of Bharuch soil specimen was observed to decrease with increase in the degree of saturation due to increment in the bulk water amongst soil particles.
- Exponential variation was obtained between swelling pressure and total suction of Bharuch expansive soil. Reduction in swell pressure with decrease in total suction was determined due to lesser affinity of water with decrease in suction.
- Collapsible nature of expansive soil was ascertained at higher suction values and higher vertical stress augmentation due to weakening of the additional inter-particle resistive force on sudden inundation and slippage among particles within the soil mass. Both swelling pressure and collapse potential of the soil would be higher for soil possessing larger clay fraction.

References

1. Fredlund, D.G., Rahardjo, H.: Soil Mechanics for Unsaturated Soils, 1st edn. Wiley, New York (1993)
2. Houston, S.L., Houston, W.N., Zapata, C.E., Lawrence, C.: Geotechnical engineering practice for collapsible soils. In: Toll, D.G. (ed.) Unsaturated Soil Concepts and Their Application in Geotechnical Practice, pp. 333–355. Springer, Dordrecht (2001)
3. Pandya, S., Sachan, A.: Effect of matric suction and initial static loading on dynamic behaviour of unsaturated cohesive soil. Int. J. Geotech. Eng. 1–11 (2017)
4. Cancela, J.J., Dafonte, J., Martinez, E.M., Cuesta, T.S., Neira, X.X.: Assessment of a water activity meter for rapid measurements of soil water potential. Biosys. Eng. **94**(2), 285–295 (2006)
5. Leong, E.C., Tripathy, S., Rahardjo, H.: Total suction measurement of unsaturated soils with a device using the chilled-mirror dew-point technique. Geotechnique **53**(2), 173–182 (2003)
6. Campbell, G.S., Smith, D.M., Teare, B.L.: Application of a dew point method to obtain the soil water characteristic. In: Schanz, T. (ed.) Experimental Unsaturated Soil Mechanics, vol. 112, pp. 71–77. Springer, Berlin (2007)
7. Cardoso, R., Romero, E., Lima, A., Ferrari, A.: A comparative study of soil suction measurement using two different high-range psychrometers. In: Schanz, T. (ed.) Experimental Unsaturated Soil Mechanics, vol. 112, pp. 79–93. Springer, Berlin (2007)
8. Malaya, C., Sreedeep, S.: Critical review on the parameters influencing soil-water characteristic curve. J. Irrig. Drain. Eng. **138**(1), 55–62 (2011)
9. Lawton, E.C., Fragaszy, R.J., Hardcastle, J.H.: Collapse of compacted clayey sand. J. Geotech. Eng. **115**(9), 1252–1267 (1989)
10. Medero, G.M., Schnaid, F., Gehling, W.Y.: Oedometer behavior of an artificial cemented highly collapsible soil. J. Geotech. Geoenviron. Eng. **135**(6), 840–843 (2009)
11. Lin, B., Cerato, A.B.: Prediction of expansive soil swelling based on four micro-scale properties. Bull. Eng. Geol. Env. **71**(1), 71–78 (2012)
12. Tan, L., Kong, L.: Study of swelling-shrinkage regularity of montmorillonite crystal and its relation with matric suction. Sci. China Ser. D-Earth Sci. **44**(6), 498–507 (2001)

On the Swelling-Shrinkage Characteristics of Weakly Expansive Soil with Different Initial Water Contents

Chen Bao(✉), Pan Yanmin, Yu Da, and Huang Yiyi

Tongji University, Shanghai, People's Republic of China
chenbao@tongji.edu.cn

Abstract. The swelling-shrinkage characteristics of expansive soils have great influence on the stability and settlement of electricity transmission towers, so that it is very important to estimate accurately the climate-dependent volume/stress change of the expansive soil foundations. In this paper, a kind of weakly expensive soil taken from a transmission tower foundation in Henan province were studied. One-dimensional swelling tests and constant volume tests were conducted to study the swelling ratio and swelling pressure of the expensive soils under different water contents which were chosen in accordance with the actual climate conditions. Results show that both swelling ratio and swelling pressure decrease linearly with the increase of initial water content, while the swelling rate increases with the decrease of water content; shrinkage ratio is linear to initial water content, while shrinkage anisotropy becomes more obvious with the increase of initial water content.

Keywords: Expansive soils · Volume change · Swelling pressure
Oedometer tests

1 Introduction

Due to the rapid urbanization progress of many areas in China, more and more transmission towers are under construction with the urgent demand of power transmission lines. Moreover, abundance of expansive soils in China [1] gives rise to the inevitability of the dependence of some towers on expansive soils. However, a transmission tower which is usually considered as a relatively tender structure is quite susceptible to the damage resulting from subsidence and stability due to the swelling-shrinkage characteristics of expansive soils. So it is very important for us to understand well how the foundation soils behave. Many researches show that the volume change behavior of an expansive soil is related to its initial state including initial water content, dry density, surcharge loads etc. [2]. During the operation period of a transmission tower, the dry density and surcharge loads of foundation soil remain constant. So initial water content, which is mainly effected by the climate and season, plays a dominant role in the settlement and heave of expensive soil foundation.

Many researchers have studied the problem of swelling and shrinkage of clays under different initial water contents. Some researchers concludes that the swelling-shrinkage

L. Hu et al. (Eds.): GSIC 2018, *Proceedings of GeoShanghai 2018 International Conference: Multi-physics Processes in Soil Mechanics and Advances in Geotechnical Testing*, pp. 156–164, 2018.
https://doi.org/10.1007/978-981-13-0095-0_18

characteristics are linear to the initial water contents. Zou et al. [3] found that the swelling ratio of expansive soils decreases linearly with the increase of water contents under the same surcharge loads. Lu et al. [4] found that the swelling pressure of Nanjing weakly expansive soils was inversely linear to initial water contents. On the other side, Ding et al. [5] found swelling pressure was negatively but not linearly correlated with the initial water contents. Yang et al. [6] used an exponential function to describe the relationship between swelling pressure and initial water contents. According to Kong et al. [7], the plastic limit could be used as a critical point which divided the relationship curves of swelling pressure and shrinkage ratio with initial water contents into two parts. Although there are so many studies on the swelling or shrinkage characteristics of expansive soils, systematic research on deformation should be further conducted.

In this paper, both swelling and shrinkage characteristics of expensive soil under different water contents were observed for a transmission tower foundation. An accurate evaluation of settlement/heave and pressure change for expensive soil foundations was made according to test results, which could be good reference to the design and remedy of transmission tower foundations in the expensive soil area.

2 Material and Specimens

2.1 Experimental Material

The experimental material used in this study was a type of weakly expansive soil obtained from a transmission tower foundation in Nanyang, China. According to the geological investigation report, the soil is alluvial and diluvial cohesive soil formed in Quaternary pleistocene, colored in yellowish-brown, rigid-plastic, containing some Fe, Mn oxide nodules. The basic physical properties of the natural soil were determined with the test methods in accordance with SL237 (1999), and the results are listed in Table 1. The particle size distribution curve obtained by hydrometer analysis is shown in Fig. 1, which presents the clay size fraction was about 33.26%. The mineral composition was analyzed by X-ray diffraction technique (D/max-rB) and the result is shown in Table 2.

Table 1. Basic physical properties of the natural soil.

Density ($g \cdot cm^{-3}$)	Dry density ($g \cdot cm^{-3}$)	Void ratio	Specific gravity	Liquid limit (%)	Plastic limit (%)	Free swelling ratio (%)
2.07	1.69	0.621	2.74	37.70	18.20	59

Table 2. Mineral composition of the soil

Mineral	Quartz	Albite	Microcline	Kaolinite	Illite	Illite-smectite	Mixed-layer ratio
Percentage (%)	50	10	6	1.36	0.68	31.96	10

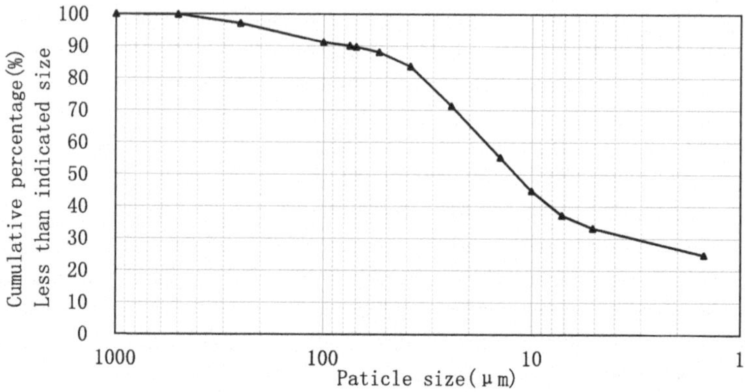

Fig. 1. Particle size distribution curve

2.2 Specimen Preparation

The specimens were prepared for the tests to obtain the vertical swelling ratio, swelling pressure and free shrinkage of the expensive soil. All the specimens were compacted into the same density of 1.7 g/cm^3 with different water contents ranging from 14% to 23%, which were chosen in accordance with the actual climate conditions [8]. At first, the naturally dried soil was grinded into powder and then passed through a 0.05 mm sieve. In order to control the initial water content of each specimen, the amount of water added was calculated according to the corresponding target water content and naturally dried water content (7.3%). The soil powder was sufficiently mixed with water, and then was sealed and stored in a plastic bag for 24 h to ensure a homogeneous water distribution. Finally the desired amount of wet soil was compacted into a specimen ring (61.8 mm in diameter and 20 mm in height) in three layers with a 2.5 kg hammer falling from a height of 30 cm.

3 Testing Methods

3.1 One-Dimensional Swelling Test

The specimens with different initial water contents were placed in oedometers under a seating load of 1 kPa. Then they were completely immersed in water and allowed to swell freely. The readings of vertical swelling deformation were recorded regularly until the deformation within 6 h did not exceed 0.01 mm. The swelling ratio is defined as the ratio of the maximum heave in specimen upon wetting to the specimen's initial height. The whole swelling process took about 3 days.

To simulate the swelling characteristics of the expensive soil under light load, the swelling ratio under a surcharge of 50 kPa was tested. The process was centrally similar to that of the free swelling test, except two differences: (a) the specimens were compressed under a 50 kPa pressure before being saturated, (b) after reaching the maximum swelling deformation, the specimens were unloaded to the seating load of 1 kPa. In this situation, the specimens would continue to swell since the load decreased. The final swelling deformation were recorded and compared with the free swelling ratio.

3.2 Constant Volume Oedometer Test

In this test, the specimens were sustained at the initial void ratio by continuously adding balance loads on the consolidation frame. The largest vertical deformations of specimens were controlled to be less than 0.01 mm. The balance loads were applied by gradually adding sands into a bucket hanged on the loading arm until no deformation was observed. Then the swelling pressure was calculated as the load required to retain zero swell divided by the specimen area.

3.3 Shrinkage Tests

The shrinkage tests of specimens with different initial water contents were performed in the free shrinkage apparatus, where the soil specimens were placed on porous plates. Then specimens were slowly dried in air at a constant temperature of 20 °C. The weight of each specimen was measured to calculate gravity water content changes during the desiccation process; while vertical and horizontal shrinkage deformation were measured by dial gauge and slide caliper respectively. The experiment was stopped when the dimensions of specimens remained constant, which took 3–4 days. Geometry factor (γ_s) [9] was analyzed to evaluate anisotropy of soils shrinkage deformation. It is a factor related to the volume shrinkage and height shrinkage:

$$\gamma_s = \frac{\ln(V_s/V_0)}{\ln(Z_s/Z_0)} \tag{1}$$

Where, V_s is the final volume of the specimen after being dried, V_0 is the initial volume of specimen; Z_s is the final height of specimen after being dried, Z_0 is the initial height of specimen; $\gamma_s = 3$: isotropic shrinkage; $1 < \gamma_s < 3$: subsidence dominates cracking; and $\gamma_s > 3$: cracking dominates subsidence.

4 Results and Discussion

4.1 Swelling Ratio

Figure 2 shows the free swelling ratio versus time of compacted specimens. Figure 3 shows swelling ratio versus time under 50 kPa, here the negative value indicates that specimens are compressed under the load before swelling. The swelling ratio is calculated by the instant specimen height divided by the initial height (20 mm). The corresponding swelling ratio versus initial water contents is shown in Fig. 4.

In Figs. 2 and 3, the curves of swelling ratio can be divided into three stages: fast swelling stage (0–3 h), slow swelling stage (3–6 h) and steady stage (more than 6 h). In the fast swelling stage, water infiltrates into specimens and reacts with clay particles

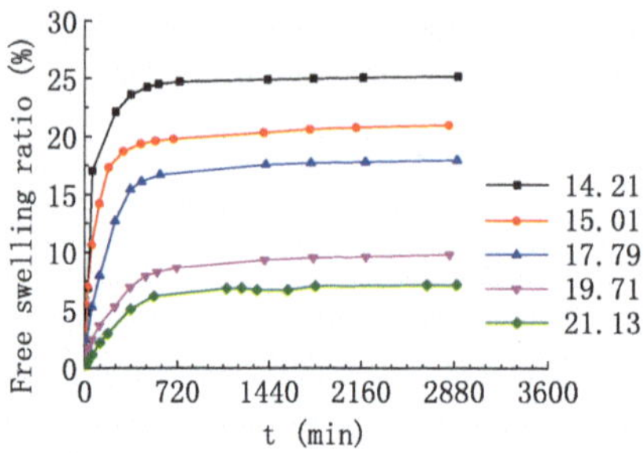

Fig. 2. Free swelling ratio versus time

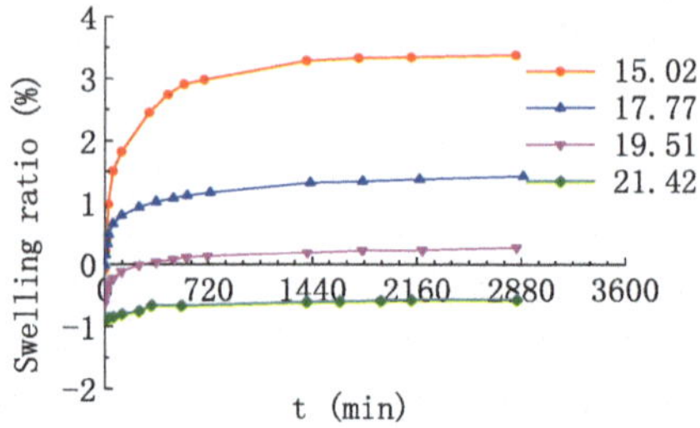

Fig. 3. Swelling ratio versus time under 50 kPa

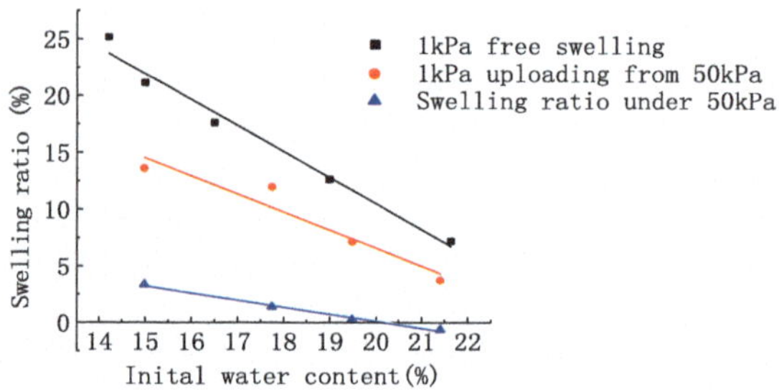

Fig. 4. Swelling ratio versus initial water contents

which increases the thickness of moisture film around soil particle. Specimens swell at a high speed, and the swelling ratio can account for 80% of the total swelling ratio. In slow swelling stage, the swelling speed slows down and the specimens only absorb little water. The swelling is deduced by the reorganization of soils' internal structure. The steady stage lasts for a relatively long time until reaching a steady state, but the swelling during this stage is small.

These three figures also illustrate that with the decrease of initial water contents, specimens swell faster and end at a higher swelling ratio. Because a lower water content leads to a higher suction, and the existence of higher suction helps soils specimens to absorb water faster when they are immersed in water. In addition, aggregates show greater swelling potential under higher initial suction during the reaction between water and illite-smectite mixed layer [10].

The middle line in Fig. 4 represent swelling ratio under 1 kPa unloading from 50 kPa which indicates that loads limit specimens from swelling, while unloading could allow specimens to continue swelling. However the final swelling ratio is less than the free swelling ratio because internal structure reorganization makes soil particles arranged more closely. This can also be understood as that the former loads 50 kPa has done negative work during the process of expansion, which consumes the swelling potential of specimens.

4.2 Swelling Pressure

Swelling pressure refers to the pressure that needed to keep the volume of expansive soils at a constant value during water infiltration. Figure 5 shows the evolution curves of swelling pressure corresponding to different initial water contents. The trend of the curves is similar to that of swelling ratio, which can be also divided into three stages. However, by comparing Fig. 5 with Figs. 2 and 3, it can be found that swelling pressure takes less time than swelling ratio to reach the steady state, probably lies in the difference of physical mechanism behind between swelling pressure and swelling ratio. Swelling ratio serves as a cumulative quantity. The swelling ratio reaches its peak level (namely steady state) after the specimen arriving at its full hydration, while the swelling pressure does not. This can be better understood by investigating the hydration process of the specimen. As shown in Fig. 6, during an early period of hydration, the specimen was partially hydrated due to the low permeability of expansive soils, where the lower part was fully hydrated but the upper part remained almost dry. As a result, the stiffness of the upper part will be much larger than that of the hydrated lower part, so that the upper part and the steel components above can be collectively treated as an effective rigid piston that transfers the axial load induced from the hydrated lower part to the consolidation frame, just as the swelling test were conducted solely on the lower part that acts as a separated specimen. In this way, the obtained swelling pressure during that period will be quite close to the maximum swelling pressure of the whole specimen and can be fairly approximated as the peak (steady) swelling pressure of it.

Figure 7 illustrates that swelling pressure, as with swelling ratio, also decreases linearly with the increase of initial water contents. This can be explained by the swelling mechanism of expansive soils. The swelling behavior of expansive soils is due to the forming of moisture film around the clay particles. For a certain soil, the maximum thickness of moisture film is fixed. However adding some water during specimen preparation can partially thicken the moisture film around clay particles, resulting in the decrease of swelling ability. So specimens compacted with a higher initial water contents would have less swelling potential.

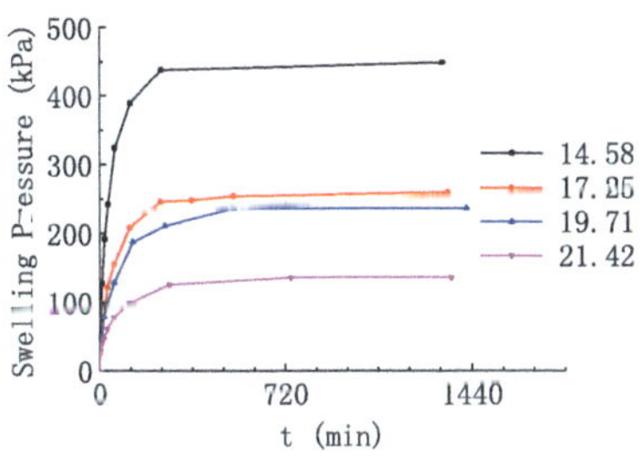

Fig. 5. Swelling pressure versus time

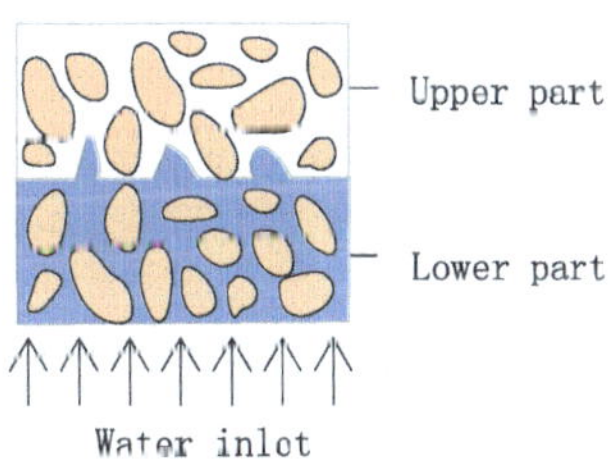

Fig. 6. Hydration process of specimen

4.3 Shrinkage Characteristics

Figure 8 illustrates that the shrinkage process due to drying can be divided into two stages. Firstly, shrinkage ratio is proportional to the variation of water contents. In this stage, the matric suction in soil specimen increases with the evaporation of pore water.

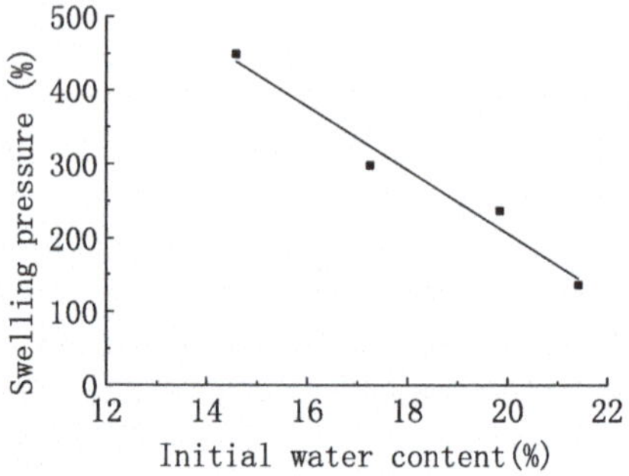

Fig. 7. Swelling pressure versus water content

Under the action of matric, the particles of soil get closer to each other gradually, which contribute to the volume shrinkage of specimens. The slopes of these curves under different water contents are nearly the same. That is to say, initial water contents have little influence on shrinkage index. The second stage is a drying process with slight volume change at the water contents less than the shrinkage limit (about 9%). In this process, although matric suction continues to increase, its action area is decreased and soils fabric has become stable. The force deduced by the matric suction has little effect on the position of soils particles. So the specimens do not show any shrinkage further. It is also worth mentioning that specimens did not crack during the whole drying process since they were dried in a relatively slow speed and the final water content was not low enough to induce cracking.

Figure 9 shows that the final horizontal shrinkage ratio increase linearly with water contents. The obtained relationship is consistent with the past observations presented by Ref. [11]. The geometry factors calculated with formula (1) are shown in Table 3. The factors of the specimens with different water contents are all greater than 3.0, which indicates that the volume change are mainly caused by horizontal shrinkage. Furthermore the anisotropy becomes greater with the increase of water contents.

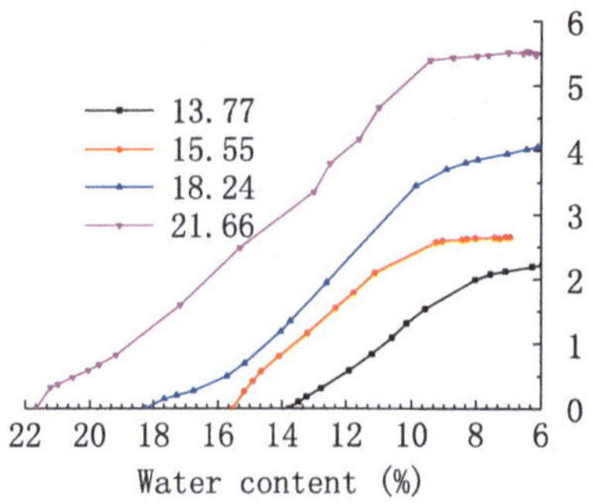

Fig. 8. Shrinkage ratio versus water content

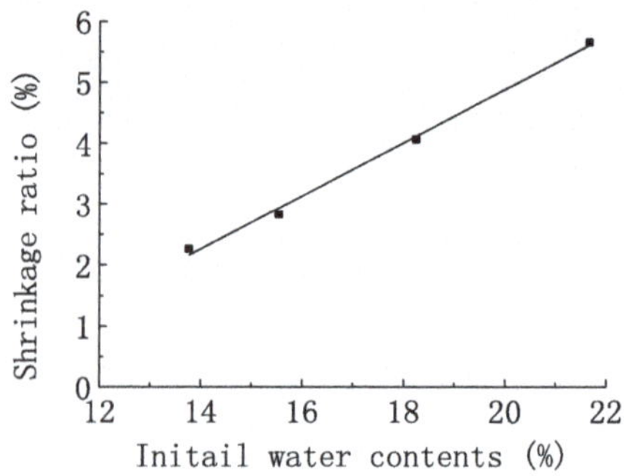

Fig. 9. Shrinkage ratio versus initial water content

Table 3. Relationships between geometry factor and initial water contents.

Water content (%)	13.77	15.55	18.67	21.66
Geometry factor	3.44	3.47	3.70	3.75

5 Conclusions

From these laboratory tests, the effect of initial water content on the swelling-shrinkage characteristics of compacted expansive soils is investigated. Following conclusions can be drawn as follows:

(1) Swelling properties including swelling ratio and swelling pressure act inversely with the initial water content. Specimens under lower water contents absorb more water and show stronger swelling potential during stabilization process.
(2) The swelling pressure or swelling ratio versus time can be divided into three stages: fast swell stage, slow swell stage and steady stage (more than 6 h).
(3) Vertical shrinkage ratio increases linearly with the initial water content. However the water contents have little influence on the shrinkage index.
(4) Geometry factors of the specimens with different water contents are all greater than 3.0, which indicates horizontal shrinkage contributes to the volume shrinkage slightly more predominantly than vertical shrinkage. The geometry factor becomes greater with the increase of initial water content, which indicates that the shrinkage anisotropy becomes more evident.

Acknowledgements. This study was financially supported by the National Grid Company Technology Project (grant no. 5217L0160001).

References

1. Shi, B.: Engineering geological characteristics of expansive soils in China. Eng. Geol. **67**(1–2), 63–71 (2002)
2. Li, Z.: Expansion characteristics and construction control of remolded Mengzi expansive soil. Chin. J. Geotech. Eng. **30**(12), 1855–1860 (2008)
3. Zou, W.: Nonlinear strength property of remolded expansive soil and its stress-strain relationship for soaking after one-dimensional consolidation. Rock Soil Mech. **33**(S2), 59–64 (2012)
4. Lu, Z.: Experimental study of the swelling force of remolded expansive soils. J. Hebei Univ. Eng. Nat. Sci. Edition **23**(3), 47–50 (2015)
5. Ding, Z.: Trial study on variation regularity of swelling force. Rock Soil Mech. **28**(7), 1328–1332 (2007)
6. Yang, Q.: Testing study on shear strength of unsaturated expansive soils. Chin. J. Rock Mechan. Eng. **23**(3), 420–425 (2004)
7. Kong, L.: Study on shear strength and swelling-shrinkage characteristic of compacted expansive soil. In: Asian Conference on Unsaturated Soils 2000, Singapore, pp. 515–519 (2000)

8. Guo, W.: On research method for air-influenced depth to expansive soil in Nan yang basin. Yangtze River **38**(9), 8–10 (2007)
9. Bronswijk, J.J.B.: Shrinkage geometry of a heavy clay soil at various stresses. Soil Sci. Soc. Am. J. **54**(5), 1500–1502 (1900)
10. Du, Y.: Swelling–shrinkage properties and soil improvement of compacted expansive soil, Ning-Liang Highway, China. Eng. Geol. **53**(3), 351–358 (1999)
11. Li, Z.: Research on expansion and contraction rules and disaster mechanism of expansive soil. In: Tenth National Symposium on Numerical Analysis and Analysis of Geotechnical Mechanics 2010 (2010)

Study on the Soil Water Characteristic Curve and Shear Strength Characteristics of Granite Soil with Different Weathered Degree

Youqian Lu[1(✉)], Changfu Wei[2], Guoqing Cai[1], and Chenggang Zhao[1]

[1] School of Civil Engineering, Beijing Jiaotong University, Beijing 100044, China
luyouqian@bjtu.edu.cn
[2] Institute of Rock and Soil Mechanics, Chinese Academy of Sciences, Wuhan 430071, Hubei, China

Abstract. In this paper, the soils of the weathered granite with double layer soil slope in the hot and humid areas are studied. The water holding characteristics of compacted samples are studied by the pressure plate technique and vapor equilibrium method with saturated salt solution, and the effect of saturation on the shear strength is studied. The experimental results show that the residual soil has better water holding capacity than the completely weathered soil, which has greater air-entry value and residual suction value, and the gradient of the transition zone is smaller. There is an optimal saturation in this two weathered soils to reach the peak of the shear strength. The influence of shear strength index is different: the effect of saturation on the internal friction angle is little, but the effect on the cohesion is significant. The different weathered degree affects the soil properties such as the mineral composition and grain size gradation of two weathered soils. The residual soil has higher clay content and residual iron oxide, which affects two weathered soils in the soil water characteristic cure (SWCC) and shear strength of the characteristics. The mechanism of these differences can be clearly elucidated by analyzing the matrix suction and the cementation between soil particles, which provides a theoretical basis for the engineering management of weathered granite slopes.

Keywords: Weathered granite soils · SWCC · Saturation · Shear strength · Cementation effect

1 Introduction

Geological disasters such as landslides and debris flows are frequent in the granite distribution area of southeastern China. Because of the different weathered degree, the weathered granite slope is diverse from the common homogeneous slope, the stability of the slope is more complicated. The weathered granite soils are formed by the weathered and laterization of granite under humid tropical climates. Numerous facts show that rainfall is the most important and most common factor affecting the shear strength of the slope, which leads to the shallow slope of the slope (Gvirtzman et al. 2008 and

L. Hu et al. (Eds.): GSIC 2018, *Proceedings of GeoShanghai 2018 International Conference: Multi-physics Processes in Soil Mechanics and Advances in Geotechnical Testing*, pp. 165–172, 2018.
https://doi.org/10.1007/978-981-13-0095-0_19

Xu et al. 2011). The stability of weathered granite slopes is studied from the aspect of grain size composition, but the effect of saturation is neglected (Zhang et al. 1997). A similar study of unsaturated water holding capacity and strength has done a lot (Baker and Frydman 2009; Gallage and Uchimura 2010). At present, many studies have found that most of the landslides in the granite landslide are located in the residual soil with low shear strength, but the typical granite landslide site investigation found that the sliding surface is mostly located in the whole weathered zone.

Although many research work has been carried out, it is still lack of a clear recognition and systematic deep study on weathered granite water holding capacity as well as the impact of water content on the shear strength of the law. Aiming at the problem of rainfall-induced landslide of weathered granite double layer soil slope, the paper uses the experiment of the soil water characteristic curve (SWCC) and direct shear test to discuss the water holding capacity and the influence of saturation on the shear strength of granite weathered soils, and then analyzes the mechanism of rainfall induced shallow landslide.

2 Test Materials

We have carried out representative sampling and testing of the residual zone (2–5 m below the surface) and the completely weathered zone (6–10 m) soil. In this paper, the particle analysis experiment of different layers of slope profile is carried out, it turns out that the content of particles in the residual soil with greater than 2 mm is less than 5%, and less than 0.5 mm particles content is greater than 50%. Its color is brick red or brown red which mixed with a small amount of white quartz particles, while the original rock structure was completely destroyed so the clay particles and silt particles are high in content, leading to a closer arrangement and higher bond strength. In the strong weathered zone, the particle content of more than 2 mm is less than 20% and the content less than 0.5 mm is 50%. The quartz content is more than the residual soil, while occasionally contains the weathered or weakly weathered granite solitary rock, which color mainly contains yellowish brown, mixes with red, gray, white and other colors. This part of soil is relatively loose and the structure is coarse clastic structure or granular structure.

According to the classification of granite residual soils as taking into account the size and structure of the soil particles (Wu 2006), the residual soil is named silty clay and the completely weathered soil is named as silty sand. The physical indicators of the two layers of soil are listed in Table 1. The comparison of the physical indices show that, all the parameters of the residual soil are higher than that of the completely weathered soil.

Table 1. Properties of soils used in this study.

Property	Residual soil	Completely weathered soil	Property	Residual soil	Completely weathered soil
Liquid limit	55.8%	43.7%	Specific gravity	2.71	2.66
Plastic limit	30.3%	28.6%	Plasticity index	24.5%	15.1%
SSA (m^2/g)	107.2	70.2	CEC (meq/100 g)	29.5	21.2

3 Test Method

The test soils were crushed after air drying then passed through a 2 mm sieve, the dry density of soil sample is taken as the average of the natural dry density of the double-layer soil, that is $\rho_d = 1.55\,\mathrm{g/cm^3}$, the size of the sample is 61.8 mm in diameter and 20 mm in height. A direct shear test with different saturation and the SWCC test were carried out.

3.1 Direct Shear Test Program

The soil samples with saturation of 20, 40, 60 and 80% were prepared respectively, and after 24 h of soil samples were sealed, the corresponding quality soil samples were taken by the static compaction method to obtain the ring-like sample. The soil samples with 100% saturation were achieved by pumping saturated soil samples with saturation of 80%. Remodeled samples were sealed and moistened for 24 h, then subjected to a direct shear test. The test adopts the ZJ-2 type strain-controlled direct shear instrument, and the ring-like sample is carried out under the vertical pressure of 100, 200, 300 and 400 kPa to do the undrained fast shear test.

3.2 The SWCC Test Program

The measurement of suction is realized by measuring the SWCC, the low suction section adopts the pressure plate method (0–1500 kPa) as shown in Fig. 1, the High suction section (3 MPa–368 MPa) adopts the saturated salt solution vapor balance method as shown in Fig. 2 and the test temperature is strictly controlled of 20 °C, the specific experimental methods could see literature Gao et al. (2016).

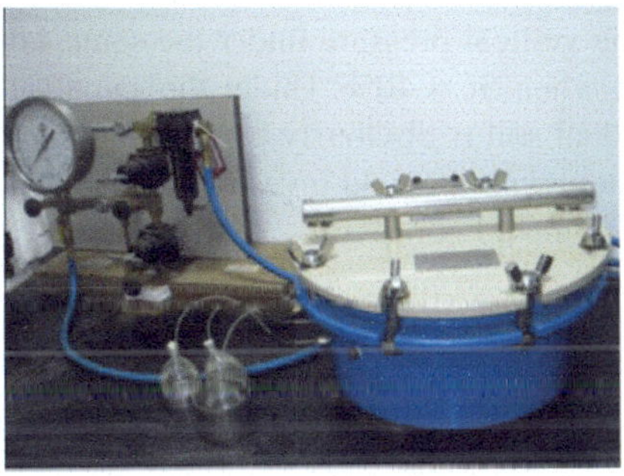

Fig. 1. Pressure plate apparatus

Fig. 2. Vapor balance container.

4 Test Results

4.1 The SWCC Test Results

The results of the SWCC test are shown in Fig. 3. It can be seen from Fig. 3 that the air-entry value of the residual soil is about 50 kPa and the residual suction value is about 3500 kPa, which is obviously higher than that of the completely weathered soil which is 4 kPa for the air-entry value and 900 kPa for the residual suction value. In the transition zone of the SWCC, the gradient of this two soil samples basically changes in a straight line, and the slope k_c is greater than k_r indicates that the completely weathered soil has a faster dehumidification rate; The saturation of the two soil samples is lower after the suction is greater than 1 MPa, and the saturation is virtually nil when the suction reaches 368 MPa, but the residual saturation of the residual soil is still higher than that of the completely weathered soil.

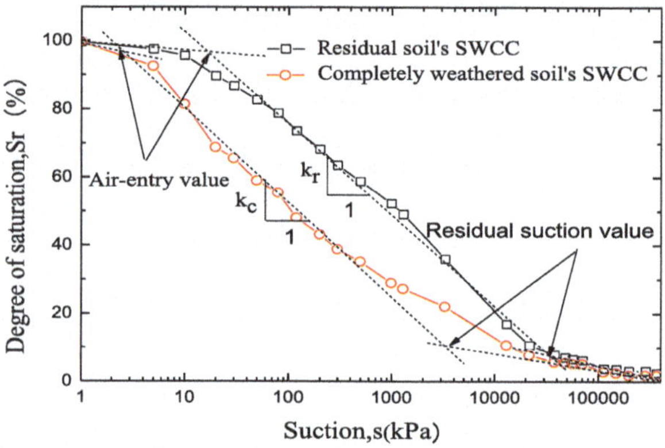

Fig. 3. The SWCCs of residual soil and completely weathered soil

4.2 The Direct Shear Test Results

Figure 4 shows the relationship between shear strength and vertical pressure of residual soil and completely weathered soil under different saturation conditions. The results show that the shear strength increases with the vertical pressure under the same saturation condition which reaches the maximum when Sr = 40%. Under the same saturation condition, the shear strength in the residual soil is obviously higher than that of the completely weathered soil, which is consistent with the field investigation results of that the sliding surface is in the completely weathered zone.

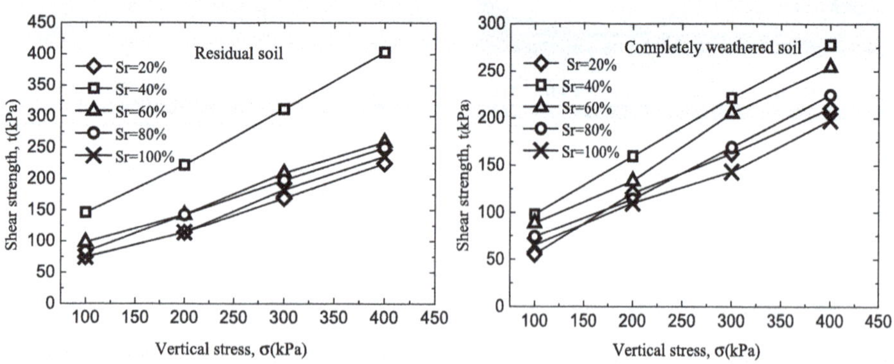

Fig. 4. Relationship between shear stress and vertical pressure under different saturation

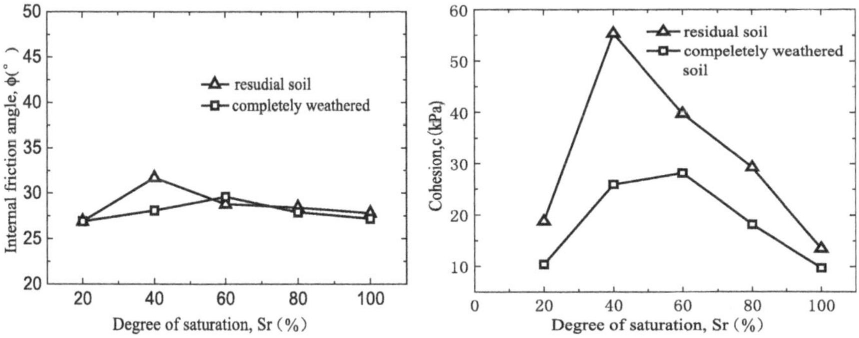

Fig. 5. Relationship between shear strength index and saturation

Figure 5 is the cohesive force c and the internal friction angle φ of the residual soil and the completely weathered soil with different saturation. The cohesive force of two kinds of soil has a distinct change under the influence of different degree of saturation, the cohesive force reaches the peak when Sr = 40%. The cohesive force shows a downward trend and the reduction rate of the residual soil was larger than that of the completely weathered soil (Fig. 5) as Sr > 40%. At the same time, the internal friction angles of the both soil are less affected by the saturation degree which only fluctuate between 26° and 30° under different saturation. However, the internal friction angle still appears a less obvious peak value near Sr = 40%. This phenomenon shows that the cohesion force c is more sensitive to the change of saturation, and it can be seen that under the same saturation condition, the cohesive force of the residual soil is obviously larger than that of the completely weathered soil and the friction angle is very small or even no difference.

5 Analysis and Discussion

The experimental results show that the cohesion force of the weathered granite soils is greatly affected by saturation, and the degree of saturation has less impact on the internal friction angle. The change of soil saturation is only a macroscopic representation, the change of microstructure and properties of soil are the direct reason leading to the change of shear strength. In the following, we analyze the influence of saturation on shear strength from matrix suction and cementation.

5.1 Effect of Matrix Suction on Shear Strength

Zhou and Fredlund (2011) has carried out a comparative analysis of the existing formulas for unsaturated strength, the results show that both the determination of model parameters or strength prediction and many other aspects are empirical and phenomenological, the most notable of which is in the high suction section over high estimated shear strength. Zhou et al. (2016) and Konrad and Lebeau (2015) use theoretical considerations and experimental observations to partition the water retention function into capillary and adsorptive parts. Since the thin film of adsorbed water should not cause effective stress, the effective stress parameters are only related to the capillary component of water holding capacity. The model explains well the peak intensity of weathered granite soil, in which only the degree of saturation based on capillary water contributes to the variation of shear strength with suction. The resulting parametric model can be written in a manner that highlights the mechanisms (capillarity and adsorption) of water holding capacity (Zhou et al. 2016):

$$S_r = S_r^{cap} + S_r^{ads} \tag{2}$$

The shear strength can be expressed as follows:

$$\tau = \left[\sigma - u_a + (u_a - u_w)S_r^{cap}\right] tan\varphi + c \tag{3}$$

Among them, S_r is the degree of saturation, S_r^{cap} the capillary component of degree of saturation, and S_r^{ads} the adsorptive component of degree of saturation. τ the shear strength, σ the apparent contact stress, u_a the air pressure, u_w the liquid pressure, φ the friction, c the inherent cohesion.

As the Fig. 6a shows that liquid saturation is separated into the adsorptive part $(S_{r,a})$ and the capillary part $(S_{r,c})$, and only the capillary part contributes to the variation of shear strength with suction, the value of the capillary stress increases with the increase of the degree of saturation first and then decreases as shown in Fig. 6b (Konrad and Lebeau 2015). So when the capillary stress increases that the water and air in the soil are connected, the reaction force of the surface tension on the shrinkage film of the water-gas interface causes pressure on the soil particles, and the cohesive force of soil is enhanced in macro performance. If the saturation of soil sample is reduced to a certain value or increased to near saturation, the capillary action will disappear which can cause the cohesion to decrease. There is a critical value in the contribution of matrix suction against shear strength, it's the shear strength and cohesion force of Sr = 40%–60% are peaked in the experimental results.

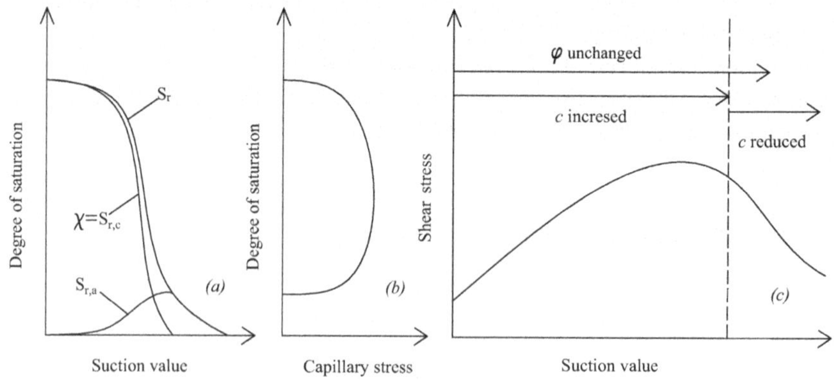

Fig. 6. Features of capillary-based model for typical soil: (a) water retention function and resulting effective stress parameter; (b) capillary stress as function of saturation; (c) shear stress as function of suction value.

5.2 Effect of Cementation on the SWCC and Shear Strength

In order to investigate the effect of cementation on the SWCC and shear strength, we carried out the SEM as shown in Fig. 7. According to the results of the SEM, the mineral crystals in residual soil have been completely transformed into fine clay granules, some of which have been formed into larger soil, the small pores between the grains are ubiquitous, and the cohesive floc of granules is more, which shows that this sample has strong bond strength (Fig. 7a and c); there are obvious large pores and fissures in the crystals, while the directional flake residues are common (Fig. 7b and d) and have higher compressibility.

a b c d

Fig. 7. SEM morphology of residual soil and completely weathered soil: a. residual soil (×100) b. completely weathered (×100) c. residual soil (×2000) d. completely weathered (×2000)

There are more iron oxides and alumina in this two soils, the free ferric oxide and alumina act as cementation to bind the soil particles together in order to form larger aggregates, which enhances the bond strength of soil particles. The residual soil contains higher free ferric oxide than the completely weathered soil, which makes the interior form more agglomeration structure. Those are containing more small pores inside the aggregates. At the same time, these factors, larger particle specific surface area, higher clay content and finer particle gradation, make the residual soil has better water holding capacity and shear strength than the completely weathered soil. In addition, there are more ions adsorbed on the surface of soil particles in residual soil which will occur ions hydration adsorption under high suction, there is an important reason for the higher water holding capacity of the residual soil, and it also strengthens the adsorption bonding between the clay particles. On the other hand, the cementation is sensitive to the saturation of the soil sample. When the saturation reaches a certain value, the water formed Sol colloid with the free iron oxide particles and fill in the pores between the particles, formed cementation with the clay minerals at last, which cause the normal cohesive force c of the soil. It represents a significant increase in the shear strength of physical and chemical forces such as Van der Waals force. The soil moisture content (saturation) is the key factor that determines the cementation, the soil will not form Sol colloid and granular colloid under the condition of low water content, and the excessive water content will lead to the increase of the film thickness, nor will it form a structure. The results show that when the saturation is between 40% to 60%, the cementation is the strongest and the greatest contributor to shear strength of soil.

6 Conclusions

Based on the comparison of physical and mechanical properties of soil in the residual soil zone and completely weather zone of granite, as well as the analysis and discussion of the results of the SWCC and direct shear test, we get the following conclusions:

The residual soil has higher water holding capacity than the completely weathered soil, such as the air-entry value and residual suction value; Only the degree of saturation based on capillary water contributes to the variation of shear strength with suction, so there is a non-linear relationship between the shear strength and saturation, as the "optimal saturation" of a saturation between 40% to 60% makes the shear strength reach the peak strength.

The composition of mineral composition and grain size gradation affect the microstructure of soil, especially the effect of oxide cements on soil structure and pore size distribution, which directly changes the soil water holding capacity and resistance shear strength.

Acknowledgement. Financial supports by the National Natural Science Foundation of China (41572293, 51722802, 51678041).

References

Baker, R., Frydman, S.: Unsaturated soil mechanics: critical review of physical foundations. Eng. Geol. **106**, 26–39 (2009)

Gallage, C.P.K., Uchimura, T.: Effects of dry density and grain size distribution on soil-water characteristic curves of sandy soils. Soils Found. **50**(1), 161–172 (2010)

Gao, Y., Sun, D.A., Zhou, A.N.: Hydro-mechanical behavior of unsaturated soil with different specimen preparations. Can. Geotech. J. **53**, 909–917 (2016)

Gvirtzman, H., Shalev, E., Dahan, O., Hatzor, Y.H.: Large-scale infiltration experiments into unsaturated stratified loess sediments: monitoring and modeling. J. Hydrol. **349**, 214–229 (2008)

Konrad, J.-M., Lebeau, M.: Capillary-based effective stress formulation for predicting shear strength of unsaturated soils. Can. Geotech. J. **52**, 2067–2076 (2015)

Wu, N.S.: Study on classification of granite residual soils. Rock Soil Mech. **27**(12), 2299–2304 (2006). (in Chinese)

Xu, L., Dai, F.C., Tham, L.G.: Field testing of irrigation effects on the stability of a cliff edge in loess, North-west China. Eng. Geol. **120**, 10–17 (2011)

Zhang, Y.B., Zhang, Y., Chen, G., et al.: The system of engineering category dividing of granite residual soils. Bull. Chin. Acad. Geol. Sci. **18**(2), 201–204 (1997)

Zhou, A.N., Fredlund, D.G.: Shear strength criteria for unsaturated soils. Geotech. Geol. Eng. **29**, 145–159 (2011)

Zhou, A.N., Huang, R.-Q., Sheng, D.: Capillary water retention curve and shear strength of unsaturated soils. Can. Geotech. J. **53**, 974–987 (2016)

Strength and Deformation Characteristics of Unsaturated Fine-Grained Tailings Soil Under Variable Moisture Content

Jianjun Dong[1,2,3(✉)] and Guofeng Zheng[2,3]

[1] Department of Civil Engineering, School of Mining Technology, Liaoning Technical University, Huludao 125105, Liaoning, China
dongjnjn@163.com

[2] State Key Laboratory of Structural Analysis for Industrial Equipment, Dalian University of Technology, Dalian 116024, Liaoning, China

[3] Department of Engineering Mechanics, Dalian University of Technology, Dalian 116024, Liaoning, China

Abstract. Previous research has not analysed the strength and deformation of unsaturated fine-grained tailings soil under variable moisture content. This paper discusses individual tests for the compressibility, direct shear strength and triaxial strength of fine-grained tailings soil, under those variable moisture content. The experimental results showed that: (1) With an increase in specimen moisture content, resistance to vertical deformation gradually declined and reached a minimum at saturation; (2) The cohesion of unsaturated fine-grained tailings soil tended to decrease with an increase of moisture content, and had an obvious peak. The angle of internal friction decreased as moisture content increased, exhibiting an approximately inverse linear relationship; (3) Moisture content exhibited a very obvious effect on the stress and strain characteristics of unsaturated fine-grained tailings soil. Under the same confining pressure, as moisture content increased, the axial stress which was applied on the specimen declined, and its ultimate strength was also reduced. As moisture content declined, the specimen damage type changed from strain hardening to strain softening. With an increase of moisture content, the volumetric strain of the specimens initially increased then decreased. Presumably, the volumetric strain should peak when the moisture content reaches an optimum level.

Keywords: Deformation · Experimental study · Moisture content Strength · Unsaturated fine-grained tailings soil

1 Introduction

Tailings are mine waste, including fine-grained sand or silt mining waste material remaining after more valuable ore has been concentrated. The tailings dam is an integral part of the tailings pond, is a special type of hydraulic construction, its stability determines the safety of the mine or not (Fan et al. 2011). A typical tailings dam would normally be in operation for a number of years to decades. During such a long period of time, the moisture content of the dam body is constantly changing, such as the increase

L. Hu et al. (Eds.): GSIC 2018, *Proceedings of GeoShanghai 2018 International Conference: Multi-physics Processes in Soil Mechanics and Advances in Geotechnical Testing*, pp. 173–180, 2018.
https://doi.org/10.1007/978-981-13-0095-0_20

of moisture content caused by rainfall and the decrease of moisture content caused by natural evaporation.

Yin et al. (2007) studied the tailings of the Diqing Mine Company, Yunnan Province, P. R. China; the results of their tests of three tailing sands showed that the compressibility and percolation of tailings are closely related to the diameter and void ratio of tailing sands. Wu et al. (2009) conducted a series of consolidation tests on saturated fine tailings in a study related to tailing dam construction with fine tailings in a copper mine in East China, and analyzed the evolutionary process of the degree of consolidation under the influence of deadweight and time factors, and the relationship of the degree of consolidation with the granulometric composition and sub-layer thickness. Qiao et al. (2015) performed a series of tests to determine the distribution of particle-size, mechanical properties and permeability of tailings in different positions on a depositional beach face, and analyzed the influence of the content of fine particles (diameter < 0.075 mm) on the engineering properties of tailing.

Wickland et al. (2010) argued that mixing waste rock and tailings can produce dense deposits with values of hydraulic conductivity that are orders of magnitude lower than those of waste rock alone, thereby limiting fluxes through the waste and lowering the associated potential for acid rock drainage. Jeeravipoolvarn et al. (2009) studied the mechanical properties of oil sands tailings, and found that fine-grained tailings should be more than 30% under the weight of variables, and the formation of effective stress is very small; however, the mixture formed could have a larger effective stress. Dimitrova and Yanfu (2011) studied the effects of moisture content, effective stress and consolidation time on the undrained shear strength of a sedimentary mine tailing ore deposit. Pi et al. (2012) analyzed the nature of silty sand tailings, and conducted a laboratory test related to particle content, physical property tests, a triaxial compression test and particle flow numerical test, then studied the physical and mechanical properties of tailings using different fines content and confining pressure.

Qin et al. (2004) analyzed the details related to the stress-strain relationship, strength and particle breakage properties of coarse aggregates under high and low confining pressures, and studied the influence of the content of mudstone on the strength behavior of rock fill. Deng et al. (2011) attained shear strength parameters based on an experimental study of tailings material from the Bai Zhi Shan mine, and determined how rules of shear strength change under different conditions such as grain size, vertical pressure, drain state and state of consolidation.

Wijewickreme et al. (2005) found that the loop resistance ratio of clay tailings increased with the initial effective confining pressure topdressing when using a circular direct shear test, and also found the cyclic resistance ratio of gold, copper, zinc mine tailings had no relationship with the changes in the initial effective confining pressure. Zhang et al. (2010) made a thorough analysis of tailings properties and the laws related to changing condition, and also investigated the law of distribution and changes of physical-mechanical characteristics and how they varied over time in the Lixi tailings fill dam of the Hua County, Shanxi Province, China. Geremew and Yanful (2012) carried out a number of cyclic triaxial tests on mine tailings and natural sediment specimens under undrained conditions to investigate their resistance to cyclical loading; they observed that axial strain and excess pore water pressure increased with the number of loading cycles while the effective stress decreased with an increasing number of loading cycles. Zhang (2012) carried out a number of dynamic property tests of saturated tailing using sand, and analyzed the dynamic deformation and dynamic strength properties of saturated tailing in sand under a dynamic load using dynamic triaxial research under the condition of undrained saturation. His study analyzed the rules controlling the development and change in dynamic pore water pressure of saturated tailing in sand under the dynamic load.

Wang et al. (2006) prepared and compacted a sand-clay mixture containing 40% fines with optimum moisture content, and conducted a series of unconventional drained compression tests. These tests showed that sand-clay mixtures showed a highly over-consolidated behavior and the Mohr-Coulomb failure envelope was nonlinear. In addition, the angle of internal friction decreased while the elastic modulus increased with an increase in the confining pressure.

However, little existing literature addresses how changes in moisture content of unsaturated tailing affect the strength and deformation characteristics of fine-grained tailings. Therefore, research related to unsaturated fine-grained tailings under different moisture content, soil strength and deformation characteristics has practical engineering significance.

2 Test Methods

The tailings soil chosen for the test was from iron ore tailings in Nanjing, Jiangsu Province, China. These tailings became uniform and brown after drying initially, then the soil got dark brown gradually as the moisture content increased. Tables 1 and 2 describe the particle composition and physical properties of the fine-grained tailings soil tested here, respectively.

Compression, direct shear and triaxial tests of unsaturated soil fine-grained tailings were employed. A study of the strength and deformation characteristics of unsaturated fine-grained tailings soil under different moisture content was conducted. The experimental design used specific moisture contents of 2.9% (moisture content of naturally air-dried soil), 5%, 10%, 15%, 20%, 25%, and 27.3% (saturated moisture content).

Table 1. Characteristics of soil particles of fine-grained tailings

The percentage (%) of variously sized particles (mm) in tailings					Characteristic value	
2 0.5	0.5 0.25	0.25 0.075	0.075 0.005	<0.005	C_u	C_c
5.01	20.49	22.08	51.19	1.23	10.77	0.24

Table 2. Physical properties of fine-grained tailings soil

G_s	$\rho_{d\max}/g.cm^{3}$	$w_{opt}/\%$	$w_P/\%$	$w_L/\%$	$I_P/\%$
3.22	1.7	15	11.3	24.2	12.9

2.1 Compression Tests of Unsaturated Fine-Grained Tailings Soil

Vertical pressure loads of 25, 100, 200, 400, 800, 1600, and 3200 kPa were employed in the compression tests. After each test the vertical deformation with slowly increasing pressure was recorded; then, the next level of loading was applied after checking for load stability. The stability criterion was consolidation for 24 h under the load at each level.

2.2 Direct Shear Tests of Unsaturated Fine-Grained Tailings Soil

Four specimens of designed moisture content were individually loaded into a direct shear apparatus and respectively applied with vertical pressures of 50, 100, 150 and 200 kPa. After applying the vertical pressure, shearing was performed immediately at a shear rate of 1.0 mm/min, and the shearing was ended when shearing deformation reached 4 mm. These series of tests are not suction-controlled tests.

2.3 Triaxial Text of Unsaturated Fine-Grained Tailings Soil

A triaxial test apparatus employed an unsaturated soil triaxial apparatus created by Dalian University of Technology (Shao et al. 2015). The method employed a consolidation drained (CD) shear test, using various levels of soil moisture content with various design specimens; the confining pressure was 100, 200, and 300 kPa under load, the control of the load rate was 0.006 mm/min, and the shear and axial strain reaches 20%.

3 Test Results and Analysis

3.1 Compressibility Test Results and Analysis

Figure 1 shows the unsaturated compressibility of fine-grained tailings soil test results.

Figure 1 shows that the vertical strain of specimens increases gradually with the increase of moisture content under the same vertical stress, and the vertical strain at saturation is the largest (Fig. 1). That is, with an increase in specimen moisture content, the resistance to vertical deformation in the specimen was gradually reduced.

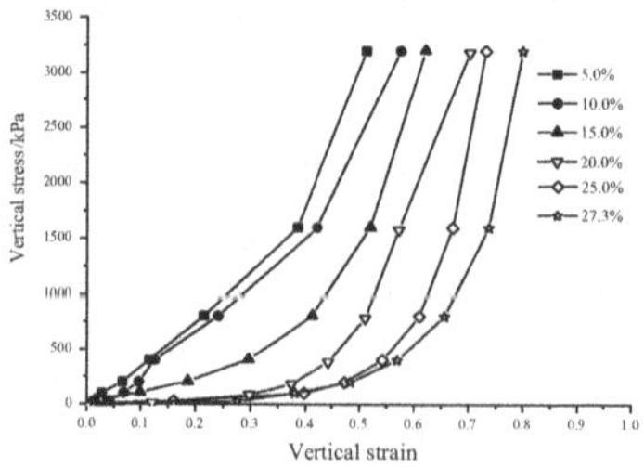

Fig. 1. Experimental results of compression tests measuring vertical strain and stress under different moisture content

The matric suction in soils with in an unsaturated state contributed to the strength of the soil; the resistance of unsaturated fine-grained tailings soil to vertical deformation was stronger than for soil in a saturated state as it related to vertical deformation. An increase in moisture content caused the matric suction in unsaturated fine-grained tailings soil to gradually decline; that is, the strength of the specimen declined with increased moisture content and the ability to resist vertical deformation also will be weakened.

3.2 Direct Shear Test Results and Analysis

Figure 2 show the results of direct shear tests for unsaturated fine-grained tailings soil.

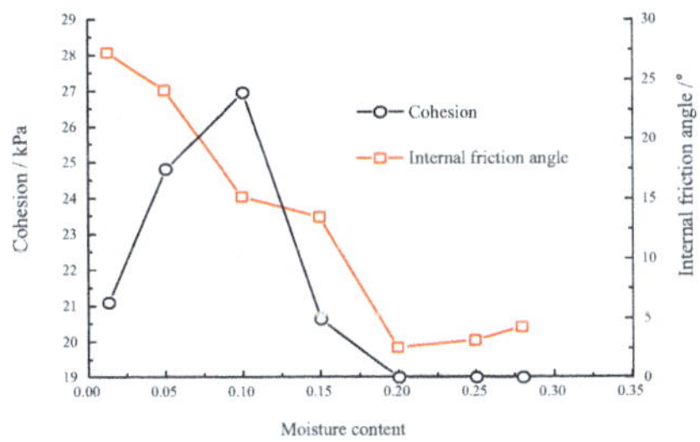

Fig. 2. The relationship curve of moisture content (w), cohesion (c) and internal friction angle (ϕ) of unsaturated fine tailings soil

The angle of internal friction (ϕ) decreases with an increase in moisture content. The internal friction angle is an important parameter that affects the relationship between moisture content and cohesion. The internal friction angle of soil reflects the frictional characteristics of soil and consists of two parts: the surface friction of soil particles, the occlusal force caused by intercalating and interlocking of particles.

The influence of moisture content on the cohesion of unsaturated fine tailings soil can be divided into three stages. The first stage occurred when the moisture content is low ($w < 10\%$), the cohesion increase with an increase in moisture content. During second stage, when the moisture content increased ($10\% < w < 20\%$), the cohesion decreased with an increase in moisture content and when the moisture content reached 20%, the cohesion decreased to zero. The third stage occurred when $w > 20\%$; here, the cohesion was always 0 and did not changed with an increase in moisture content. Generally, the cohesion of unsaturated fine tailings soil showed a trend of initially increasing and then decreasing with an increase in moisture content, and has an obvious peak.

This analysis of the change in the cohesion of unsaturated fine-grained tailings soil with a change in moisture content shows that the cohesion of fine tailings soil was mainly generated by capillary action. That is, the surface tension on the gas-water shrinkage film interface generated the cohesion. When this capillary action force has an effect on soil particles, it produced cohesion in the soil that is caused by compressional stress. Therefore, when the soil moisture content gradually rose within a certain range, the surface tension on the gas-water shrinkage film interface also improved accordingly. This caused the role of total pressure in the soil body stress to gradually increase so that cohesion gradually became large. When the moisture content continued to increase, the gas in the soil was gradually surrounded by water, and formed a closed gas system. This part of the surface tension of the gas water shrinkage film interface no longer affected the soil directly. Therefore the cohesion created by gas and capillary action was reduced, and with an additional increase in moisture content, soil particles became completely surrounded by water creating a lubrication effect, which led to an increase in cohesion as the moisture content declined. One can see that the moisture content of unsaturated fine-grained tailings soil has a strong influence on the cohesion (*c*) and the internal friction angle (ϕ) is one of the important parameters related to shear strength.

3.3 Triaxial Test Results and Analysis

Figures 3 and 4 show the triaxial test results for unsaturated fine-grained tailings soil.

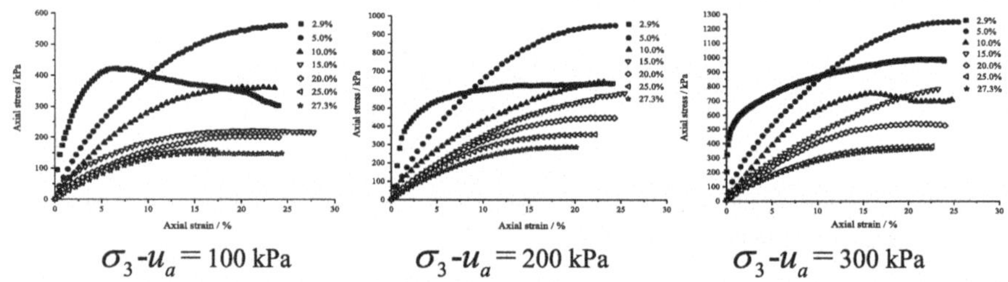

Fig. 3. Axial stress-strain curves for unsaturated fine-grained tailings soil with different moisture content under confining pressure

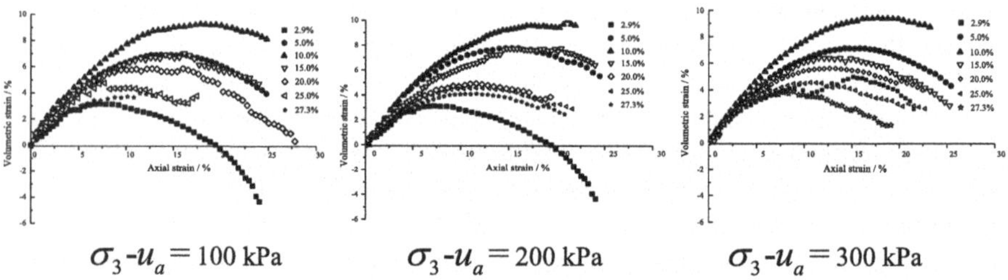

Fig. 4. Volumetric-axial strain curves for unsaturated fine-grained tailings soil with different moisture content under confining pressure

Based on the variations in specific moisture content under different confining pressure stress-strain curves one can see that moisture content had very obvious effects on the stress and strain characteristics of unsaturated fine-grained tailings soil. Under the same confining pressure, with an increase in moisture content, the axial stress which was applied on the specimen declined at the same level of strain, and its ultimate strength was also reduced. With the exception of the specimen with the lowest moisture content (2.9%), the remaining specimens exhibited a strain hardening model, and no obvious peak was observed in the stress-strain curves. Under a confining pressure of 100 kPa, the specimen with a moisture content of 2.9% had a stress-strain curve with an obvious peak, and matched the strain softening model. When the confining pressure was 200 kPa and 300 kPa, the stress-strain curve of the specimen of a moisture content of 2.9% gradually became horizontal in the absence of a significant peak. This may be caused because a failure surface had emerged in the specimens internal composition before the peak value of the stress-strain curve was reached. Thus, the moisture content also had an effect by damaging the form of the specimens.

The relationship of the volumetric and axial strain curves can be divided into two stages. In the first stage, 10% of the volumetric strain grew as the moisture content increased. In the second stage >10% of the volumetric strain decreased with an increase in moisture content. Next, the moisture content continued to increase, the volumetric strain of the specimens initially increased then decreased. This may have occurred because w = 10% was closest to the optimum moisture content, and once the optimum moisture content was exceeded (w > 15%) the volumetric strain decreased. Presumably, the volumetric strain should peak when the optimum moisture content is reached.

4 Conclusions

The moisture content is an important factor related to the strength and deformation characteristics of fine-grained tailings soil. The specific findings related to the performance of this type of soil follow:

(1) With an increase in moisture content of the specimen, resistance to vertical deformation gradually declined. Resistance to compression was weakest when the specimen was saturated.
(2) The cohesion (c) of unsaturated fine tailings soil showed a decreasing trend after initially increasing with the moisture content rising, and had an obvious peak value. The angle of internal friction (ϕ) decreased with an increase in moisture content, and had an approximate linear relationship.
(3) Moisture content had very obvious effects on the stress and strain characteristics of unsaturated fine-grained tailings soil. Under the same confining pressure, with an increase in moisture content, the axial stress which was applied on the specimen declined at the same rate, and its ultimate strength was also reduced. With a loss of moisture content, the specimen damage type changes from where strain hardens the specimen to where strain softens the specimen. With an increase in moisture content, the volumetric strain of the specimens initially increased then decreased. Presumably, the volumetric strain should peak when optimum moisture content is reached.

Acknowledgments. The National Natural Science Foundation of China (Grant No. 51479023), the Natural Science Foundation of Liaoning Province, China (Grant No. 201602349), the Education Department of Liaoning Province, China (Grant No. L2014126) and the Liaoning Technical University, China (Grant No. 13-1061) supported this research.

References

Fan, P.F., Hu, G.S.: Study on Test of the Dynamic Properties About the Fine-grained Tailings Soil. Dissertation Chang'an University, Xi-an, China (2011)

Yin, G.Z., Yang, Z.Y., Wei, Z.A., Tan, Q.W.: Physical and mechanical properties of YangLa-copper's tailings. J. Chongqing Univ. **30**(9), 117–121 (2007). (Natural Science Edition)

Wu, X.G., Wang, B., Xiang, H.H., Zhou, Y.X.: Research and application of the large strain consolidation of saturated fine-grained tailings in tailings reservoir. Metal Mine **392**(2), 53–56 (2009)

Qiao, L., Qu, C.L., Cui, M.: Effect of fines content on engineering characteristics of tailings. Rock Soil Mech. **36**(4), 923–927 (2015)

Wickland, B.E., Wilson, G.W., Wijewickreme, D.: Hydraulic conductivity and consolidation response of mixtures of mine waste rock and tailings. Can. Geotech. J. **47**(4), 472–485 (2010)

Jeeravipoolvarn, S., Scott, J.D., Chalaturnyk, R.J.: 10 m standpipe tests on oil sands tailings: long-term experimental results and prediction. Can. Geotech. J. **46**(8), 875–888 (2009)

Dimitrova, R.S., Yanfu, E.K.: Undrained strength of deposited mine tailings beds: effect of moisture content, effective stress and time of consolidation. Geotech. Geol. Eng. **29**, 935–951 (2011)

Pi, Q.H., Yang, C.H.: Test Study and Application on Fines Content and Confining Pressure Influence the Mechanical Properties of Tail Silty Sand. College of Resources and Environment Science of Chongqing University, Chongqing, China (2012)

Qin, H.Y., Liu, H.L., Gao, Y.F., Dai, P.F.: Research on strength and deformation behavior of coarse aggregates based on large-scale triaxial tests. Rock Soil Mech. **25**(10), 1575–1580 (2004)

Deng, T.H., Zhu, J.W., Cao, Y.J.: Experimental study on shear strength of tailing materials. West-China Explor. Eng. **23**(2), 10–16 (2011)

Wijewickreme, D., Sanin, M.V., Greenaway, G.R.: Cyclic shear response of fine-grained minetailings. Can. Geotech. J. **42**(5), 1408–1421 (2005)

Zhang, Y.X., He, J.G., Guo, Z.S.: Study on tailings accumulation characteristics of high tailings dam. China Molybdenum Ind. **34**(5), 8–12 (2010)

Geremew, A.M., Yanful, E.K.: Laboratory investigation of the resistance of tailings and natural sediments to cyclic loading. Geotech. Geol. Eng. **30**, 431–447 (2012)

Zhang, B.: Dynamic Properties of the Saturation Tailing and Dynamic Stability Analysis of the Tailing Dam. Dissertation Shijiazhuang Tiedao University, Shijiazhuang, China (2012)

Wang, J.P., Ling, H.I., Mohri, Y.: Stress-Strain behavior of a compacted Sand-Clay mixture. Soil Stress Strain Behav. Meas. Model. Anal. **11**(2), 491–502 (2006)

Shao, L.T., Guo, X.X., Liu, G., Liu, X.: Application of digital image processing technique to measuring specimen deformation in triaxial test. Rock Soil Mech. **36**(S1), 669–684 (2015)

The Effect of Suction on Shear Strength of Silty Clay

Wenjing Tian(✉) and Fuguo Tong

China Three Gorges University, Yichang, China
934859484@qq.com

Abstract. The shear strength is one of the main mechanical performance indexes of soil, which is affected by many factors, such as the types of soil, the structure and the water saturation. The effect that soil water saturation exerts on shear strength is caused by the changing of the matric suction. As a result, It is important to research relationship between matric suction and the shear strength in engineering field. Therefore, the following study presented in this article takes silty clay as the object. In this paper, the matric suction and shear strength of soil under different saturation conditions are measured by pressure plate extractor and the triaxial compression apparatus, and the effect of matrix suction on the shear strength of silty clay is explored. Also, the functional relationship between the matric suction and shear strength of silty clay indicators that include cohesion and internal friction angle is obtained by the method of least square. Meanwhile, the formula of Coulomb is modified based on the above studies. The result shows that the cohesion of silty clay increases with the increase of matrix suction in the form of a power function, and the internal friction angle increases with the increase of matrix suction as a logarithmic function.

Keywords: Matric suction · Shear strength · Internal friction angle Cohesion

1 Introduction

Shear strength of silty clay, as the standard of shear resistance, is an important parameter used in evaluating stability of embankment slopes such as canal, roadbed and earth dam. As we all know, degree of compaction, water saturation all have an effect on the shear strength of silty clay. Above all, water saturation plays an important role, which is a mark that discriminates the saturated soils and unsaturated soils. Accurate calculations for the shear strength formula has a valuable significance in practical engineering.

So far, the formulas of shear strength have been continued to be developed and modified (Fredlund et al. 1978; Sheng et al. 2011; and Zhou et al. 2016). There are two famous formulas being used by scholars all over the world widely (Fredlund et al. 1978; Bishop and Blight 1963). However, the two formulas all emphasize the importance of the suction to the shear strength and it is difficult to measure the matric suction in laboratory test, let alone to the field test. It is easy to measure the water saturation of soil in laboratory test, so the shear strength formula about water saturation instead of matric suction has a valuable significance in practical engineering.

L. Hu et al. (Eds.): GSIC 2018, *Proceedings of GeoShanghai 2018 International Conference: Multi-physics Processes in Soil Mechanics and Advances in Geotechnical Testing*, pp. 181–189, 2018.
https://doi.org/10.1007/978-981-13-0095-0_21

Therefore, the main objective of the research presented in this paper is to investigate the relationship between shear strength and matric suction as to modify the traditional formula of shear strength. In this paper, the soil-water characteristic curve test and triaxial shear test are designed as to get the matric suction and shear strength under different water saturation conditions. Then the effect of matrix suction on the shear strength of silty clay is explored as to obtain the functional relationship between the matric suction and shear strength of silty clay indicators that include cohesion and internal friction angle. Meanwhile, the formula of Coulomb is modified based on the above studies.

2 Shear Strength of Silty Clay

The shear strength of silty clay can be described by Coulomb, given by:

$$\tau_f = c + \sigma \tan \varphi \tag{1}$$

Where τ_f is the shear strength of saturated soil, c and φ are the shear strength parameters (cohesion and internal friction angle, respectively), σ is the total stress.

By the above equation, shear strength scales in proportion to total stress, the shear strength has two part, the first part is frictional resistance (which scales in proportion to total stress), the second part is the cohesion (which is caused by many factors, including soil cementation and electrostatic interaction).

3 Experimental Program and Results

3.1 Physical Property of Experimental Soil

The soil from the district of the Three Gorges Dam in Yichang of Hubei, China. Table 1 gives the physical properties of the soil.

Table 1. Physical properties of the soil.

Types	Density (g/cm^3)	Dry density (g/cm^3)	Water saturation (%)	PSD
Silty clay	2.07	1.68	0.274	0.003

3.2 Soil Water Characteristic Curve Test

The soil-water characteristic curve is very important for studying the physical and mechanical characteristics of unsaturated soils. Several models that relate soil suction and water saturation are found in the literature (Brooks and Corey 1964; Van Genuchten 1980; Fredlund and Xing 1994). Above all, VG numerical model is adopted in this paper because the VG model is suitable for all kinds of soil. This paper adopts pressure plate extractor as the test instrument in order to obtain the suction under different saturation. Figure 1 gives the soil sample which is put into the cutting ring.

Figure 2 gives the simple device which is used to saturate soil and Fig. 3 shows the equipment of soil-water characteristic curve test. Table 2 shows the experimental data of soil-water characteristic curve test. The relationship between matric suction and water content is notably nonlinear. Figure 4 gives the fitting curve of VG model. The result shows that the curve of VG model fit well with the experimental data. VG numerical model is given as follows:

$$P_c = P_0[S_e^{-1/m} - 1]^{1/n} \quad (2)$$

Where P_c is the matric suction; P_0 is the air entry value; m and n are the parameters, m = 1–1/n; S_e is the effective saturation, $S_e = \frac{S_r - S_{rw}}{1 - S_{rw}}$, S_r is water saturation, S_{rw} is the residual staturation.

Fig. 1. The soil sample in cutting ring

Fig. 2. The simple device for saturating soil

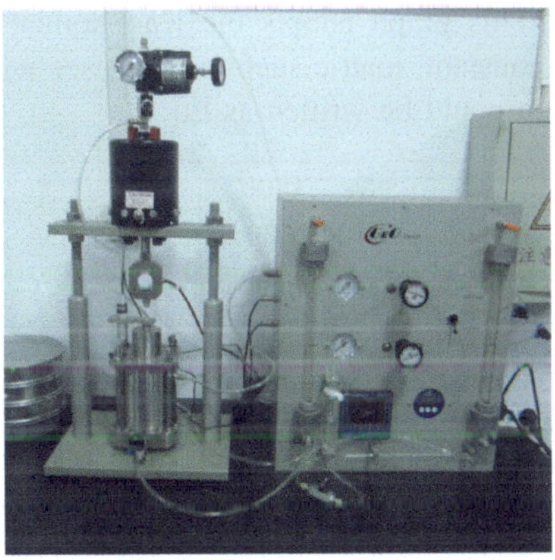

Fig. 3. Suction plate apparatus

Table 2. Relationship between water saturation and suction

Saturation	Matric suction (m)	Matric suction (kPa)
0.877	1.3	13
0.837	2.4	24
0.814	3.4	34
0.753	7.5	75
0.722	10.5	105
0.601	20.5	205
0.5272	35.6	356
0.4753	45.6	456

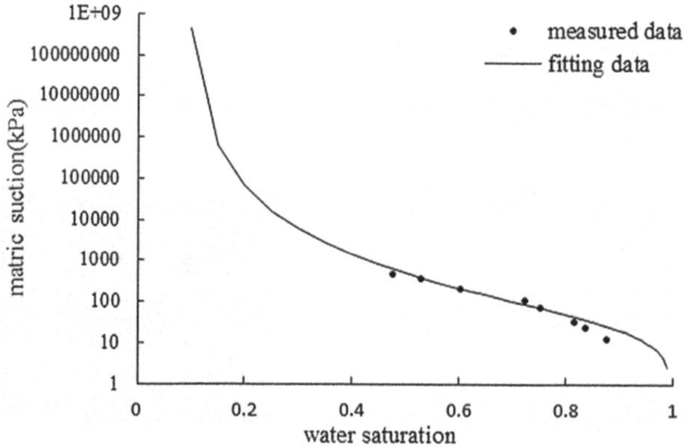

Fig. 4. A fitting curve of soil water characteristic curve test

As shown in Fig. 4, This graph adopts the least square method by MATLAB. When the dry density is constant, matric suction decreases with the increase of water saturation. Fitting equation could be written as Eq. 3.

$$P_c = -0.0262[(\frac{S_r - 0.09}{1 - 0.09})^{-1/0.216} - 1]^{1-0.216} \tag{3}$$

3.3 Triaxial Shear Strength Test

When the soil is sieved by the sieve that the diameter is 0.5 mm, the sieved soil is put into the drying box as to get the soil dried. Then the soil of different water saturation is prepared and the water saturation include 35%, 45%, 55%, 65%, 85% and 100%. Next the soil of different water saturation is put into a container which is sealed to make the soil be blend well. Afterwards, the soil is put into a saturator which is made of iron by compaction method and the saturator is lifted slowly as to separate the saturator and moulded soil. Then the moulded soil is wrapped by rubber. The saturator's shape is cylindrical and the diameter is 6.12 cm, the height is 12.5 cm. Figure 5 shows the

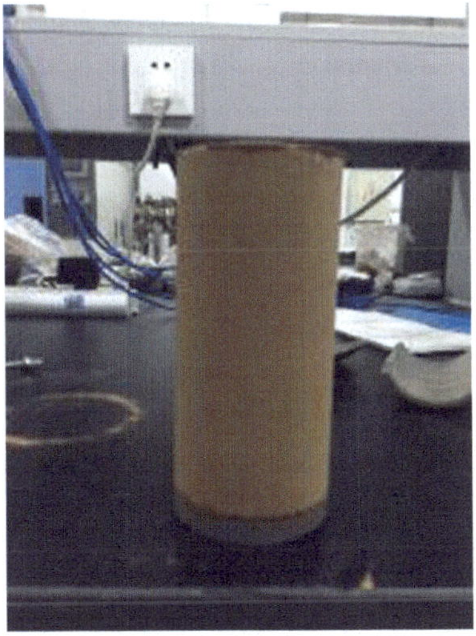

Fig. 5. Moulded soil

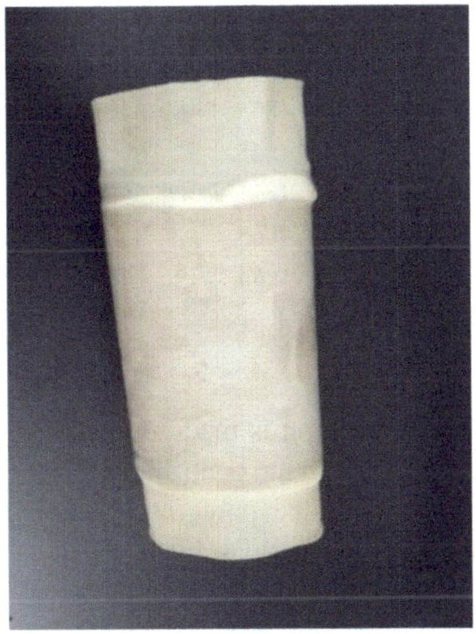

Fig. 6. Moulded soil wrapped by rubber

Fig. 7. Triaxial shear equipment

moulded soil and Fig. 6 shows the moulded soil which wrapped by rubber. Finally put the shaped soil into triaxial compression apparatus to carry a series of triaxial compression test under different confining pressure including 50 kPa, 100 kPa, 150 kPa, 200 kPa and 250 kPa. Triaxial shear equipment is shown as Fig. 7.

Based on the experimental data and analysis results, the shear strength envelope curve is drawn as to get the internal friction angle and cohesion. The relationship between cohesion and water saturation and the relationship between the internal friction angle and water content are given in Table 3.

Figure 8 shows the relationship between cohesion and water saturation and Fig. 9 shows the relationship between internal friction angle and water saturation. As shown in the Fig. 8, cohesion decreases with the increase of water saturation. Microscopically

Table 3. Relationship between saturation and cohesion, internal friction angle

Saturation	Cohesion (kPa)	Internal friction angle (°)
0.35	81.36	25.81
0.45	69.24	25.21
0.55	54.94	22.55
0.65	35.24	15.56
0.85	17.78	8.27
1	15.27	7.85

speaking, cohesion is made up of effective cohesion and capillary cohesion. Capillary cohesion refers to the contribution of the matric suction in shear strength. When the soil is saturated, all the pores are occupied by water so that the soil lose the ability of absorbing water as to lead to the decrease of the matric suction. In conclusion, only effective cohesion provides the cohesion when the soil is saturated as to lead to downtrend of cohesion.

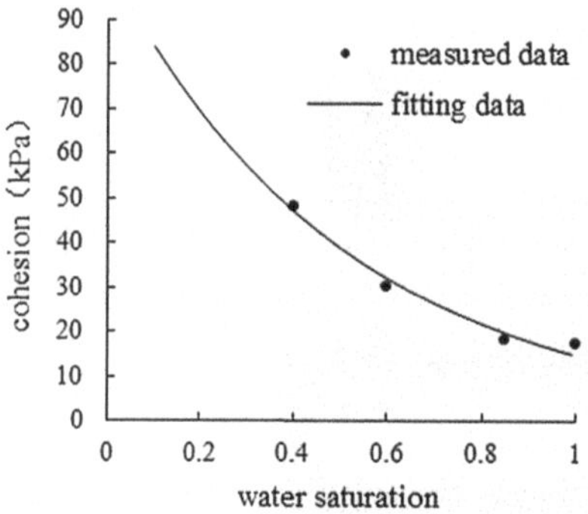

Fig. 8. The relationship between cohesion and water saturation

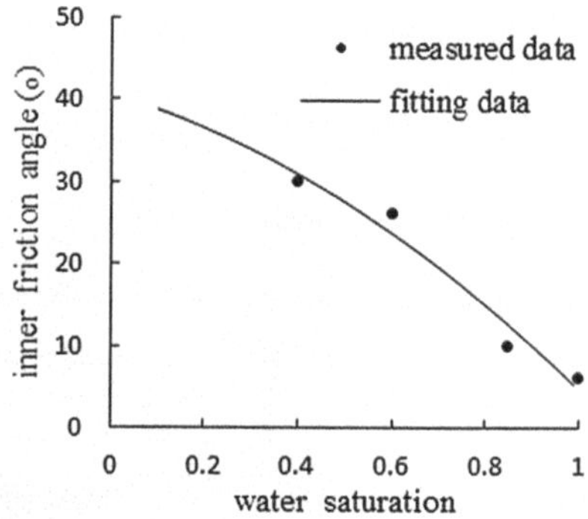

Fig. 9. The relationship between internal friction angle and water saturation

As shown in the Fig. 9, internal friction angle decreases as the water saturation increases. The objects are existed the relative movement trends under the action of friction and rough interface which hold back the moving of objects. When the saturation increase, water film is formed on the surface of soil particles as to divide the soil particles as a function of lubrication and cause the decrease of friction angle.

Adopt the least square method by MATLAB, the relationship between cohesion, internal friction angle and the saturation can be given as Eqs. 4 and 5. Coefficient correlation are 0.99 and 0.97 respectively.

$$c = 101.84e^{-1.94S_r},\ R^2 = 0.99 \tag{4}$$

$$\varphi = -19.16S_r^2 - 16.74S_r + 40.68,\ R^2 = 0.97 \tag{5}$$

Where c is the cohesion, S_r is the water saturation, φ is the internal friction angle. The cohesion grows exponentially with the increase of water saturation and internal friction angle has a cubic function to the water saturation. The Eqs. 4 and 5 could describe the experimental data better.

4 Result and Discussion

4.1 Shear Strength and Matric Suction

Based on the above research, the relationship between shear strength and matric suction is obtained by the soil-water characteristic curve test and triaxial shear test. Figure 10 shows the relationship between cohesion and matric suction and Fig. 11 shows the relationship between internal friction angle and matric suction.

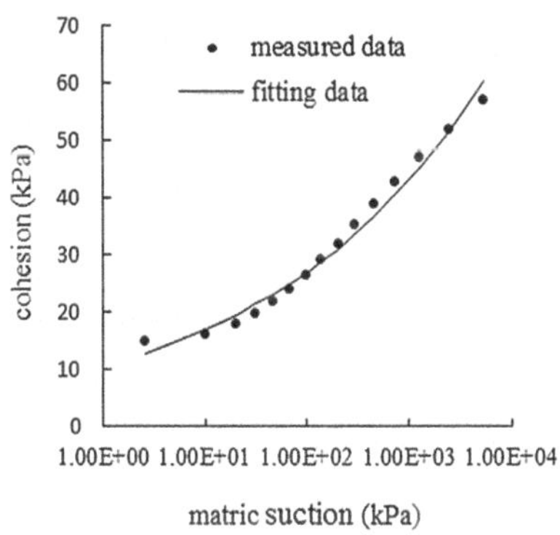

Fig. 10. The relationship between cohesion and matric suction

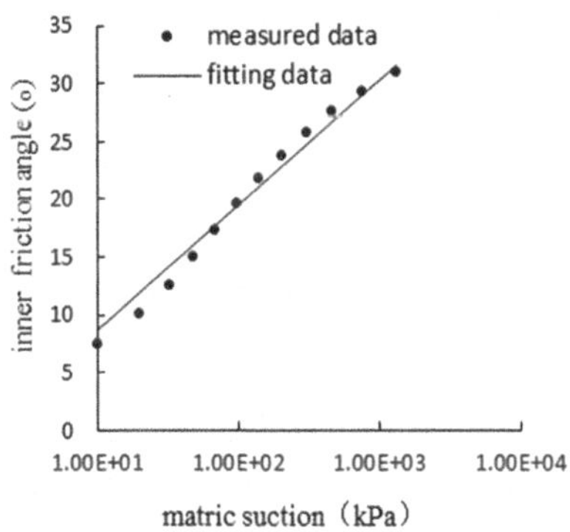

Fig. 11. The relationship between internal friction angle and matric suction

As shown in Figs. 10 and 11, fitting curve has a high goodness fit with experimental data. The functional relationship is obtained by the least square method. The result shows that the cohesion of silty clay increases with the increase of matrix suction in the form of a power function, and the internal friction angle increases with the increase of matrix suction as a logarithmic function. The equations are shown as follows:

$$c = 10.67P_c^{0.20}, R^2 = 0.99 \quad (6)$$

$$\varphi = 4.69 \ln P_c - 2.11, R^2 = 0.97 \quad (7)$$

4.2 Modified Shear Strength Formula

It is difficult to ensure the shear strength of the soil in practical engineering. And it is difficult to measure the matric suction. If the shear strength is established by water saturation instead of matric suction, the shear strength considered water saturation could be applied in practical engineering. The modified shear strength formula which considered water saturation is shown as follows:

$$\tau_f = 101.84e^{-1.94S_r} + \sigma \tan(19.16S_r^2 - 16.74S_r + 40.68) \tag{8}$$

Where τ_f is the shear strength, S_r is the water saturation, σ is the total stress.

In order to analyze the relationship between shear strength and matric suction, the formula of shear strength could be expressed as Eq. 9:

$$\tau_f = 10.67P_c^{0.2} + \sigma \tan(4.69 \ln P_c - 2.11) \tag{9}$$

Where τ_f is the shear strength, P_c is the matric suction, σ is the total stress.

5 Conclusion

In order to explain and quality the effect of slope stability on natural rainfall infiltration, an shear strength model from silty clay which considered water saturation needs to be developed. In this study, Laboratory tests on soil water characteristic curve and triaxial strength are performed as to get the shear strength and matric suction under different water saturation. The least square method is adopted in order to get the fitting curve between shear strength and water saturation and the fitting curve between matric suction and water saturation. So that the formula of shear strength is modified. Main conclusions are shown as follows:

(1) The soil water characteristic curve test shows that, when the dry density is constant, matric suction decreases with the increase of water saturation. The experimental data fit well with the VG model.
(2) Triaxial shear test shows that, when the dry density is constant, the cohesion grows exponentially with increase of water saturation and internal friction angle has a cubic function to the water saturation. The fitting equation could describe the relationship between shear strength and water content better.
(3) The relationship between parameters of shear strength and water saturation is further analyzed. Cohesion increases at a power function with the increase of matric suction, internal friction angle increases at a logarithmic function with the increase of matric suction.

References

Fredlund, D.G., Morgenstern, N.R., Widger, A.: Shear strength of unsaturated soils. Can. Geotech. J. **2**(15), 313–321 (1978)

Sheng, D., Zhou, A.N., Fredlund, D.G.: Shear strength criteria for unsaturated soils. Geotech. Geol. Eng. **29**(2), 145–159 (2011)

Zhou, A., Huang, R.Q., Sheng, D.: Capillary water retention curve and shear strength of unsaturated soils. Can. Geotech. J. **4**(53), 974–987 (2016)

Bishop, A.W., Blight, G.E.: Some aspects of effective stress in saturated and partly saturated soils. Geotechnique **13**(3), 177–197 (1963)

Brooks, R.J., Corey, A.T.: Hydraulic properties of porous media. Hydrol. Pap. **5**(3), 151–161 (1964)

Van Genuchten, M.T.: A closed-form equation for predicting the hydraulic conductivity of unsaturated soils. Soil Soc. Am. J. **44**(5), 892–898 (1980)

Fredlund, D.G., Xing, A.: Equations for the soil-water characteristic curve. Can. Geotech. J. **31**(4), 521–532 (1994)

Seepage and Porous Mechanics

Micro-Mechanism of Pressure Variation of Grout in Porous Media Considering Filtration

Zilong Zhou[1], Xueming Du[1,2(✉)], Shanyong Wang[2], and Xin Cai[1]

[1] School of Resources and Safety Engineering, Central South University, Changsha 410083, China
2007-dxm@163.com
[2] Faculty of Engineering and Built Environment, ARC Centre of Excellence for Geotechnical Science and Engineering, The University of Newcastle, Callaghan, Australia

Abstract. The pressure variation of cement grout is an important issue when it is used to eliminate the standard for grouting at the ending of grouting. To better understand the effects of grouting parameters on the pressure variations considering filtration, a series of laboratory tests were carried out under the condition of different grouting flowrates, water/cement ratios and mass fractal dimensions. After a grouting flowrate was performed on sand samples, the pressures in each measurement point were monitored by transducers. The experimental results show that: (1) the grouting pressure presents a nonlinear decreasing law with the increase of the distance from the grouting hole, whereas the grouting pressure at the same measuring point gradually increases with time or increases firstly and then decreases; (2) too large grouting flow or too small grouting flow is unfavorable to the diffusion of grout; (3) the grouting pressure decreases as the mass fractal dimension increases.

Keywords: Grouting pressure · Micro-mechanism · Filtration
Fractal dimension

1 Introduction

As known, the grouting is an effective method for improving the soils stabilization without disturbing the original soil structure. The cement-based grout has been widely used in civil engineering applications [1]. However, due to the concealment nature of the grouting engineering, the effect of grouting treatment cannot be directly visualized, and the standard for grouting at ending cannot be directly obtained. In engineering practice, the variation rule of diffusion pressure can be used as the standard to judge the distribution of slurry and whether or not the end of the grouting. For this reason, some researchers have carried out some research on diffusion pressure. For example, Yang et al. [2] deduced a formula for calculating the diffusion radius of power-law and Bingham based on generalized Darcy's law and the theory for spherical diffusion model, and analyzed the influence of slurry properties on diffusion pressure and

L. Hu et al. (Eds.): GSIC 2018, *Proceedings of GeoShanghai 2018 International Conference: Multi-physics Processes in Soil Mechanics and Advances in Geotechnical Testing*, pp. 193–200, 2018.
https://doi.org/10.1007/978-981-13-0095-0_22

diffusion radius. Zou et al. [3] and Kelessidis et al. [4] studied on the distribution of pressure attenuation along the diffusion radius of slurry.

Previous studies have mainly focused on the change in the characteristics of the slurry, without considering the change of pore structure caused by the change characteristics of the slurry. As a matter of fact, during penetration into the porous media, grouting could be stopped due to building a stable arch caused by cement grains. This process, defined as the filtration of grout, has been studied by Widmann [5]. The filtration process results in a reduction of the ability of the grout to flow the porous media and then makes the rule of the diffusion pressure become unambiguous. Therefore, it is of great significance to explore the micro-mechanism for the pressure variation of grout in porous media based on filtration.

This paper presents a laboratory study devoted to investigate the effect of grouting parameters including different grouting flowrates, water/cement ratios, as well as fractal dimensions on the grouting pressure during the grouting process. In order to accomplish the purpose, a serial of column transport experiments were conducted on fractal characteristics' sand with different parameters during the injection process. By using a series of transduces, the variation of pressure at the different point/time was investigated under constant flow condition.

2 Materials and Methods

2.1 Materials

2.1.1 Particle-Size Distributions by Fractal Dimension

Fractal theory is well suited to describe particle size distribution on soil structure, and it has been widely used in many fields. A lot of studies showed that the mass fractal dimension of soil particles is one of the inherent properties of soil, as well as the particle size distribution of the soils could be described with a mass-based relationship by correlating the measured mass with the sieve diameter. The fractal theory can be expressed by [6]:

$$\frac{M(\delta < \overline{d_i})}{M_T} = \left(\frac{\overline{d_i}}{\overline{d}_{\max}}\right)^{3-D_m} \tag{1}$$

where $\overline{d_i}$ is the average diameter of the i particle size, $\overline{d}_{\max}$ is the diameter of the maximum particle of the sample, $M(\delta < \overline{d_i})$ is the cumulative mass of the soil particles with size δ smaller than $\overline{d_i}$, M_T is the total mass of the sample of sand, D_m is the fractal dimension of particle size distribution of soils, $3 - D_m$ can be obtained by the slope of the fitted line on a log $(M(\delta < \overline{d_i})/M_T)$ against log $(\overline{d_i}/\overline{d}_{\max})$ graph.

2.1.2 Materials Preparation and Properties

In this study, sand with particle sizes ranging from 0.05 mm to 1.2 mm was collected from the south of Changsha, Hunan province of China. Considering the demands in the test, the sand was firstly put through a series of grade sieves to get sandpiles with specific particle sizes, i.e., 0.05 mm, 0.15 mm, 0.2 mm, 0.4 mm, 0.6 mm and 1.2 mm, respectively. After that, five sand samples with mass fractal dimensions (D_m) of 1.5, 1.7, 1.9, 2.1 and 2.3 were further prepare. The percent contents by weight of particles for the experiment were determined according to Eq. (1) and listed in Table 1 [7].

Table 1. Particle size proportion in sand samples

Sample No.	Particle size proportion %						Density	D_m
	0.05 mm	0.15 mm	0.2 mm	0.4 mm	0.6 mm	1.2 mm	P g/cm^3	
S-1	5.4	15.1	20.6	25.4	21.6	11.9	18.19	2.3
S-2	3.5	8.0	16.6	28.4	18.6	24.9	18.16	2.1
S-3	2.8	7.8	14.3	16.4	23.6	35.1	17.50	1.9
S-4	1.9	3.6	9.6	15.4	29	40.5	17.40	1.7
S-5	1.2	1.3	8.6	14.4	28	46.5	17.10	1.5
S-6	0	0	0	0	0	100	18.09	*

2.2 Experimental Procedures

2.2.1 Apparatus

The diffusion of cement-based tests were carried out by a homemade diffusion test device, as shown in Fig. 1. The device consists of a grout container and a diffusion system: (1) The grout container is composed of a pneumatic agitator, a pressure regulator, a safety valve, flow regulator, flowmeter, lock rings, a steel barrel and some valves. The volume of the steel barrel is 80 L and the maximum pressure in the steel drum is less than 1 MPa. The pressure in the barrel can be controlled by adjusting the flow regulator. Besides, in order to improve the experimental precision in the grouting process, the slurry is stirred by the pneumatic agitator to keep the uniformity of

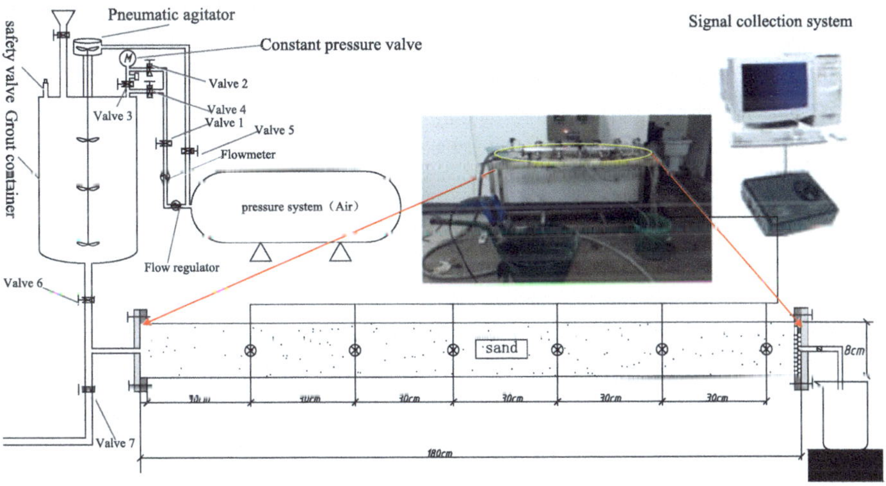

Fig. 1. Diagram of the experimental set up

the slurry. (2) The diffusion system consists of a stainless steel pipe with a length of 1800 mm and an internal diameter of 80 mm and six transduces installed on the steel pipe in the form of every 30 cm from the inlet to the outlet of stainless steel.

2.2.2 Test Procedures

The test procedures included several steps: (1) the cement slurry was produced according to the design of water/cement ratio, and was filled in the grouting container from the inlet; (2) the column was packed in 5 cm increments by pouring the sand into the column, mixed with deionized water (saturated condition) to avoid trapping air bubbles; (3) Pressing the switch of the (pressure system), opening the valve 1, valve 4, valve 5 and valve 6, and then adjusting the flow regulator until the reading of flow meter reached the design value. After that, the cement slurry was injected into the porous media at a constant flow; (4) the pressure variation were measured by the pressure transduces, and then the effects of grouting parameters, e.g., grouting flowrate, water/cement ratios and mass fractal dimension, on the pressure variations considering filtration were obtained.

3 Description of Filtration Mechanism

As the grouting proceeds, cement particles are captured or accumulated in the porous media, leading to the progressive formation of a plug. Subsequently, the accumulation continues and the plug grows, and then a complete plug has formed. The consequence is a possible blockage of the pores of porous media and a reduction in the porosity of the media. What is more, the plug starts near the injection point and stops the flow [8]. Above behavior has a great influence on the change of diffusion pressure due to the variation of pore structure in porous media.

Figure 2 shows the variation of the pressures P_0, P_1, P_2, P_3, P_4, P_5, and P_6 during grouting in the cases without filtration and with filtration when the cement particles were filtered under the constant flow. In the case without filtration, all pressures increase with as time. In the case with filtration, only the pressure P_0 increases as time, while the other pressures P_1, P_2, P_3, P_4, P_5, and P_6 firstly increase and subsequently

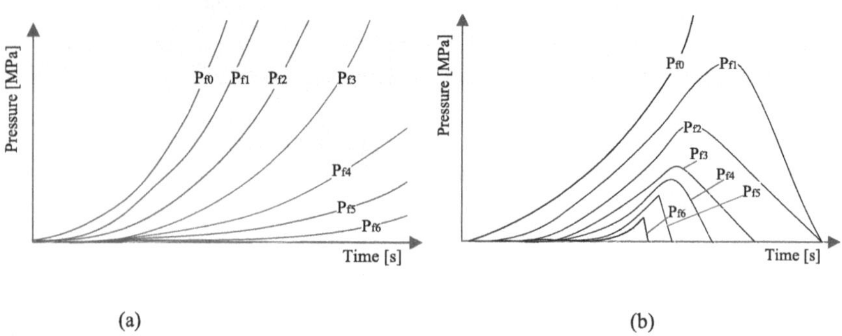

Fig. 2. The characteristic of pressure during grouting in cases without filtration (a) and with filtration (b)

decrease with time. At the beginning of the grout, the porous media has a relatively large penetration. The pressure at each measured point gradually increases due to the diffusion of slurry. With the continuous injection of grout, the cement particles are filtered and agglomerated in the sand pores, leading to the progressive formation of an impermeable cake. Subsequent filtration results in the increasing deposition of cement particles on the surface of the cake. Consequently, a possible block of the access pores and a reduced penetration of the grout results in that the pressures of the test points in front of the plug continue to increase under constant flow condition and the pressures of the test points behind the plug decrease sharply until it becomes zero.

4 Results and Discussion

The variation curves of diffusion pressure with the distance are plotted in Fig. 3. It can be seen that, on the whole, the diffusion pressure decreases with the increase of distance. Specifically, in the initial stage of grouting, the variation of diffusion pressure with the distance presents approximately convex decreasing law. After about 30 s, it decreases concavely with the increase of distance. Especially, in the later period of grouting (50 s), the diffusion pressure is sharply decreased in the 0.3–0.9 m, and it drops to zero at about 1.2 m.

Generally, at the grouting begin, because the distance from diffusion pressure source is near and the resistance of the slurry in the diffusion process is smaller due to the porosity of the porous media is larger, which causes the diffusion pressure changes slowly with the diffusion distance between 0.2 m and 0.9 m. As far away from the grouting source, due to the pressure front in the slurry diffusion is smaller than that in the initial stage and the viscous force between the soil and rock particles, the diffusion pressure decreases rapidly with the increase of diffusion distance.

Figure 4 shows the variation curves of diffusion pressure at each measuring point with the grouting time. It can be seen that the diffusion pressure at 0/3 m (P0/P1)

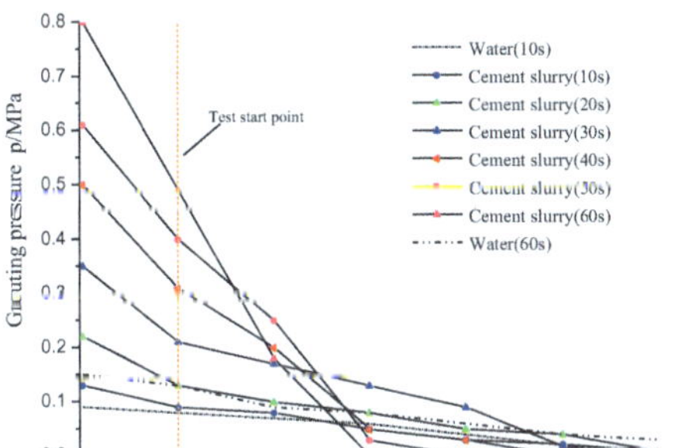

Fig. 3. The variation curves of diffusion pressure with grouting distance (he average particles diameter is 1.2 mm, Q = 1.4 L/min, w:c = 1.5:1)

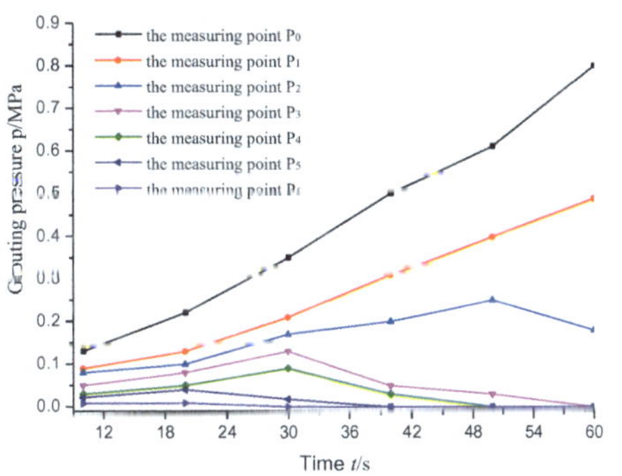

Fig. 4. The variation curves of diffusion pressure at each measuring point with the grouting time (the average particles diameter is 1.2 mm, Q = 1.4 L/min, and w:c = 1.5:1)

gradually increases, whereas the diffusion pressure at 0.6–1.8 m (P_2, P_3, P_4, P_5, and P_6) increase firstly and then decreasing with time until it drops to zero. The reason is that, when the grouting begins, plug-building can start due to the cement particles stick the skeleton of porous media, leading to a decrease of porosity of porous media. According to the literature [9], in the case of constant flow, with the decrease of porosity, the starting pressure gradient increases slowly firstly; below a certain value, the starting pressure gradient increases rapidly, so the diffusion pressure increases; with the continuous injection of slurry, a complete plug has formed, the pressures of the test points in front of the plug continue to increase under constant flow and the pressures of the test points behind the plug decreases sharply until it becomes zero.

The relationship of the diffusion pressure versus distance is shown in Fig. 5. The results show that, the diffusion pressure of slurry in the media with higher fractal dimension sharply decreases with the increase of distance, and drops to zero at 0.6 m. However, that in the media with lower fractal dimension decreases slowly with the increase of distance; until the end of grouting, the diffusion pressure at 1.8 m is still 0.15 MPa.

According to formula (1) and the literature [10], if the fractal observing scale is unvaried, the contents of fine particle increases and the proportion of fine pore increases with the increase of mass fractal dimension, leading to a reduction in the porosity. Therefore, the particles of cement-based are trapped by the relatively dense skeleton of the higher mass fractal dimension easily, which makes a block of the access pores and a reduced penetration of the grout. Once a complete plug has formed, the pressures of the test points in front of the plug continue to increase under constant flow condition and that behind the plug decrease sharply until it becomes zero.

Figure 6 presents the variation of diffusion pressure with grouting distance under the different grouting flowrates. It can be seen that, when the grouting flow rate is 1.0 L/min, the diffusion pressure reaches zero at 0.9 m, i.e., the diffuse stoppage of the grout occurs; the diffusion pressure becomes zero at 1.2 m when the grouting flow rate

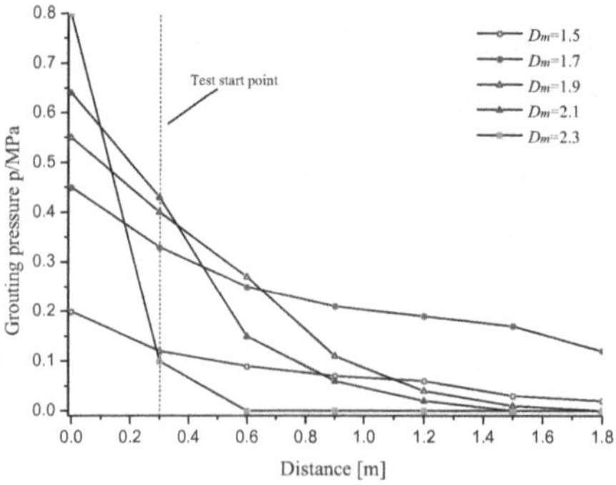

Fig. 5. The variation curves of diffusion pressure with grouting distance for the considered porous media (w:c = 1.5:1, Q = 1.4 L/min, t = 10 s)

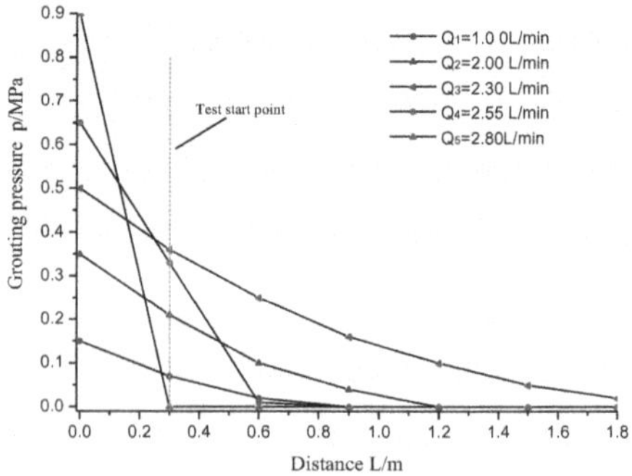

Fig. 6. The variation curves of diffusion pressure with grouting distance for the different grouting flowrate (w:c = 1.5:1, and t = 10 s)

is 1.2 L/min; when the grouting flow continues to increase, the diffusion pressure is still above zero until the test points at 1.8 m; however, when the grouting flow increases further (Q = 2.8 L/min), the diffusion pressure drops rapidly and it goes to zero at 0.3 m. The results indicate that too large grouting flow or too small grouting flow is unfavorable to the diffusion of grout. This is can be interpreted that at low grouting flow rate, the cement-based particles are prone to precipitation and be captured by skeleton easily owing to the low flow velocity, which easily leads to the decrease of porosity and velocity in the media and then makes the cement-based particles can be captured by skeleton. As a comparison, for the case of the high grouting flow rate, due to the increase in the amount of particles within the same cross-section in unit time, cement particles are filtered and agglomerate in the sand pores, leading to the progressive formation of an impermeable cake. The consequence is a block of the access pores and a reduced penetration of the grout, which causes the diffusion pressure drops rapidly and it goes to zero at a short distance from the grouting role.

5 Conclusions

To study the effects of grouting parameters on the pressure variations considering filtration, a series of laboratory tests were carried out to investigate the response of the fractal characteristics of sand in column. Based on the experimental results, the specific findings are highlighted as follows.

The filtration has a great influence on the diffusion pressure. Under constant flow condition, in the case without filtration, all pressures always increase with time. In the case with filtration, the pressure P_0 increases with time, while the pressures P_1, P_2, P_3, P_4, P_5, and P_6 increase firstly and then decrease.

The mass fractal dimension of porous media also controls the diffusion pressure during the injecting process. The diffusion pressure of slurry decreases sharply with the increase of distance from the grouting hole under the higher fractal dimension. However, the diffusion pressure of slurry decreases slowly with the increase of distance from the grouting hole under lower fractal dimension.

References

1. Yoon, J., Mohtar, C.S.E.: A filtration model for evaluating maximum penetration distance of bentonite grout through granular soils. Comput. Geotech. **65**, 291–301 (2015)
2. Yang, X.Z., Lei, J.S., Xia, L.N., Wang, X.H.: Study on grouting diffusion radius of exponential fluids. Rock Soil Mech. **26**(11), 1803–1806 (2005)
3. Zou, J.F., Liang, L.I., Yang, X.L.: Penetration radius and pressure attenuation law in fracturing grouting. J. Hydraul. Eng. **37**(3), 314–319 (2006)
4. Kelessidis, V.C., Dalamarinis, P., Maglione, R.: Experimental study and predictions of pressure losses of fluids modeled as Herschel–Bulkley in concentric and eccentric annuli in laminar, transitional and turbulent flows. J. Petrol. Sci. Eng. **77**(3–4), 305–312 (2011)
5. Widmann, R.: International society for rock mechanics commission on rock grouting. Int. J. Rock Mech. Min. Sci. Geomech. Abs. **33**(8), 803–847 (1996)

6. Fazeli, M.: A fractal approach for estimating soil water retention curve. Environ. Manag. **1**, 177–183 (2010)
7. Zhou, Z.L., Du, X.M., Wang, S.Y., et al.: Cement grout transport within sand with fractal characteristics considering filtration. Eur. J. Environ. Civil Eng. (2017). https://doi.org/10.1080/19648189.2017.1358217
8. Draganović, A., Stille, H.: Filtration of cement-based grouts measured using a long slot. Tunn. Undergr. Space Technol. Incorporating Trenchless Technol. Res. **43**(7), 101–112 (2014)
9. Wang, Q.L., Tang, H., Lv, D.L., et al.: An experimental study on threshold pressure gradient in low permeability reservoir. Petrol. Geol. Recovery Effi. **18**(1), 97–100 (2011)
10. Zhang, J.R., Huang, L., Zhu, J., Huang, W.Z.: Sem analysis of soil pore and its fractal dimension on micro scale. Acta Pedol. Sin. **45**, 207–215 (2008)

Numerical Evaluation on the Filtration and Clogging Behavior of Porous Pavement

Guoyang Lu[1(✉)], Guoxiang Zhou[2], Dawei Wang[1,3], Jing Zhong[2], and Markus Oeser[1]

[1] Institute of Highway Engineering, RWTH Aachen University, Mies-van-der-Rohe-Street 1, 52074 Aachen, Germany
lu@isac.rwth-aachen.de

[2] School of Civil Engineering, Harbin Institute of Technology, Harbin 150090, People's Republic of China

[3] School of Transportation Science and Engineering, Harbin Institute of Technology, Harbin 150090, People's Republic of China

Abstract. To avoid the floods in urban pavement system, the use of permeable pavement is increasingly considered as an effective way to optimize the existing sealed pavement. Porous Asphalt (PA) has constituted the permeable pavement infrastructure for years, which can realize the fast infiltration and precipitation of surface water through layers. But the poor durability is the main obstacle inhibits the application of conventional PA. The substitution of the innovative Polyurethane binder can perform excellent mechanical property as well as the permeability. Apart from it, the filtration property is found effectively to reduce the quantity of particulate matter (PM) flooded by the rainfall. In this study, the filtration and clogging behavior of PU-bonded pavement under different concentration gradient of PM are quantified. Both filtration property and hydraulic conductivity are predicted by numerical modelling and adopted as comparative indexes. Results indicate that PU-Bonded permeable pavement has significant hydraulic conductivity and the effective absorptive-filtration performance for the contaminant of flow. The numerical modelling can well conduct the prediction of filtration and clogging performance of permeable pavement.

Keywords: Filtration · Porous Asphalt · Permeable pavement
Hydraulic conductivity · Porosity · Polyurethane

1 Introduction

1.1 Permeable Pavement

Conventional pavements are designed to be sealed to avoid moisture penetrating into the pavement structure and causing damage to it. However, with the rapid expansion of sealed area constructed by urban development and industrial activity, the natural retention capacity experiences a significant decrease. To recover the natural hydrological cycle and relief the urban flood risk, Permeable Asphalt Pavement (PAP) as a Low Impact Development (LID) facility is implemented by directly allowing the rainwater seeping through the pavement surface. By hydraulic restoration and

L. Hu et al. (Eds.): GSIC 2018, *Proceedings of GeoShanghai 2018 International Conference: Multi-physics Processes in Soil Mechanics and Advances in Geotechnical Testing*, pp. 201–209, 2018.
https://doi.org/10.1007/978-981-13-0095-0_23

constituent load reduction, the in-situ management of storm water can be achieved [1]. In case of rain event, water can quickly infiltrate through the pavement structure into subsoil, which consequently reduces the pressure of urban drainage and recharges the water supply of natural cycle. Furthermore, PAP was also proved with high capacity of tire noises absorption and advantages for driving safety [2].

1.2 Pavement Runoff and Filtration Property

Pavement runoff has long been recognized as a main source of pollution by washing out large amount of pollutants from urban areas and carrying them with water flow [3]. According to the report, urban runoff attributes to 46% surface water pollution, and the infiltration of pollutants may cause the contamination in the subsoil water [4]. Therefore, the assessment and control of quantity and quality of urban runoff permeating into the subgrade is vital to the ecological environment of road sections. Based on the aim of achieving high hydraulic conductivity, the PAP is designed with different level of porosity which constitutes the pavement structure. Such pore characteristic enable PAP equipping the properties of gravitational infiltration, evaporation, absorption as well as the filtration of urban runoff [5].

From various field investigation and laboratory assessment of the penetrated water quality, PAP presents a constant absorption capability of PM, heavy metals as well as the nutrients under different boundary conditions [6]. With comparing to the conventional pavements, a reduction of 75% heavy metal and suspended soils can be achieved by the PAP system, lower lead and copper concentration was also found in the porous structure of this kind of pavement [7].

Previous researches conducted to evaluate the hydraulic and filtration performance of PAP were mainly concentrated on the analysis and description from the field tests as well as the laboratory observation [8]. These researches are mostly based on macroscopic phenomenological level. The microscopic performance such as the inner water permeation, storage and release, pollutant distribution and migration are not clear, which cannot provide theoretical support for the understanding of the hydraulic and filtration mechanisms. As a result, a systematic investigation of the PAP must be carried out with multi-scales together with fully modelling processes. In which case, the long-term performance of the PAP can be accurately simulated for determine the varied design purpose and mitigation level [7].

To realize the fast infiltration and precipitation of surface water through layers, a void-rich pavement structure is currently the most feasible and effective way, such as Porous Asphalt (PA), Porous Concrete (PC) and Permeable Interlocking Concrete Pavement (PICP) which mixed by a low content of fine aggregate to reach high void content. Hence, PAP is often considered as a porous structure media when conducting the simulation on the hydraulic conductivity as well as the quantitation of the relationship to the porosity [3].

As regarding to the prediction of filtration and clogging behavior in PAP system, Albrecht theory using unviscous fluid potential flow equation and Ideal fluid equation [9]. Kaufmann (1936) coupled Brownian movement and inertia sedimentation on fiber filtration theory and elicited mathematical expression of filtration efficiency [10]. A new theory describes filtering quality by coupling three mechanisms: diffusion, intercept and inertia was established [11]. Tien [12] proposed simulation of aerosol filtration considering particle size and distribution mode. Recently, the numerical method combining discrete element method based on molecular dynamics theory and simulated the deposition track of particles on fiber was developed [13]. And the influence of media packing density on filtration efficiency and pressure loss was analyzed. However, there are less literature was found based on the filtration and clogging modelling in the PAP system. The construction and attempt of prediction of the filtration and clogging behavior concerning to the urban runoff, is vital important to fully understand the hydraulic property of PAP and optimize the design of it.

1.3 Polyurethane Bonded Permeable Asphalt Pavement (PPAP)

To achieve high permeability, a void-rich pavement structure is currently the most feasible and effective way, such as Porous Asphalt (PA) and Porous Concrete (PC) pavements mixed by a low content of fine aggregate to reach 10–25% void content [8]. Apart from the high hydraulic conductivity, the porous structure can also make contribution to the noise adsorption of the transportation system. However, the adhesion failure and short life-span of porous mixture has been an obstacle restricting the application of permeable pavement. In this case, innovative bound permeable material is being developed to gain high mechanical property [14]. Polyurethane (PU) bounded pavement initial researches shows that the combination of high functionality and mechanical property can be both achieved based on the excellent physical performance of PU. Hence, based on the manufacture and measurement of Polyurethane-bonded Permeable Asphalt Pavement (PPAP), the present research aims to investigate a modified model for the hydraulic conductivity as well as the filtration and clogging mechanism. Based on the modified theoretical framework, the simulation will be solved by finite element method (FEM). With revealing the deterioration mechanism of the hydraulic and filtration properties, the basic law of optimizing PAP design, construction and maintenance will be established.

2 Experimental Program

2.1 Material and Specimen Manufacture

Preparing polyurethane bounded asphalt specimens is based on a precise preparing process as the steps build up on each other. Generally, the mixing procedure is equal to the process of mixing hot mix asphalt, regardless of high temperatures. This is because of the viscosity of polyurethanes which is not affected by temperatures.

In the present study, diabase aggregates and limestone filler were selected to form the mixtures, the designed grain size distribution is presented as Fig. 1.

Regarding to the specimens of permeability test and the optimized binder ratio obtained, the proposed proportions of aggregates (94%), the binder content (6%) and the air voids (26%) are calculated concerning the strived bulk density. Based on the determination of a special type of grading curve (Fig. 1), all components are weighed out into its components to fit into a plate mold. To obtain the specimens in a same characteristic of mixture, a test plate was formed, where the test specimens were cut from.

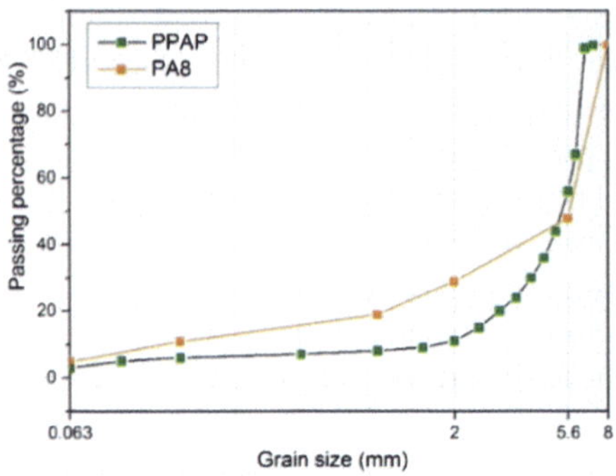

Fig. 1. Grain size distribution of the PPAP specimen

2.2 Void Ratio

The void system in PAP ensures the infiltration, absorption and all other proposed properties of porous structures. Additionally, the mechanical characteristics such as strength, stiffness and fatigue performance are also influenced by the total void content, which is decided by the compaction process and gradation of aggregates.

To measure the total void content, conventional way is by measuring the solid density ρ_s and bulk density ρ_b (2.1). Where the solid density can be calculated by proportion and density of dried aggregates and binder respectively. And the bulk density can be determined by dried mass of specimen M_s to the volume V_s.

$$e_t = \frac{\rho_s}{\rho_b} - 1 \tag{2.1}$$

However, for the void system is inner-connected and could partially access from the outside, the conventional way of void content measuring is hard to accurately describe void characteristic of porous material. Hence, in the present study, the hydraulically effective void ratio e_{eff} is applied to conduct the prediction of permeability as well as the filtration and clogging behavior, which can be calculated as follow Eq. (2.2).

$$e_{eff} = \frac{V - V_w}{V} \tag{2.2}$$

Where, V was measured from total specimen's volume, and V_w was measured from displaced water of the immersed specimen after being placed in a steam room for 24 h.

2.3 Permeability Test

Permeability is mainly denoted as the ability of water conductivity in material or other pore structure, which property is highly depended on the hydraulic characteristics of the material (such as void content and the pore structure) and the viscosity of moisture. If a high permeability was assumed, the material can quickly drain the surface moisture into the porous media below. The drainage performance of a material can be evaluated by coefficient of permeability K_f which denotes the speed of given quantity of water flowing through a porous structure area at a given hydraulic gradient.

According to the fact that Darcy's law has been widely adopted to describe the moisture flow distribution in asphalt mixtures as proposed by [15], the constant falling head test was adopted to investigate the permeability performance of PU-bonded PAP specimens. The coefficient of permeability, $K_{f,v}$, in vertical direction can be determined by quantifying the geometry of specimens and the water flow Q_v (m^3/s) through the specimen in a defined time at the constant effective hydrostatic head h of 300 mm (2.3).

$$K_{f,v} = \frac{4 \times Q_v \times l}{h \cdot \pi D^2} \tag{2.3}$$

With regarding to the FGSV [16], the coefficient of permeability can be classified into 5 levels from 'very slightly permeable' to 'very highly permeable'. Pavement can be determined as permeable if the coefficient of permeability $K_f \geq 5.4 \times 10^{-5} m/s$ is achieved according to German permeable pavement standards.

3 Mathematical Model Development

3.1 Permeability Model

Pore structure characteristics are usually adopted for estimating the hydraulic conductivity k in porous media. Most existing relationship of these two parameters is established based on the pore factors in terms of grain size distribution (GSD) or the measured indices [17]. Due to the hardly measurements of different pore indices regarding to the pore structure, pore shape factor and tortuosity, this kind of modeling are mostly constructed on varied assumptions, while the model integrated from GSD presents more empirical and convenient [15]. Among which, two famous approaches for predicting hydraulic conductivity k based on the GSD are Hazan equation and Kozeny-Carman model (KCM). Contrary to the Hazan model based on the effective aggregate diameter D_{10}, the KCM is the integration of the entire GSD as well as the void ratio and shape factor to achieve the k prediction. In which case, the accuracy of KCM (3.1) turned out highly surpasses that of Hazan equation. As a result, The KCM is widely adopted for modeling the hydraulic conductivity in porous material and pavements.

$$k^*_{wr11} = \frac{\gamma_w}{\mu_w \cdot C_{k-c}} \cdot \left(\frac{100\%}{\sum \left(f_i / \left(D_{lk}^{0.5} \cdot D_{sk}^{0.5} \right) \right)} \right)^2 \frac{1}{S_F^2} \frac{e_{eff}^3}{1 + e_{eff}} \tag{3.1}$$

Where, μ_W denotes the viscosity of water, which is approximately 1×10^{-6} KPa s at 20 °C; γ_W represents unit weight of water, which is 9810 N/m^3; C_{K-C} is the Kozeny–Carman empirical coefficient for which a value of 5 is suggested; S_F represents the shape factors of aggregate: (a) spherical grains: 6.0; (b) rounded grains: 6.1; (c) worn grains: 6.4; (d) sharp grains: 7.4–7.5 and (e) angular grains: 7.5–8.4.

3.2 Filtration and Clogging Model

Laminar Flow is used to compute the velocity and pressure fields for the flow of a single-phase fluid in porous media. Particle Tracking was added to compute the motion of particles in a background fluid which was calculated by Laminar Flow. A Fluid-Particle Interaction multi-physics coupling computes a volume force that is equal in magnitude and opposite in direction to the total drag force exerted on particles in each mesh element in the selected domains. This volume force contributes to the total force acting on the fluid in the Laminar Flow interface. The filtration property of PU-bonded pavement was realized by a stick boundary set on the surface of fluid pathway. Stick boundary was selected to fix the particle position at the instant the wall is struck, the particle velocity is set to zero. The interaction force between fluid and PM was described as Stokes Drag law which was detail described by as follows [17]:

$$F = \frac{1}{t_p} m_p (u - v) \tag{3.2}$$

$$t_p = \frac{\rho_p d_p^2}{18\mu} \tag{3.3}$$

Where, F is the interaction force between fluid and PM, m_p, ρ_p, d_p are particle mass, particle density, particle diameter respectively; u, v are the velocity of fluid and particle; μ is the dynamic viscosity of fluid.

An inlet velocity boundary at the surface of permeable pavement was set for the fluid based on raining conditions. In addition, the uniform distribution of particles with the velocity of fluid are released at the same boundary. It is important to note that the particle number at the inlet boundary was calculated by 160 cm^3 of PM 2.5 at the concentration of 115 $\mu g/m^3$. The outlet boundary for the fluid and particles was set at the bottom of fluid passageway. The interaction force between fluid and PM was defined as Drag Force which is loaded on the entire fluid region.

4 Results and Discussions

4.1 Permeability

Based on the determination on the average hydraulically effective void ratio as 32% for the Pu-bonded PAP specimens, the coefficient of permeability was obtained by the conduction of constant falling head test. The result turns out a high coefficient $K_{f,v}$ (10^{-4} m/s) of vertical permeability can be achieved by the PU-bonded PAP specimen, with reaching the value of 88 when comparing to the conventional PA8 with only valued 23 approximately.

The prediction of vertical permeability in the present study is based on the modified KCM model. By determination of the shape factor of aggregate, grain size distribution as well as the hydraulic effective void ration, the prediction of permeability can be obtained by the aid of commercial program software Matlab®. It is notable to find that

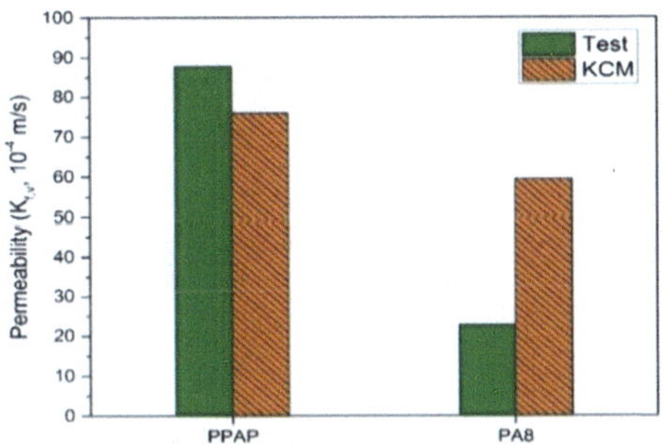

Fig. 2. Validation of KCM model

the tested value of permeability in PU specimen is higher than the KCM value, while the tested value on PA8 is much lower than the prediction (Fig. 2). The difference can be considered as the application of different binders. The polyurethane binder can form a smoother coat for the aggregates, which promote the infiltration process of the runoff. On the contrary, the bitumen retards the infiltration to some extent, according to the high viscosity. In this case, based on the hypothesis of KCM model and the instinct property of polyurethane, the error between tested and predicted coefficient of permeability in PU specimens 13% is acceptable.

4.2 Prediction of Filtration and Clogging Behaviour

Due to the high accordance with the permeability prediction, factors of PPAP are also applied to conduct the primary prediction of filtration and clogging behavior in PAP structure. Both Laminar Flow and Particle Tracking in COMSOL Multiphysics 5.2 were employed to analyze the path for PM. A 2D geometric model was built using a dimension of 25 mm in width and 25 mm in depth, where the blue area in the geometry were the fluid flow passage.

After inlet velocity boundary depends on rain condition was loaded, an initial distribution of flow field without PM particles was computed in as control group. In order to analyze the relationship between filtration efficiency and PM concentration, five levels of PM with concentration of 2.5 mg/m^3, 5 mg/m^3, 7.5 mg/m^3, 10 mg/m^3 were under consideration respectively. The PM particles was dragged by the fluid with the same velocity, but once it hit the wall of binder coat it will be adsorbed by the wall. Through a stick boundary loaded on asphalt surface as shown as blackspots in Fig. 3. The filtration efficiency was defined by the stacked particles number to the whole particle number. After all the particles flow past the pavement structure, part of them will stick on the pavement structure and the rest of them will flow always. Since the particles stacked on the pavement, the velocity of the particles was zero which leads to a resistance to the fluid. The clogging behavior was performed by the comparison between the fluid flow with PM.

It is obvious that the higher the concentration of PM more PM particles were caught but the lower of filtration efficiency was acquired. However, the least filtration efficiency of 11.75% was still positive. What's more it is easy to find that the length of

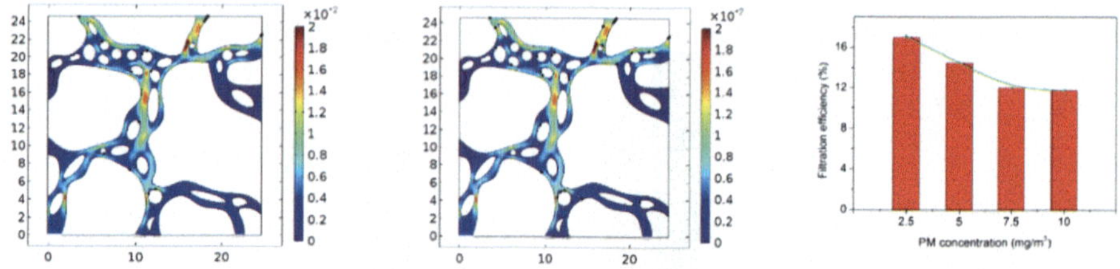

(a) (b) (c)

Fig. 3. Filtration efficiency and clogging behaviour at different particle density. (a) 17% filtration efficiency at 2.5 mg/m^3 particle density; (b) 14.3% filtration efficiency at 5 mg/m^3 particle density. (c) Filtration and clogging behaviour under different PM concentration

particle pathway is a key factor to the filtration efficiency. The micro structure of a geometry with 25 mm * 25 mm was built as a consideration of limit elements for less calculated amount since the model contains large calculated amount. With this mind, we can see that the filtration efficiency maybe better in the actual pavement with the size of 40 mm * 40 mm.

However, this was an initiatory prediction model of filtration and clogging performance of permeable pavement. More related experiment need to be carried out to update our model for better prediction.

5 Conclusion

The quantitative analysis on hydraulic properties of PAP is vital important to the construction of LID and sustainable drainage system. To optimize the design of PAP and understand the deterioration mechanism it, the simulation of hydraulic properties was established based on the effective void ration and validated with the innovative PPAP. Further, the prediction of filtration and clogging behavior was attempted based on the hydraulic properties. The main achievements in this paper can be concluded as follow:

- Based on the specific property of binder, the innovative PPAP was proved having higher hydraulic properties when comparing to that of conventional PAP;
- Based on the effective void ratio, the modified KCM model presents a high accordance to the hydraulic properties of PPAP;
- The simulation on the filtration and clogging behaviour can be established based on the FEM, where the PPAP presents a high absorption capability of PM subjected to the urban runoff;
- The filtration and clogging behaviour varied by the concentration of PM, specifically the high filtration efficiency can be achieved by smaller PM density, while high PM density leads to a lower filtration efficiency but higher clogging phenomenon.

Acknowledgement. The work underlying this project was carried out under at the request of the German Research Foundation (DFG), under research project No. OE 514/4-1.

References

1. Kuang, X., Sansalone, J.: Cementitious porous pavement in stormwater quality control: pH and alkalinity elevation. Water Sci. Technol. **63**(12), 2992–2998 (2011)
2. Wang, D., Schacht, A., Leng, Z., Leng, C., Kollmann, J., Oeser, M.: Effects of material composition on mechanical and acoustic performance of poroelastic road surface (PERS). Constr. Build. Mater. **135**, 352–360 (2017)
3. Sansalone, J., Teng, Z.: In situ partial exfiltration of rainfall runoff. I: quality and quantity attenuation. J. Environ. Eng. **130**(9), 990–1007 (2004)
4. Teng, Z., Sansalone, J.: In situ partial exfiltration of rainfall runoff. II: particle separation. J. Environ. Eng. **130**(9), 1008–1020 (2004)
5. Rushton, B.T.: Low-impact parking lot design reduces runoff and pollutant loads. J. Water Resour. Plan. Manag. **127**(3), 172–179 (2001)
6. Pagotto, C., Legret, M., Le Cloirec, P.: Comparison of the hydraulic behaviour and the quality of highway runoff water according to the type of pavement. Water Res. **34**(18), 4446–4454 (2000)
7. Kamali, M., Delkash, M., Tajrishy, M.: Evaluation of permeable pavement responses to urban surface runoff. J. Environ. Manag. **187**, 43–53 (2017)
8. Oeser, M., Hovagimian, P., Kabitzke, U.: Hydraulic and mechanical properties of porous cement-stabilised materials for base courses of PICPs. Int. J. Pavement Eng. **13**(1), 68–79 (2012)
9. Albrecht, F.: Theoretische untersuchungen über bie ablegerung staub und luft und ihre anwendung auf die theorie der staubfilter. Physik. Zeitschr **32**, 48 (1931)
10. Kaufmann, A.: Die faserstoffe für atemschutzfiltern. Z. Ver. Dtsch. Ing. **8**, 593–597 (1936)
11. Davies, C.N.: Air Filtration. Academic Press, London (1973)
12. Tien, C., Wang, C.S., et al.: Chainlike formation of particle deposits in fluid-particle separation. Chem. Eng. Sci. Technol. **196**(7), 983–985 (1977)
13. Leung, W., Hung, C.: Investigation on pressure drop evolution of fibrous filter operating in aerodynamic slip regime under continuous loading of sub-micron aerosols. Sep. Purif. Technol. **63**(11), 691–700 (2008)
14. Renken, L., Oeser, M., Milatz, M., Grabe, J.: Measurement of hydraulic properties of unsaturated permeable polyurethane bound asphalt materials. In: Unsaturated Soil Mechanics-from Theory to Practice: Proceedings of the 6th Asia Pacific Conference on Unsaturated Soils, Guilin, China, 23–26 October 2015, p. 407. CRC Press, 4 November 2015
15. Cooley Jr., L.A.: Permeability of superpave mixtures: evaluation of field permeameters. National Center for Asphalt Technology, NCAT Report No. 99-1, February 1999
16. Forschungsgesellschaft für das Straßen- und Verkehrswesen (FGSV): Handbuch für die Bemessung von Straßenverkehrsanlagen (HBS) (2009)
17. Vukovic, M., Soro, A.: Determination of Hydraulic Conductivity of Porous Media from Grain-Size Distribution. Water Resources Publications, Littleton (1992)

Study on the Influence of Impermeable Membrane's Blanket Length on the Seepage of a Dam

Lu Yu, Zhenying Zhang, Chenyang Tu, Jincheng Zhang, and Dazhi Wu(✉)

Zhejiang Sci-Tech University, Hangzhou, China
wudz@zstu.edu.cn

Abstract. Geotechnical membrane is often used in dam engineering to prevent the seepage. But the research of the influence of impervious membrane on the seepage of a dam is not enough. Combining the indoor model experiment and numerical analysis methods, the influence of the impermeable membrane's blanket length on the seepage of a dam had been carried out in present paper. By analyzing the results from these two methods, it was found that the seepage quantity of the dam linearly changed with the increase of the blanket length. Besides, bases on the numerical analysis method, the influence of different horizontal impervious blanket length on the seepage of a dam had been investigated, and the calculation formula for different thickness of the foundation had been fitted. Finally, the optimum length of horizontal blanket for different dam foundation was analyzed.

Keywords: Impermeable membrane · Blanket length · Dam · Seepage

1 Introduction

China has a long coastline, and a large number of offshore and estuary dams have been constructed along the coastline. It was confirmed that there is a close relationship between the common diseases of the dam and the permeation. It was reported that 40.5% of the dam destruction were caused by dam seepage [1].

During the flood in 1998, many dangerous conditions were happened in dams along Yangtze River, and 54.5% of these dangerous conditions were the result of seepage failure. In history, more than 90% of the Yangtze River inrushes were caused by seepage failure [2]. The "Katrina" hurricane brought heavy damage to America, and the disaster survey report pointed out that the direct factor of dams' destruction in New Orleans was seepage failure of dam foundation [3]. Studies indicated that the engineering reinforcement of embankment had met the requirements of national standard. However, due to lack of understanding on the dam foundation seepage failure, there are still some problems that need to be studied. Impervious geotechnical membrane can adapt to dam's deformation due to its soft characteristics, so it is widely used in engineering. For example, Sichuan Tianwanhe Renzonghai rockfill dam, Sichuan Ganzi Huashan core film/gravelly soil core rockfill dam and Shandong Dezhou Datun

L. Hu et al. (Eds.): GSIC 2018, *Proceedings of GeoShanghai 2018 International Conference: Multi-physics Processes in Soil Mechanics and Advances in Geotechnical Testing*, pp. 210–219, 2018.
https://doi.org/10.1007/978-981-13-0095-0_24

plain reservoir are all representative membrane impermeable dam [4]. Whitfield [5] introduced the application of geotechnical impervious material in Roller compacted concrete dam. Messerklinger [6] introduced a seepage failure case of a geomembrane lining dam. Yuan *et al.* [7] studied the influence of blanket length to infinite deep pervious foundation dam with boundary element method, and revealed the relationship between the blanket length of the infinite deep pervious foundation dam and the upstream head. Yin *et al.* [8] analysed the length selection of impervious blanket in non homogeneous infinite deep pervious dam foundation, and gave the computing method about seepage of non homogeneous infinite deep pervious dam foundation by using model experiment. Combining with the engineering example, Wen *et al.* [9] analyzed the influence of the depth of cutoff wall, blanket length and permeability of overburden soil on the seepage control of a dam foundation, and concluded that the blanket could not effectively control the overburden foundation seepage. Xu *et al.* [10] used model experiment and numerical analysis methods to carry out the research on the selection of the impervious blanket length of the infinite deep pervious foundation. By developing a model experiment, Mao *et al.* [11] calculated the effective depth of impermeable membrane on earth rockfill dam.

Combining indoor model experiment and numerical analysis method, the influence of impermeable membrane's blanket length on the seepage of a dam were systematically carried out, the conclusions could provide a reference for the design, construction and management of a dam.

2 Model Experiment

2.1 Experiment Method

The indoor experiment was carried out by using a self-made seepage apparatus, which was shown in Fig. 1, and the tank's size were 3.0 m × 0.5 m × 1.2 m. The water flew into the tank through the adjustable bin. The upstream water level could be adjusted by changing the bin's height. There was a drainage outlet in the right wall of the tank. To observe the seepage process, the front face of the tank was made of organic glass. To gain the distribution of pore water pressure, a row of piezometric tubes were installed 200 mm above the bottom of the tank.

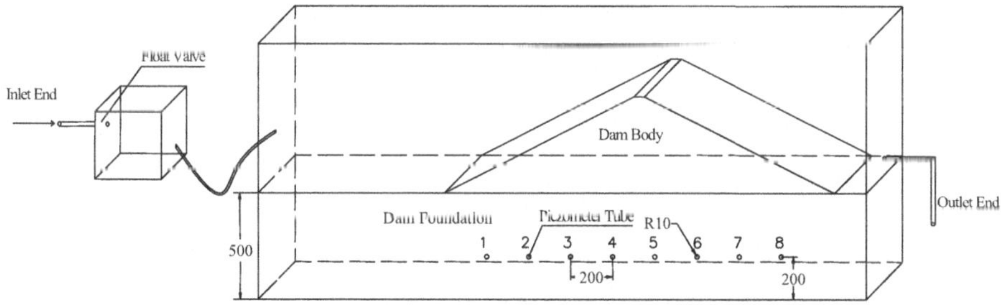

Fig. 1. Diagrammatic sketch of self-made seepage apparatus (mm).

Sand was used to build the dam foundation to simulate the pervious characteristics of the foundation. In order to simulate the actual engineering, dam body was filled with silt. The soil was carried from a building foundation pit in Hangzhou. Basic physical property indexes of the filling materials were shown in Table 1. Common plastic film was selected as the material of impervious blanket.

Table 1. Basic physical property indexes of the materials.

Position	Soil type	Plastic limit (%)	Liquid limit (%)	Optimum moisture content/(%)	Specific gravity	Void ratio	Water content/(%)	Permeability coefficient/ ($cm\cdot s^{-1}$)
Dam body	Sand				2.65	0.68	8.7	2.8×10^{-2}
Dam foundation	Silt	20.27	29.83	22.2	2.70	1.04	17.6	1.46×10^{-5}

The width of the dam bottom was 185 cm, the width of the dam crest was 5.0 cm, the height of the dam foundation was 50 cm, the height of the dam body was 45 cm, and the slope ratio of the upstream slope and the downstream slope were 1:2. Hierarchical construction method was used for the filling of dam foundation and dam body, each layer thickness was 10 cm, and the filling compaction degree was controlled at 88%.

The upstream water level and downstream water level were selected as 20 cm and 7 cm, respectively. In order to avoid the water oozing out from the leakage of blanket and slope, membrane was laying on the surface from dam body to foundation. Membrane was 50 cm larger than the tank's width. The extra membranes in two sides were pasted on the inner wall of the tank firmly by using vaseline, and there was no large air bubble under the membrane. Besides, the junction of membrane and foundation was covered by sand to prevent the membrane floating and water seeping through the underside of the blanket. It was shown in Fig. 2.

Fig. 2. Laying of impermeable membrane.

After finish laying membrane on the dam, water was slowly poured from the inlet end. For the difference of the water levels between the upstream and downstream, seepage occurred. The piezometric tubes' water level and the drainage quantity from outlet end could be measured directly. The readings of piezometric tubes would not recorded until the readings did not change and the drainage quantity was kept stable during every three minutes. Finally, the flow rate was calculated based on the seepage quantity.

2.2 Experiment Results

The lengths of the membrane were selected as 20 cm, 40 cm, 60 cm and 80 cm, which were 1–4 times of the upstream water level. The experiment without membrane was also carried out for contrast. The seepage quantity of these 5 groups experiments were list in Table 2.

Table 2. Seepage quantity of dam with different length blanket.

Blanket length (cm)	Seepage quantity ($ml \cdot s^{-1} \cdot m^{-1}$)
0	7.67
20	7.22
40	6.94
60	6.50
80	6.11

3 Numerical Analysis

3.1 Numerical Analysis Method

The numerical analysis was carried out by using seep/w module of the software Geo-Studio. Seep/w was developed to analyze porous material from simple saturated steady-state problems to complex, saturated unsaturated variable problem. In this paper, saturated material was selected to simulate the permeable dam foundation, and unsaturated material was selected to simulate the dam body.

3.2 Compared with the Model Experiment

In order to make a comparative analysis of indoor model test and numerical calculation, the parameters of the numerical analysis were selected as 100 times of the sizes of the laboratory model. Those were, the width of the dam bottom was 185 m, the width of the dam crest was 5 m, the height of the dam foundation was 50 m, the height of the dam body is 45 m, and the slope ratio of the upstream slope and the downstream slope were 1:2. The saturated permeability coefficient of the dam body and dam foundation were 5.68×10^{-7}m/s and 1×10^{-14}m/s. The calculation results were shown in Table 3.

Table 3. Seepage quantity of dam with different length blanket.

Blanket length (m)	Seepage quantity ($ml \cdot s^{-1} \cdot m^{-1}$)
0	6.2941
20	6.2262
40	6.1407
60	6.0715
80	6.0125

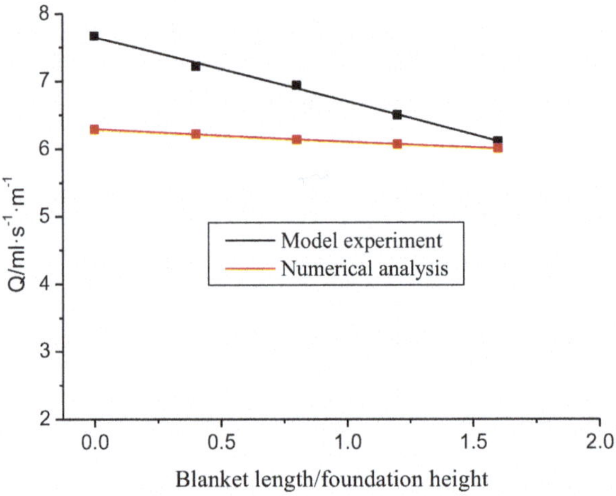

Fig. 3. Comparison between numerical analysis and indoor model test.

Compared the results of indoor model test to numerical analysis, the comparative figure was plotted in Fig. 3.

From Tables 2 and 3, it could be found that, the seepage quantity of the dam was linearly changed with the blanket length. This indicated both the indoor model test and numerical analysis method were correct and reliable. The difference values between these two methods were 17.94%, 13.76%, 11.52%, 6.59%, 1.60% for the membrane length were 0, 1, 2, 3, 4 times of the upstream water level, respectively. The difference values of two methods decreased with the increase of the upstream blanket length. This was probably due to the limitation of experimental conditions, membrane and foundation could not fit tightly enough when the membrane was short.

3.3 Numerical Analysis Results

Selected the head difference of upstream and downstream H as a benchmark. The height of dam foundation was selected in the range of 1H to 5H, the blanket length was in the range of 1H to 12H. The seepage results were list in Table 4.

Table 4. Seepage quality for different blanket length.

Blanket length	Dam foundation				
	H	2H	3H	4H	5H
0H	4.4556	7.4105	9.2946	10.469	11.216
1H	4.3928	7.2855	9.1558	10.318	11.070
2H	4.3466	7.1395	8.9616	10.096	10.848
3H	4.3228	7.0371	8.8139	9.9035	10.646
4H	4.3085	6.9561	8.6951	9.7348	10.471
5H	4.2989	6.9123	8.6090	9.6088	10.319
6H	4.2922	6.8723	8.5474	9.4973	10.188
7H	4.2871	6.8408	8.4932	9.4032	10.074
8H	4.2831	6.8155	8.4540	9.3226	9.9735
9H	4.2799	6.7946	8.4229	9.2521	9.8825
10H	4.2773	6.7771	8.3979	9.1883	9.8005
11H	4.2752	6.7619	8.3768	9.1267	9.7255
12H	4.2734	6.7478	8.3570	9.0591	9.6565

3.4 Analysis and Discussion

Based on Table 4, following figure could be drawn.

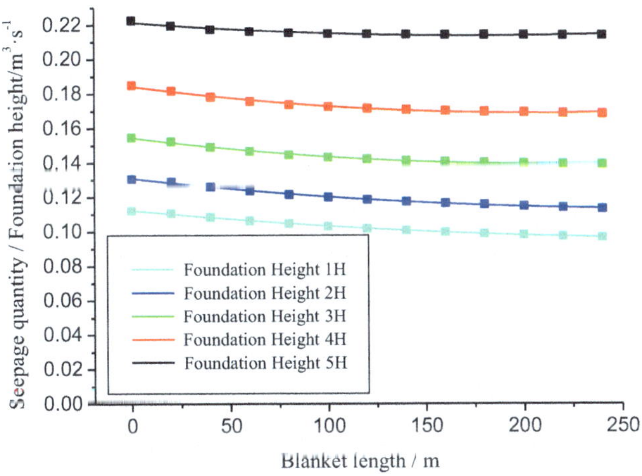

Fig. 4. Curve of seepage quantity versus blanket length.

From Fig. 4, it was found the seepage quantity decreased with the increase of blanket length, and the curve gradually tended to be gentle. The curves could be considered as linear change when the blanket length was smaller than 100, this was in consistent with Fig. 3. Comparing the 5 curves in Fig. 4, it could also be found that, with the increase of dam foundation height, the seepage quantity decreased rapidly and achieved the gentle range quickly.

Table 5. Ratio of seepage quantity with blanket to that of without blanket.

Blanket length	Dam foundation				
	H	2H	3H	4H	5H
0H	0.9859	0.9831	0.9851	0.9856	0.9870
1H	0.9755	0.9634	0.9642	0.9644	0.9672
2H	0.9702	0.9496	0.9483	0.9460	0.9492
3H	0.9670	0.9387	0.9355	0.9299	0.9336
4H	0.9648	0.9328	0.9262	0.9178	0.9200
5H	0.9633	0.9274	0.9196	0.9072	0.9084
6H	0.9622	0.9231	0.9138	0.8982	0.8982
7H	0.9613	0.9197	0.9096	0.8905	0.8892
8H	0.9606	0.9169	0.9062	0.8838	0.8811
9H	0.9600	0.9145	0.9035	0.8777	0.8738
10H	0.9595	0.9125	0.9013	0.8718	0.8671
11H	0.9591	0.9106	0.8991	0.8653	0.8610
12H	0.9859	0.9831	0.9851	0.9856	0.9870

By divided the dam seepage quantities with the seepage quantity without blanket, following ratios of seepage quantity with blanket to that of without blanket could be written.

Based on Table 5, the ratio curve of seepage quantity with blanket to that of without blanket could be drawn as Fig. 5.

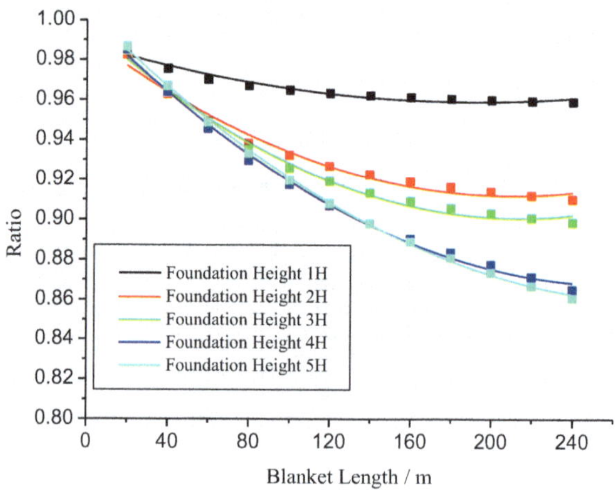

Fig. 5. Ratio curve of seepage quantity with blanket to that of without blanket.

The relationships in Fig. 5 could be fitted as a polynomial:

$$k = ah_l^2 + bh_l + c \tag{1}$$

Where k was the proportion of dam seepage quantity with blanket to that of without blanket, h_l was blanket length, and a, b, c were fitting coefficients.

Table 6. Fitting coefficients of *a*, *b*, *c*.

Foundation height	*a*	*b*	*c*	Correction coefficient
1H	3.20×10^{-4}	−0.0060	0.9880	0.9377
2H	2.88×10^{-3}	−0.0302	0.9921	0.9785
3H	7.65×10^{-3}	−0.0544	0.9979	0.9886
4H	1.23×10^{-2}	−0.0812	1.0019	0.9959
5H	1.73×10^{-2}	−0.1006	1.0046	0.9988

The fitted coefficients *a*, *b* and *c* were list in Table 6.

After gained the relationship of the ratio of seepage quantity with blanket to that of without blanket for different foundation height, the minimum value and the corresponding blanket length could be computed. The minimum seepage quantities and optimum membrane blanket lengths for five foundation heights were list in Table 7 and plotted in Fig. 6.

Table 7. Minimum seepage quantity and optimum membrane blanket length.

Foundation height	Minimum value	Optimum blanket length/*H*
1H	0.960	9.38
2H	0.913	5.24
3H	0.902	3.56
4H	0.867	3.30
5H	0.859	2.91

The relationship in Fig. 6 could be formulated as:

$$K_{\min} = 0.0039h_r^2 - 0.0479h_r + 1.0016 \quad (2)$$

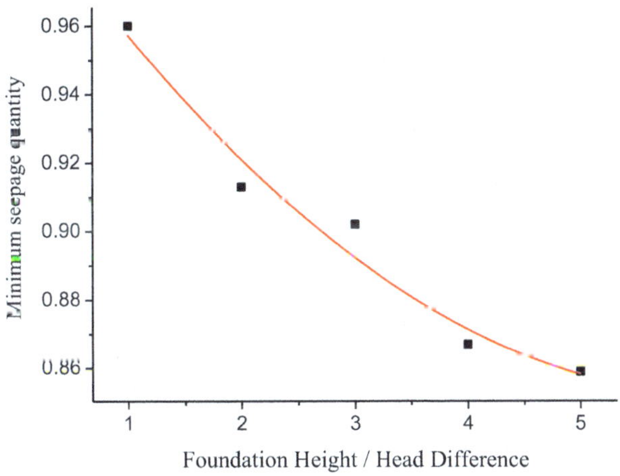

Fig. 6. Minimum seepage quantity versus dam foundation height.

Where K_{min} was the minimum value of the ratio of seepage quantity with blanket to that of without blanket, and h_t was the thickness of the dam foundation.

The fitting curve correlation coefficient 0.9437 showed that the fitting was reliable. From Eq. (2), it could be seen that, with the increase of the dam foundation thickness, the minimum seepage quantity of the dam was gradually reduced.

4 Conclusions

Combining indoor model experiment and numerical analysis method, the influence of impermeable membrane's blanket length on the seepage of a dam were studied. The results of these two methods were in accordance with each other well.

The seepage quantity decreases with the increase of blanket length, there is a linear relationship among them when the blanket length is small. With the increase of the height of dam foundation, the seepage quantity decreased rapidly and achieved the gentle range quickly.

The relationship between the ratio of seepage quantity with blanket to that of without blanket and the blanket length can be expressed by a polynomial.

The relationship of minimum seepage quantity and dam foundation height can also be formulated as a polynomial. With the increase of the thickness of dam foundation, the minimum seepage quantity of the dam was gradually reduced.

Acknowledgments. This paper is funded by National Natural Science Foundation of China under the contract No. 51678533.

References

1. Gu, W.: Principle and Application of Seepage Calculation. Chinese Building Materials Industry Press, Beijing (2000)
2. Water Resources Department of Construction and Management: Analysis of dam danger in the Yangtze River and the Songhua River happened in 1998 flood. China Flood Control Drought Relief **19**(4), 20–24 (1998)
3. Seed, R.B., Bea, R.G., Abdelmalak, R.I., Athanasopoulos, A.G., Boutwell, G.P., Bray, J.D., Briaud, J.-L., Cheung, C., Cobos-Roa, D., Cohen-Waeber, J., Collins, B.D., Ehrensing, L., Farber, D., Hanemann, M., Harder, L.F., Inkabi, K.S., Kammerer, A.M., Karadeniz, D., Kayen, R.E., Moss, R.E.S., Nicks, J., Nimmala, S., Pestana, J.M., Porter, J., Rhee, K., Riemer, M.F., Roberts, K., Rogers, J.D., Storesund, R., Govindasamy, A.V., Vera-Grunauer, X., Wartman, J.E., Watkins, C.M., Wenk Jr., E., Yim, S.C.: Investigation of the Performance of the New Orleans Flood Protection System in Hurricane. University of California-Berkeley, State of California (2006)
4. Shu, Y., Wu, H., Jiang, X.: Technical progress of pervious membrane used in seepage control of reservoir dam in China. Geotech. Eng. J. **38**(S1), 1–9 (2016)
5. Whitfield, B.L.: Geomembrane application for an RCC dam. Geotext. Geomembr. **14**, 253–264 (1996)
6. Messerklinger, S.: Failure of a geomembrane lined embankment dam-case study. Geotext. Geomembr. **42**, 256–266 (2014)

7. Yuan, Y., Shi, K., Li, Y., et al.: The influence of blanket length to seepage of infinite deep pervious foundation. Hydroelectric Power **33**(1), 34–36 (2007)
8. Yin, H., Shi, K., Mao, H.: The selection of impervious membrane length used in non homogeneous infinite deep dam foundation. J. Water Resour. Water Eng. **22**(2), 62–65 (2011)
9. Wen, L., Fan, Y., Chai, J., et al.: Numerical analysis on the effect of seepage control measures of deep overburden foundation. J. Water Resour. Water Eng. **25**(1), 127–132 (2014)
10. Xu, Y., Shi, K., Xu, Q., et al.: The influence of micro porous membrane to seepage of infinite deep permeable foundation. J. Water Resour. Water Eng. **24**(1), 89–92 (2013)
11. Mao, H., Wang, X., He, Y.: A Pilot study of the method choice about the blanket length over earth-rock dam on the infinite deep pervious foundation. In: International Conference on Electric Technology and Civil Engineering, pp. 5698–5701 (2011)

Inter-porosity Exchange in Saturated Double-Porosity Hollow Cylinder Subject to Axisymmetric Load

Zhu Song(✉) and Yanqiu Xiang

Nanjing Tech University, Nanjing 210009, China
zhusong@njtech.edu.cn

Abstract. Drilling a borehole in saturated geotechnical materials often results in the excavation disturbed zone constituted by the original porous matrix and the fractures. In order to study the hydro-mechanical behaviors of such double-porosity medium, the analytical solution to the hollow cylinder subject to axisymmetric load is derived with Laplace transform. Emphases have been placed on the inter-porosity behaviors in terms of transport coefficient between the matrix and fractures and the results show a higher value of transport coefficient will accelerate the dissipation in the matrix while increase the pore pressure significantly in the fractures due to the fluid from the matrix.

Keywords: Double porosity · Hydro-mechanical behaviors · Hollow cylinder

1 Introduction

After drilling in the geotechnical materials, the cylindrical borehole was usually surrounded by an excavation disturbed zone (EDZ) [1]. The disturbed structure, following with the hydro-mechanical behaviors such as permeability and stiffness, departs from the original porous medium. Such alterations greatly affect the mass transfer for engineering purposes, including injection/extraction of water or gas for heat exchange, carbon sequestration and oil production [2]. The investigations of the altered characteristics in the EDZ had been implemented by assuming the disturbed or undisturbed zones of different material properties, where the development of fractures is neglected. Nevertheless, the sudden imposed penetration increases the fractures in the formation [3], and fractures interconnect as an additional system of flow channels where the capacity of water movement is much larger than the original porous matrix. Such heterogeneous medium can be described in terms of the overlapping of two different types of porosity [2]: (1) Storage porosity, often in the form of matrix of porous medium, holds the most volume of fluid while the permeability is rather low; (2) Transport porosity, often in the form of fractures in the reservoir or joints in rock mass, is low in volume but high in permeability. Two systems of porosity are coupled by a mass transport between pores and fractures and such inter-porosity effect is estimated as inter-porosity flow coefficients.

As an enrichment of Biot description [4] to present two fields of pore pressure in interaction, the theory of double-porosity porous medium has been expanding in the

L. Hu et al. (Eds.): GSIC 2018, *Proceedings of GeoShanghai 2018 International Conference: Multi-physics Processes in Soil Mechanics and Advances in Geotechnical Testing*, pp. 220–227, 2018.
https://doi.org/10.1007/978-981-13-0095-0_25

over three decades [5, 6]. The governing equations of the two parallel-model porosities consider the interaction between matrix, fractures and pores in both the momentum balance and the mass conservation. Numerical methods had been applied for field-scale examples [7]. Meanwhile, the analytical solutions for the comparison between the double-porosity media and the single-porosity counterpart is highlighted [2]. The afore-mentioned analytical solutions, however, are limited to the cylinder case which are not suitable for the excavation disturbed zone around the borehole.

In this analysis, the analytical solution to the double porosity medium is extended into a hollow cylinder case which is sandwiched by two rigid plates and subject to a vertical load. After the application of Laplace transform for time variable and the boundary conditions, the hydro-mechanical behaviors in the matrix and fractures are analyzed and emphases have been placed on the transport coefficient between the matrix and fractures. The results show that with a high transport coefficient, the dissipation in the matrix differs from the single-porosity Biot's solution due the absence of Mandel-Cryer effect and the pore pressure in the fractures can increase significantly due to the fluid from the matrix.

2 Governing Equations

In this section, a saturated doubled-porosity cylinder with an inner radius r_i and an outer radius r_o is studied which is sandwiched between two rigid, frictionless and impervious plates. The linear isotropic constitutive equations are written as [6]:

$$\sigma_{ij} = \left[2\bar{G}\varepsilon_{ij} + \frac{2\bar{v}\bar{G}}{1-2\bar{v}}\varepsilon_{kk}\delta_{ij}\right] + \left(\bar{\alpha}^m p^m + \bar{\alpha}^f p^f\right)\delta_{ij} \tag{1}$$

$$\zeta^m = -\bar{\alpha}^m \varepsilon_{kk} + \frac{p^m}{\bar{M}^m} + \frac{p^f}{\bar{M}^{m,f}};\ \zeta^f = -\bar{\alpha}^f \varepsilon_{kk} + \frac{p^m}{\bar{M}^{m,f}} + \frac{p^f}{\bar{M}^f} \tag{2}$$

where σ_{ij}, ε_{ij}: total stress and strain tensor, respectively; $\bar{K}$, $\bar{v}$: overall drained elastic bulk modulus and Poisson's ratio, respectively; ε_{kk}: overall bulk volumetric strain, $\varepsilon_{kk} = \Delta V/V$; δ_{ij}: Kronecker delta; $^-$: overall double-porosity poroelastic material coefficients; ^{m}, ^{f}: superscripts representing for the matrix and fractures, respectively; $\bar{\alpha}$: effective pore-pressure coefficient; p: pore pressure; ζ: volumetric change, $\bar{M}^m$, $\bar{M}^f$, $\bar{M}^{m,f}$: the effective coupled Biot's moduli which are defined in [8].

Equilibrium and strain-displacement relation

$$\frac{\partial \sigma_{ij}}{\partial x_j} = 0;\ \varepsilon_{ij} = \frac{1}{2}\left(\frac{\partial u_i}{\partial x_j} + \frac{\partial u_j}{\partial x_i}\right) \tag{3}$$

Conservation of the fluid phases

$$\frac{\partial \zeta^m}{\partial t} + \frac{\partial q_i^m}{\partial x_i} = \Gamma\left(p^f - p^m\right);\ \frac{\partial \zeta^f}{\partial t} + \frac{\partial q_i^f}{\partial x_i} = -\Gamma\left(p^f - p^m\right) \tag{4}$$

Darcy's law

$$q_i^m = -\frac{v^m k^m}{\mu^m}\frac{\partial p^m}{\partial x_i};\ q_i^f = -\frac{v^f k^f}{\mu^f}\frac{\partial p^f}{\partial x_i} \tag{5}$$

where u_i and x_i: displacement vector and the spatial coordinate, respectively; q_i^m, q_i^f: specific fluid discharge vector in the matrix and fractures; v^m, v^f: bulk volume fraction occupied by the matrix and fractures, $v^f + v^m = 1$; k^m, k^f: intrinsic permeability of porous matrix; μ^m, μ^f: fluid dynamic viscosity; Γ: transport coefficient.

Combining the equilibrium Eq. (3) with the stress-strain-pressure Eq. (1), we can obtain the compatibility relation of the system as

$$\nabla^2\left(\varepsilon_{kk} + \frac{\bar{\eta}^m}{\bar{G}}p^m + \frac{\bar{\eta}^f}{\bar{G}}p^f\right) = 0 \tag{6}$$

where $\bar{\eta} = \bar{\alpha}(1-2\bar{v})/[2(1-\bar{v})]$; $\bar{G} = 3\bar{K}(1-2\bar{v})/[2(1+\bar{v})]$; $\nabla^2 = \partial^2/\partial r^2 + (\partial/\partial r)/r$.

The diffusion equations in the matrix and fractures can be obtained by the substitution of Eq. (2) into Eq. (4), and combining the Darcy's law from Eq. (5):

$$-\bar{\alpha}^m\frac{\partial\varepsilon_{kk}}{\partial t} + \frac{1}{\bar{M}^m}\frac{\partial p^m}{\partial t} + \frac{1}{\bar{M}^{mf}}\frac{\partial p^f}{\partial t} = \frac{v^m k^m}{\mu^m}\nabla^2 p^m + \Gamma(p^f - p^m) \tag{7}$$

$$-\bar{\alpha}^f\frac{\partial\varepsilon_{kk}}{\partial t} + \frac{1}{\bar{M}^{mf}}\frac{\partial p^m}{\partial t} + \frac{1}{\bar{M}^f}\frac{\partial p^f}{\partial t} = \frac{v^f k^f}{\mu^f}\nabla^2 p^f - \Gamma(p^f - p^m) \tag{8}$$

The solution to Eq. (6) can be given as

$$\varepsilon_{kk} + \frac{\bar{\eta}^m}{\bar{G}}p^m + \frac{\bar{\eta}^f}{\bar{G}}p^f = C_0 \tag{9}$$

where C_0 is the integration constant.

Substitution of Eq. (9) into Eqs. (7) and (8) can eliminate ε_{kk} term, and the variations of coupled equations can be compacted as

$$\left\{[\mathbf{A}_1]\frac{\partial}{\partial t} + [\mathbf{A}_2] - [\mathbf{A}_3]\left(\frac{\partial^2}{\partial r^2} + \frac{1}{r}\frac{\partial}{\partial r}\right)\right\}\begin{bmatrix} p^f \\ p^m \end{bmatrix} = \begin{bmatrix} \bar{\alpha}^f \\ \bar{\alpha}^m \end{bmatrix}\frac{\partial C_0(t)}{\partial t} \tag{10}$$

where $[\mathbf{A}_1] = \begin{bmatrix} \frac{\bar{\alpha}^m\bar{\eta}^m}{\bar{G}} + \frac{1}{\bar{M}^m} & \frac{\bar{\alpha}^m\bar{\eta}^f}{\bar{G}} + \frac{1}{\bar{M}^{mf}} \\ \frac{\bar{\alpha}^f\bar{\eta}^m}{\bar{G}} + \frac{1}{\bar{M}^{mf}} & \frac{\bar{\alpha}^f\bar{\eta}^f}{\bar{G}} + \frac{1}{\bar{M}^f} \end{bmatrix}$, $[\mathbf{A}_2] = \begin{bmatrix} \Gamma & -\Gamma \\ -\Gamma & \Gamma \end{bmatrix}$, and $[\mathbf{A}_3] = \begin{bmatrix} \frac{v^m k^m}{\mu^m} & 0 \\ 0 & \frac{v^f k^f}{\mu^f} \end{bmatrix}$.

After applying the Laplace transform to Eq. (10) on the time variable (denoted by the overbar $\sim$) results in the following equation:

$$\left\{[\mathbf{A_1}]s+[\mathbf{A_2}]-[\mathbf{A_3}]\left(\frac{\partial^2}{\partial r^2}+\frac{1}{r}\frac{\partial}{\partial r}\right)\right\}\begin{bmatrix}\tilde{p}^m\\ \tilde{p}^f\end{bmatrix}=\begin{bmatrix}\bar{\alpha}^m\\ \bar{\alpha}^f\end{bmatrix}s\tilde{C}_0 \tag{11}$$

The solution to Eq. (11) is obtained from theory of system of ordinary differential equations from [9] and can be stated as follows:

$$\begin{bmatrix}\tilde{p}^m\\ \tilde{p}^f\end{bmatrix}=\begin{pmatrix}1 & 1\\ \chi_1 & \chi_2\end{pmatrix}\left\{\begin{pmatrix}B_1 I_0\left(r\sqrt{\xi^{\mathrm{I}}}\right)\\ B_2 I_0\left(r\sqrt{\xi^{\mathrm{II}}}\right)\end{pmatrix}+\begin{pmatrix}B_3 K_0\left(r\sqrt{\xi^{\mathrm{I}}}\right)\\ B_4 K_0\left(r\sqrt{\xi^{\mathrm{II}}}\right)\end{pmatrix}\right\}+\begin{bmatrix}\psi_1\\ \psi_2\end{bmatrix}\tilde{C}_0 \tag{12}$$

where I_0 and K_0 are the modified Bessel function of the first kind of order 0, and the modified Bessel function of the second kind of order 0, respectively. B_1, B_2, B_3, and B_4 are arbitrary integration; ξ^{I} and ξ^{II} are eigenvalues of the characteristic matrix $[\mathbf{E}]=[\mathbf{A_3}]^{-1}\{s[\mathbf{A_1}]+[\mathbf{A_2}]\}$ of the system of ODEs in Eq. (11), where $\xi^{\mathrm{I}}, \xi^{\mathrm{II}}=\left[E_{11}+E_{22}\pm\sqrt{(E_{11}-E_{22})^2+4E_{12}E_{21}}\right]\Big/2$, respectively; χ_1 and χ_2 stands for: $\chi_1=\left(\xi^{\mathrm{I}}-E_{11}\right)/E_{12}$ and $\chi_2=\left(\xi^{\mathrm{II}}-E_{11}\right)/E_{12}$, respectively; $[\psi_1 \quad \psi_2]^{\mathrm{T}}$ is the particular solution to Eq. (11) and can be expressed as $[\psi_1 \quad \psi_2]^{\mathrm{T}}=s\{s[\mathbf{A_1}]+[\mathbf{A_2}]\}^{-1}\cdot[\bar{\alpha}^m \quad \bar{\alpha}^f]^{\mathrm{T}}$.

The integration of $r\tilde{\varepsilon}_{kk}$ based on $r\tilde{\varepsilon}_{kk}=\partial(r\tilde{u}_r)/\partial r+r\tilde{\varepsilon}_{zz}$, together with integrations of I_0 and K_0 in [10], the radial displacement is written as

$$\begin{aligned}\tilde{u}_r=&-b_2\frac{I_1\left(r\sqrt{\xi^{\mathrm{I}}}\right)}{\bar{G}\sqrt{\xi^{\mathrm{I}}}}B_1-b_3\frac{I_1\left(r\sqrt{\xi^{\mathrm{II}}}\right)}{\bar{G}\sqrt{\xi^{\mathrm{II}}}}B_2+b_2\frac{K_1\left(r\sqrt{\xi^{\mathrm{I}}}\right)}{\bar{G}\sqrt{\xi^{\mathrm{I}}}}B_3\\ &+b_3\frac{K_1\left(r\sqrt{\xi^{\mathrm{II}}}\right)}{\bar{G}\sqrt{\xi^{\mathrm{II}}}}B_4+\frac{r}{2\bar{G}}b_1\tilde{C}_0+\frac{1}{r}\tilde{C}_1-\frac{r}{2}\tilde{\varepsilon}_{zz}\end{aligned} \tag{13}$$

where $b_1=\bar{G}-\bar{\eta}^m\psi_1-\bar{\eta}^f\psi_2$, $b_2=\bar{\eta}^m+\bar{\eta}^f\chi_1$ and $b_3=\bar{\eta}^m+\bar{\eta}^f\chi_2$.

The radial stress can be obtained from Eq. (1) as:

$$\begin{aligned}\tilde{\sigma}_{rr}=&\left[\frac{3(1-\bar{\nu})\bar{K}}{1+\bar{\nu}}-b_1\right]\tilde{C}_0+2b_2I_1\left(r\sqrt{\xi^{\mathrm{I}}}\right)\Big/\left(r\sqrt{\xi^{\mathrm{I}}}\right)B_1+2b_3I_1\left(r\sqrt{\xi^{\mathrm{II}}}\right)\Big/\left(r\sqrt{\xi^{\mathrm{II}}}\right)B_2\\ &-2b_2K_1\left(r\sqrt{\xi^{\mathrm{I}}}\right)\Big/\left(r\sqrt{\xi^{\mathrm{I}}}\right)B_3-2b_3K_1\left(r\sqrt{\xi^{\mathrm{II}}}\right)\Big/\left(r\sqrt{\xi^{\mathrm{II}}}\right)B_4-\frac{2\bar{G}}{r^2}\tilde{C}_1-\bar{G}\tilde{\varepsilon}_{zz}\end{aligned} \tag{14}$$

The vertical force is obtained by integrating the vertical stress:

$$2\pi\int r\tilde{\sigma}_{zz}dr = 2\pi r^2\bar{G}\varepsilon_{zz} + \pi r^2 b_4\tilde{C}_0 + 2\pi rb_5\left[I_1\left(r\sqrt{\xi^{\mathrm{I}}}\right)/\sqrt{\xi^{\mathrm{I}}}B_1 - K_1\left(r\sqrt{\xi^{\mathrm{I}}}\right)/\sqrt{\xi^{\mathrm{I}}}B_3\right] + 2\pi rb_6\left[I_1\left(r\sqrt{\xi^{\mathrm{II}}}\right)/\sqrt{\xi^{\mathrm{II}}}B_2 - K_1\left(r\sqrt{\xi^{\mathrm{II}}}\right)/\sqrt{\xi^{\mathrm{II}}}B_4\right] \tag{15}$$

where $b_4 = \bar{\alpha}^m\psi_1 + \bar{\alpha}^f\psi_2 + \left(\bar{G} - \psi_1\bar{\eta}^m - \psi_2\bar{\eta}^f\right)\frac{2\bar{v}}{1-2\bar{v}}$;
$b_5 = \bar{\alpha}^m + \bar{\alpha}^f\chi_1 - \left(\bar{\eta}^m + \chi_1\bar{\eta}^f\right)\frac{2\bar{v}}{1-2\bar{v}}$; $b_6 = \bar{\alpha}^m + \bar{\alpha}^f\chi_2 - \left(\bar{\eta}^m + \chi_2\bar{\eta}^f\right)\frac{2\bar{v}}{1-2\bar{v}}$.

3 Boundary Conditions

As for the hollow cylinder with a height of 2H subject to a confining load f from the top, the boundary conditions can be described as follows:

$$r = r_o,\ \tilde{\sigma}_{rr} = \tilde{p}^{\mathrm{I}} = \tilde{p}^{\mathrm{II}} = 0;\ r = r_i,\ \tilde{\sigma}_{rr} = \tilde{p}^{\mathrm{I}} = \tilde{p}^{\mathrm{II}} = 0;\ z = \pm H,\ \int_{r_i}^{r_o} r\tilde{\sigma}_{zz}dr = \frac{\tilde{f}}{2\pi}$$

The unknowns are solved in the Laplace transform domain, and the real solution can be obtained by using the numerically inversed method proposed by [11].

4 Numerical Analysis

In this section, we present the numerical analysis of the characteristics of pore pressure in a hollow cylinder and focus on the effect of transport coefficient. The parameters of porous medium are listed in the Table 1 and the outer and inner radius are r_{o} = 1.0 m and r_i = 0.5 m, respectively.

Table 1. Poroelastic parameters

	Matrix	Fractures
Bulk volume fraction	$v^m = 0.98$	$v^f = 0.02$
Intrinsic permeability (m/s)	$k^m = 1\times10^{-10}\,\mathrm{m/s}$	$k^f = 1\times10^{-8}\,\mathrm{m/s}$
Fluid dynamic viscosity (Pa·s)	$\mu^m = 0.01\,\mathrm{Pa\cdot s}$	$\mu^f = 0.01\,\mathrm{Pa\cdot s}$
Bulk modulus (MPa)	$K^m = 110\,\mathrm{MPa}$	$K^f = 22\,\mathrm{MPa}$
Bulk modulus of solid grain (MPa)	$K_s^m = 27.6\,\mathrm{GPa}$	$K_s^f = 27.6\,\mathrm{GPa}$;
Porosity	$\phi^m = 0.14$	$\phi^f = 0.95$
Bulk modulus of fluid (MPa)	$K_f = 1744\,\mathrm{MPa}$	
External force on the top (N)	$f = 10000\,\mathrm{N}$	

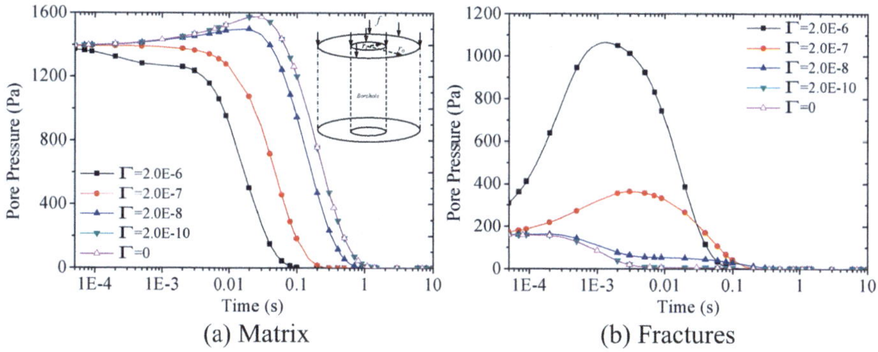

Fig. 1. Pore pressure in the matrix and fractures at $r = 0.75$ m varied with time ($\Gamma = 2\times10^{-6}$ $(\text{Pa·s})^{-1}$, 2×10^{-7} $(\text{Pa·s})^{-1}$, 2×10^{-8} $(\text{Pa·s})^{-1}$, 2×10^{-10} $(\text{Pa·s})^{-1}$ and 0, respectively).

Figure 1 depicts the pore pressure varied with time in the matrix and fractures at the r = 0.75 m, and five transport coefficients are adopted ($\Gamma = 2 \times 10^{-6}$ $(\text{Pa·s})^{-1}$, 2×10^{-7} $(\text{Pa·s})^{-1}$, 2×10^{-8} $(\text{Pa·s})^{-1}$, 2×10^{-10} $(\text{Pa·s})^{-1}$ and 0). With a larger value of transport coefficient, the fluid in the matrix can be transported in a faster rate into the fractures. The pore pressure in the matrix dissipates without Mandel-Cryer effect

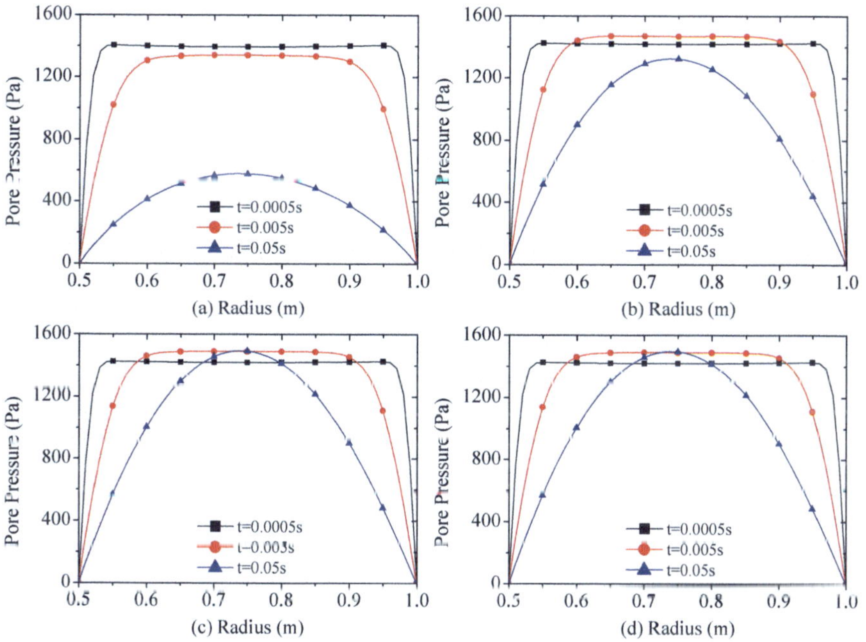

Fig. 2. Pore pressure in the matrix along the radius at time = 0.0005 s, 0.005 s and 0.05 s, respectively. (a) $\Gamma = 2 \times 10^{-7}$ $(\text{Pa·s})^{-1}$; (b) $\Gamma = 2 \times 10^{-8}$ $(\text{Pa·s})^{-1}$; (c) $\Gamma = 2 \times 10^{-10}$ $(\text{Pa·s})^{-1}$; (d) $\Gamma = 0$.

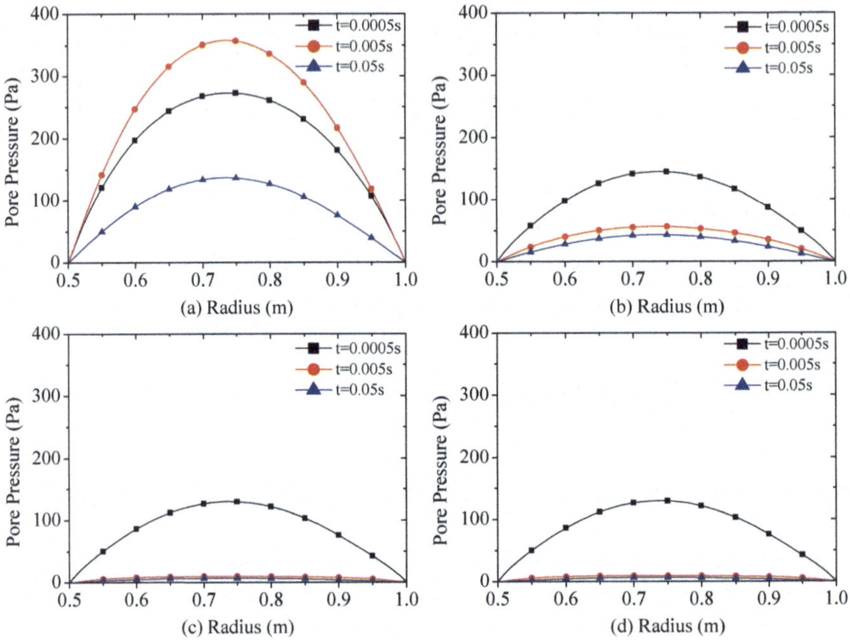

Fig. 3. Pore pressure in the fractures along the radius at time = 0.0005 s, 0.005 s and 0.05 s, respectively. (a) $\Gamma = 2 \times 10^{-7}$ (Pa·s)$^{-1}$; (b) $\Gamma = 2 \times 10^{-8}$ (Pa·s)$^{-1}$; (c) $\Gamma = 2 \times 10^{-10}$ (Pa·s)$^{-1}$; (d) $\Gamma = 0$.

($\Gamma = 2 \times 10^{-6}$ (Pa·s)$^{-1}$), whilst the pore pressure in the fractures climbs obviously and the peak pressure can be 3 times higher than the initial one. As the transport coefficient being close to zero, the pore pressure in the matrix and fractures behaves as the single-porosity way in Biot's theory with two different values of permeability.

Figures 2 and 3 depict the pore pressure along the hollow cylinder with four transport coefficients ($\Gamma = 2 \times 10^{-7}$ (Pa·s)$^{-1}$, 2×10^{-8} (Pa·s)$^{-1}$, 2×10^{-10} (Pa·s)$^{-1}$ and 0) at different points in time (t = 0.0005 s, 0.005 s and 0.05 s). The maximum pore pressure is located in the middle of hollow cylinder and declines due to the drained outer and inner boundaries. As suggested in Fig. 1, Mandel-Cryer effect in the matrix is less obvious with a high transport coefficient since pore pressure in Fig. 2(a) declines monotonously, and that in fractures shown in Fig. 3(a) increases obviously due to the fluid from the matrix before it later decreases into final zero value. However, the effect of transport coefficient has a limit as little differences between the case of $\Gamma = 2 \times 10^{-10}$ (Pa·s)$^{-1}$ and that of $\Gamma = 0$ can be observed.

5 Conclusions

In this paper, the analytical solutions to the saturated hollow cylinder with double porosities are derived to investigate the hydro-mechanical behaviors of fractured formation around a borehole. Laplace transform for the time domain is adopted to de-couple the governing equations and after the application of boundary conditions, the real solution is obtained by the corresponding Laplace transform inversion. The pore pressure in the matrix and fractures is studied and emphases have been placed on the influence of the transport coefficients between the matrix and fractures. The results show that due to the mass transfer between the matrix and fractures with a high transport coefficient, the dissipation in the matrix can declines monotonously while the pore pressure in the fractures increases significantly due to the fluid from the matrix. However, the hollow cylinder simulating the excavation disturbed zone is a simplified model since the outer boundary conditions in the real production problem is governed by the compatibility between the disturbed zone and the undisturbed zone. Such case involves the radial heterogeneity and a radially multilayered model should be presented for the further analysis.

References

1. Kaewjuea, W., Senjuntichai, T.: Poromechanical response of borehole in excavation disturbed zone. Comput. Geotech. **56**, 148–159 (2014)
2. Mehrabian, A., Abousleiman, Y.: Gassmann equations and the constituitive relations for multi-porosity and multi-permeability poroelasticity with applications to oil and gas shale. Int. J. Numer. Anal. Meth. Geomech. **39**, 1547–1569 (2015)
3. Liu, C., Mehrabian, A., Abousleiman, Y.: Poroelastic dual-porosity/dual-permeability after-closure pressure-curves analysis in hydraulic fracturing. Soc. Petrol. Eng. **22**(1), 1–20 (2017)
4. Biot, M.: General theory of three-dimensional consolidation. J. Appl. Phys. **12**(2), 155–164 (1941)
5. Wilson, R., Aifantis, E.: On the theory of consolidation with double porosity. Int. J. Eng. Sci. **20**, 1009–1035 (1982)
6. Berryman, J.: Extension of poroelastic analysis to double-porosity materials: new technique in microgeomechanics. J. Eng. Mech. **128**(8), 840–847 (2002)
7. Salimzadeh, S., Khalili, N.: Three-dimensional numerical model for double-porosity media with two miscible fluids including geomechanical response. Int. J. Geomech. **16**(3), 1–14 (2016)
8. Berryman, J., Pride, S.: Models for computing geomechanical constants of double-porosity materials from the constituents' properties. J. Geophys. Res. **107**, 1–15 (2002)
9. Edwards, C., Penny, D.: Differential equations and boundary value problem. Prentice Hall, New Jersey (1996)
10. Abramowitz, M., Stegun, I.: Handbook of Mathematical Function. National Bureau of Standards, Washington (1964)
11. Talbot, A.: The accurate inversion of Laplace transforms. J. Inst. Math. Appl. **23**, 97–120 (1979)

Rapidly Determining the Unsaturated Permeability by Transient Two-Steps Outflow Experiment

Tiande Wen[1,2], Longtan Shao[1,2(✉)], and Xiaoxia Guo[1,2]

[1] State Key Laboratory of Structural Analysis for Industrial Equipment, Dalian University of Technology, Dalian 116024, China
shaolt@hotmail.com, shaolt@126.com
[2] Department of Engineering Mechanics, Dalian University of Technology, Dalian 116024, China

Abstract. Numerous methods have been developed to measure the unsaturated permeability in multi-step outflow experiments. Silicon micro-powder (SMP) and Guangxi Guiping clay (GGC) have been conducted by transient outflow methods in this paper. The water release upon two-steps increase in matric suction. For each step of the experiment, the water outflow from the soil specimen is weighed in real time in order to determine the change of the water content in the specimen. As a gas pressure is exerted, the difference of the matric potential between the transient water content curve (TWCC) and the soil water retention curve (SWRC) at the same water content is calculated. This difference is considered to have the potential to drive the flow of pore water. Then, the unsaturated permeability can be calculated by using Darcy's law. This two-steps measurement method agrees with the Gardner outflow analysis method (1956), which have the same ranges of variation for unsaturated permeability. Additionally, the test time is greatly reduced, and generally takes 10 days.

Keywords: Unsaturated soil · Two-steps outflow experiment
Unsaturated permeability

1 Introduction

The unsaturated permeability is an important parameter for studying seepage, unsaturated soil consolidation and contaminant transport. It is mainly affected by the mineral composition of soil particles, the size and distribution of pores, the water content and different ion concentrations. The unsaturated permeability is generally measured by experiment or is estimated from the soil water retention curve.

Techniques for measuring the unsaturated permeability, including the horizontal infiltration method, the centrifuge method, the multi-step outflow method, and the instantaneous profile method (e.g., ASTM D7664 2010; Gardner 1956; Klute 1972; Elzeftawy and Mansell 1975; Raimbault 1986; Abuhejleh et al. 1993; Meerdink et al. 1996; Benson and Gribb 1997; Znidarcic et al. 2002; Vanapalli et al. 2007; Zornberg and Mccartney 2010; Znidarcic and Mccartney 2010; Parks et al. 2012), have been well developed. The multi-step outflow method has become one of the most promising and

L. Hu et al. (Eds.): GSIC 2018, *Proceedings of GeoShanghai 2018 International Conference: Multi-physics Processes in Soil Mechanics and Advances in Geotechnical Testing*, pp. 228–235, 2018.
https://doi.org/10.1007/978-981-13-0095-0_26

practical methods for estimating unsaturated permeability. The original outflow method for measuring the unsaturated permeability via a pressure plate instrument was proposed by Gardner (1956). In the multi-step outflow method, matric suction is applied on a soil specimen by increasing the gas pressure. Then, the outflow rate and cumulative outflow volume corresponding to each step of matric suction are measured. The linear equation of the volume of the pore water is used to derive the water diffusivity of unsaturated soil, and the water diffusivity D is calculated from the slope of the linear relationship in Eq. (1). The unsaturated permeability for matric suction is given as shown in Eq. (2):

$$\ln(V_0 - V_t) = \ln(\frac{8V_0}{\pi^2}) - \frac{D\pi^2 t}{4L^2} \tag{1}$$

$$k_u = D\frac{\Delta\theta}{\Delta\psi} \tag{2}$$

where D is the water diffusivity, V_0 is the cumulative outflow volume for a specific step of matric suction, V_t is the cumulative outflow volume at elapsed time t from the initial application, L is the thickness of the specimen, and $\Delta\theta/\Delta\psi$ is the change in water content for an increment in matric suction; this is the slope of the SWRC.

The multi-step outflow method is a transient measurement method for estimating the unsaturated permeability in the laboratory. Because this method tends to produce bubble accumulation close to the ceramic disk, diminishing the accuracy of measurement. Therefore it improved by Miller and Elrick (1958), Rijtema (1959), Kunze and Kirkham (1962), Van Dam et al. (1994), Crescimanno and Iovino (1995), Hwang and Powers (2001), and Figueras et al. (2006). Kool et al. (1985) developed an inverse method to estimate the parameters of the SWRC from cumulative outflow measurements. They demonstrated that the outflow data obtained from a multi step outflow experiment provided sufficient information to measure the unsaturated permeability. Benson and Gribb (1997) presented a comprehensive review of inverse modeling approaches used to estimate the hydraulic conductivity, including one-step and multi-step outflow methods.

Shao et al. (2017) presented a method to measure the unsaturated permeability in multi-step outflow experiments, and it aggred with Gardner (1956) analysis method and van Genuchten-Mualem model (1980). However, the multistep outflow experiment is time consuming, it generally takes two or three months. In this paper, a new method is proposed to calculate the unsaturated permeability in two-steps outflow experiment, which took approximately 10 days. Silicon micro-powder (SMP) and Guangxi Guiping clay (GGC) have been conducted by transient two-steps outflow experiments.

2 Measurement Method

The level of matric suction conditions can be considered to have reached a stable state, when the pore water is no longer discharged from the soil specimen. At the same time, the pore water is in a state of static equilibrium corresponding to the previous step's outflow. It can be said that the gas pressure potential increment applied is completely transformed into the matric potential (Shao et al. 2017). The SWRC is described the relationship between the matric suction and volumetric water content, as shown in Fig. 1. The matric potential of the soil water can be increased instantaneously with the application of a gas pressure potential increment. The gas pressure path is [0, ψ_B, ψ_D] or A-B-C-D-E. It is a transient two-steps outflow experiment. As shown in Fig. 1, Δu_a is the gas pressure potential increment, therefore the matric potential satisfies $\psi_E = \psi_D = \psi_B + \Delta u_a$ at point E. When the gas pressure potential increment is applied, the pore water in the soil specimen outflows gradually from D to E during this gas pressure potential increment. The hydraulic gradient driving pore water seepage gradually decreases until it is close to 0 at point E.

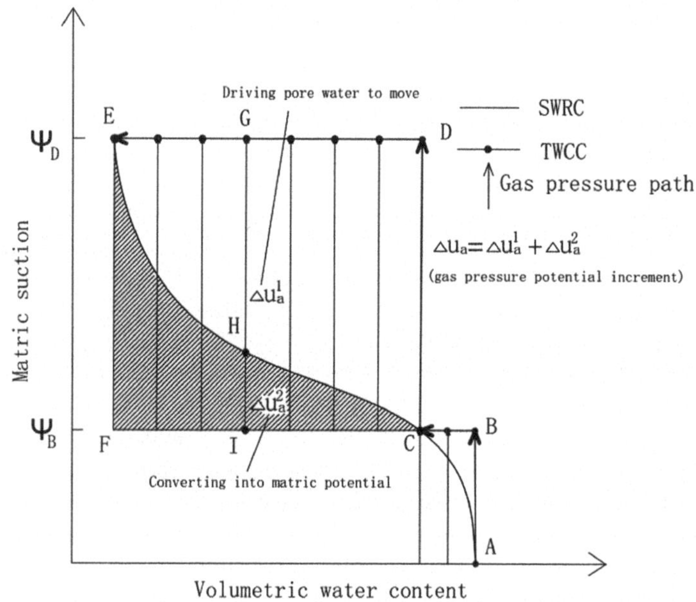

Fig. 1. A diagram of the direct measurement method

As shown in Fig. 1, the test data curve is the transient water content curve (TWCC). The first step outflow is $\overline{ABC}$, the gas pressure increment at this step is little larger than the air-entry value of soil. When the gas pressure increment at second step outflow is applied ($\overline{CDE}$), and the limit of total matric suction must less than the air entry pressure of the ceramic disk. Here, G is one test data point on the TWCC under the condition of matric suction ψ_D. H is the point at which the volumetric water content equals that at point G on the SWRC, which mean that $\theta_G = \theta_H$. During this step, the path of gas pressure increment is $\overline{CDE}$, the change path of volumetric water content on the TWCC is $\overline{DGE}$, and the corresponding SWRC is $\widehat{CHE}$. In this method, the increment of the gas pressure potential, Δu_a, is divided into two parts: Δu_a^1 and Δu_a^2. Where Δu_a^1 overcomes the osmotic resistance and drives pore water; Δu_a^2 balances the surface tension in the hydrostatic equilibrium condition and converts it into matric potential. At the moment of gas pressure potential is applied, Δu_a^1 equals Δu_a and gradually decreases with the outflow of pore water. When the stable stage of matric suction is reached, Δu_a^1 equals zero. Conversely, Δu_a^2 behaves in a completely different way. When the gas pressure potential is first applied, $\Delta u_a^2 = 0$ and then gradually increases with the outflow of pore water. When matric suction reaches a steady state, Δu_a^2 equals Δu_a. Finally, the gas pressure potential is completely transformed into the matric potential.

It should be noted that Δu_a^1 is the part of the gas pressure potential that overcomes the osmotic resistance and drives pore water. Therefore, the hydraulic gradient can be written as follows:

$$\frac{\partial h}{\partial z} = 1 + \frac{1}{\gamma_w}\frac{\partial \psi}{\partial z} = 1 + \frac{1}{\gamma_w}\frac{\Delta u_a^1}{\Delta z} \tag{3}$$

The flow of pore water in unsaturated soil can be expressed by Darcy's law. The ratio of flow velocity to the hydraulic gradient is called the unsaturated permeability:

$$k_u = -\frac{\Delta V}{A\Delta t}\left(\frac{\partial h}{\partial z}\right)^{-1} = -\frac{\Delta V}{A\Delta t}\left(1 + \frac{\Delta u_a^1}{\gamma_w \Delta z}\right)^{-1} \tag{4}$$

where $\partial h/\partial z$ is the hydraulic gradient, $v = \frac{\Delta V}{A\Delta t}$ is the flow velocity, ΔV is the cumulative outflow volume at the time interval Δt, A is the cross-sectional area of the specimen, Δz is the thickness of the specimen, and γ_w is the unit weight of water.

3 Instrument, Materials and Experimental Procedures

3.1 Instrument

The pressure plate instrument is adopted an evaporation compensation system in contrast to the conventional instrument. The specimen is a cylinder with a diameter of 54.7 mm and a height of 20 mm. The specimen is tightly attached to the two ceramic disks during the experiments. The ring cutter guarantees that the gas pressure is applied around the soil so that the water is trapped in the soil. The increased gas pressure leads to elevated gas pressure while maintaining the pore water pressure constant; the real-time volume of cumulative outflow water and the water content of the soil specimen with high-precision balance are then measured. Notably, the evaporation compensation system is vital for compensating for the evaporative loss in the cumulative outflow volume measurement and has a great impact on the test results (Shao et al. 2017).

3.2 Materials and Experimental Procedures

Silicon micro-powder (SMP) and Guangxi Guiping clay (GGC) were used in this experiment. Table 1 lists their properties. The experimental data of SMP and GGC

Table 1. Property parameters of the tested soils

Soil	G_s	Grains composition, %						ρ_d g/cm^3
		k_s m/s	0.5–0.25 mm	0.25–0.075 mm	0.075–0.025 mm	0.025–0.005 mm	<0.005 mm	
SMP	2.68	1.09e−7	0.90	15.34	43.38	31.45	8.93	1.6
GGC	2.74	4.72e−9	0.00	5.32	30.64	43.51	20.53	1.35

obtained from the two-steps outflow experiments are recorded. The saturated soil specimen in this experimental study is preprocessed: (1) flush the bubbles from the exhaust pipe and water pipe, and fill them with air-free water; (2) saturate the ceramic disk and measure the saturated permeability of the ceramic disk; and (3) install the saturated soil specimen, and start the dehydration experiment. It should be noted that the gas pressure can be controlled to apply matric suction step by step. The matric suction path of SMP is [0 kPa, 40 kPa, 450 kPa]; the matric suction path of GGC is [0 kPa, 40 kPa, 450 kPa].

4 Results

It can be obtained the relationship between the cumulative outflow volume and time during this two-steps experiment. And we also can get the relationship between the volumetric water content and time, and the flow velocity of pore water. The TWCC can be obtained from this two steps of the outflow process. Comparing the TWCC with the SWRC, the hydraulic gradient can be obtained by using Eq. (3). Then, the unsaturated permeability can be derived from Eq. (4).

Figure 2 shows the relationship between the cumulative outflow volume and time for SMP and GGC. The experimental time is longer with the increase in the matric suction. Comparison of these two types of soils shows that the testing time of SMP, it took about 75 h, is much faster than the GGC, which took approximately 217 h.

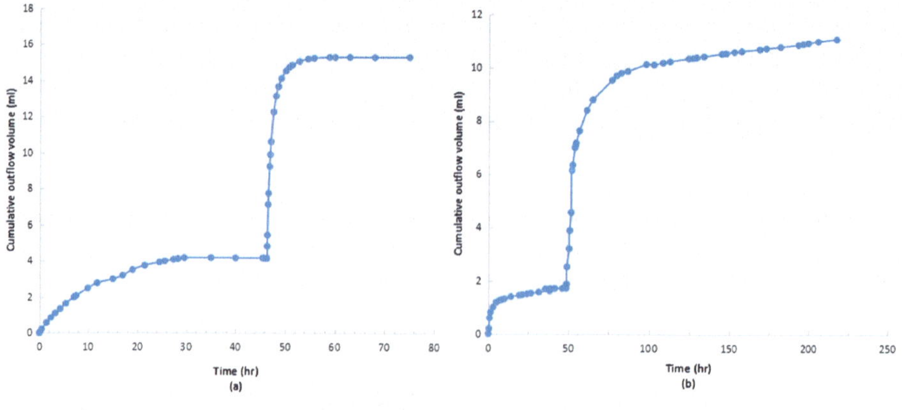

Fig. 2. Relationships between cumulative outflow volume and time for (a) silica micro-powder and (b) Guangxi Guiping clay

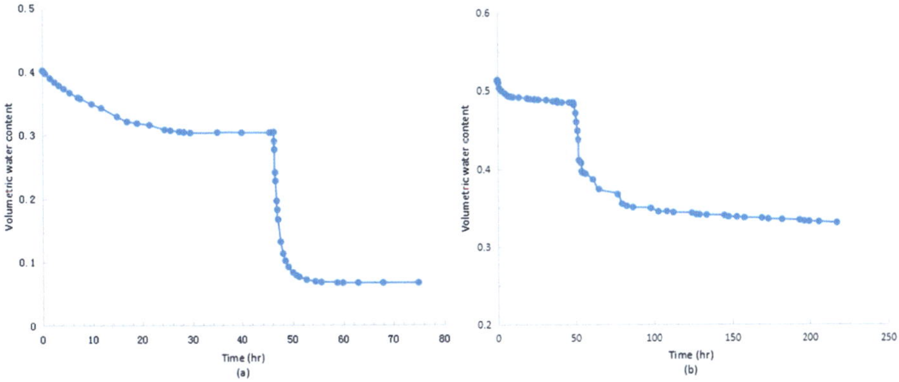

Fig. 3. Relationships between volumetric water content and time for (a) silica micro-powder and (b) Guangxi Guiping clay

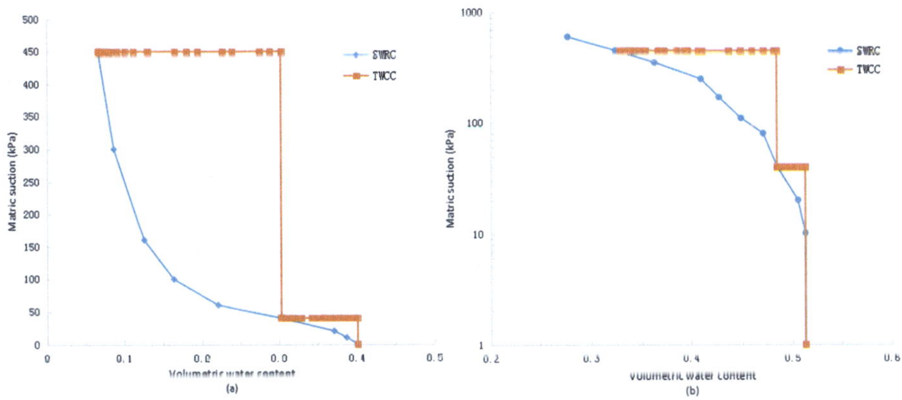

Fig. 4. Transient volumetric water content curves (TWCC) and static soil water retention curves (SWRC) for (a) silica micro-powder and (b) Guangxi Guiping clay

Figure 3 shows the relationship of the volumetric water content and time for SMP and GGC. The saturated volumetric water content is 0.401 for SMP, and 0.513 for GGC. The volumetric water content decrease during each step of the applied constant matric suction, and tend to stable finally.

Figure 4 shows the TWCC, which demonstrates a decreasing volumetric water content during each step of the applied constant matric suction. Additionally, the SWRC is below the TWCC. From the data shown in Fig. 4, we can obtain Fig. 5 by using Eq. (4), which clearly demonstrates that the range of variation in unsaturated permeability during each step of the outflow process is positively correlated to the volumetric water content. The unsaturated permeability is obtained with the two-steps measurement method proposed by using Eq. (4) in this paper. In the meantime, the unsaturated permeability can be obtained by Gardner outflow analysis method (1956) in multistep (more than 2-steps) outflow experiments, the detailed content could be

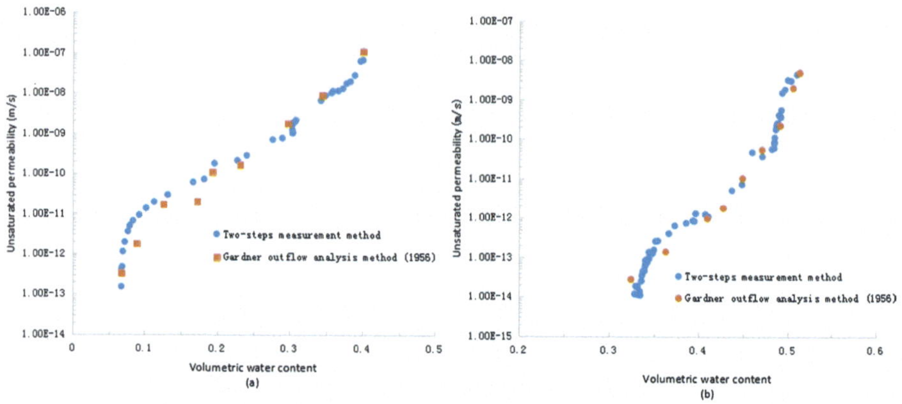

Fig. 5. The hydraulic conductivity versus volumetric water content curves for (a) silica micro-powder and (b) Guangxi Guiping clay

seen in Shao et al. (2017). This two-steps measurement method agrees with the Gardner outflow analysis method, which have the same ranges of variation for unsaturated permeability.

5 Conclusions

The data points obtained by the two-steps measurement method in this paper and the Gardner outflow analysis method have the same ranges of variation. Thus, it can be concluded that the use of the two-steps measurement method is feasible. Additionally, it is important to note that it is relatively easy to measure the unsaturated permeability according to the TWCC and SWRC directly, and test time is greatly reduced compared with Gardner outflow analysis method in multistep outflow experiments.

Acknowledgment. This research was funded by the National Natural Science Foundation of China (No. 51479023).

References

ASTM D7664: Standard Test Methods for Measurement of Hydraulic Conductivity of Unsaturated Soils. Annual Book of ASTM Standards, vol. 4.09. ASTM International, West Conshohocken (2010)

Abuhejleh, A.N., Znidarčić, D., Illangasekare, T.H.: Permeability determination for unsaturated soils. In: Unsaturated Soils, pp. 163–174. ASCE (1993)

Benson, C.H., Gribb, M.M.: Measuring unsaturated hydraulic conductivity in the laboratory and field. Geotech. Spec. Publ. **68**, 113–168 (1997)

Crescimanno, G., Iovino, M.: Parameter estimation by inverse method based on one-step and multi-step outflow experiments. Geoderma **68**(4), 257–277 (1995)

Elzeftawy, A., Mansell, R.S.: Hydraulic conductivity calculations for unsaturated steady-state and transient-state flow in sand. Soil Sci. Soc. Am. J. **39**(4), 599–603 (1975)

Figueras, J., Gribb, M.M., McNamara, J.P.: Design of a new automated multi-step outflow test apparatus. In: AGU Fall Meeting. AGU Fall Meeting Abstracts (2006)

Gardner, W.R.: Calculation of capillary conductivity from pressure plate outflow data. Soil Sci. Soc. Am. J. **20**(3), 317–320 (1956)

Hwang, S., Powers, S.E.: Using a multi-step outflow technique to estimate unsaturated hydraulic conductivity function for quartz sands with variable interfacial properties. American Geophysical Union (2001)

Kunze, R.J., Kirkham, D.: Simplified accounting for membrane impedance in capillary conductivity determinations. Soil Sci. Soc. Am. J. **26**(5), 421–426 (1962)

Kool, J.B., Parker, J.C., Van Genuchten, M.T.: Determining soil hydraulic properties from one-step outflow experiments by parameter estimation: I. Theory and numerical studies. Soil Sci. Soc. Am. J. **49**(6), 1348–1354 (1985)

Klute, A.: The determination of the hydraulic conductivity and diffusivity of unsaturated soils1. Soil Sci. **113**(4), 264–276 (1972)

Miller, E.E., Elrick, D.E.: Dynamic determination of capillary conductivity extended for non-negligible membrane impedance1. Soil Sci. Soc. Am. J. **22**(6), 483–486 (1958)

Meerdink, J.S., Benson, C.H., Khire, M.V.: Unsaturated hydraulic conductivity of two compacted barrier soils. J. Geotech. Eng. **122**(7), 565–576 (1996)

Parks, J.M., Stewart, M.A., McCartney, J.S.: Validation of a centrifuge permeameter for investigation of transient infiltration and drainage flow processes in unsaturated soils. ASTM Geotech. Test. J **35**(1), 182–192 (2012)

Rijtema, P.E.: Calculation of capillary conductivity from pressure plate outflow data with nonnegligible membrane impedance. Institute for Land and Water Management Research (1959)

Raimbault, G.: Diffusivite et conductivite hydrauliques de materiaux ou sols non satures en eau - mesure en laboratoire. Bull. Liaison Lab. Ponts Chauss **145**, 61–68 (1986)

Shao, L.T., Wen, T.D., Guo, X.X., Sun, X.: A method for directly measuring the hydraulic conductivity of unsaturated soil. Geotech. Test. J. **40**(6), 20160197 (2017)

van Dam, J.C.: Inverse method to determine soil hydraulic functions from multi-step outflow experiments. Soil Sci. Soc. Am. J. **58**(3), 647–652 (1994)

van Genuchten, M.T.: A closed-form equation for predicting the hydraulic conductivity of unsaturated soils. Soil Sci. Soc. Am. J. **44**(5), 892–898 (1980)

Vanapalli, S.K., Garga, V.K., Brisson, P.: A modified permeameter for determination of unsaturated coefficient of permeability. Geotech. Geol. Eng. **25**(2), 191–202 (2007)

Znidarcic, D., Mccartney, J.S.: Testing system for hydraulic properties of unsaturated nonwoven geotextiles. Geosynth. Int. **17**(5), 355–363 (2010)

Znidarcic, D., Hwang, C. Bicalho, K. V.: Experimental determination of hydraulic characteristics for unsaturated soils. In: Juca, J.F.T., de Campos, T.M.P., Marinho, F.A.M. (eds.) Proceedings of the 3rd International Conference on Unsaturated Soils, vol. 3, pp. 1137–1141. Swets & Zeitlinger, Lisse (2002)

Zornberg, J.G., McCartney, J.S.: Centrifuge permeameter for unsaturated soils I: theory and experimental philosophy. ASCE J. Geotech. Geoenviron. Eng. **136**(8), 1051–1063 (2010)

Pore-Network Model for Geo-Materials

Liming Hu[1(✉)], Haohao Guo[1], Pengwei Zhang[1], and Dongming Yan[2]

[1] Department of Hydraulic Engineering, Tsinghua University, Beijing, China
gehu@tsinghua.edu.cn
[2] National Laboratory of Pattern Recognition (NLPR), Institute of Automation, Chinese Academy of Sciences, Beijing, China

Abstract. Pore-network model is a convenient tool to investigate the micromechanics of seepage in porous media. Geo-materials are typical porous media, including the different types of soils and rocks from rock-fill with mm-scale pores with high connectivity to gas shale with nm-scale pores and little connectivity. Based on the 2D image from CT or micro-CT technology, the 3D image of soil aggregates/rock matrix and pore structures for different types of geo-materials were obtained by the advance computational graphics technology. The pore size distribution and connectivity were derived from the developed 3D model, which agreed well with the experiment result. The seepage process was also simulated numerically via the developed micro-mechanics seepage model, and the pore-scale phenomena was revealed such as preferential flow. The hydraulic conductivities for various types of geo-materials from numerical simulation agreed well with the laboratory testing data, demonstrating the potential capability of pore-network model in hydraulic properties study for geo-materials.

Keywords: Porous materials · CT images reconstruction
Pore-network model

1 Introduction

In recent years, seepage phenomenon in porous media had been extensively studied due to its influence in many engineering fields. Water seepage phenomenon caused damages in most part of mine, rock slope and dam [1], with an annual loss of over 10 billion yuan because of the roads waterlogging phenomenon [2]. Hence, the physical properties of the porous media and the micromechanics of seepage were thoroughly studied.

As a typical porous media, geo-materials include different types of soil and rock, and the size of them covers from rock-fill with mm-scale pores and high connectivity to gas shale with nm-scale pores and little connectivity. To study the microstructure of the geo-materials, the existing instruments mainly acquired the micro feature of the materials by using the technology of CT, MRI or SEM etc. Among the various technologies, CT technology was fit for extracting the internal characteristics of geo-materials relying on its high resolution with a pixel resolution of μm-level [3], quasi real-time property which could be used to create the quasi dynamic changing process [4], lossless properties that would not affect the geo-material sample when

L. Hu et al. (Eds.): GSIC 2018, *Proceedings of GeoShanghai 2018 International Conference: Multi-physics Processes in Soil Mechanics and Advances in Geotechnical Testing*, pp. 236–243, 2018.
https://doi.org/10.1007/978-981-13-0095-0_27

scanning [5], and relatively lower cost compared to other technologies. Reconstruction of the 3D structure of the geo-materials according to 2D CT-images serial would reveal the micro features of the geo-material obviously, which could be transformed to a pore-network model further. Pore network, a 3D network consisting of connected pores/nodes and throats/bonds [6], is widely used to study the macroscopic transport properties including capillary pressure, relative permeability and residual saturation, etc. A wide range of the applications such as wettability, multiphase flow dynamics, hysteresis could all use this model to make further research [7].

2 Existing Research

Since the invention of the first CT machine by Hounsfield in 1972, this technology had been widely used in the field of medicine, and the scanning time decreased rapidly from hours to seconds [8]. By 1990 s, the CT technology was used in the geo-mechanics researches in China [9], and many achievements had been accomplished in the fields of structure characteristics [10], deformation characteristics [11, 12], fracture development [13] and pore network [14] of the geo-materials, and pore-network model was going popular due to its high resolution.

On the other side, pore-network model was considered as an available method to simulate the real geo-material micromechanics. Lochmann et al. [15] directly analyzed the changing rules of statistic characteristics when ideal spheres were packed randomly. Dai Hongqin [16] had designed an algorithm to simulate the real sphere packing. From the view of reconstructing the pore-network model, Bryant et al. [17] presented one method to choose the appropriate geometry used in pore-network model construction. Kantzas and Chatzis [18] proposed a flux equation that was relevant to the porous ratio and the geometric shape of the porous. Gao et al. [7] presented a hydraulic numerical simulation of the pore network model, and a comparison between the prediction permeability values the published experimental data had shown good agreement.

3 2D CT-Images Serial Processing

3.1 CT Images Acquirement

CT machine, a scanning and imaging system usually consist of transmitting part, receiver part and processor part, is able to acquire the inner structure of the sample by scanning and transmitting rays. The transmitting part is responsible to transmit rays, mainly X-ray that can be received by receiver part [19]. According to Radon's proof, the attenuation ratio called CT number of all the positions could be back-derived based on the CT images of different angles [20]. After scanning the sample, a serial of CT images could be obtained, and generally a mapping from 12-bit CT images to 8-bit gray-scale images was proceeded. The Fig. 1(a) is one typical CT image [21].

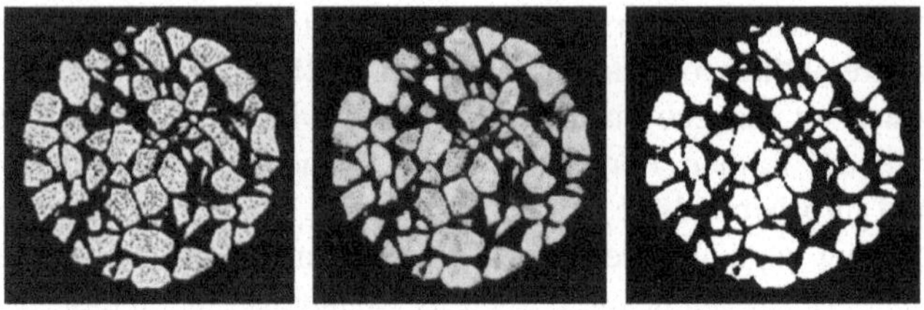

Fig. 1. (a) The origin image outputted from CT. (b) The image proceeded by bilateral filter. (c) The image proceeded by close operation and thresholding.

3.2 CT Images Pre-processing

By eliminating noises from data and revising its structure, we can thus further analyze the data. This procedure is done by applying filtering to the close operation as well as adjusting a threshold.

Filtering is an operation to retain the specific information of the image. Different filters can be divided into linear filters and non-linear filters, and generally the image processed by linear filter such as average filter and Gaussian filter will gain blur edges due to smoothing of pixels. Hence, an improved Gaussian filter called bilateral filter was proposed to solve this problem, which could reserve the edges of the image and remove the noises of it at the same time by dividing the core of the filter into two parts named spatial kernel and range kernel, described as Eq. 1 [22]. $\omega(x, y)$ is the spatial kernel meaning the spatial distance, and $\phi(g(x), g(y))$ is the range kernel meaning the intensity differences. f and g are input and guidance image separately.

$$\tilde{f}(x) = \frac{\int_{\Omega} \omega(x, y)\phi(g(x), g(y))f(y)dy}{\int_{\Omega} \omega(x, y)\phi(g(x), g(y))dy} \tag{1}$$

The effect of the operation is shown at Fig. 1(b). This is not hard to see that the frequency noises, which are shown as black dots in white particles, have not been cleared completely. One of the morphology operation called close operation was efficient to exclude the tiny hole in the image. Close operation is a compound operation combined with a dilation operation followed by an erosion operation. After processing this operation, tiny holes had been excluded mostly. Then a thresholding was proceeded to the image, which gave the result shown as Fig. 1(c).

When the image serial had been proceeded entirely, a 3D model of the sample could be reconstructed easily by interpolation. Figure 2(a) displays a 3D model of the CT data from Blunt et al. [23].

4 Pore-Network Model Construction

4.1 Segmentation of Model

The particles are sticky in the image shown in Fig. 2(a) due to the reason of particles deformation and projection direction, which will lead to a failure to divide different particles by common distinguish methods such as connected domain detection. This article used weight reserved erosion-dilation method to process the segmentation, and it bases on normal erosion-dilation method.

The weight reserved erosion-dilation method appended a weight for each pixel that represented the core-level, in other words the erosion times of the pixel. Larger was the weight, more central was the position. According to this weight, one pixel could be easily determined which domains it belonged. The segmentation result was shown in Fig. 2(b).

It was quite clear that different particles had been segmented to corresponding region, which meant the segmentation result was acceptable. The flow diagram of the method is shown in Fig. 3.

Fig. 2. (a) The 3D structure reconstructed by the CT images serial. (b) One sliced layer after segmentation. (c) Two sample tetrahedron produced by Delaunay tessellation.

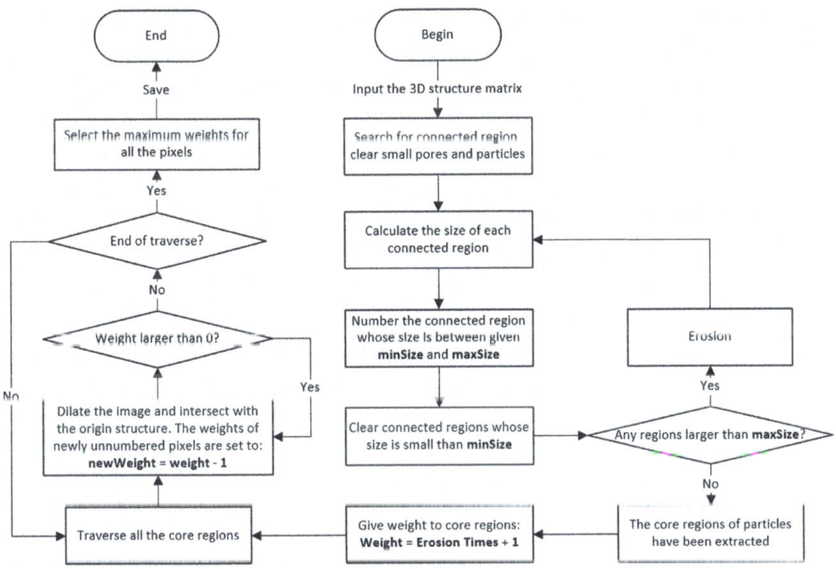

Fig. 3. The flow diagram of the weight reserved erosion-dilation method.

4.2 Pore-Network Structure Extracting

Since all the particles had been segmented, it could be considered as the packing of the geo-material particles. Here, the Delaunay tessellation was used to describe and simplify the structure, which gave a subdivision of the total space into a set of tetrahedral volumes. The volumes represented the void of the packing and the vertices represented the particles, here the centroids of particles were used (see Fig. 2(c)).

Delaunay tessellation produced a convex tetrahedron tessellation, and as the result there were lots of invalid tetrahedrons on the surface of the tessellation, where "invalid" meant voids were directly connected to outer space. Hence, a subtraction from the tetrahedrons set was proceed after the tetrahedrons were divided. Once the invalid tetrahedrons had been eliminated, the pores of the structure could be extracted right away. Here an approximation was used in the model: the pores of the structure could be abstracted as the maximum balls which was able to be inscribed into the void of the tetrahedrons, and the other part of the void be abstracted into a tube to connect pores. Due to the irregularity of the particles shape, the extraction algorithm used a global optimum method instead of a local optimum method to search for the maximum inscribed spheres. Here SCE-UA method [24] that represented the global optimum method and simplex method [25] that represented the local optimum method were both used to find the maximum inscribed balls, and the result showed the local optimum method could not find the maximum value with very high probability, which revealed the limitation of the simplex method.

It could be recognized that some pores were close enough to merge them together. According to Gao et al. [7], The connection situations between two neighbor pores were divided into 3 levels (see Fig. 4). The first situation gave two separate pores and a pore throat to connect them; The second situation two pores were connected directly, while there was still a narrow throat between two pores; The third situation showed two pores that seemed like one pore because the centers of which was close enough to combine with each other. The third situation made the pores own a higher coordination number more than 4, which was closer to the reality. Three situations met the Eqs. 2–4 respectively, in which R_1 and R_2 are the radii of two pores, and d is the distance of the centers.

$$d > R_1 + R_2 \tag{2}$$

$$\sqrt{R_1^2 - R_2^2} < d \leq R_1 + R_2 \tag{3}$$

$$d \leq \sqrt{R_1^2 - R_2^2} \tag{4}$$

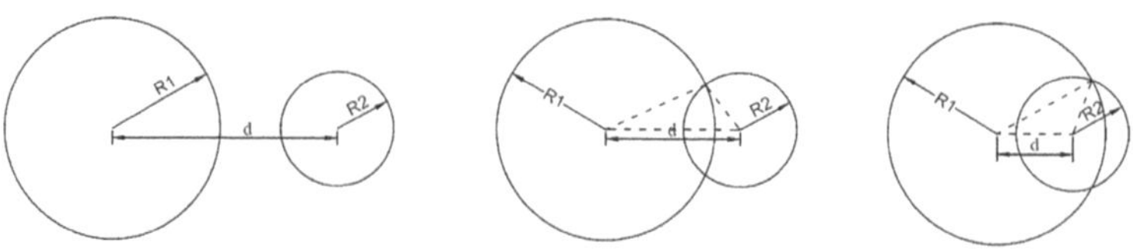

Fig. 4. Three typical connection situations for two pores.

5 Analysis of the Model

To analyze the distribution of the pore radius and the coordination numbers, the CT image data from Blunt et al. [23] was used to extract the pore-network structure. Figure 5(a) shows one typical pore radius distribution, which accorded well with the gamma-distribution fitting line. Using PFC^{3D} to simulate the randomly packing of the spheres, the pore radius distribution accorded with the same fitting line as well (see Fig. 5(b)). Similarly, the coordination number of the structure was shown as Fig. 5(c). Here, the coordination number means the amount of connections one pore connected to others.

As it is shown, the differences between the statistic coefficients of the reconstruction model and the simulation model was small. According to Blunt et al. [26], the water-mechanics properties would vary small if the geometry statistic parameters of the pores were determined. Hence, the developed micromechanics seepage model could be used to process the seepage simulation.

Simulating the single-phase flow in the model gave the predicted permeability of the structure, which could be indirectly compared by published experiment data. The permeability results predicted by the pore-network models reconstructed by CT images from Blunt et al. [23] are shown in Table 1.

Some researchers had given results for the relationship between the permeability and pore size, like Wyllie et al. [27] and Chu et al. [28]. The experiments were proceeded in packing structure, and a constant head permeability experiment was conducted. Plot the result with the CT images models together (see Fig. 6), and one fitting line appears with $R^2 = 0.9976$, which indirectly proved the correctness of the model to some extent, demonstrating the potential capability of pore-network model in hydraulic properties study for geo-materials.

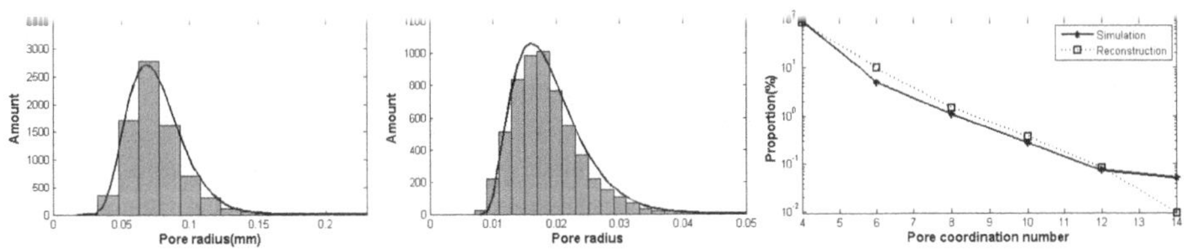

Fig. 5. (a) Pore radius distribution of reconstructed model. (b) Pore radius distribution of simulated model. (c) Pore coordination number distribution of two models.

Table 1. Parameters of the pore-network models and corresponding permeability results.

ID	Ave. particle radius (mm)	Porosity rate (%)	Ave. pore radius (mm)	Ave. pore Coordination Num.	Permeability (μm^2)
F42A	0.2077	32.72	0.0805	4.26	152.48
F42B	0.1970	33.80	0.0773	4.27	146.98
F42C	0.1934	33.31	0.0743	4.26	144.54
LV60A	0.1394	37.85	0.0560	4.29	85.57
LV60C	0.1331	38.57	0.0546	4.29	77.62

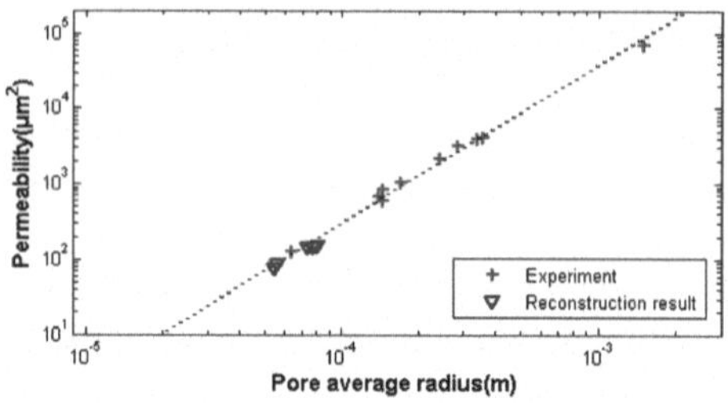

Fig. 6. Relationship between permeability and pore average radius.

6 Summary and Conclusion

A construction method to extract the pore-network model from CT images serial was presented, and the single-phase flow simulation was proceeded to predict the permeability of the structure, which agreed well with the laboratory testing data. A segmentation method named weight reserved erosion-dilation method was presented to solve the problem of sticky particles, and a global optimization method was found required when searching for the maximum inscribed spheres as the abstract pores in pore-network model. Therefore, the model could be used to simulate the single phase flow to predict the permeability of the structure. The simulated permeability results were compared with the published experiment results, and an indirect validation was proceeded by determining the relationship between the permeability and pore radius. The comparison showed the credibility of the model, and more rules could be found by making use of the pore-network model to other fields.

References

1. Zhou, H., et al.: Modeling research on the response of geoelectric fields in a porous media seepage process. J. Geophys. Eng. **14**(2), 408–416 (2017)
2. Tao, Y.: Seepage and stability analysis of coarse grained soil embankment slope under the condition of rainfall. Changsha University of Science and Technology, Changsha (2013)
3. Vicent, V., et al.: A new method developed to characterize the 3D microstructure of frozen apple using X-ray micro-CT. J. Food Eng. **212**, 154–164 (2017)
4. Qin, Y., et al.: A quasi real-time approach to investigating the damage and fracture process in plain concrete by X-Ray tomography. J. Civ. Eng. Manage. **22**(6), 792–799 (2016)
5. Thali, M.J., et al.: VIRTOPSY - Scientific documentation, reconstruction and animation in forensic: individual and real 3D data based geo-metric approach including optical body/object surface and radiological CT/MRI scanning. J. Forensic Sci. **50**(2), 428–442 (2005)
6. Song, W., et al.: Assessing relative contributions of transport mechanisms and real gas properties to gas flow in nanoscale organic pores in shales by pore network modelling. Int. J. Heat Mass Transf. **113**, 524–537 (2017)
7. Gao, S., et al.: Two methods for pore network of porous media. Int. J. Numer. Anal. Meth. Geomech. **36**(18), 1954–1970 (2012)

8. Shuangli, T.: New advances of multislice spiral computed tomography. CT Theory Appl. **14**(4), 50–53 (2005)
9. Li, X.: A research for reprocessing the data of rock and soil material CT text for the degree of master of engineering. Changjiang River Scientific Research Institute, Wuhan (2012)
10. Li, C., et al.: 3D mesh generation for soil-rock mixture based on CT scanning. Rock Soil Mech. **35**(9), 2731–2736 (2014)
11. Jiang, J., et al.: CT triaxial rheological test on coarse-grained soils. Rock Soil Mech. **35**(9), 2507–2514 (2014)
12. Cheng, Z., et al.: Application of CT technology in geotechnical mechanics. J. Yangtze River Sci. Res. Inst. **28**(3), 33–38 (2011)
13. Sun, H., et al.: 3D identification and analysis of fracture and damage in soil-rock mixtures based on CT image processing. J. China Coal Soc. **39**(3), 452–459 (2014)
14. Cheng, Y., et al.: Three-dimensional reconstruction of soil pore structure and prediction of soil hydraulic properties based on CT images. Trans. Chin. Soc. Agric. Eng. **28**(22), 115–122 (2012)
15. Lochmann, K., et al.: Statistical analysis of random sphere packings with variable radius distribution. Solid State Sci. **8**(12), 1397–1413 (2006)
16. Hongqin, D.: The Research on Sphere Random Packings and Packing Structure. Soochow University, Suzhou (2011)
17. Bryant, S.L., et al.: Network model evaluation of permeability and spatial correlation in a real random sphere packing. Transp. Porous Media **11**(1), 53–70 (1993)
18. Kantzas, A., Chatzis, I.: Network simulation of relative permeability curves using a bond correlated-site percolation model of pore structure. Chem. Eng. Commun. **69**, 191–214 (1988)
19. Gao, L., Chen, W.: The Application and Prospect of CT. CT Theory Appl. **18**(1), 99–109 (2009)
20. Xianchao, W.: Research on Local Reconstruction Algorithm of CT Images. PLA Information Engineering University, Zhengzhou (2013)
21. Wenli, Y.: Research on the Techniques for 3D Reconstruction and Visualization of Slicing Image Sequence of Granular Soil. Huazhong University of Science and Technology, Wuhan (2012)
22. Yang, K., et al.: Fast bilateral filtering using the discrete cosine transform and the recursive method. Optik **126**(6), 592–595 (2015)
23. Imperial College London. http://www.imperial.ac.uk/earth-science
24. Kan, G., et al.: Accelerating the SCE-UA global optimization method based on multi-core CPU and many-core GPU. Adv. Meteorol. (2016)
25. Lagarias, J.C., et al.: Convergence properties of the nelder-mead simplex method in low dimensions. SIAM J. Optim. **9**(1), 112–147 (1998)
26. Bryant, S., Blunt, M.: Prediction of relative permeability in simple porous-media. Phys. Rev. A **46**(4), 2004–2011 (1992)
27. Wyllie, M.R.J., Gregory, A.R.: Fluid flow through unconsolidated porous aggregates - effect of porosity and particle shape on kozeny-carman constants. Ind. Eng. Chem. **47**(7), 1379–1388 (1955)
28. Chu, C.F., Ng, K.M.: Flow in packed tubes with a small tube to particle diameter ratio. AIChE J. **35**(1), 148–158 (1989)

Laboratory Dissolution Experiments of Internal Erosion in Sandy Soil: Underground Cavities and Piping

Yang Yang[1,2(✉)], Chao Xu[1,2], and Samanthi Indiketiya[3]

[1] Department of Geotechnical Engineering, Tongji University, Shanghai 200092, China
2011yang@tongji.edu.cn

[2] Key Laboratory of Geotechnical and Underground Engineering of Ministry of Education, Tongji University, Shanghai 200092, China

[3] Institute of Industrial Science, The University of Tokyo, Tokyo 153-8505, Japan

Abstract. Loss of soil particles due to mineral dissolution, degradation or internal erosion can lead to a change in the microstructure of soil, which as a result, affects the strength and stiffness of soil. This paper presented an approach to investigate the behavior of granular soil subjected to local particle removal by introducing a certain amount of erodible particles. Through sufficiently long period of water flushing, internal erosion was simulated by dissolution of glucose shaped in block/pipe, in representing underground cavities and piping. A series of model tests were conducted, and the formation of cavity/piping was recorded by digital images from the front of the soil chamber. Extent of loosening during the propagation of erosion and changes in local strength at post-erosion state were obtained by conducting penetration tests.

Keywords: Internal erosion · Dissolution · Particle loss · Cavity
Penetration test

1 Introduction

During the last few decades, disasters of soil erosion associated with ground cave-ins and piping have been increased remarkably. On Oct. 2016, a massive sinkhole accident occurred in front of station in Fukuoka, spanning the entire length of the four-lane road. According to the reports, deterioration of the sewer systems underground has been considered as one of the crucial reason for the cave-in. It is said that a large portion of sewer networks in Japan, which covers 460,000 km nationwide, are exceeding 40 years of service life and could cause enormous problems such as road subsidence and ground cave-ins in near future due to self-defects [1]. On the other hand, piping-triggered hydraulic failures and landslides have also been commonly observed in plenty of countries, particularly in rainy season [2, 3]. Such internal erosion in the form of an enlarged preferential channel through which continuous migration of soil particles takes place under seepage force is responsible for over half of dam failures, based on the investigation by Foster et al. [4].

L. Hu et al. (Eds.): GSIC 2018, *Proceedings of GeoShanghai 2018 International Conference: Multi-physics Processes in Soil Mechanics and Advances in Geotechnical Testing*, pp. 244–251, 2018.
https://doi.org/10.1007/978-981-13-0095-0_28

In both of the sink holes and piping phenomena, internal erosion of soil is initiated by locally concentrated particle loss. Previous laboratory studies concerning soil erosion could be divided into two types. One is to trigger the erosion inside the specimen through particle dissolution by mixing water soluble material such as salt or sugar [5, 6]. The other is to flush particles with certain size from the modified apparatus. For example, by replacing the traditional pedestal in triaxial apparatus with the one on which small holes are distributed, fine particles could be washed out during water infiltration [7, 8]. In these experimental studies, the weakened mechanical behavior of soil was repeatedly stated, which was mainly attributed to the increase in void ratio and decrease of particle contact. However, it should be noted that erosion consequences in specimen of uniform sand-salt mixtures or specimen suffered with fine particle loss are different from that in cavity/piping. When soil is subjected to local erosion, significantly anisotropic fabric would be generated due to the discontinuities along the erosion path.

Up till now, little research work has been done addressing properties of non-uniform soil at post erosion state, and questions still arise as to how the concentrated particle removal would affect the mechanical behavior of soil. In this study, in order to simulate internal erosion in the form of underground cavity and piping, glucose in the shape of cube and column was installed into Toyoura sand respectively. By injecting enough water into the soil, erosion was reproduced. Through a series of model test, migration of soil particles during erosion and variations in strength properties of soil after erosion were evaluated.

2 Laboratory Model Tests

2.1 Test Apparatus

Figure 1 shows the apparatus for laboratory experiments that allows visual observation of the model ground through the transparent acrylic wall of the soil chamber. The chamber is 300 mm in width, 200 mm in height and 80 mm in thickness. Besides, there are two side chambers and a bottom acrylic chamber, consisting of the main part of the apparatus.

The main chamber and the side chamber are separated by the steel wall, on which small holes in diameter of 4 mm are distributed uniformly. These holes would be partially covered by stick tape during the test, and those open ones serve as the exit of the seepage. The drained water, coming out from the opened holes, is then collected in the side chambers. Outlet tubes are fixed at the bottom of the side chambers in order to discharge the water.

The bottom acrylic chamber is connected to the bottom plate of the soil chamber, with an input control valve connected to the water supply. After the bottom chamber is full filled with water, drainage is then allowed through a small aperture in the base plate with width of 5 mm and length of 50 mm. Upon the aperture filter paper in the right size is covered to prevent washing out of sand particles.

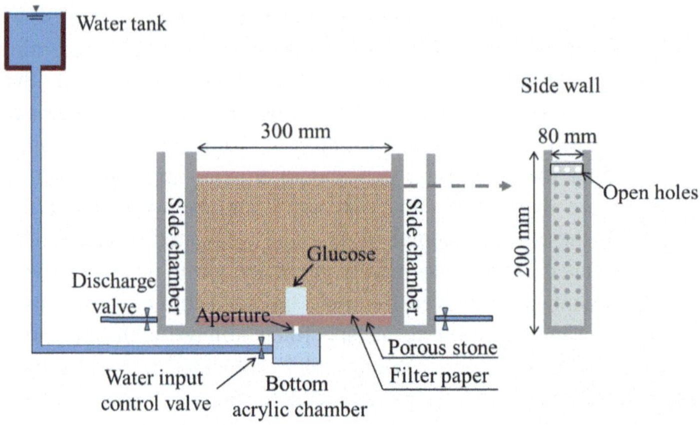

Fig. 1. Apparatus for laboratory model test simulating cavity/piping erosion.

2.2 Test Procedure

Toyoura sand was used in this study, which was fine-grained and uniformly graded sand with sub-angular particles. The index properties of the sand are $G_s = 2.64$, $e_{max} = 0.992$, $e_{min} = 0.678$.

For test cases simulating underground cavity, a cylindrical glucose block was placed at the bottom the chamber firstly, which was casted by compacting glucose paste in a mould. The paste was produced by mixing the dry powder of glucose with around 5% of water (by weight). After that, dry Toyoura sand was poured into the soil chamber using a funnel with appropriate opening. By adjusting the pouring height between top surface of the sand and the tip of the funnel opening, the density of the specimen could be varied. It is noted that filter paper together with the porous stone was put at both bottom and top of the sand specimen so as to achieve a more uniform seepage.

As for the cases simulating piping erosion, a plastic straw with the outer diameter of 14 mm was placed at the center of the cross-section of the soil chamber, with the top end impaled by a fine wire. Then the top opening of the pipe was covered and the wire was fixed to the side chambers through the holes on it. Similarly, the sand was pluviated until the decided height. Then the plastic straw was open and dry glucose powder was filled into the pipe with great care. Finally, the straw was slowly so as to minimize disturbance to the surrounding sand.

After the completion of the model ground, the inlet valve was open and water was infiltrated into the soil chamber from the bottom plate at the rate around 20 ml/min, in order to dissolve the glucose and induce the erosion. After water level reached the surface of the soil, constant water head was maintained by draining out the overflow water from the open holes on both the side walls. Deformation of sand was recorded by digital images taken from the front of the soil chamber. The penetration stopped till near 10000 ml of water discharge, and the soil model was left for 20 min to stabilize. Then drainage was allowed from the bottom acrylic chamber.

After leaving the soil samples for one day, laboratory penetration tests were conducted to evaluate the variation in stiffness of the model ground. Figure 2 illustrated the

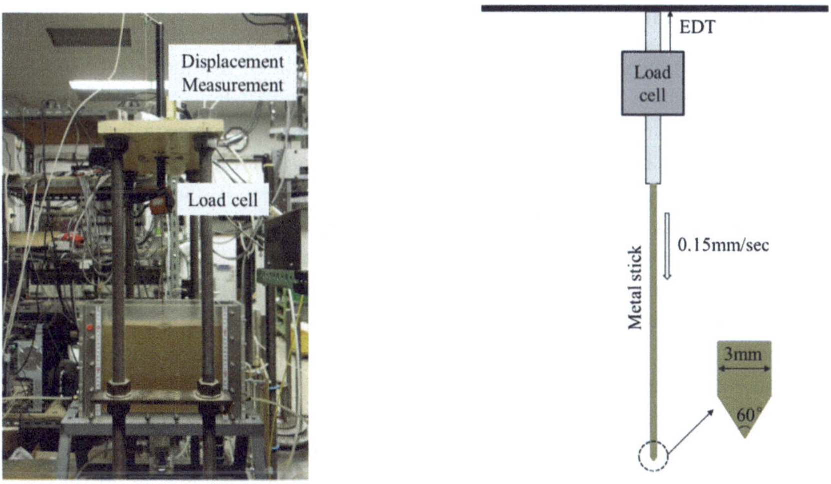

Fig. 2. Apparatus for penetration test

testing device. It mainly consists of an external displacement transducer (EDT), a load cell and a metal stick (3 mm in diameter) mounted below. During the test, the metal stick was penetrated from the surface to the bottom of the soil samples slowly and automatically, and the penetration resistance as well as depth was measured by the load cell and EDT respectively. The results were recorded automatically at 0.1 s intervals. In total five points with different distance from the center were tested, and all the measured points were located along the center line of the cross-section of the specimen, seen in Fig. 3.

2.3 Test Conditions

In total five test cases were performed. In the first two cases, glucose pipe was placed in the center to study the effect of piping erosion. The relative densities D_r in each scenario were around 35% and 80% for representing the loose and dense ground. In the other two cases, underground cavity was simulated by introducing the glucose block in different sizes into the model ground with D_r around 80%. Besides, another experiment was performed in dense colored sand mixed with 20% DL clay, with glucose pipe placed next to the front wall of the chamber in order to observe the migration of fines during erosion.

3 Test Results and Discussions

Through the above model tests, the new experimental approach of developing artificial local erosion in uniform grain sand was achieved. Meanwhile, important factors including the soil density, erodible particle volume and fines content on the extent of disturbance and effect of ground loosening were discussed.

3.1 Piping Simulation

Figures 3 and 4 showed the test results along the height of the model ground for the two test cases simulating piping erosion. The penetration resistance was acquired at five points as illustrated in the figures, and the distances from the measuring points to the center point (C) are 18.75 mm, 37.50 mm, 75.00 mm and 113.50 mm for points R1, L1, R2 and L2, respectively. By comparing the two cases, it could be observed that penetration resistance in dense sand showed nearly twice times larger value than that in loose sand.

In both cases, accompanying the progress of penetration, resistance started to increase from ground surface till the bottom, except for point C which was in the center of the model ground. The rather limited resistance measured in point C implied that erosion was generated successfully, i.e. concentrated voids were located following the glucose pipe. On the other hand, it was observed that there was some difference in the resistance between L2 and R2, where the ground loosening was thought to be negligible. The possible reasons for this result might lie in the fact that at the moment of penetration test the moisture distribution was non-uniform.

By defining the reduction ratio *R* as the proportion of difference in penetration resistance between the measuring point and the point nearest to the chamber wall (L2), the range of piping-influenced area due to loss of glucose could be evaluated. It was found that *R* for point R1, L1 and R2 were around 46%, 21% and 11% in loose sand, and in dense sand, they were 34%, 35% and 21%. Given the same amount of initial glucose, such difference indicated that during the generation and propagation of piping erosion, the scope of disturbance was affected by the sand density. In specimen with lower density, loosening effect might spread to a wider area due to the weaker internal contacts, i.e., sand particles were more likely to migrate or collapse into the eroded voids.

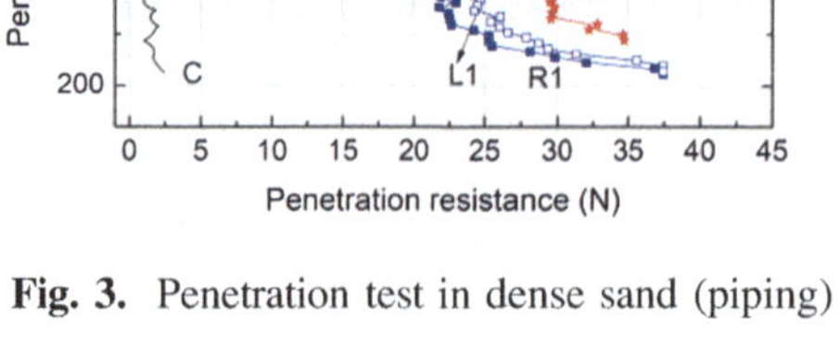

Fig. 3. Penetration test in dense sand (piping)

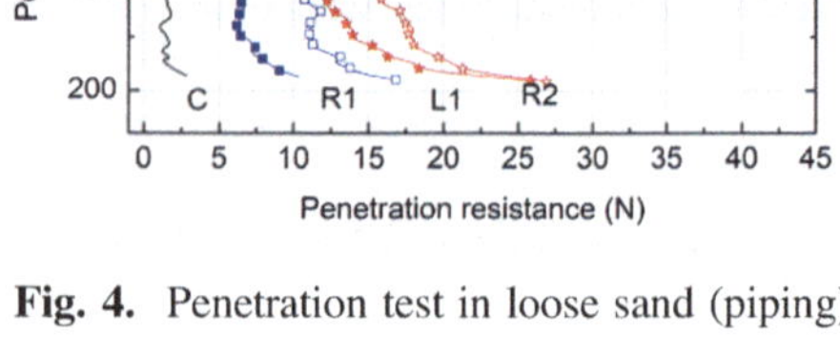

Fig. 4. Penetration test in loose sand (piping)

3.2 Cavity Simulation

For observation purpose, thin colored sand layers were used to separate each 2 cm Toyoura sand layer in the two dissolution tests on cavity simulation. Similarly, the

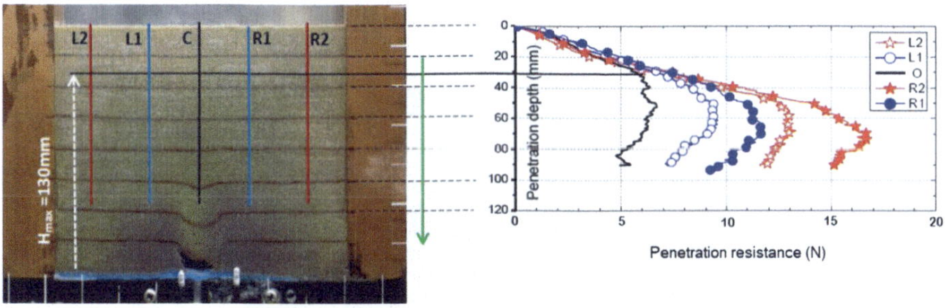

Fig. 5. Penetration test in dense sand with larger glucose block (cavity simulation)

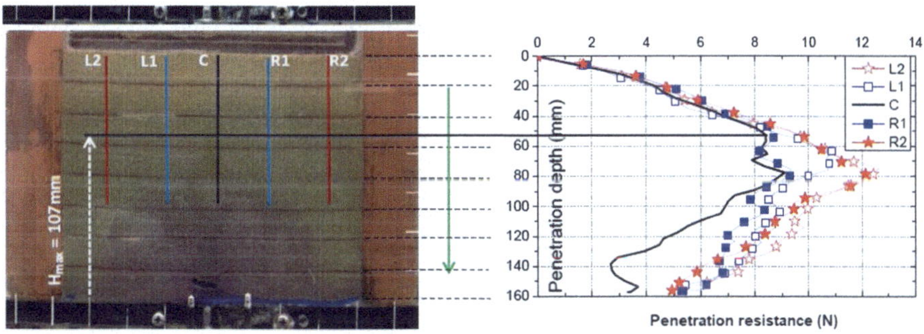

Fig. 6. Penetration test in dense sand with smaller glucose block (cavity simulation)

model was tested for penetration resistance through five different profiles representing the center, left and right side of the cavity to evaluate the strength variation. Figures 5 and 6 show the model ground aft post-erosion state and the penetration results of the two cases where glucose block in different sizes were placed in the dense Toyoura sand. In Fig. 5, the rectangular cross section for the glucose block was 20 mm in height and 30 mm in length, while in Fig. 6, it was 10 mm in height and 20 mm in length.

In both test cases, penetration resistance increased from ground surface and then decreased. By defining H_{max} as the height from bottom where differences in penetration resistance were observed among all measuring points, the extent of the disturbance caused due to the loss of glucose block could be compared. As could be seen in the following figures, the value of H_{max} was larger in the model with larger glucose block than the smaller one. Besides, the value of resistance showed more scatter in the test cases with larger glucose block. These facts implied the more severe migration and rearrangement of sand particles when the ground was suffered large portion of particle loss. It could also be found in Fig. 5 that obvious subsidence of the sand reached over 60 mm in height, which was 3 times larger than the height of initial glucose block.

Figure 7 shows the dissolution test conducted with sand containing fine particles. The migration of fines after the disappearance of glucose was rather obvious, along with certain settlement of the model ground. Compared with Figs. 5 and 6, it could be observed that the loosened effect in cavity erosion transferred and propagated in a

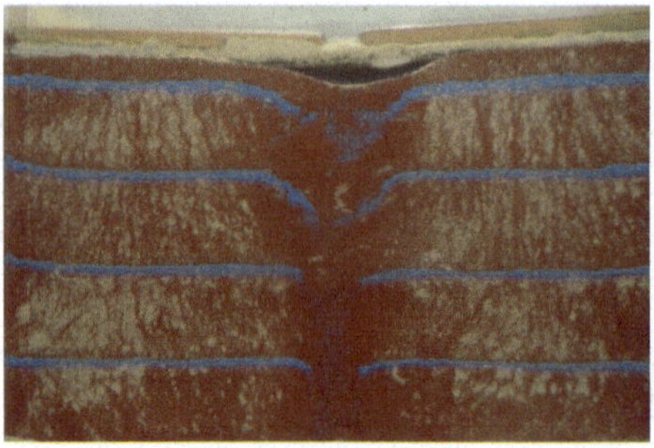

Fig. 7. Dissolution test in dense Toyoura sand containing 20% of fines with glucose pipe

wider range in the dense sand in vertical direction than in horizontal, due possibly to the influence of direction of water seepage and self-weight.

4 Conclusions

This paper presents a preliminary experimental approach for investigation of soils subjected to internal erosion of concentrated particle loss. As a first step, the glucose was just applied within a segment of the soil specimen in this study, representing a relative simple case. Still, the spatial distribution of glucose to the mechanical behavior of soil must be further studied.

Based on the current research, it was observed that propagation and expansion of internal erosion of soil were influenced by the soil density and the amount of soluble material. A larger extent of loosening effect and disturbance were observed in loose soil than dense soil. Penetration resistance showed great reduction due to particle loss. Piping erosion migrated mainly in the horizontal direction while loosening transferred vertically in the case of underground cavity. During water infiltration, abrupt collapses of the soil sitting upon cavities due to self-weight and the relocation of soil particles under the seepage force were commonly observed, indicating an unpredictable structure of soil subjected to local particle loss.

Acknowledgements. The authors acknowledge the final support from the National Natural Science Foundation of China under Grant No. 51609171. Experiments presented in this paper were performed in the Department of Civil Engineering at the University of Tokyo.

References

1. Kuwano, R., Horii, T., Kohashi, H., Yamauchi, K.: Defects of sewer pipes causing cave-ins in the road. In: Proceedings of Fifth International Symposium on New Technologies for Urban Safety of Mega Cities in Asia, Phuket, pp. 347–353 (2006)
2. Jones, J.A.A.: The effects of soil piping on contributing areas and erosion patterns. Earth Surf. Proc. Land. **12**(3), 229–248 (1987)

3. Richards, K.S., Reddy, K.R.: Critical appraisal of piping phenomena in earth dams. Bull. Eng. Geol. Env. **66**(4), 381–402 (2007)
4. Foster, M.A., Fell, R., Spannangle, M.: The statistic of embankment dam failures and accidents. Can. Geotech. J. **37**(5), 100–124 (2000)
5. Shin, H., Santamarina, J.C.: Mineral dissolution and the evolution of k_0. J. Geotech. Geoenviron. Eng. **134**(8), 1141–1147 (2009)
6. Truong, Q.H., Eom, Y.H., Lee, J.S.: Stiffness characteristics of soluble mixtures. Géotechnique **60**(4), 293–297 (2010)
7. Sato, M., Kuwano, R.: Effects of internal erosion on mechanical properties evaluated by triaxial compression tests. In: Proceedings of the 15th Asian Regional Conference on Soil Mechanics and Geotechnical Engineering, pp. 1056–1059 (2016)
8. Luo, Y.L., Qiao, L., Liu, X.X., Zhan, M.L., Sheng, J.C.: Hydro-mechanical experiments on suffusion under long-term large hydraulic heads. Nat. Hazards **65**(3), 1361–1377 (2013)

Selecting Optimum Cutoff Wall Position for Rehabilitation of an Inclined Core Earthfill Dam

Hamed Farshbaf Aghajani(✉), Mahsa Mousavi Anzabi, Zahra Sheikhi, and Rahele Shokri

Department of Civil Engineering, Faculty of Engineering, Azarbaijan Shahid Madani University, Kilometer 35 of Tabriz/Azarshahr Road, P.O. Box 53714-161, Tabriz, Iran
h.farshbaf@azaruniv.ac.ir

Abstract. Over the last few decades, a particular number of existing dams in Iran have required rehabilitation to improve the water sealing system performance. This paper aims to thoroughly investigate the effect of the cutoff wall position on the seepage behavior of an existing earthfill dam with an inclined core. Several conditions are considered for the permeability and depth of a dam foundation. Then, for each condition, the remedial cutoff wall is imposed in dam at various locations and the seepage pattern and magnitude through the foundation is analyzed via finite element GeoStudio SEEP/W software.

The analyses results show that if the cutoff wall is not connected to a lower impervious layer of foundation, the effects of the cutoff wall position variations on the seepage value are insignificant. However, if the cutoff wall is constructed at the dam heel, minimum seepage occurs. When the cutoff wall covers the whole of pervious foundation, the seepage through foundation is fully corresponded to the cutoff wall position. Minimum seepage is observed in cases where the cutoff wall is constructed from the dam crest. Besides, in foundation with higher permeability, the effect of the cutoff wall position on the seepage is more significant.

Keywords: Remedial cutoff wall · Inclined clay core · Earthfill dam Seepage analysis · Rehabilitation

1 Introduction

Today, dams play a prominent role in human cavitation. They supply water and energy, and account for major infrastructures. Any defects and shortages in dams and associated hydraulic structures can result in dam malfunctions and catastrophic consequences. Despite its long history in construction and operation, dams face unfortunate problems. One of the current problems in the earthfill dam operation process is the presence of extra seepage in the dam's body and foundation. This problem arises from either, shortages in the seepage barrier system during design stages or other construction problems [1].

L. Hu et al. (Eds.): GSIC 2018, *Proceedings of GeoShanghai 2018 International Conference: Multi-physics Processes in Soil Mechanics and Advances in Geotechnical Testing*, pp. 252–260, 2018.
https://doi.org/10.1007/978-981-13-0095-0_29

In some cases, defects in the seepage barrier system of a dam body and foundation emerge as a result of phenomena [2] like the existence of evaporative materials in a dam clay core or foundation. The high susceptibility of these materials to the solution may lead to dissolution and enhance the permeability of the dam core or foundation [3, 4]. Other infective factors, such as the internal erosion of clay core material [5] and fracturing can lead to extra seepage or relevant problems [6]. There are many cases worldwide where the presence of extra seepage in the dam body and foundation caused a total collapse of the dam [1].

In order to solve the extra seepage problem in clay core earthfill dams, a new remedial cutoff wall is constructed through the dam body and foundation. Remedial cutoff wall selection in an earthfill dam with an inclined core becomes more critical due to the unsymmetrical shape of the existing clay core. This paper aims to thoroughly study seepage behavior in an inclined core earthfill dam, which is rehabilitated by a remedial cutoff wall. To this end, the remedial cutoff wall effect on seepage behavior in the foundation and body of the dam is determined by conducting a 2-D finite element seepage analysis.

2 Numerical Modeling

The GeoStudio 2004 software is used to conduct a 2-D finite element analysis of seepage. This software is a powerful means to analyze saturated and unsaturated flow in porous media, and is widely used in the seepage analysis of dams [7].

The numerical model consists of a zoned earthfill dam with a defected inclined clay core. Weakness in the clay core increases the permeability of the clay core to about 1000 times greater than the common value of the intact impervious condition. The selected outer face slope of the dam body is 1V:3H. The height and crest width of the dam are about 40 and 5 m, respectively.

The dam foundation consists of a one-layer alluvium that is 60 m deep and is located on an impervious bedrock. At its current state, there is no seepage barrier system in the foundation.

A 1-m thick remedial cutoff wall is constructed inside the dam body, and extended to the foundation to improve the current earth fill dam performance, and decrease seepage in the dam body and foundation. Figure 1 illustrates that the remedial cutoff wall in the dam body, denoted by x, is varied and located at a different distance from the downstream corner of the dam crest. The values include x = 0, 5, 15, 25 and 45 m. In the cases x = 0 and x = 5, the remedial cutoff wall is constructed from the dam crest and across the current defected clay core. However, for other cases of x, the remedial cutoff wall is constructed from the upstream face of the dam. Complete water sealing of the dam body is achieved by covering the dam slope face between the remedial cutoff wall and the dam crest with an impervious 1-m thick slab. In cases where the distance of x is greater than 15 m, the remedial cutoff wall is only crossed through the dam shell and out of the inclined clay core. As a result, there is no interaction between the current clay core and the remedial cutoff wall. The cutoff wall has different depth in foundation to address the embedment effect of the cutoff wall. The embedment depth of the cutoff wall from a horizontal natural ground level (denoted by d) is 20, 30, 50 and 60 m.

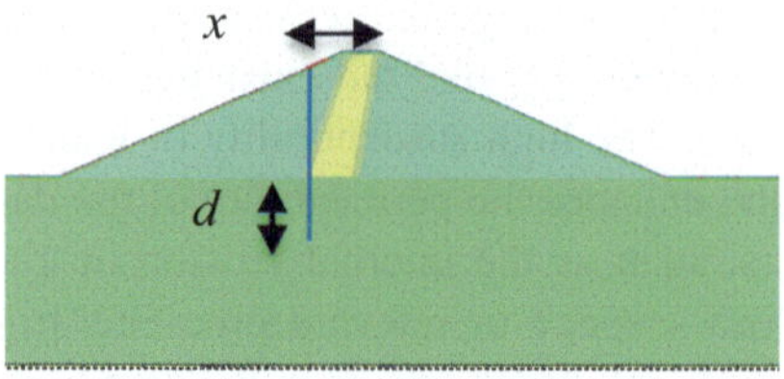

Fig. 1. The definition of cutoff wall location (x) and embedment depth (d)

Table 1. The permeability of materials in numerical model

Material name	Permeability (m/s)
Foundation	10^{-4}
Inclined core in defected condition	10^{-5}
Remedial cutoff wall	10^{-8}
Filter	10^{-4}
Upstream impervious slab	10^{-7}
Shell	10^{-2}

In the case of the d = 60 m embedment depth, the remedial cutoff wall not only covers the entire alluvial layer, but is also connected to the lower impervious bed rock layer. However, in cases where the embedment depth (d) is less than 60 m, the cutoff wall only partially covers the foundation, and floats within the foundation. The values of permeability of materials used in the numerical model are summarized in Table 1.

3 The Analysis Results

Figure 2 illustrates the seepage analysis results, including seepage paths and phreatic lines of three cutoff wall locations for dams rehabilitated by the 20 m deep embedment remedial cutoff wall. The hydraulic head fall occurs in the cutoff wall when the remedial cutoff wall is constructed at the top of the dam crest (i.e. $x = 0$) and overlapped with the current clay core. As a result, the inclined clay core at the upstream of the cutoff wall is completely submerged and hydraulic head in clay core placed under the equilibrium with the reservoir. In other words, the existing clay core does not contribute as much to the hydraulic dissipation of the pore water pressure raised from seepage.

In contrast, when the remedial cutoff wall is constructed at the upstream face of the dam ($x = 15$ m), the phreatic line pattern in the dam body is also somewhat changed. As seen in Fig. 2-b, most of the hydraulic head dissipation occurs in both the impervious face and the cutoff wall. However, despite weakness area in the dam core, the current clay core material is relatively impervious compared to the shell material. Hence, some of the hydraulic head fall occurred in the current clay core. The move of the cutoff wall location towards the dam heel increases the hydraulic head dissipation in the upstream impervious slab. Most of the hydraulic head fall occurred in the upstream

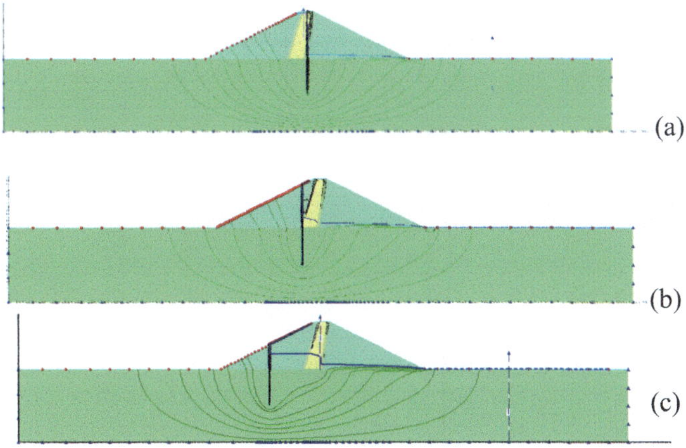

Fig. 2. The phreatic line and flow path for analysis cases with partially embedded cutoff wall of d = 20 m and various locations for cutoff wall; (a) x = 0, (b) x = 15 m and (c) x = 45 m.

face rather than the cutoff wall. Moving the cutoff wall position along the upstream slope of the dam does not significantly influence the general shape of the phreatic line within the dam body. Furthermore, the form and height of the hydraulic head in the remedial cutoff wall was similar for the different remedial cutoff wall locations. Due to the fact that the cutoff wall embedment depth was less than the full depth of the foundation layer, some discharge was observed in the foundation zone beneath the cutoff wall. Parts of the seepage discharge in the foundation flowed upward and entered the dam shell after passing the cutoff wall axis. This caused a slight flow in the downstream shell, and sketched the flow path from the dam foundation to the downstream shell. The water flow in the downstream shell causes that the hydraulic head falls at a slight rate until it reaches the zero pressure head in the downstream ground. As a result, the cutoff wall floats in the alluvial layer and is not connected to the lower impervious layer. A horizontal drainage layer should be constructed to separate the downstream shell from the foundation, and avoid extra foundation seepage into the downstream shell.

Figure 3 illustrates the seepage analysis results of cases where the remedial cutoff wall fully covers the foundation ($d = 60$ m). Similar to previous analysis cases, the upstream impervious slab and remedial cutoff wall play a very distinctive role in pore water pressure dissipation of the seepage flow. However, in contrast to the partially embedded cutoff wall, when the cutoff wall fully seals the foundation depth, seepage does not occur in the dam core and downstream shell, making the phreatic line horizontal in the clay core. Moreover, the seeped water in the foundation directly flows towards the downstream natural ground, and water is prevented from entering the downstream shell.

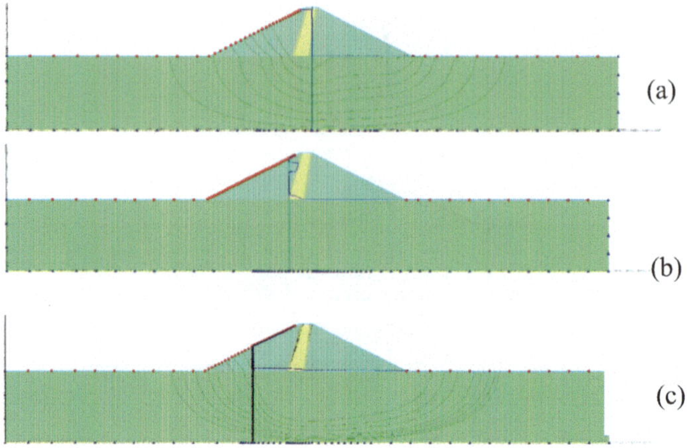

Fig. 3. The phreatic line and flow path for analysis cases with fully embedded cutoff wall of d = 60 and various locations for cutoff wall; (a) x = 0, (b) x = 15 m and (c) x = 45 m.

4 The Effect of Cutoff Wall Location on the Seepage

Figures 4 and 5 illustrate the seepage discharge graph as calculated for all the analysis cases of the dam body and foundation, respectively. As seen, the seepage graph for the full depth embedded cutoff wall has different form from the seepage graph of partially cutoff walls. In cases where the remedial cutoff wall did not extend to the lower impervious bedrock foundation layer, the minimum seepage through the dam body occurred when the cutoff wall was constructed at the center of the dam crest (x = 5 m). Moving the cutoff wall upstream thereafter, considerably decreased the effectiveness of the remedial cutoff wall, and allowed the flow of more water through the dam. If the cutoff wall is constructed at the downstream corner of the crest (x = 0), the cutoff axis is located outside of the clay core, and the overlap between the two elements greatly reduces. Thus, remediation is less effective, and water flux is increased through the dam. Increasing the cutoff distance by more than 5 m leads to hydraulic head dissipation in the existing clay core, and water flow in the dam core. As earlier mentioned, some of the seepage discharge in the foundation is entered into the downstream shell, which in effect enhances the flux of the dam body. The partially embedded cutoff wall gives maximum performance at x = 5 m.

However, when the cutoff wall fully covers the foundation layer, there is less water flow through the dam body. In this case, the minimum discharge in the dam body occurs when the cutoff wall is constructed at mid- height of the upstream dam slope (i.e. x = 25 m). The remedial cutoff wall of other locations has an insignificant flux variation through the dam body, and is not considerably affected by the cutoff wall location.

The full or partial embedment of the remedial cutoff wall directly affects the seepage flux through the foundation. As seen in Fig. 5, full embedment of the cutoff wall in the foundation significantly decreases the flux. The minimum flux occurs in a

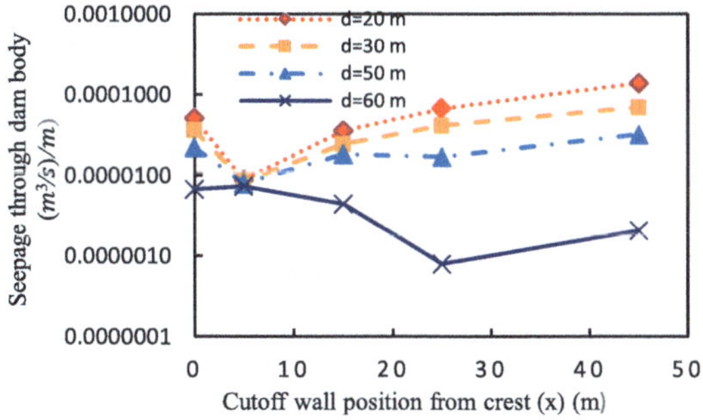

Fig. 4. Seepage discharge through dam body for all analysis cases, where the remedial cutoff wall has different embedment depths (d) and distance from the dam crest (x)

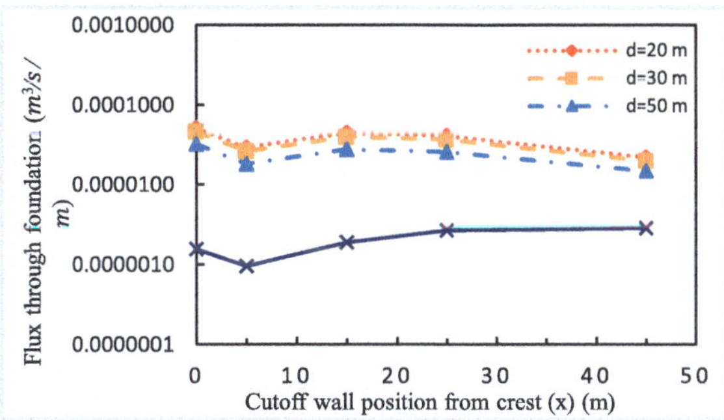

Fig. 5. Seepage discharge through the foundation for all analysis cases, where the remedial cutoff wall has different embedment depths (d) and distance from dam crest (x)

case that cutoff wall locates of x = 5 m from the dam crest. There is a slight increase in the foundation flow due mainly to a decrease in the water flow path of the foundation.

Contrary to full embedment, there is more water flow when the remedial cutoff wall only partially covers the foundation. In this research, due to the partial seal of the foundation with the water barrier system, the water flows freely in the foundation. As a result, neither the embedment depth nor the cutoff wall location significantly influences the seepage flux.

5 The Critical Gradient in the Remedial Cutoff

Due to the large contribution of the remedial cutoff wall in the hydraulic water head dissipation, the cutoff wall suffers a larger gradient. If the maximum gradient imposed on the cutoff wall exceeds the critical limit, the likelihood of erosion in the cutoff wall material increases, and may be lead to defects in the cutoff wall. It is therefore essential to determine the maximum value and location of the critical gradient in the cutoff wall when designing cutoff wall specifications. Figure 6 illustrates the gradient profile along the remedial cutoff walls with a 30 m deep embedment. The trend on the gradient graph is similar for all cutoff wall locations, except the maximum gradient depends on the cutoff wall location. The gradient within a part of the cutoff wall, located inside the dam body is linearly increased, and the maximum gradient is imposed on the cutoff wall at an elevation equivalent to the interface between the dam body and foundation top. By increasing the embedment depth of the cutoff wall inside the foundation, the gradient gradually decreases. The maximum gradient varied between 27 for x = 25 m and 33 for x = 5 m. Since the maximum performance of the remedial cutoff wall in reducing the flow through the dam body occurred at the x = 5 location, the cutoff wall experienced a greater fall in the hydraulic head, and thus the maximum gradient is attained.

Figure 7 illustrates the gradient profile of fully embedded cutoff walls. Similar to the partial embedded cutoff wall, the gradient in the cutoff wall located inside the dam body linearly increases with depth. However, in contrast to the partial embedded cutoff wall, the gradient is same value in a zone that the cutoff wall is located inside the foundation. It is noteworthy that the situation of the cutoff wall from the dam crest does not significantly impact the maximum gradient in the fully embedded cutoff wall. This implies that if the cutoff wall totally covers the foundation, the cutoff wall undergoes uniform gradient along the embedment depth in the foundation, regardless of the cutoff wall location.

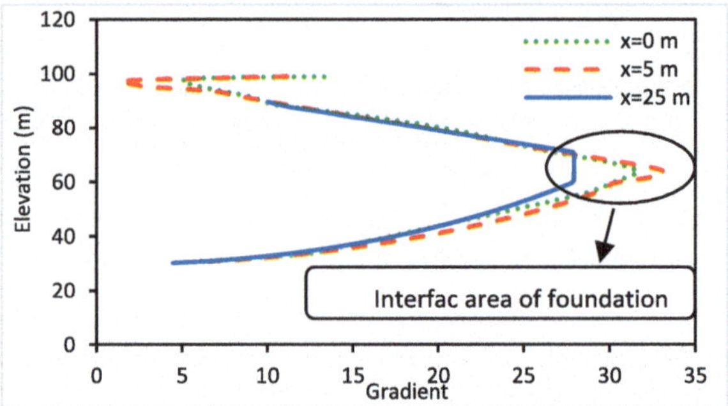

Fig. 6. The gradient graph along the remedial cutoff wall with an embedment depth of d = 30 m

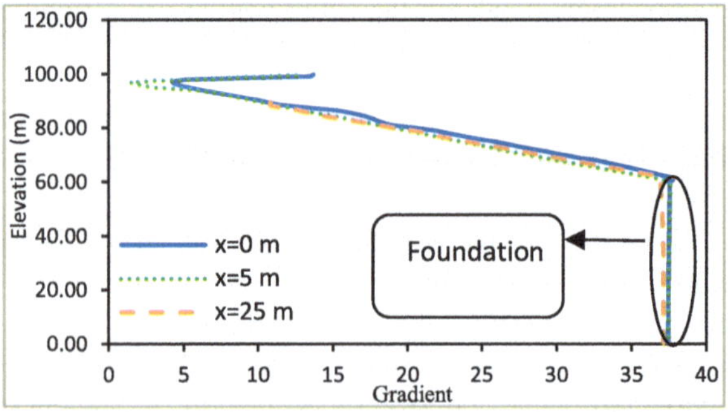

Fig. 7. The gradient graph along the remedial cutoff wall with embedment depth of d = 60 m

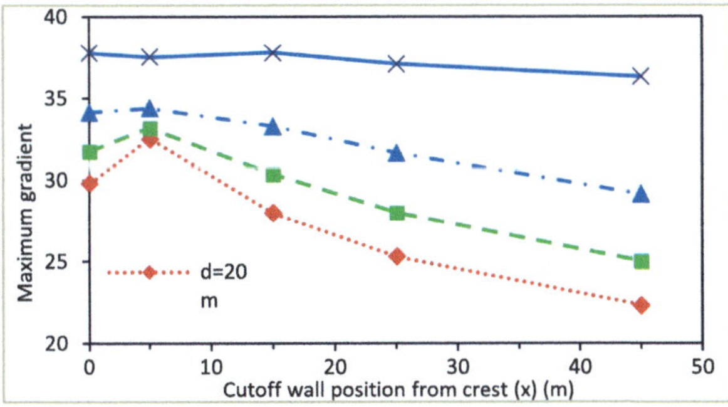

Fig. 8. The variation of maximum gradient in the cut off wall for all analyses cases.

Figure 8 compares the maximum gradient of the cutoff walls and their different embedment depths and locations for all analysis cases through a maximum gradient graph. As seen in the figure, fully embedded cutoff walls, as opposed to partially embedded cutoff walls, are completely sealed against water flow in the foundation, and the large gradient is imposed in the cutoff wall. Increasing the cutoff wall depth among partially embedded cutoff walls on the other hand, enhances the sealing performance of the cutoff wall, and increases the gradient within the cutoff wall. The effect of the cutoff wall location on the maximum gradient is more observed in the cutoff family with the shallowest embedded depth (i.e. d = 20 m). Similarly, the greatest gradient imposed in the cutoff wall is observed in the x = 5 m location. As earlier stated, the water flux through the dam foundation in this cutoff wall location is minimum because of a greater gradient of water in the cutoff wall. Hence, it can be said that among the partially embedded cutoff walls, the cutoff wall constructed at x = 5 m has the best sealing performance.

6 Conclusion

This paper attempted to thoroughly investigate the effects of embedment and remedial cutoff wall location on sealing performance. The findings of this paper are summarized as follows:

- A greater part of the hydraulic head dissipation in both the impervious face and cutoff wall occurred when the remedial cutoff wall was partially embedded in the foundation. However, a slight fall of the pore water pressure occurred in the clay core. As a result, parts of the seepage flow in the foundation entered the dam shell after passing the cutoff wall axis.
- If the remedial cutoff wall fully covers the foundation, the entire hydraulic head dissipates in the remedial cutoff wall, and seepage does not occur in the dam core and downstream shell.
- The minimum seepage for the partially embedded remedial cutoff wall occurs through the dam body when the cutoff wall is located at the center of the dam crest (x = 5 m). Directing the cutoff wall upstream considerably decreases the effectiveness of the remedial cutoff wall.

- The minimum flux in the dam body occurs when the cutoff wall fully covers the foundation layer, and the cutoff wall is constructed mid-height of the upstream dam slope.
- In all cases, regardless of the cutoff location and embedment depth, the minimum flux in the foundation is related to the x = 5 m cutoff wall location.
- The maximum gradient along the cutoff wall in the partially embedded type occurred at interface area between the dam and foundation. In contrast, fully embedded cutoff walls have the gradient inside a portion of the cutoff wall, which is located inside the foundation. In this case, the gradient is almost equal and holds its highest value.
- Among the partial embedded cutoff walls, the best sealing performance was observed in the cutoff wall constructed at the x = 5 location.

References

1. Rice, J.D., Duncan, J.M.: Findings of case histories on the long-term performance of seepage barriers in dams. J. Geotech. Geoenvironmental Eng. **136**(1), 2–15 (2009a)
2. Anderson, E.I.: Exact and approximate solutions for seepage through semipermeable cutoff walls. Int. J. Geomech. **15**(6), 04014087 (2014)
3. Turkmen, S.: Treatment of the seepage problems at the Kalecik Dam. Eng. Geol. **68**(3–4), 159–169 (2003)
4. Aghajani, H.F.: Deciding for remediation of the seepage barrier system in foundation of an earthfill dam based on the monitoring data and numerical analysis. In: Shehata, H., Rashed, Y. (eds.) Numerical Analysis of Nonlinear Coupled Problems: Proceedings of the 1st GeoMEast International Congress and Exhibition, Egypt 2017 on Sustainable Civil Infrastructures, pp. 1–12. Springer, Cham (2018). https://doi.org/10.1007/978-3-319-61905-7_1
5. Foster, M., Fell, R., Spannagle, M.: A method for assessing the relative likelihood of failure of embankment dams by piping. Can. Geotech. J. **37**(5), 1025–1061 (2000)
6. Rice, J.D., Duncan, J.M.: Deformation and cracking of seepage barriers in dams due to changes in the pore pressure regime. J. Geotech. Geoenvironmental Eng. **136**(1), 16–25 (2009b)
7. GEOSLOPE International, GeoStudio software (2004). http://www.geo-slope.com/

Model Test of Seepage Erosion Around Defective Buried Pipelines

Dong-Mei Zhang[1,2](✉), Wei-Wei Du[1], and Cheng-Peng Gao[1]

[1] Department of Geotechnical Engineering, Tongji University, Shanghai 200092, China
dmzhang@tongji.edu.cn

[2] Key Laboratory of Geotechnical and Underground Engineering of Ministry of Education, Tongji University, Shanghai 200092, China

Abstract. In recent years, the cave-in on roads frequently occurs in urban area to threaten the city safety, which is usually initiated by the ground cavity due to internal erosion around defective buried pipelines. In order to study the mechanism of internal erosion inducing cavity, the model tests are performed using gap grading soil samples. The effects of soil saturation, hydraulic head, initial soil grading and defect size of the pipeline on the internal erosion are studied with eight model tests. The erosion is characterized in terms of the erosion area and cavity formation during erosion. The test results show that the erosion can get stable due to the support of coarse particle skeleton in gap grading soil. The erosion areas can be classified into three forms of gravel area, flow area and cavity according to erosion state. Furthermore, the erosion area especially the cavity in saturated soil is more extensive than that in unsaturated soil. Besides, no hidden cavity forms in unsaturated soil, but the surface settlement. The higher hydraulic head induces bigger erosion area and cavity forms at the water table. More initial fine-sand content means more sand loss, leading to more significant soil deformation. While bigger defect size may cause a little more erosion amount, the effects on cavity is not significant. Finally, the mechanism of internal erosion and cavity formation are interpreted from the microscopic and macroscopic aspects.

Keywords: Pipeline leakage · Internal erosion · Model test
Influential factors

1 Introduction

In recent years, the settlement or cave-in of the urban road surface frequently occurs, which is mainly due to the hidden cavity under the ground [1]. The cavity induced by internal erosion has been a big threat to city safety. Therefore, most of the researchers focus on the control measures such as GPR or other methods to detect the underground cavity [2]. However, it is more important to clarify the mechanism of the appearance of the underground cavity to prevent the accident fundamentally.

The original version of this chapter was revised: Incorrect second author name has been corrected. The erratum to this chapter is available at https://doi.org/10.1007/978-981-13-0095-0_61

L. Hu et al. (Eds.): GSIC 2018, *Proceedings of GeoShanghai 2018 International Conference: Multi-physics Processes in Soil Mechanics and Advances in Geotechnical Testing*, pp. 261–270, 2018.
https://doi.org/10.1007/978-981-13-0095-0_30

For the seepage failure of soil layer, most studies focus on internal erosion. Foster et al. [3] analyzed the causes of accidents in embankment dams, finding that almost half of the failures were caused by the piping failure. Kenney et al. [4] presented that the soil gradation is the most critical factor of seepage failure. The confining pressure of soil, size of the filter and other factors also affect the internal stability of soil [5, 6]. All these factors are internal causes of the seepage erosion and soil deformation.

Hu et al. [7] found the defects of sewer pipes is the direct cause leading to erosion through 120 cases of road cave-in accidences. The soil particles may flow into the pipelines with water through the defects, which produces and expands the ground cavity until ground cave-in occurs. Mukunoki et al. [8] revealed that the forms of pipeline defect, as well as hydraulic conditions, influence the formation of the cavity by X-ray CT method in cylindrical soil tank. Kuwano et al. [9] investigated the formation characteristics of cavity to indicate the dangerous pattern of cavity by model tests in which the water enters and scours the soil from the opening at the bottom of soil tank. Sato et al. [10] presented that the location of underground structures changes the seepage path, leading to local erosion around structures due to high hydraulic conductivity.

Previous studies mostly focused on the seepage failure without considering the cavity. The mechanism of the cavity around defective pipeline were studied fully [8–10], while, the model tests were performed in uniform grading sand whose erosion is different from the gap grading sand where fine particles migrate through the pores of coarse particles. Thus, this study aims to clarify the development of internal erosion in gap grading sand caused by the pipeline defects and reveal the effects of soil saturation, hydraulic head, initial soil grading and pipeline defects dimension on the erosion using model tests. The mechanism of erosion and cavity formation is finally discussed.

2 Test Methods

2.1 Test Apparatus

A new apparatus is designed that allows soil particles flow into pipeline through the defect under the hydraulic head. The apparatus is made of transparent acrylic plates which allow for visual observation of erosion. This 1-g model test is applicable because cavity can be formed in the shallow ground [9]. Figure 1 shows the schematic diagram of test apparatus which consists of a sand tank, two water tanks, and a detective pipeline.

The center sand tank is 400 mm long, 200 mm wide and 750 mm high, used for holding soil sample with 500 mm depth for seepage erosion. The water tanks are on both sides of the sand tank with 100 mm wide, having water inlets and overflow holes to control the hydraulic head. The plates made of permeable porous stones are set between the water tank and sand tanks to allow water flow only. A pipeline with 100 mm in diameter can be inserted into the bottom of the sand tank, and defects are on the top of the pipeline. The pipeline extends the sand tank by 50 mm to collect the drained water and soil.

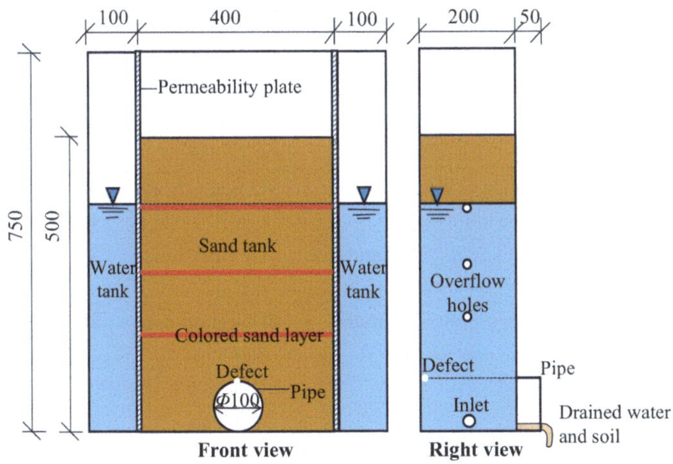

Fig. 1. Schematic diagram of test apparatus (Unit: mm)

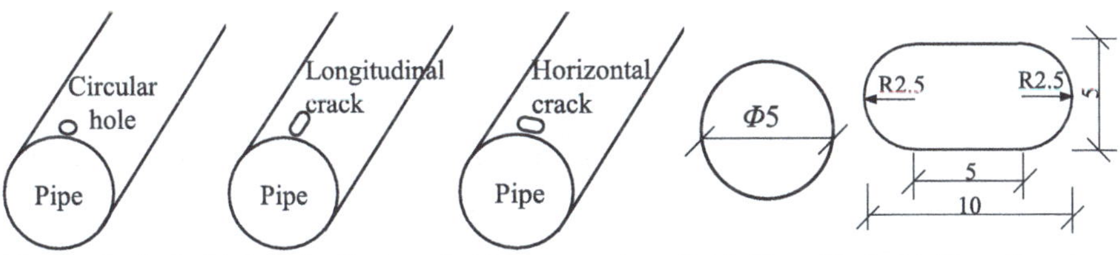

(a) Circular hole (b) Longitudinal crack (c) Horizontal crack (d) Size of hole (e) Size of crack

Fig. 2. Types and size of defect (Unit: mm)

As shown in Fig. 2, there are three types of pipelines defects. The defect hole is 5 mm in diameter, and the defect crack is 10 mm long and 5 mm wide. The defect can be blocked before the test and opened during the test.

2.2 Test Conditions

In order to clarify the influence of four factors on the seepage erosion, eight model tests are performed. The specific test cases and conditions are shown in Table 1.

Among cases, Case1 is unsaturated soil of about 75% saturation, while Case3 is saturated soil by pumping method to explore the effect of soil saturation. Case2–Case4 have different hydraulic heads. Case1, Case5, and Case6 have different gradations. Case2, Case7, and Case8 have different size pipeline defects.

Table 1. Test conditions

Factors	Soil gradation (Sand: gravel)	Saturation	Pipeline defect	Hydraulic head	Case
Saturation	6:4(S6:4)	Unsaturated	Circular hole	30 cm	Case1
				20 cm	Case2
Hydraulic head	6:4(S6:4)	Saturated	Circular hole	30 cm	Case3
				40 cm	Case4
Soil gradation	4:6(S4:6)	Unsaturated	Circular hole	30 cm	Case5
	8:2(S8:2)				Case6
Pipeline defect	6:4(S6:4)	Saturated	Longitudinal crack	20 cm	Case7
			Horizontal crack		Case8

(Note: The record of S6:4 means fine-sand content of 60% and fine gravel content of 40%)

Besides, the tests use gap grading soil. The particle size of fine-sand is 0.1–0.25 mm to simulate the fine silty-sand layer. Binner [6] found the particle size of the skeleton is about 4 mm, so gravel with the particle size of 2–5 mm is selected to form the soil skeleton. The width of pipeline defect is 5 mm to ensure the loss of fine-sand only.

2.3 Test Procedure

The basic concept of tests is to irritate the seepage erosion through the pipeline defect. Therefore, the detailed test procedure is presented as follows:

(1) Check the instruments, install the pipeline and block the defect before the test.
(2) Mix the soil sample according to the proposed soil gradation, then strew soil on the tank freely. Stratified compaction to reach the relative density of 80%. Compact the ground by moist tamping into four layers and set a thin layer of colored soil at the interface of each layer for observing the ground deformation more clearly.
(3) Open the defect and trigger the erosion test. Use a camera to record the erosion process and collect the loss of water and sand during the test.
(4) Collect the soil sample of each layer to analyze the gradation change after the test.

3 Test Results and Analysis

3.1 The Seepage Erosion Around Defective Pipeline

The development process of seepage erosion caused by pipeline defect obtained from Case3 is presented in Fig. 3, and other cases show the similar characteristics.

The erosion is triggered by the defect and develops upwards gradually. The fine-sand closed to the defect begins to flow into the pipeline, and upper sand transports

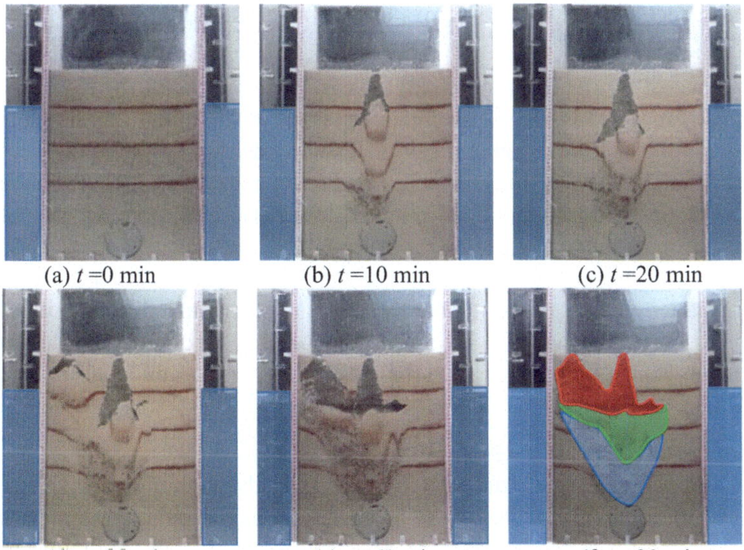

(a) t =0 min (b) t =10 min (c) t =20 min

(d) t =30 min (e) t =60 min (f) t =90 min

Fig. 3. Process of erosion development

downwards, leading to loose and deformation of eroded soil. The cavity forms at the height of the water level and expands gradually over time. In fact, the formation of cavities is also affected by saturation as well, which will be interpreted later. Besides, the closer to the defect, the more fine-sand loses. Therefore, the eroded area can be classified into three types: gravel area where fine-sand loses severely and the skeleton of gravel is visible (shown as blue in Fig. 3(f)), flow area where the settlement of soil occurs due to compression deformation of gravel area, showing relative rigid movement downwards (shown as green in Fig. 3(f)), the cavity (shown as red in Fig. 3(f)). The erosion develops rapidly at the beginning and then converges gradually. Erosion doesn't develop after 60 min anymore and the whole soil gets stable.

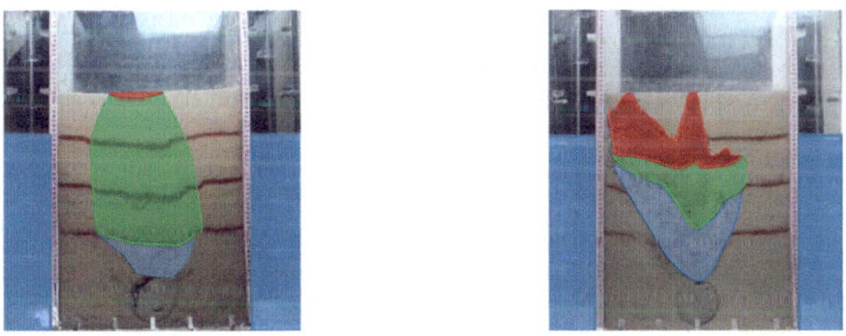

(a) Final state of unsaturated soil (Case1) (b) Final state of saturated soil (Case3)

Fig. 4. The final state of eroded areas in soil with different saturation

3.2 The Effect of Influential Factors on Seepage Erosion

The Influence of Saturation on Seepage Erosion

The final erosion state in different saturated soil is presented in Fig. 4. The saturation of soil has great effects on erosion and cavity. The eroded area of saturated soil is large, which means more loss of fine-sand. It can be explained that the permeability and seepage force of the unsaturated soil is much lower. Besides, pore air may block the flow channel that fine particles migrate through. Therefore, the erosion amount is reduced.

In addition, the hidden cavity hardly forms in unsaturated soil, but settlement occurs on the surface due to soil suction and low seepage force, which is different from saturated soil. Hence, the risk of the ground cave-in is relatively small in unsaturated soil.

The Influence of Hydraulic Head on Seepage Erosion

The final erosion state with different hydraulic heads is presented in Fig. 5. It is obvious that the location of the cavity is closely related to the hydraulic head. The cavities almost form at the water table. Besides, the higher the hydraulic head, the larger the eroded areas, especially the cavity and gravel area, which means more loss of fine-sand. The gravel area develops upward constantly along with the rising of hydraulic head, even extends to the surface when water table reaches the ground. Therefore, the high hydraulic head can induce large seepage force to drag the fine-sand and to scour the channel that particles migrate through, making much soil loss and deformation.

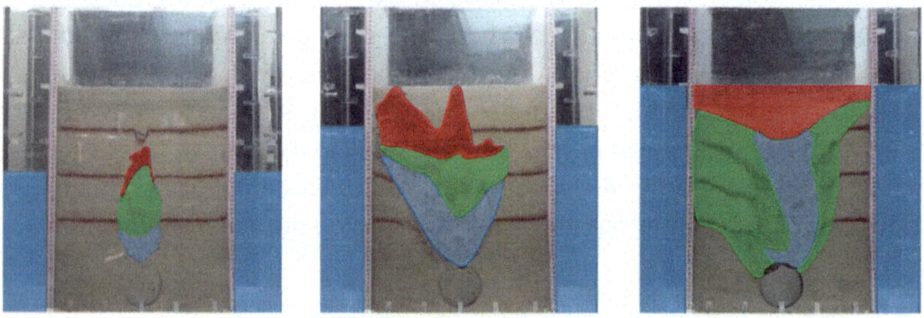

(a) 20cm head (Case2) (b) 30cm head (Case3) (c) 40cm head (Case4)

Fig. 5. The final state of eroded areas with different hydraulic heads

The Influence of Soil Gradation on Seepage Erosion

The final state of erosion areas with different soil gradation is presented in Fig. 6. It is obvious that with more gravels forming the more stable skeleton, the soil deformation hardly occurs, as shown in Case 5 of S4:6. With no much gravels forming the stable skeleton, the fine-sand migrates and soil deformation occurs constantly. The skeleton deformation may change the particle migrating channel, and result in large eroded area and ground settlement, as shown in case 6 of S8:2. After the test, the sieving test was

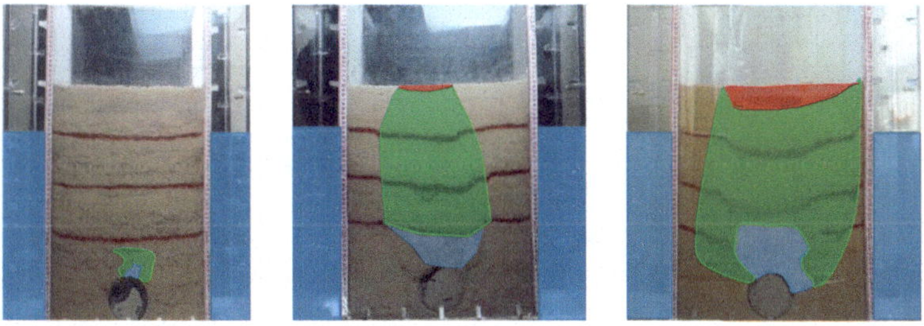

(a) S4:6 (Case5) (b) S6:4 (Case1) (c) S8:2 (Case6)

Fig. 6. The final state of eroded areas with different soil gradation

done on the soil sample around the defects. The contents of fine-sand of Case5, Case1, and Case6 after erosion are 39%, 42%, and 45%, respectively. It is indicated that different initial graded soil samples can achieve the similar state regarding the ratio of fine-sand and gravel after internal erosion with the same condition. Therefore, it can be speculated that if the content of fine-sand is less than the threshold, erosion won't occur.

The Influence of Pipeline Defect Size on Seepage Erosion

The final state of eroded areas with different pipeline defect is presented in Fig. 7. The cavities are all located at the water table. Besides, three tests can reach stability quickly, generally in the first 30 min. They also show the similar distribution of eroded areas. Case7 and Case8 have same defect size, showing the same eroded area, which implies that crack direction has no effect on erosion. The gravel area and cavity in Case7 and Case8 are a little larger than that in Case2 due to increased defect size, which means more loss of soil. Because bigger defect induces more channels that soil migrate to the pipeline, but the final increased amount of sand loss is not much due to the same seepage force. On the whole, the defect size of the pipeline has a little effect on seepage erosion when only fine sand migrates through the defect. If the defect is so large that gravel loses together with sand, the defect size would have great effects on erosion.

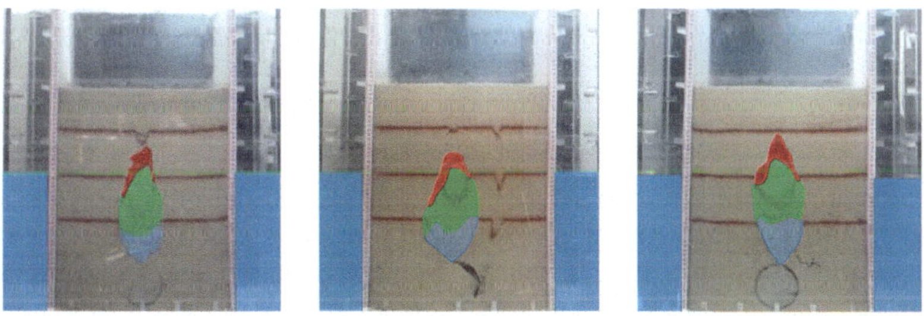

(a) Circular hole (Case2) (b) Longitudinal crack (Case7) (c) Horizontal crack (Case8)

Fig. 7. The final state of erosion areas with different pipeline defects

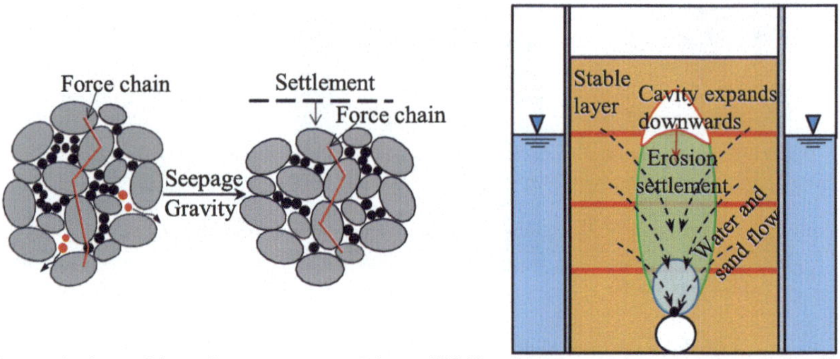

(a)The evolution of the microstructure of the soil[11] (b) The settlement of eroded soil

Fig. 8. The diagrammatic sketch of erosion settlement

4 Discussion

Figure 8 presents the mechanism of settlement and cavity formation microscopically and macroscopically. Microscopically, as shown in Fig. 8(a), coarse particles form the soil skeleton and force chains. Fine particles migrate through the pore of the skeleton with the seepage to erosion, while the skeleton structure remains stable. However, the skeleton becomes loose, and the force chains become unstable due to the loss of lateral support by fine particles. Then the skeleton would be compressed with the influence of seepage force and gravity. Hence, deformation and settlement occur constantly. But the deformation of skeleton won't occur under low hydraulic gradient, so the hydraulic gradient threshold of skeleton-deformation will be investigated soon.

Macroscopically, in saturated soil, the settlement of eroded soil below the water table occurs as long as skeleton deformation [12]. Meanwhile, the soil above the water table has great suction to remain stable without internal erosion. So the cavity is generated near the water table ordinarily. With the development of erosion and settlement, the cavity expands downwards. Finally, all the channels that fine particles migrate through are blocked due to compression of the skeleton. Besides, the content of fine-sand decreases to a certain threshold. Then the erosion will cease as well as soil deformation.

5 Conclusions

Model tests are carried out to simulate the erosion around defective pipeline considering soil saturation, hydraulic head, soil gradation and defect size. The process of cavity formation is measured. Based on the results, the main findings are presented as follows:

(1) The internal erosion is triggered by the pipe defect and develops upwards. The erosion can converge and get stable finally due to the support of soil skeleton.

(2) Regarding the erosion state, the eroded areas can be classified as gravel area, flow area, and cavity. The loss of fine-sand and soil deformation mainly occur in the gravel area, which leads to the upper soil move downwards in the flow area.
(3) From a microscopic perspective, the erosion breaks the stable structure of the soil. When soil skeleton becomes loose, it is rearranged due to loss of fine-sand that provides lateral support to skeleton, which leads to settlement of eroded soil below the water table. Hence, cavity forms at the water table in saturated soil.
(4) The erosion area especially the cavity in saturated soil is larger than that in unsaturated soil. No hidden cavity forms in unsaturated soil, but the surface settlement. Besides, the higher hydraulic head means stronger seepage force, inducing larger erosion mass and eroded area. The water table also affects the location of the cavity. In addition, more initial fine-sand content means more sand loss, leading to more extensive soil deformation. While bigger defect size may cause a little more erosion amount, the effects on cavity is not significant when only fine sand migrates through the defect.

Acknowledgments. This study is financially supported by National Natural Science Foundation of China (Grants No. 41772295). The support is gratefully acknowledged.

References

1. Tao, L.J., Yuan, S., An, J.H.: Development mechanism of cavity damage under urban roads and its influence on road surface subsidence. J. Heilongjiang Univ. Sci. Technol. **3**, 289–293 (2015). (in Chinese)
2. Mochales, T., Casas, A.M., et al.: Detection of underground cavities by combining gravity, magnetic and ground penetrating radar surveys: a case study from the Zaragoza area, NE Spain. Environ. Geol. **53**(5), 1067–1077 (2008)
3. Foster, M., Fell, R., Spannagle, M.: The statistics of embankment dam failures and accidents. Can. Geotech. J. **37**(5), 1000–1024 (2000)
4. Kenney, T.C., Lau, D.: Internal stability of granular filters. Can. Geotech. J. **22**, 215–225 (1985)
5. Chang, D.S., Zhang, L.M.: A stress-controlled erosion apparatus for studying internal erosion in soils. Geotech. Testing J. **34**(6), 579–589 (2011)
6. Binner, R., Homberg, U., et al.: Identification of descriptive parameters of the soil pore structure using experiments and CT data. In: The Fifth International Conference on Scour and Erosion, San Francisco, pp. 397–407 (2010)
7. Hu, Y.H., et al.: Analysis of reasons for urban road collapse prevention and control countermeasures in recent decade of China. Highway **9**, 130–135 (2016). (in Chinese)
8. Mukunoki, T., Kumano, N., et al.: Visualization of three dimensional failure in sand due to water inflow and soil drainage from defective underground pipe using X-ray CT. Soils Found. **49**(6), 959–968 (2009)
9. Kuwano, R., Sato, M., Sera, R.: Study on the detection of underground cavity and ground loosening for the prevention of ground cave-in accident. Jpn. Geotech. J. **5**, 219–229 (2010). (in Japanese)

10. Sato, M., Kuwano, R.: Influence of location of subsurface structures on development of underground cavities induced by internal erosion. Soils Found. **55**(4), 829–840 (2015)
11. Chang, D.S., et al.: Critical hydraulic gradients of internal erosion under complex stress states. J. Geotech. Geoenviron. Eng. **139**(9), 1454–1467 (2013)
12. Maeda, K., Kondo, A.: Micro and macro modeling of ground depression due to internal erosion. In: Geo-Shanghai 2014, Shanghai, pp. 445–455 (2014)

Numerical Analysis of Wall Deformation of PBFC Anti-seepage Slurry in Landfill

Guozhong Dai, Jia Zhu(✉), Guicai Shi, and Weicheng Shi

School of Civil Engineering and Architecture,
Changzhou Institute of Technology, 666 Liaohe Road, Changzhou 213032,
People's Republic of China
{daigz, shigc, shiwc}@czu.cn, zhujiagz@163.com

Abstract. This research is based on an anti-seepage project of landfill in Jiangsu province, China. Finite element method in ANSYS is used to investigate numerically the stress and strain of diaphragm wall in the landfill. The pulp ratio of diaphragm wall is determined by orthogonal experimental method. Results show that the maximum stress (absolute value) of the wall is only 0.24–0.38 MPa and no tensile stress is produced with the wall. The maximum horizontal displacement is in the bottom of the wall with the value about 0.19% of the height of the wall. The maximum vertical displacement is in the top of the wall with the value about 2.5% of the thickness of the wall. In addition, no plastic deformation occurs in the diaphragm wall and the deformation of wall is proportional to that of its surrounding soil, which meets the operational requirements of landfill.

Keywords: Landfills · Diaphragm wall · Stress

1 Introduction

For the vertical diaphragm wall in the landfill, the wall mainly bears the load of the gravity, the earth pressure and the water pressure, which enlarges the deformation of wall [1]. The wall does not crack during its service life, because the deformation modulus of the vertical diaphragm wall is close to that of surrounding soil. Thus, the deformation of the wall and the surrounding soil can be coordinated [2–4]. Experiments show that the concretion body of PBFC anti-seepage slurry has a good anti-seepage performance and the average permeability coefficient is about 0.53–1.86 × 10^{-8} cm/s. It can prevent effectively the diffusing of leachate to the surrounding soil [5]. And elastic modulus of the wall made of PBFC material is about 50–800 MPa. Its stiffness is also close to that of the surrounding soil. All these features lead the anti-seepage slurry to meet the requirements of mechanical impervious wall.

This paper is based on an diaphragm wall of a landfill project in Jiangsu Province, China. The stress and deformation of the diaphragm wall is analyzed by ANSYS finite element software. The distributions of the maximum principal stress and the minimum principal stress are estimated accurately. The change law of horizontal displacement and vertical displacement is summarized. The results can provide theoretical basis for the design of vertical diaphragm wall around the landfill [6–8].

L. Hu et al. (Eds.): GSIC 2018, *Proceedings of GeoShanghai 2018 International Conference: Multi-physics Processes in Soil Mechanics and Advances in Geotechnical Testing*, pp. 271–279, 2018.
https://doi.org/10.1007/978-981-13-0095-0_31

2 Engineering Survey and Calculation Parameter Selection

2.1 Engineering Survey

The landfill is in the alluvial plain of the Yangtze River Delta, which belongs to the floodplain terraces unit. The terrain of landfill is relatively flat, the natural ground elevation of the Yellow Sea is about 3.7–5.8 m and the average thickness of ground cover layer is 51 m. The maximum length of the landfill (East to West) is about 440 m and the maximum width (North to South) is about 180 m. The vertical permeability coefficient is 2.4×10^{-6}–1.3×10^{-7} cm/s and the horizontal permeability coefficient is 5.0×10^{-6}–2.4×10^{-7} cm/s. The depth of groundwater table varies with topography, which is 0.60–1.30 m away from the surface. The thickness of the diaphragm wall is 0.5 m, the average depth of the wall is 15 m, and the maximum depth is 25 m. The engineering geological conditions of landfill (top-down hierarchical) and related parameters of the diaphragm wall are shown in Table 1.

Table 1. Engineering geological conditions in landfill and related parameters of the diaphragm wall.

Material	Compression modulus E /MPa	Poisson ratio ν	Saturation severity γ/KN/m^3	Cohesive force C/kPa	Friction angle $\phi/°C$	Average thickness of soil /m
Plain fill	6.1	0.26	18.7	8	8.2	1.45
Clay	9	0.32	19.5	56	12.8	3.53
Silty clay plus silt	8.2	0.31	19.3	36	12.3	3.45
Silty clay	5.6	0.32	19.3	27	12.9	3.45
Clay	12	0.27	19.7	60	13.5	9.28
Silty clay	8.2	0.32	19.6	37	14.5	10.50
Diaphragm wall	210	0.26	20.2	780	27	15–25

2.2 Radio of Anti-seepage Slurry

The slurry with different ratio has various impervious performance and mechanical performance [9]. By statistical analysis of the existing anti-seepage material of landfill [10], we synthesize the preliminary test of this subject. The orthogonal test of 4 factors and 3 levels is adopted to optimize the slurry formula. The factors and the levels are shown in Table 2.

The impermeability and mechanical properties of slurry under different formulations are analyzed by orthogonal test [11, 12]. The variation of permeability coefficient about the slurry concretion body at different ages is presented in Fig. 1. The unconfined compressive strength of the slurry concretion body with different proportions is shown in Fig. 2.

By analysis of experiment data, the optimum formulation is selected as: 20% cement, 22% bentonite, 0.2% polyvinyl alcohol, 0.03% superplasticizer, 18% fly ash

Table 2. The factors and the levels of orthogonal test.

Level	Factors /%			
	Cement	Organic bentonite	Water reducer	Polyvinyl alcohol
1	16	18	0.01	0.2
2	20	22	0.02	0.5
3	24	26	0.03	0.8

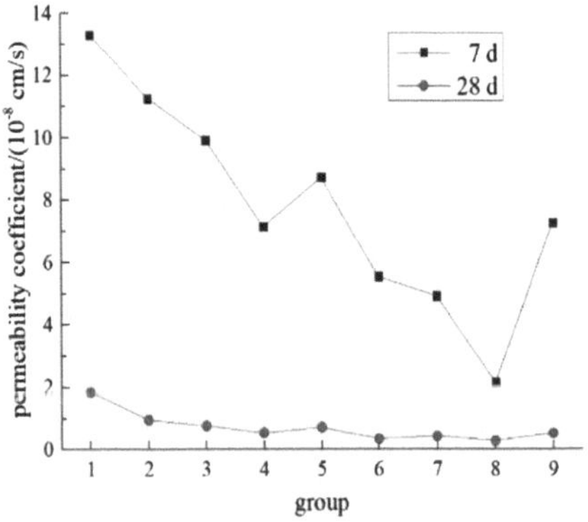

Fig. 1. Variation of permeability coefficient with age

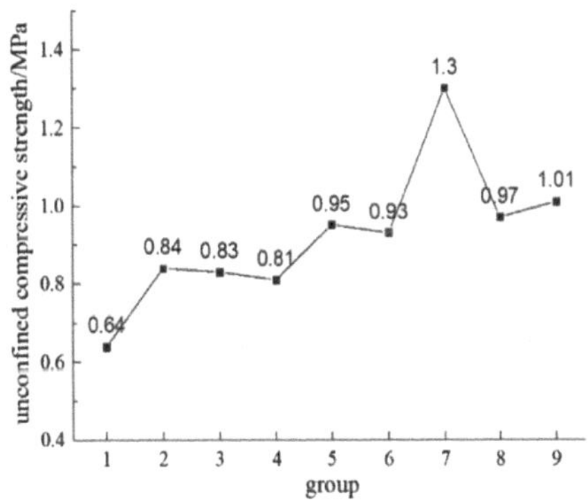

Fig. 2. Variation of unconfined compressive strength with the ratio

and 0.5% sodium carbonate. The impermeability of slurry on this radio is stable and the average permeability coefficient of the concretion body after 28 days curing is 0.73×10^{-7} cm/s. The unconfined compressive strength of concretion body after 28 days curing is 0.95 MPa. It can resist the erosion of acid liquid and groundwater [13].

2.3 Calculation Model Selection

The appropriate constitutive model needs to be considered first before analyzing the wall deformation [14]. For most engineering materials, the stress lower than the proportional limit are generally in linear relationship with strain and when the stress is greater than the proportional limit, nonlinear relationships emerges [15]. In general, the proportional limit is close to the yield limit, so we treated the two values as the same point in ANSYS. Considering that the unconfined compressive strength of concretion body after 28 days is 0.5–2.0 MPa, far greater than the tensile strength, and the concretion body expands subjecting to shear. Therefore, the Drucker-Prager (DP) plastic model is used to analyze the stress and deformation of wall and surrounding soil [16]. This model considers the influence of volume expansion caused by the material yield and ignores the influence of temperature change. It is suitable for granular materials, such as concrete, rock, soil and other materials [17–19]. For the analyzed material, the DP model applies both the relevant flow criterion and the irrelevant flow criterion. In the process of material yield, the yield surface does not change, so there is no reinforcement criterion. Its yield criterion is as follows [20]:

$$F = 3\beta\sigma_m + \left[\frac{1}{2}\{s\}^T[M]\{s\}\right]^{\frac{1}{2}} - \sigma_y \tag{1}$$

Where σ_m is hydrostatic pressure, β is material constant. Material parameters β and material parameters σ_y is defined as follows:

$$\beta = \frac{2\sin\phi}{\sqrt{3}(3 - \sin\phi)};\ \sigma_y = \frac{6(c)\cos\phi}{\sqrt{3}(3 - \sin\phi)} \tag{2}$$

To calculate the DP plastic model, three parameters are needed: viscous force (shear yield stress) C, internal friction angle and dilatancy angle (used to control the volume expansion). To evaluate whether the wall meets the requirements of stress and deformation, a small volume expansion model is used to calculate the distribution of compressive stress, tensile stress and displacement [21]. The most unfavorable situation is the landfill without waste, the wall under soil pressure and water pressure. The slope of inside the wall (h: b) is 1:0.5. The boundary constraint conditions are as follows: the both sides of vertical boundary is horizontal constraint, the bottom is fixed constraint, the other is free boundary. The quadrilateral mesh is used to process the diaphragm wall, which is divided into 3 rows according to the thickness of the wall. The mesh section of the diaphragm wall and formation section is shown in Fig. 3.

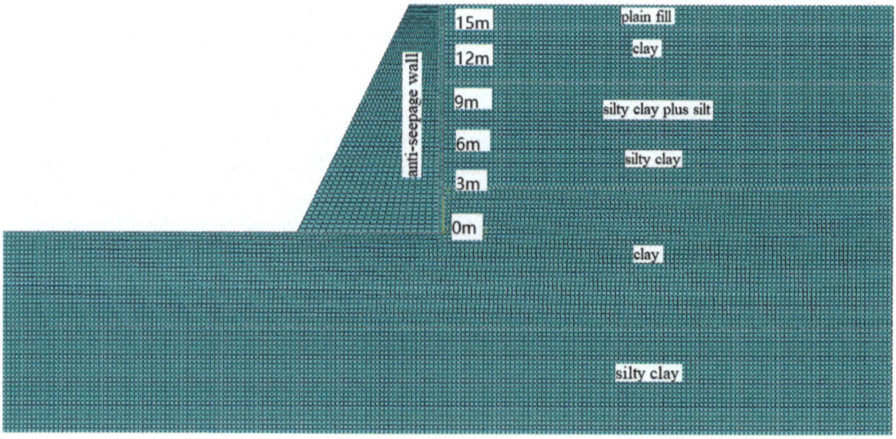

Fig. 3. Strata profile and mesh generation for diaphragm wall of 15 m height

3 Calculation Result Analysis

3.1 Stress Distribution of the Wall

According to the calculation, the principal stress distribution of diaphragm wall and surrounding soil is shown in Fig. 4. Calculation and analysis shows that the maximum principal stress decreases along with the height of the wall (from the wall bottom elevation date) and the turning point emerges in the 2–4 m of the wall. The varying trend of the minimum principal stress is the same as that of the maximum principal stress: the stress value increases in the 2–3 m of the wall. The maximum value of the minimum principal stress, 0.24 MPa, is located at the bottom of the wall.

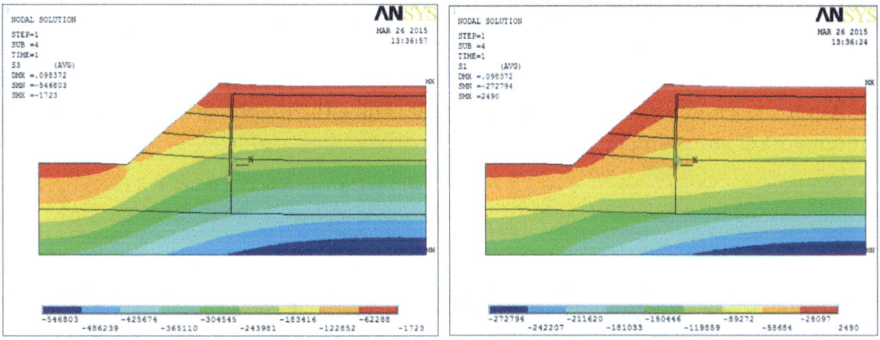

(a) The maximum main stress (b) The minimum principal stress

Fig. 4. Principal stress distribution of the diaphragm wall and surrounding soil (wall depth is 15 m)

To analysis the influence of depth of diaphragm wall, diaphragm wall with three depths: 15 m, 20 m and 25 m. The distribution curves of the maximum principal stress and the minimum principal stress along the wall are shown in Fig. 5. The maximum stresses (absolute value) of three walls are 0.24 MPa, 0.30 MPa, 0.38 MPa. No tensile

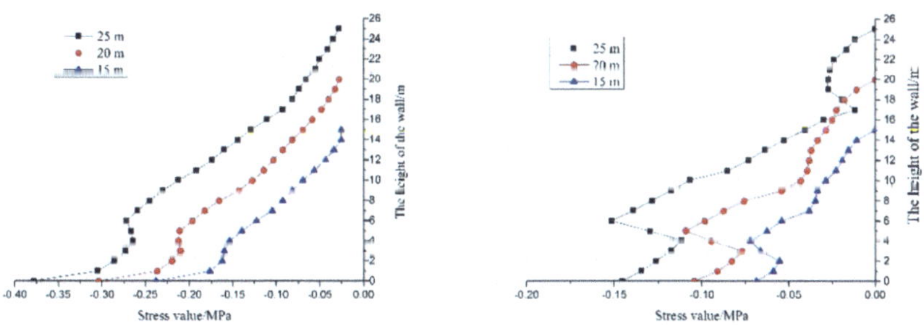

(a) The maximum main stress (b) The minimum principal stress

Fig. 5. Curves of the principal stress distribution of the wall (pressure stress is negative)

stress is detected for the three models due to the small stress value [22]. The unconfined compressive strength of plastic concrete diaphragm wall after 28 days curing is 0.95 MPa, which fully meets the requirements of stress and ensures that the stress concentration does not occur [23].

3.2 Stress Intensity Distribution of the Wall

The stress intensity distributions of three depths are similar and the stress intensity distribution of the 15 m cutoff wall and surrounding soil is shown in Fig. 6. The results show that the wall stress intensity gradually reduces along the wall height. The maximum stress strength of the three walls, located at the bottom of the wall, are 0.17 MPa, 0.20 MPa and 0.23 MPa. With the increase of height of the wall, the soil pressure, hydrostatic pressure and thus stress intensity increase.

The distribution curve of stress intensity along the wall is shown in Fig. 7. The change range of stress intensity of the wall is small and the maximum stress strength is less than the strength limit of the wall material. Thus, the diaphragm wall can meet the requirement on stress intensity.

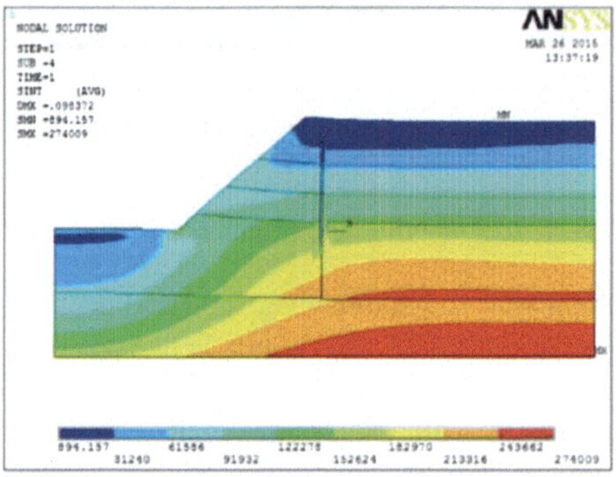

Fig. 6. Stress intensity distribution of wall (15 m wall)

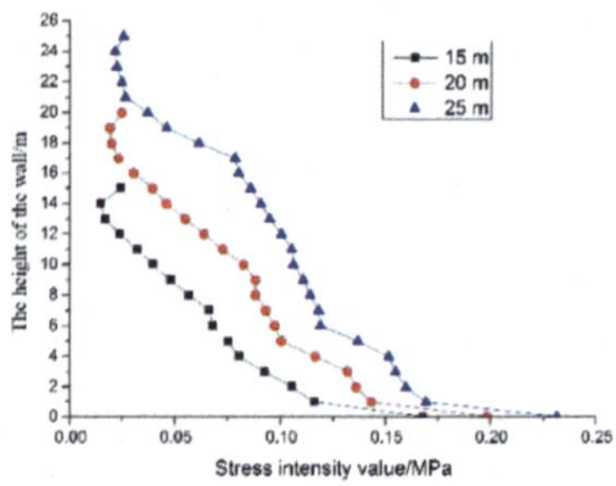

Fig. 7. Distribution curve of stress intensity distribution

3.3 Horizontal Displacement Distribution of the Wall

The horizontal displacement distribution of the wall and surrounding soil is shown in Fig. 8. The horizontal displacement distribution shows the maximum horizontal displacement at the bottom of the wall. For the diaphragm wall with depth of 15–25 m, the maximum horizontal displacement value is 1.0–4.8 cm and the overall level of diaphragm wall displacement is small. The maximum horizontal displacement is about 0.19% of the wall height and the horizontal displacement of the wall along the wall height changes as shown in Fig. 9.

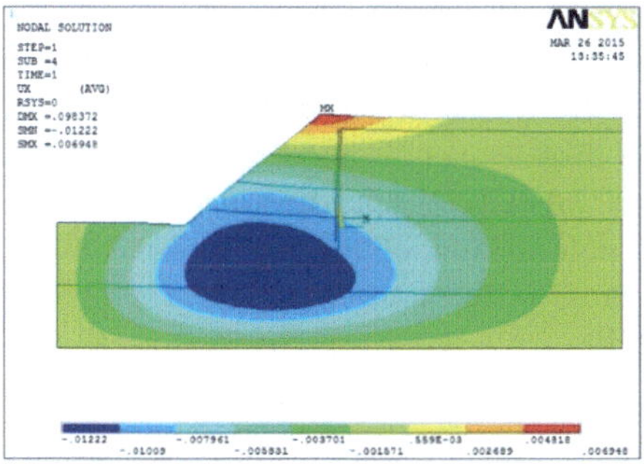

Fig. 8. Horizontal displacement distribution of the wall (its depth is 15 m)

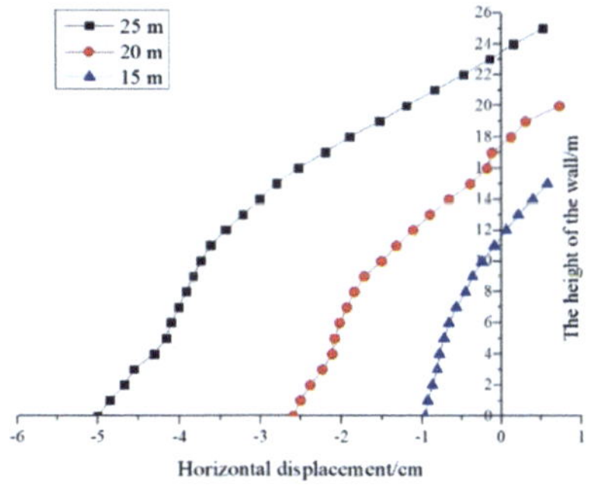

Fig. 9. Changing curves of the horizontal displacement

3.4 Vertical Displacement Distribution of the Wall

The vertical displacement of the diaphragm wall reaches the maximum value at the top of the wall. And the maximum value for the three walls are around about 2.5% of the wall thickness, which are 7.28 mm, 9.56 mm, 12.55 mm, respectively. The vertical displacement of the wall is relatively small and it also increases with the increase of the depth of wall. The settlement of the diaphragm wall is basically the same as that of the surrounding soil. In fact, the PBFC concretion body has stronger ability to resist deformation and the ultimate strain under unconfined condition can reach 6%. This means that the plastic deformation will not occur and can meet the requirements of the use of landfill [24, 25].

By test in situ, the distribution of the maximum principal stress, the distribution of the minimum principal stress, the horizontal displacement of the diaphragm wall and the vertical displacement of the diaphragm wall are measured directly. And the tested results are basically consistent with the calculated values by ANSYS, which fully demonstrates that the chosen model is reasonable.

4 Conclusion

(1) The PBFC slurry has the characteristics of high seepage resistance, low elastic modulus, moderate compressive strength and so on, so it is very suitable for the construction of the vertical diaphragm wall of the landfill.

(2) The stress ratio of diaphragm wall is basically the same as that of the corresponding wall height, and this change is linearly decreasing along the wall. The maximum principal stress of the wall increases at the root of the wall (2–4 m from the bottom of the wall). For the depth of 15–25 m in the diaphragm wall, the maximum stress (absolute stress) of the wall is only 0.24–0.38 MPa, the stress value is small, and there is no tensile stress.

(3) The maximum horizontal displacement of the diaphragm wall is located at the bottom of the wall and the maximum horizontal displacement is about 0.19% of the wall height. The maximum vertical displacement of the wall is located at the top of the wall and the maximum vertical displacement is about 2.5% of the wall thickness. The diaphragm wall does not produce plastic deformation and the deformation of soil around the plastic deformation of the diaphragm wall is coordinated. The wall does not show stress concentration and crack problem, which ensure the safe operation of landfill.

Acknowledgments. The national natural science foundation of China (51678083) is gratefully acknowledged.

References

1. Li, S., Qin, Z.: The finite element analysis of the diaphragm wall's deformation and stress of Lin Xi reservoir's main dam. Yangtze River (19), 72–74 (2012)
2. Ma, X., Liang, G., Zheng, M., Xu, C., Mou, R.: The numerical analysis of concrete diaphragm wall with low elastic modulus in earth-rockfill dam. Hydroelectr. Power **5**(4), 51–54 (2011)
3. Su, Y.: The analysis plastic concrete diaphragm wall effects on the stability of earth-rockfill dam. Shandong University (2013)
4. Ding, Y., Zhang, Q., Zhang, B.: The finite element analysis of diaphragm wall's deformation and stress characteristics in high core wall of rockfill dam. J. Hydroelectr. Power **32**(3), 162–167 (2013)
5. Wang, Y., Dai, G., Shi, G.: Numerical analysis of the deformation and stress of waste landfill's diaphragm wall. J. Yangtze River Sci. Res. Inst. **37**(4), 89–93 (2015)
6. Amadi, A.A., Eberemu, A.O.: Delineation of compaction criteria for acceptable hydraulic conductivity of lateritic soil-bentonite mixtures designed as landfill liners. Environ. Earth Sci. **67**, 999–1006 (2012)
7. Wang, R., Dou, T., Xiong, H., Feng, Y., Lu, Y.: The finite element analysis of diaphragm wall's deformation and stress of Beijing, Daning reservoir dam. Water Resour. Hydropower Technol. **9**(3), 46–49 (2010)
8. Chen, Y., Shi, W.: The research on influence of the bedrock fault rupture to core-wall rockfill dam. J. Rock Mech. Eng. **25**(3), 578–583 (2006)
9. Evans, J.C., Asce, F., Ruffing, D.G., et al.: Design and construction of an experimental soil-bentonite cutoff wall. In: Geofrontiers (2017)
10. Abbaslou, H., Ghanizadeh, A.R., Amlashi, A.T.: The compatibility of bentonite/sepiolite plastic concrete cut-off wall material. Construct. Build. Mater. **124**, 1165–1173 (2016)
11. Guo, C., Wang, F.: Mechanism study on the construction of ultra-thin antiseepage wall by polymer injection. J. Mater. Civ. Eng. **24**(9), 1183–1192 (2012)
12. Asada, M., Ishikawa, A., Horiuchi, S.: Large-scale cutoff wall model test using ethanol bentonite slurry. J. Mater. Civ. Eng. **17**(6), 719–724 (2005)
13. Bryukhovetskiy, O.S., Borovkov, Y.A., Naydenko, I.Y.: Construction of anti-seepage curtains during jet mining. Gornyi Zhurnal (6), 44–47 (2017)
14. Amiri, S.A.G., Grimstad, G.: Constitutive model for long-term behavior of saturated frozen soil. In: Biot Conference on Poromechanics, pp. 1005–1012 (2017)

15. Isbuga, V., Regueiro, R.A.: Finite element analysis of finite strain micromorphic Drucker-Prager plasticity. Comput. Struct. **193**, 31–43 (2017)
16. Asada, M., Horiuchi, S.: High-Density bentonite slurry for seepage barriers. J. Mater. Civ. Eng. **17**(2), 178–187 (2005)
17. Zhu, Y., Hao, Z., Yang, Z.: The application of ANSYS in the numerical simulation of dam. Rock Soil Mech. **27**(6), 965–972 (2006)
18. Nguyen, T.B., Lee, C., Choi, H.: Slug test analysis in vertical cutoff walls with consideration of filter cake. J. Geotech. Geoenviron. Eng. **137**(8), 785–797 (2011)
19. Chen, L., Zhu, J., He, S., Wu, X., Yin, J.: Effects of different constitutive models on the calculation of stress of diaphragm wall. J. Chongqing Univ. **36**(10), 120–125 (2013)
20. Fiedler, J., Koudelka, T.: Plasticity calculation of plates using layered model. Appl. Mech. Mater. **825**(2), 111–118 (2016)
21. Beretta, E., Bonnetier, E., Francini, E., et al.: Small volume asymptotics for anisotropic elastic inclusions. Inverse Prob. Imaging **6**(1), 1–23 (2017)
22. Liu, X., Zhang, F.: Study on the influence of tensile stress on the stress intensity factor of ANSYS. Shanxi Architecture (2016)
23. Shi, J., Luan, J.: Stability analysis method for composite failure through base liner and waste filling. Rock Soil Mech. **34**(9), 2576–2582 (2013)
24. Hamdi, N., Srasra, E.: Hydraulic conductivity study of compacted clay soils used as landfill liners for an acidic waste. Waste Manag. **33**, 60–66 (2013)
25. Qiu, G., Liang, L., Sun, H.: Slope stability of landfill under biodegradation. J. Northeastern Univ. (Nat. Sci.) **34**(10), 1495–1498 (2013)

Supercritical CO_2 Permeability in Rock: An Experiment Study

Bin Ye(✉), Xueqian Ni, Yadong Zhang, and Weimin Ye

Tongji University, Siping Road 1239, Shanghai 200092, China
yebin@tongji.edu.cn

Abstract. Carbon capture and storage projects have traditionally targeted deep sedimentary basins where CO_2 is maintained as supercritical state ($scCO_2$). In order to study the permeability of $scCO_2$, a series of tests were conducted by independent research and development of the instrument with related environment variables, including temperature, confining pressure, injection pressure and different types of rocks. Through normalization of the experiment data, the results showed that the permeability of $scCO_2$ increased with the increment of temperature and injection pressure, but with the reduction of confining pressure and effective stress. Besides, silt rock was more sensitive for temperature, while was less sensitive for confining pressure than sandstone. The conclusions summarized above possess instructional significance and referenced value for the siting selection of geological storage.

Keywords: $ScCO_2$ · Permeability · Temperature · Confining pressure Inlet pressure

1 Introduction

With the increasing development of modern industry, people are facing serious pressure from greenhouse effect for which scientists believe that the increasing emission of carbon dioxide (CO_2) into the atmosphere is the main causative factor. Nevertheless, traditional fossil energy, such as coal, petroleum, and natural gas, is still occupying the main energy market, and this situation might last for a long period. The long-term reliance on fossil fuel prompts governments to look for ways to reduce the emission of anthropogenic CO_2 to the atmosphere. Carbon Capture and Storage (CCS) is thought to be one of the most promising approaches to achieve this goal [1, 2].

CCS is a process through which CO_2 is captured from industrial or energy-related sources and then transported and injected into underground geological formations. Generally, CO_2 can be isolated in oil/gas field, saline aquifer, nonproductive deep coal seam or other geological isolation forms [3]. Among all possible CO_2 storage sites, saline aquifer has the greatest storage potential up to 400–10,000 Gt worldwide, and can be found in almost all of the sedimentary basins [4]. Thus, it has gained extensive attention in the world. For CO_2 storage in saline aquifers, understanding the permeability of CO_2 in porous rocks is of great significance in predicting the migration and evaluating the long-term stability of injected CO_2. The injected CO_2 are normally stored in geological media up to several kilometers below the ground surface. In this

L. Hu et al. (Eds.): GSIC 2018, *Proceedings of GeoShanghai 2018 International Conference: Multi-physics Processes in Soil Mechanics and Advances in Geotechnical Testing*, pp. 280–288, 2018.
https://doi.org/10.1007/978-981-13-0095-0_32

context, the CO_2 is in a supercritical state, i.e., its pressure and temperature are beyond the critical point (7.38 MPa and 31.8 °C). In addition, the environmental temperature and pressure will increase with the underground storage depth. The variation of environmental pressure and temperature can greatly influence the permeability of $scCO_2$ in rock because the properties of $scCO_2$, such as the density and viscosity, greatly depend on its pressure and temperature. Furthermore, the confining pressure for the reservoir rock is also high due to the thick overburden ground layers, and this high confining pressure can influence the porosity and skeleton of the reservoir rock, which can also influence the permeability of $scCO_2$ in porous rock. Therefore, to understanding the permeability of CO_2 in porous rocks, it is vital to investigate how the permeability of $scCO_2$ in porous rock varies with the environmental temperature, the confining pressure of reservoir rock and even the injecting pressure.

Many works about the permeability of CO_2 through rock have been conducted. There are two usual methods that can be used to test CO_2 flowing in laboratory, namely, the steady-state method and unsteady-state method. For unsteady-state method, the flow-rate and the pressure of CO_2 are not kept constant during the test, and therefore the permeability can only be evaluated empirically based on a given mathematical model. The relative permeability of CO_2 was related with the interfacial tension of rock, which is dependent on pressure, temperature and salinity tested by unsteady-state experiments ([5, 6]). In addition, the permeability of $scCO_2$ decreased with the increasing injecting pressures at low temperatures while increases at higher temperatures (more than 50 °C) using high pressure triaxial equipment (unsteady-state method) [7]. In contrast, the steady-state method, in which the flow-rate and the pressures are in a steady state, is more suitable for measuring the permeability of $scCO_2$. However, because of the difficulties in experimental equipment and techniques, only a few studies were reported by using the steady-state method to investigate the permeability of $scCO_2$ in rocks [8–12].

In this study, we focused on finding out how permeability of $scCO_2$ varies with the change of influencing factors, including temperature, confining pressure and inlet pressure, by conducting series of core-flood experiments on two specific types of rock, namely silt rock and sandstone. So a self-made apparatus [13] was established to meet the experiment condition. It was found that the permeability coefficient of CO_2 increases with the increment of temperature and inlet pressure, while decreases with the increment of confining pressure.

2 Tested Rock Samples

Two types of rocks were tested in this study, i.e. a silt rock extracted from BOSO peninsula, Japan and sandstone extracted from Shandong Province, China respectively. The physical properties of the rocks are shown in Table 1. The tested samples were prepared with the size of 5 cm in diameter and 1 cm in height. Four silt rock samples and four sandstone samples were used for studying the influences of confining pressure and temperature on the permeability of $scCO_2$. Meanwhile, another one sandstone sample was used to investigate the influences of inlet pressure on the permeability of $scCO_2$. Before conducting the CO_2-flooding tests, the samples were saturated by

Table 1. Physical properties of the silt rock and the sandstone

Physical properties	Unit	Silt rock	Sandstone
Soil particle density, ρ_s	g/cm^3	2.64	2.27
Dry unit weight, γ_d	kN/m^3	14.4	0.047
Porosity, n	-	0.44	-
Natural moisture content, w	%	25.9	-

injecting sufficient water slowly into the samples. The injecting process lasted for more than 5000 min, and when the outflow water equaled to the inflow water, the sample was considered to be fully saturated.

3 Experimental Setup and Procedures

3.1 Apparatus and Test Procedure

The test system used in the study was developed by Ye et al. [13]. The schematic diagram and panorama of the whole system are shown in Fig. 1. The device primarily comprises a pressure cell, a syringe pump for CO_2, a pressure amplifier, a heating unit, and a measurement system. Two buffer tanks are set at the inlet and outlet of the seepage chamber to maintain stable CO_2 injection and back pressures during testing. More details can be referenced to Ye et al. [13].

High purity CO_2 was reserved in the refrigerator to 15 °C below. Before injecting CO_2 from refrigerator, sample was fixed in the pressure cell. Temperature in the pressure cell was controlled by heating unit which was connected to pressure cell by two plastic pipes. CO_2 was injected into inlet buffer tank by syringe pump and outlet buffer tank by adjusting regulator 1 before injecting CO_2 into pressure cell. The value of both inlet and outlet pressure were shown on the measurement system. CO_2 wasn't injected into the pressure cell until both the inlet and outlet pressure near to the

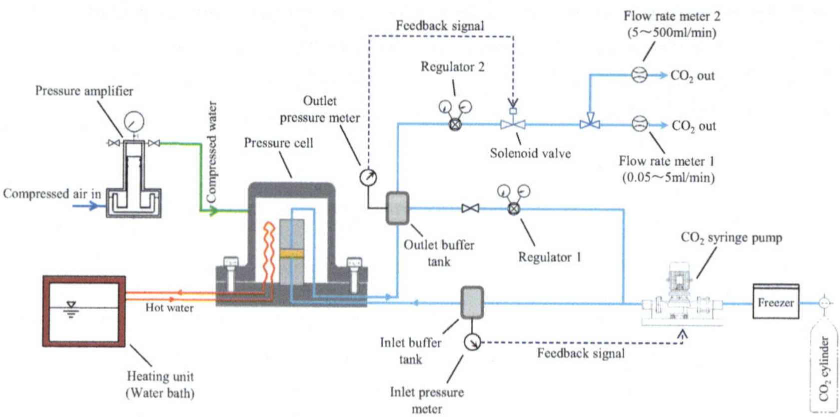

Fig. 1. Schematic diagram of the test system

anticipated value. Pressure amplifier could press water into the pressure cell and control the confining pressure in the pressure cell. The mechanism of the apparatus is similar to the steady state technique, except only $scCO_2$, rather than the CO_2-brine mixture, is injected into the core during the test. In this testing system, the syringe pump can provide the inlet CO_2 pressure up to 15 MPa to ensure the supercritical state of CO_2. The available back pressure (outlet pressure) is in the range of 0–9 MPa.

3.2 Experimental Conditions

Set Confining Pressure and Temperature as Variables. To study the influence of confining pressure and temperature on the permeability of supercritical CO_2, we conducted a series of permeability tests on the samples under different confining pressures and temperatures. For silt rock and sandstone, we used four samples for testing respectively. For each sample, we conducted the permeability tests under five combined conditions of confining pressures and temperatures. The experimental conditions are summarized in Table 2, where *T* refers to temperature, while *CP* refers to confining pressure. Although every sample was tested along a path different from each other, their temperatures and confining pressures were set the same in the inception stage and the final stage. In these tests, the inlet pressure and outlet pressure were set as 10.0 MPa and 9.0 MPa respectively.

Table 2. Specific confine pressures and temperatures for each sample

Sample	Stage									
	1		2		3		4		5	
	T (°C)	CP (MPa)	T (°C)	CP (MPa)	T (°C)	CP (MPa)	T (°C)	CP (MPa)	T (°C)	CP (MPa)
1	45	11	55	11	65	11	65	13.5	65	15
2	45	11	55	11	55	13.5	55	15	65	15
3	45	11	45	13.5	55	13.5	65	13.5	65	15
4	45	11	45	13.5	45	15	55	15	65	15

Set Inlet Pressure as Variable. Besides, another test was performed on a sandstone sample to investigate the influence of inlet pressure on the permeability of $scCO_2$. In the test, the inlet pressure varied from 10 MPa to 15 MPa and the outlet pressure kept constant as 9 MPa. The temperature and the confining pressure were set as 65 °C and 15 MPa respectively (Table 3).

Table 3. Specific inlet pressure for each sample

Sample	Stage				
		1	2	3	4
	T(°C)	Inlet pressure MPa	Inlet pressure MPa	Inlet pressure MPa	Inlet pressure MPa
5	65	10	11	13.5	15

3.3 Calculation of the Permeability Coefficient

The permeability of $scCO_2$ can be calculated based on Darcy Law and the equations are described as follow [13]:

$$k = \frac{\bar{q}_{sc}}{A \cdot i} = \frac{\bar{q}_{sc} \cdot \rho_w g h}{A(\bar{p}_{in} - \bar{p}_{out})} \quad (1)$$

$$\bar{q}_{sc} = c \cdot \bar{q}_{gas} \quad (2)$$

$$c = \frac{\rho_{gas}}{\rho_{sc}} \quad (3)$$

where $\bar{p}_{in}$ and $\bar{p}_{out}$ are the average inlet and outlet pressures, A is the section area of sample, h is the height of the sample, ρ_w is the density of water, g is the gravitational acceleration, $\bar{q}_{gas}$ and $\bar{q}_{sc}$ are the average flow rates in gaseous and supercritical states, and c is the ratio of the gaseous CO_2 density ρ_{gas} to the $scCO_2$ density ρ_{sc}. Specifically, ρ_{gas} equals to a constant value $1.9\,\text{kg/m}^3$. ρ_{sc} varies with temperature and pressure, and could be queried from Peace Software [14].

4 Results and Discussions

4.1 The Influences of Confining Pressure and Temperature on the Permeability of $ScCO_2$

As described above, the confining pressure and temperature effect on CO_2 permeability were investigated for eight samples in different paths (Table 2). According to the test results, the curves of k versus confining pressure and k versus temperature are plotted in Fig. 2 for silt rock and Fig. 3 for sandstone respectively.

As shown in Figs. 2(a) and 3(a), when temperature increases from 45 °C to 65 °C, there is a clear increase in the permeability of $scCO_2$. For instance, at 11 MPa confining pressure in silt rock samples, when the temperature increased by 20 °C, the permeability increased by 1.7 times, and the permeability increases by 50% at 15 Mpa confining pressure. It indicates that the speed of permeability decreases with the

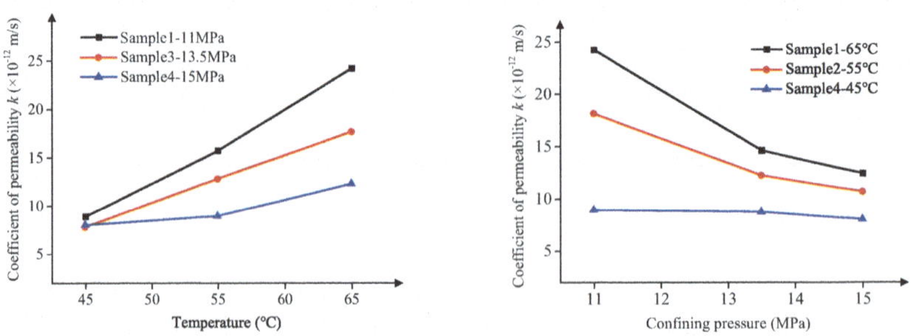

Fig. 2. k varies with temperature (a) or confining pressure (b) (Silt rock).

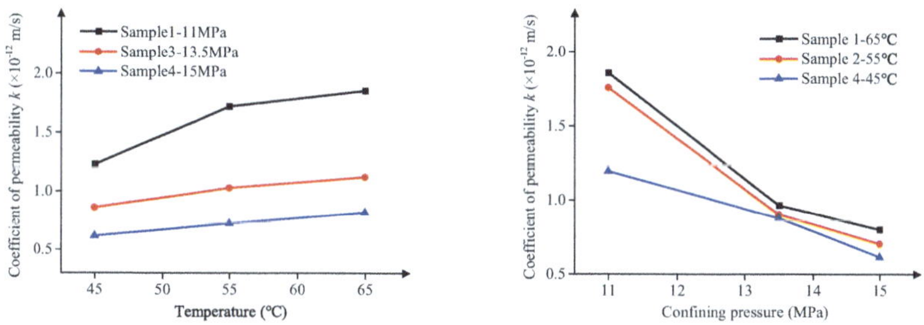

Fig. 3. *k* varies with temperature (a) or confining pressure (b) (Sandstone)

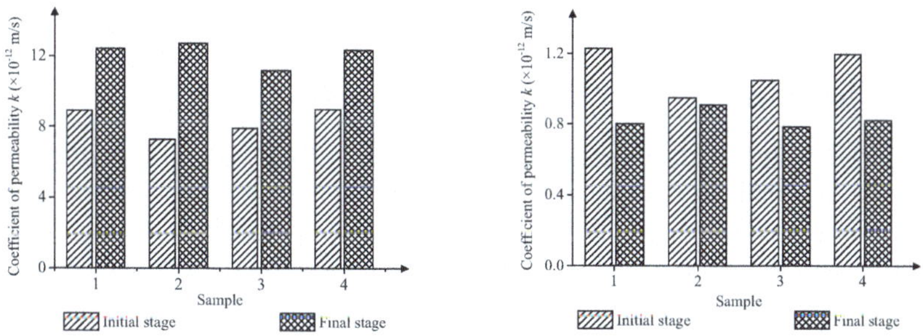

(a) Silt rock (b) Sandstone

Fig. 4. *k* under the initial and final stage of each sample

increasing of confining pressure. Though there is not a significant change in sandstone samples, the trend is in agreement with silt rock samples. On the other hand, the lower the temperature is, the higher the permeability of $scCO_2$ under the same confining pressure. The tendency is also demonstrated by Figs. 2(b) and 3(b), which illustrate the permeability *k* decreases with the increment of confining pressure. Though samples subjected to different temperature in each group, the trend of *k* is the same. These conclusions can be interpreted as the molecular movement. When the temperature goes up, the internally molecules heat of $scCO_2$ move faster and faster, then a greater average molecular kinetic each molecule has. This could lead to more $scCO_2$ passing through a certain cross section when the rest of the variables are the same. However, if the confining pressure enhanced, due to compression effect of fissures inside samples, molecules are not easy to pass through the cross section. In other words, there is less $scCO_2$ passing through the section when the rest variables are the same.

Figure 4 displayed the measured coefficients of permeability *k* of each sample in the initial and final stages. The experimental conditions are identical both the initial (T45-P11) and the final stage (T65-P15) for each sample. The value of *k* of the last stage is higher than that of the first stage for silt rock samples, while for sandstone samples, the situation is reverse. As presented above, there is a positive correlation between temperature and *k*, and a negative correlation between confining pressure and *k*. Therefore, it can be concluded that the permeability of silt rock is more sensitive to temperature, while less sensitive to confining pressure than that of sandstone.

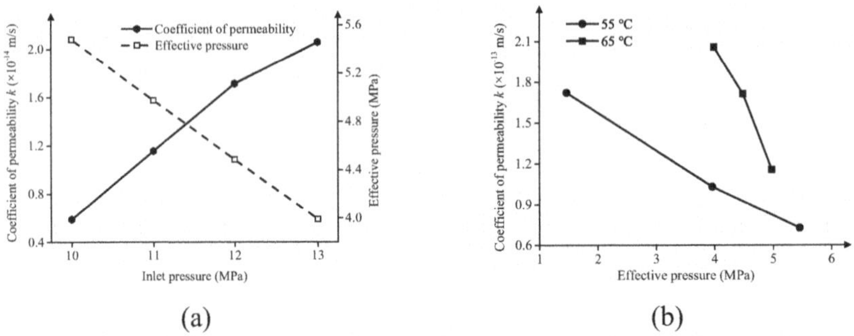

Fig. 5. (a) Relationship between inlet pressure and k and effective pressure versus injection (b) Relationship between effective pressure and k

4.2 The Influences of Inlet Pressure on *k* of $ScCO_2$

The curve of k versus inlet pressure is presented in Fig. 5(a). The permeability of $scCO_2$ increases with increasing inlet pressure, and this trend is mainly related to the effective pressure effect on the pore space of sandstone sample. According to Ranjith and Perera [15], the effective pressure is a function of fluid pressure and expressed as $p_{effective} = p_{cp} - 0.5 * (\bar{p}_{in} + \bar{p}_{out})$, where p_{cp} is the confining pressure. With the increment of the inlet pressure, the pore pressure inside the sandstone sample increases and leads to a decrease in the effective pressure, as shown by the dash line in Fig. 5(a). As a result, the pore space of rock expands and the flow channels extends, which leads to an increase in permeability of $scCO_2$.

Figure 5(b) shows the relation between effective pressure and the permeability of $scCO_2$ more obviously. The square-line is sample 5 mentioned above, while the dot-line is sample subjected to a constant injection pressure 10 MPa in constant 55 °C but changed with confining pressure. According to the results, three conclusions may be summarized. First, the permeability of $scCO_2$ decreases with the increase of effective pressure which has discussed in Fig. 5(a). Second, the higher temperature exhibits the greater permeability under the same effective stress, such as when the effective pressure is 4 MPa. Finally, though the confining pressure and inlet pressure are changing, the tendency between effective pressure and k is the same.

5 Conclusions

Four silt rock samples and five sandstone samples were tested to investigate the influence of temperature, confining pressure and inlet pressure on the permeability of $scCO_2$. The main conclusions are drawn as follows:

(1) The permeability of $scCO_2$ (k) increases with the increment of temperature. When the temperature goes up, the internally molecules of $scCO_2$ exercise more and more quickly, then each molecular has a greater average molecular kinetic.

(2) k decreases with the increment of confining pressure. Due to compression effect of fissures inside samples, molecules are not easy to pass through the cross section so that less $scCO_2$ pass through.
(3) k increases with increasing injection pressure. Because the increasing internal pore pressure results in reduced effective stress, the pore space of sample expand, then the flow channels extended and more $scCO_2$ pass through.
(4) Silt rock is more sensitive for temperature, while is less sensitive for confining pressure. On the contrary, sandstone is more sensitive for confining pressure but less sensitive for temperature.

In conclusion, engineer should select a site with great confining pressure but lower temperature as much as possible. Meanwhile, the effective stress and inlet pressure are also considered during the process.

References

1. US Climate Change Technology Program [EB/OL] (2002). http://www.climatetechnology.gov
2. CSLF Forum [EB/OL] (2004). http://www.cslfforum.org
3. Koide, H., Tazaki, Y., Noguchi, Y., Nakayama, S., Iijima, M., Ito, K.: Subterranean containment and long-term storage of carbon dioxide in unused aquifers and in depleted natural gas reservoirs. Energy Convers. Manag. **33**(5), 619–626 (1992)
4. Bert, M.: IP CC Special Report on CO_2 Capture and Storage. Cambridge University Press, Landon (2005)
5. Xue, Z., Ohsumi, T., Koide, H.: Seismic wave velocity monitoring of CO_2 migration in porous sandstones saturated with water. Chin. J. Rock Mech. Eng. **22**(6), 1002 (2003)
6. Bachu, S., Bennion, B.: Effects of in-situ conditions on relative permeability characteristics of CO_2-brine systems. Environ. Geol. **54**(8), 1707–1722 (2008)
7. Perera, M.S.A., Ranjith, P.G., Choi, S.K., Airey, D.: Investigation of temperature effect on permeability of naturally fractured black coal for carbon dioxide movement: an experimental and numerical study. Fuel **94**(1), 596–605 (2012)
8. Demessie, E.S., Hassan, A., Levein, K.L., Kumar, S., Morrell, J.J.: Supercritical carbon dioxide treatment: effect on permeability of Douglas-fir heartwood. Wood Fiber Sci. **27**(3), 296–300 (1995)
9. Chang, C., Zhou, Q., Xia, L., Li, X., Yu, Q.: Dynamic displacement and non-equilibrium dissolution of supercritical CO_2 in low-permeability sandstone: An experimental study. Int. J. Greenh. Gas Control **14**, 1–14 (2013)
10. Chang, C., Zhou, Q., Guo, J., Yu, Q.: Supercritical CO_2 dissolution and mass transfer in low-permeability sandstone: Effect of concentration difference in water-flood experiments. Int. J. Greenh. Gas Control **28**, 328–342 (2014)
11. Meng, Q.: A numerical study of supercritical carbon dioxide migration in porous media under conditions of saline aquifers. University of Science and Technology of China, China (2014)
12. Schanaar, G., Digiulio, D.C.: Computational modeling of the geologic sequestration of carbon dioxide. Vadose Zone J. **8**(2), 389–403 (2009)

13. Ye, B., Ye, W.M., Zhang, F., Xu, L.: A new device for measuring the supercritical CO_2 permeability in porous rocks under reservoir conditions. Geotech. Test. J. **38**(3), 338–345 (2015)
14. Peace software [EB/OL]. http://www.peacesoftware.de/einigewerte/co2_e.html
15. Ranjith, P.G., Perera, M.S.A.: A new triaxial apparatus to study the mechanical and fluid flow aspects of carbon dioxide sequestration in geological formations. Fuel **90**(8), 2751–2759 (2011)

A Parallelized Water-Soil-Coupled SPH Model Considering the Effect of Permeability and Its Application in the Piping Simulation of Dike

Weijie Zhang[1,2](✉), Yufeng Gao[1,2], and Kenichi Maeda[3]

[1] Key Laboratory of Ministry of Education for Geomechanics and Embankment Engineering, Hohai University, Nanjing 210098, China
zhangwj2016@hhu.edu.cn

[2] Geotechnical Engineering Research Center of Jiangsu Province, Hohai University, Nanjing 210098, China

[3] Department of Civil Engineering, Nagoya Institute of Technology, Nagoya 466-8555, Japan

Abstract. The Smoothed Particle Hydrodynamics model has been widely used in geotechnical engineering for the research of flow-like landslide, flowing liquefied soil and dike or slope failure. However, the efficiency is still an issue when the simulation has a large number of particles and only a few research focused on the parallel optimization of the water-soil-coupled SPH model. In this study, an SPH model considering the water-soil coupling was proposed based on the basic principle of SPH method, Drucker-Prager model, and mixture theory of porous media. The OpenMP, as a widely used parallelling framework, was adopted to accomplish the parallel implementation and improve the efficiency. Besides, an improved particle searching method, compatible with the OpenMP parallel optimization, has been proposed. Using the proposed SPH model, the effect of different permeability on the seepage failure of dike was simulated. Meanwhile, by comparing the time consumptions of different numbers of CPU thread, it has been proven that the parallel implementation can greatly reduce the time-consumption.

Keywords: Piping simulation · Permeability · OpenMP
Water-soil coupling · SPH

1 Introduction

Extreme weather events caused by the global warming have induced many geological disasters worldwide, among which seepage failures due to water level-up pose a great threat to the peoples' lives and property. Therefore, more attention should be focused on this topic to identify the failure mechanism, and to propose related countermeasures. For such disasters, an analysis of the entire process, including the coupling and the large deformation, is very essential to more precisely identify the mechanism [1, 2]. In the field of dike failure analysis, the Smoothed Particle Hydrodynamics (SPH) method has unique advantages for the simulation of free surfaces, deformation boundaries, and large deformations [3]. Peng et al. introduced a hypoplastic constitutive model into the

L. Hu et al. (Eds.): GSIC 2018, *Proceedings of GeoShanghai 2018 International Conference: Multi-physics Processes in Soil Mechanics and Advances in Geotechnical Testing*, pp. 289–298, 2018.
https://doi.org/10.1007/978-981-13-0095-0_33

SPH method to simulate the large deformation of sand collapse and slope failure, demonstrating the advantage of SPH method coupled with complex soil model [4]. Maeda and Sakai [5] introduced air dynamics into the SPH framework to investigate the air effect in the seepage process. Huang et al. [6] conducted an analysis of large-deformation flows of liquefied soils using a water-soil-coupled SPH method. Bui and Fukagawa [7] adopted the two-phase SPH method to analyze the stability of slope. Zhang and Maeda [8, 9] have presented preliminary results of a slope failure test and coupled SPH simulations. These researchers have found that the SPH method is suitable for the simulation of seepage failures, however, the numerical simulation of seepage failure often involves the multi-phase modelling resulting in a lot of particles, therefore the improvement of efficiency is an essential issue that should be focused on.

To improve the efficiency, many parallelized SPH models have been proposed and applied in practice. These parallel implementations are mainly adopting the OpenMP framework, MPI (Message Passing Interface) framework, and GPGPU including CUDA (Compute Unified Device Architecture) and OpenCL (Open Computing Language). The typical examples are DualSPHysics based on the GPU and OpenMP solving the free surface problem and LAMMPS (Large-scale Atomic/Molecular Massively Parallel Simulator) base on distributed memory using MPI solving the atomic or molecular dynamic problem. Besides, Wróblewski and Boryczko conducted the parallel implementation of SPH using the OpenMP environment MPI environment [10]. Jin et al. [11] used an OpenMP-parallelized SPH model for the incompressible fluid to study the wave flume stability. For the explosion-induced fragmentation problem, Fan and Li [12] developed a coupled method of state-based peridynamics and SPH. From these studies, it can be concluded that previous parallel implementations are mainly in the field of free surface problem and atomic/molecular dynamic problem, while the parallel implementation of SPH model for geotechnical problems and multi-phase modelling is rarely found in literatures. In order to improve the efficiency of the SPH model in geotechnical engineering, this research proposed a SPH model considering the water-soil coupling and conducted the parallel implementation using the OpenMP environment, aiming at benefiting other numerical methods in geotechnical engineering. The accuracy and efficiency of the proposed SPH model were carefully checked by the analysis of a piping failure test with different permeability. At last, some conclusions were derived from the analysis and the parallel efficiency was discussed as well.

2 The Proposed Water-Soil-Coupled SPH Model

2.1 Concepts of Water-Soil-Coupled SPH Model

In the proposed SPH model, the water phase and soil phase are simulated on different layers, which can be combined by the interaction force, based on the mixture theory of Biot [13] and Zienkiewics et al. [14]. Particles of different phase are assigned to different layers by using different material type: type of boundary is 0, type of water is 1, type of soil is 2 and type of water inlet is 7. After that, we can obtain the components of density and stress on different phase layers. The dynamic behavior of particles in the proposed SPH model is governed by Navier-Stokes equations, including the continuity equation, the momentum equation, and Equation of State [15]. The continuity equation is derived from the conservation of mass as,

$$\frac{\mathrm{d}\rho_i}{\mathrm{d}t} = \sum_{j=1}^{N} m_j(v_{im} - v_{jm})\frac{\partial W_{ij}}{\partial x_{im}} \tag{1}$$

where W_{ij} is the smoothing function of supporting particle j evaluated at particle i, N is the total number of supporting particles, v is the velocity, m, n are coordination indices, and i, j are particle indices. This is the original continuity equation in the SPH method and normalization correction can be introduced to obtain more precise simulation. The momentum equation is derived from the Newton's second law, and related to the stress and interaction forces. For the water phase, the momentum equation is

$$\frac{dv_{im}^{f}}{dt} = -\sum_{j=1 j\in water}^{N} m_j\left(\frac{\sigma_{imn}^{f}}{\bar{\rho}_i^{f2}} + \frac{\sigma_{jmn}^{f}}{\bar{\rho}_j^{f2}}\right)\frac{\partial W_{ij}}{\partial x_{in}} + F_i + \frac{R_{im}^{sf}}{\bar{\rho}_i^{f}} \tag{2}$$

For the soil phase, it is

$$\frac{dv_{im}^{s}}{dt} = \sum_{j=1 j\in soil}^{N} m_j\left(\frac{\sigma_{imn}^{s}}{\bar{\rho}_i^{s2}} + \frac{\sigma_{jmn}^{s}}{\bar{\rho}_j^{s2}} - \delta_{mn}\Pi_{ij}\right)\frac{\partial W_{ij}}{\partial x_{in}} + F_i + \frac{R_{im}^{fs}}{\bar{\rho}_i^{s}} \tag{3}$$

where F_i is the external force, R^{sf} is the interaction force on the water phase, and R^{fs} is the interaction force on the soil phase. $\bar{\rho}^{f}$ and $\bar{\rho}^{s}$ are the summarized densities of water phase and soil phase, respectively, calculated from the continuity equation. The artificial viscosity Π_{ij} is used to avoid the stress oscillation [9].

The water phase is regarded as a quasi-compressible liquid, and its dynamic pressure is calculated from the equation of state for water [15]. For the soil phase, stress increment is described by an incremental plasticity model proposed by Bui and Fukagawa [7], combining the Drucker-Prager (DP) yield criterion and the associated flowing rule.

In order to describe the coupling of water and soil, frictional forces [9] are used to simulate the interaction effect, considering the fluid density ρ^{w} (1000 kg/m^3), gravitational acceleration g, porosity n, permeability k, and velocity difference. The B-spline smoothing function, providing a smoothing curve and high efficiency, is used as the smoothing kernel function. To realize a better physical meaning of the boundary, this research adopted the boundary treatment method by Takeda et al. [16] and Morris et al. [17]. It assumed that the boundary particle has a virtual velocity, and its effect on the moving particle is determined according to the relative distance of two particles. The second order Runge Kutta method is employed as the time integration method, where the time step Δt is constrained by the Courant-Friedrichs-Levy (CFL) condition [9].

2.2 Parallel Implementation

The OpenMP environment has been selected to conduct the parallel implementation for its simplicity and flexibility. OpenMP is a programming API that supports a multi-threading platform with shared-memory architecture, using a fork-join model to allocate a master thread that forks a specified number of slave threads running concurrently on different processors [4]. OpenMP provides a set of compiler directives and callable runtime library routines that extend C, C++, or FORTRAN, to express parallelization. The section of code that is meant to run in parallel is marked with a preprocessor directive, *#pragma omp parallel* or *#pragma omp parallel for*, that causes a thread to be spawned before the section is executed. For the nature of shared memory, when one memory that is used by a thread is updated with new information, the change needs to be reflected to other threads, otherwise different threads will be working with incoherent data, yielding wrong results.

The Nearest Neighbor Particle Search (NNPS) subroutine contributes the largest part of time consumption and its parallel implementation is of great difficult. Different threads will access a same address simultaneously and incorrect computing will appear if memory consistency is not strictly guarantied. To deal with this problem, the Linked-Cell method is adopted as the NNPS method to obtain a higher efficiency. It firstly places cells on the problem domain, and then neighboring cells will be determined, if given the total number of cells at each coordinate (*nXm*, *nYm*, and *nZm*). It is easy to find out that the number of neighboring cells for an internal cell is nine for two dimensions and 27 for three dimensions. When searching nearest neighbor particles,

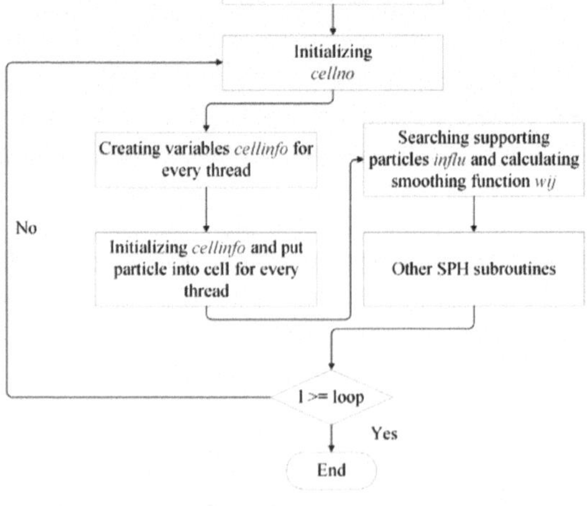

Fig. 1. Flow chart of improved Linked-Cell searching algorithm.

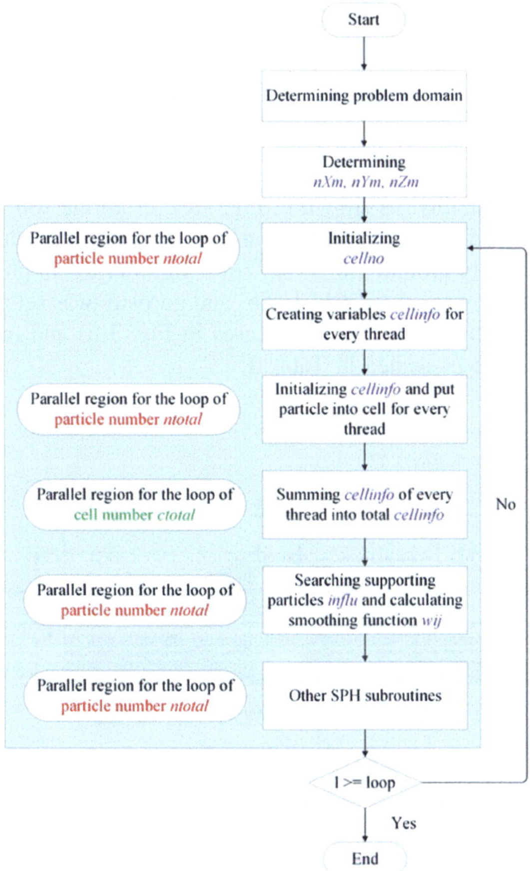

Fig. 2. Flow chart of parallelized Linked-Cell searching algorithm.

only particles in adjacent cells are selected as candidates. The flow chart of Linked-Cell searching algorithm is shown in Fig. 1, where *celllink* is a variable storing neighboring cell IDs, *cellno* is the cell ID that a particle belongs to, *cellinfo* stores particle IDs of cell, *influ* stores supporting particle IDs, and *wij* stores smoothing function values.

To make the parallel implementation of the NNPS subroutine, the flow of Linked-Cell searching algorithm has been improved as Fig. 2. We defined independent variable *cellinfo* for each thread, which had no storage association with the same variable outside, and after the execution of parallel region for the loop of particle number *ntotal*, independent *cellinfo* was summed to the total *cellinfo* using the parallel region for the loop of total cell number *ctotal*. This improvement avoided the accessing confliction and guaranteed the memory consistency. After that, a fully parallelized SPH algorithm has been proposed and a computing program was written using the C++ programming language.

3 Simulation Cases of Dike Seepage Failure

Simulation cases of dike seepage were based on a model test, which was to reveal the failure process of Yabe River in Japan caused by the water level-up of rainfall. The dimension of the model test is shown in Fig. 3(a). In the test, dike body was mainly consisting of Fujimori clay. Silica sand was used in the base layer of dike to form a cushion of high permeability, as shown in Fig. 3(c). From the soil experiments, Fujimori clay had a very small permeability around 1.0×10^{-7} m/s while the Silica sand was of high permeability around 1.0×10^{-5} m/s. In order to simplify the calculation, density of soil particles was set to 2644 kg/m^3 and porosity was set to 0.53. According to the different arrangement of sand, two cases in Fig. 3(b) and (c) were simulated. Detailed parameters are presented in Table 1.

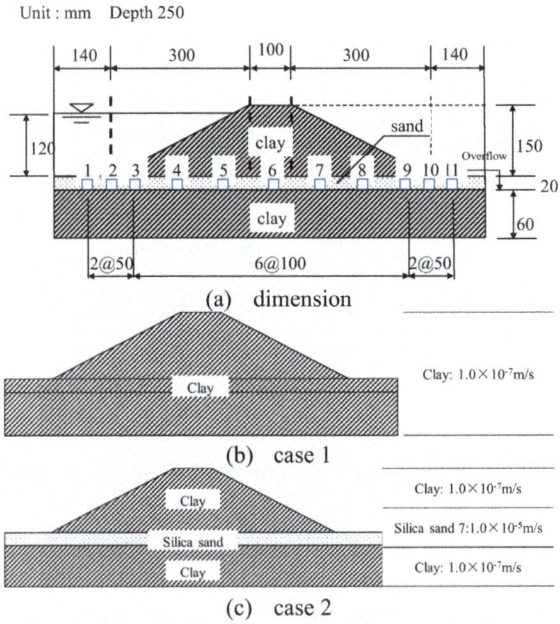

Fig. 3. Simulation cases of dike piping failure test.

Table 1. Simulation parameters.

Parameter	Value	Parameter	Value
Density	2644 kg/m^3	Internal friction angle	29°
Void ratio	0.53	Cohesion	1250 Pa
Elastic module	1.0 MPa	Compression index	0.008
Poisson ratio	0.25	Time steps	1.2×10^{-5} s
Total particles	13881	Total steps	9×10^6 steps

In the simulation, soil particles and water particles were initially stationary, but could move under gravity, stress and interaction forces. Rising water level was produced in the way that inlet particles were changed to water particles. Initial geo stress was generated using the elastic model, and after that, soil behavior was described by the D-P model.

4 Analysis Results and Parallel Efficiency

The SPH simulated second invariant of strain was showed in Fig. 4 at 4.5 million and 9.0 million steps, respectively. It demonstrated that the invariant value increased with time. At the end of simulation, an area of high strain has appeared at the left part of both cases, because of the accumulated pore water pressure and the reduced effective

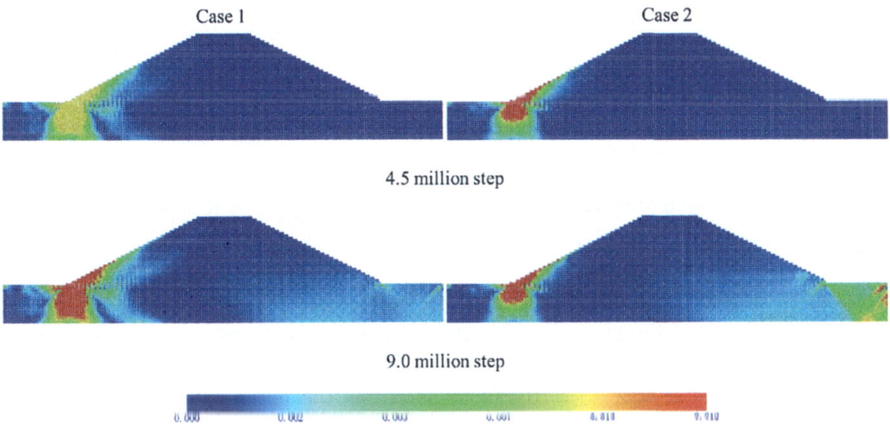

Fig. 4. Simulated second invariant of strain.

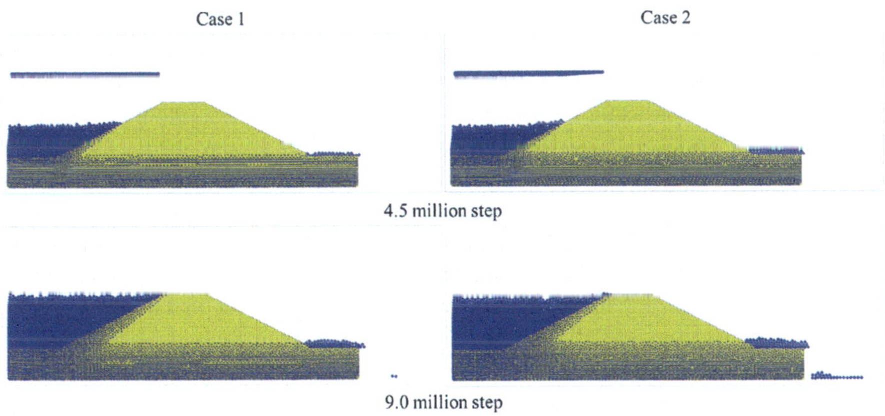

Fig. 5. Simulated piping process (yellow particles: soil and blue particles: water)

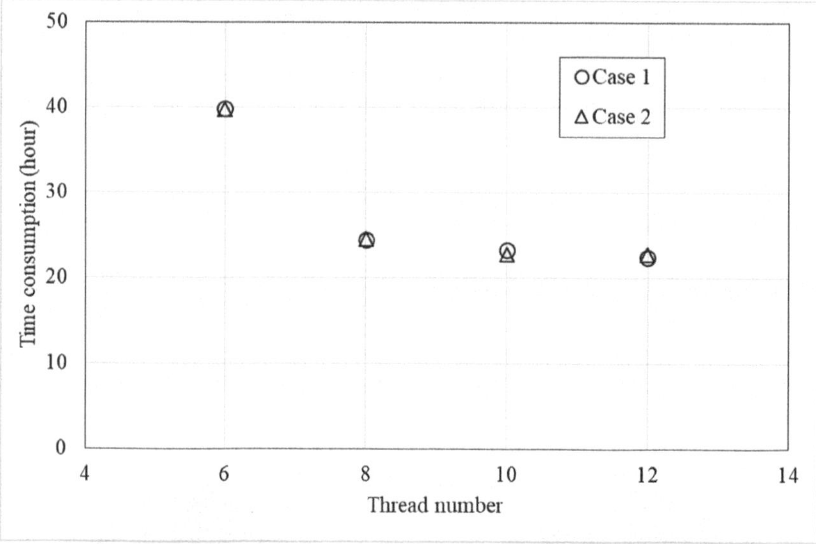

Fig. 6. Time consumption with different number of threads

stress. It is also found that Case 2 had a larger shear zone in the right part for the influence of high permeable base layer. The seepage process from SPH simulations is presented in Fig. 5. It is obvious that Case 2 had more drained water than Case 1 in the right side of dike, demonstrating the effect of high permeable layer in Case 2.

In terms of the parallel efficiency shown in Fig. 6, simulations of Case 1 and Case 2 using 12 threads of Intel Xeon 2690 V3 took about 22 h, while using 6 threads the consumed time was almost 40 h. It can be seen that the parallel implementation can make good use of multi-core technology of CPU and greatly improve the efficiency.

5 Conclusions

To conduct the parallel implementation of the water-soil-coupled SPH analysis, this work proposed a SPH model considering the water-soil coupling and conducted the parallel implementation using the OpenMP environment, based on the basic principle of SPH method, Drucker-Prager model, and the theory of porous media. In order to improve the efficiency, an improved neighboring particle searching algorithm was proposed based on the linked-cell searching method. Using the proposed SPH model, we conducted the simulations of seepage failure test of dike, and from the simulations, the effect of permeability on the seepage has be revealed to some extent. Meanwhile, the parallel implementation has been proven to be an effective way that makes good use of multi-core technology of CPU and benefits other numerical methods in geotechnical engineering.

Acknowledgements. This work is supported by the National Key Research and Development Program of China (grant No. 2016YFC0800205); National Natural Science Foundation of China (NSFC grant No. 41630638); Natural Science Foundation of Jiangsu Province (grant No. BK20170887); the Fundamental Research Funds for the Central Universities (grant No. 2015B25914) and China Postdoctoral Science Foundation.

References

1. Mergili, M., Moreiras, S.M., Fellin, W., Stötter, J.: Preliminary results of slope stability simulations for the prediction of debris flows in the Central Andes (Mendoza, Argentina). In: Proceedings of Geomorphology for the Future, Obergurgl, Austria, 2–7 September 2007, pp. 145–152 (2007)
2. Pastor, M., Hadda, B., Sorbino, G., Cuomo, S., Drempetic, V.: A depth-integrated, coupled SPH model for flow-like landslides and related phenomena. Int. J. Numer. Anal. Methods Geomech. **33**(2), 143–172 (2009)
3. Liu, M.B., Liu, G.R.: Smoothed particle hydrodynamics (SPH): an overview and recent developments. Arch. Comput. Methods Eng. **17**(1), 25–76 (2010)
4. Peng, C., Wu, W., Yu, H.S., Wang, C.: A SPH approach for large deformation analysis with hypoplastic constitutive model. Acta Geotechnica **10**(6), 703–717 (2015)
5. Maeda, K., Sakai, H.: Seepage failure and erosion of ground with air bubble dynamics. In: Proceedings of GeoShanghai, Shanghai, China, June 2010. Geoenvironmental Engineering and Geotechnics (Geotechnical Special Publication No. 204), pp. 261–266 (2010)
6. Huang, Y., Zhang, W.J., Dai, Z.L., Xu, Q.: Numerical simulation of flow processes in liquefied soils using a soil-water-coupled smoothed particle hydro-dynamics method. Nat. Hazards **69**(1), 809–827 (2013)
7. Bui, H.H., Fukagawa, R.: An improved SPH method for saturated soils and its application to investigate the mechanisms of embankment failure: case of hydrostatic pore-water pressure. Int. J. Numer. Anal. Methods Geomech. **37**(1), 31–50 (2013)
8. Zhang, W.J., Maeda, K.: The model test and SPH simulations for slope and levee failure under heavy rainfall considering the coupling of soil, water and air. In: Proceedings of GeoShanghai, Shanghai, China, May 2014. Soil Behavior and Geomechanics (Geotechnical Special Publication No. 236), pp. 538–547 (2014)
9. Zhang, W.J., Maeda, K., Saito, H., Li, Z.Q., Huang, Y.: Numerical analysis on seepage failures of dike due to water level-up and rainfall using a water–soil-coupled smoothed particle hydrodynamics model. Acta Geotech **11**(6), 1401–1418 (2016)
10. Wróblewski, P., Boryczko, K.: Parallel simulation of a fluid flow by means of the SPH method: OpenMP vs. MPI comparison. Comput. Inform. **28**, 139–150 (2009)
11. Jin, Z., Ren, B., Xu, Z.L., Wang, G.Y.: Numerical wave flume based on parallelized SPH. In: Proceedings of the 11th National Conference on Hydrodynamics, Wuxi, pp. 332–339 (2012)
12. Fan, H., Li, S.: Parallel peridynamics–SPH simulation of explosion induced soil fragmentation by using OpenMP. Comput. Part. Mech. **4**(2), 199–211 (2017)
13. Biot, M.A.: General theory of three-dimensional consolidation. J Appl Phys **12**, 155–164 (1941)
14. Zienkiewicz, O.C., Chan, A.H.C., Pastor, M., Shrefler, B.A., Shiomi, T.: Computational Geomechanics. Wiley, Chichester (2000)

15. Liu, G.R., Liu, M.B.: Smoothed Particle Hydrodynamics: A Mesh-Free Particle Method. World Scientific Press, Singapore (2003)
16. Takeda, H., Miyama, M., Sekiya, M.: Numerical simulation of viscous flow by smoothed particles hydrodynamics. Prog. Theor. Phys. **92**(5), 939–960 (1994)
17. Morris, J.P., Fox, P.J., Zhu, Y.: Modeling low Reynolds number incompressible flows using SPH. J. Comput. Phys. **136**(1), 214–226 (1997)

Analysis of the Underground Water Level Considering the Block Effect of Underground Structure on Seepage

Ji-wen Zhang[1,2], Linlong Mu[3,4(✉)], Jie Li[3,4], Jie Cao[2], and Le Wang[3,4]

[1] School of Human Settlements and Civil Engineering, Xi'an Jiaotong University, Xi'an 710049, Shanxi, China
[2] China Jikan Research Institute of Engineering Investigations and Design, Co., Ltd., Xi'an 710043, Shanxi, China
[3] Key Laboratory of Geotechnical and Underground Engineering of the Ministry of Education, Tongji University, Shanghai 200092, China
mulinlong@tongji.edu.cn
[4] Department of Geotechnical Engineering, Tongji University, Shanghai 200092, China

Abstract. The hydrostatic pressure calculation method, which cannot consider the block effect of the underground structure on seepage, is used to calculate the buoyancy of the underground structures in practice nowadays. This will lead to inaccurate calculation of buoyancy. A series of finite element analysis are conducted to investigate the water level distribution around the structure under the condition of the seepage. The influence factors and the distribution rules of the underground water level around the structures are revealed based on the results from FEM. Then an empirical method is proposed to calculate the underground water level around the structure considering the block effect of the underground structure on the seepage. Through validation, the proposed method is proved to be reasonable to calculate the underground water level for calculation of buoyancy on structures considering the block effect of structures on seepage.

Keywords: Seepage · Underground structure · Underground water level

1 Introduction

Plenty of underground structures are constructed as the development of the cities. Anti-floating becomes one of the key ingredients of the design of the underground structures, especially at those places where the underground water level is high. According to the design method, the buoyancy on the underground structures depends on the underground water level. However, the underground water level will be affected by the block effect of underground structures on seepage [1]. In order to evaluate the buoyancy on the underground structures accurately, it is necessary to estimate the underground water level considering the block effect of underground structure on seepage firstly.

L. Hu et al. (Eds.): GSIC 2018, *Proceedings of GeoShanghai 2018 International Conference: Multi-physics Processes in Soil Mechanics and Advances in Geotechnical Testing*, pp. 299–306, 2018.
https://doi.org/10.1007/978-981-13-0095-0_34

Many studies have been carried out to study the effects of underground structures on the seepage of the underground water [2–8]. Pope and Ho [3] pointed out that the underground water level would be raised due to the block effect of underground foundations on seepage through FEM analysis. Brassington [5] and Cao et al. [9] pointed out that even the presents of the piles would change the underground water level. Ding et al. [6] found that the constructions of the underground structures changed the underground water level of Hongkong. Xu et al. [10–12] found that the alternation of different soil layers influenced the block effect of the underground structures on seepage. Fan [13] and Zheng et al. [14] carried out several tests to investigate the influence of underground structures and the dimensions of the underground structures on the seepage. Although the influence is significant, the influence rules are still not clear due to too less tests are carried out. The influence of the underground structures on the underground water level considering the seepage still cannot be estimated based on the existing researches.

In order to find an empirical method to estimate the underground water level considering the effect of the underground structures on seepage, a serious of FEM studies are carried out to investigate the underground water level considering the block effect of underground structure on seepage.

2 Numerical Analysis

2.1 Finite Element Model

The mesh of the calculation model is shown in Fig. 1. The pore pressure element is used in the model. The relative parameters are as following, the soil gravity γ = 1700 kN/m^3, the soil void ratio n = 0.7, the soil permeability coefficient $k = 5\times10^{-5}$ m/s, the soil modulus E = 10 MPa, the Poisson Ratio v = 0.35. The hydraulic gradient is set as 2%. The length of the underground structure is set as 10 m, the width of the underground structures is set as 40 m, the depth of the underground structures is set as 30 m. Herein, dimension of the model is set as 200 m × 200 m × 100 m(width × length × depth).

Fig. 1. Mesh of the model

2.2 Influence of Underground Structure on Underground Water Level in Homogeneous Soil

Firstly, the dimensions of the underground structures are changed in different calculation cases to investigate the influence of the dimensions of the underground structures on the underground water level considering the seepage. As shown in Fig. 2, the width of the structures influences the underground water level significantly. Based on the same hydraulic gradient, the difference of the underground water levels at the upstream of the underground structure increases linearly with the width of the underground structure. The influence of the structure length on the underground water level keep constant when the length is larger than 30 m. The influence of the structure depth on the underground water level also keep constant when the depth is larger than 0.3 m.

Then, the hydraulic gradient is changed in different calculation cases to investigate the influence of the hydraulic gradient on the underground water level around underground structures considering the seepage. As shown in Fig. 3, the difference of the water level at the upstream of the structure increases linearly with the increase of the hydraulic gradient. While the difference of the water level at the downstream of the structure decreases linearly with the increase of the hydraulic gradient.

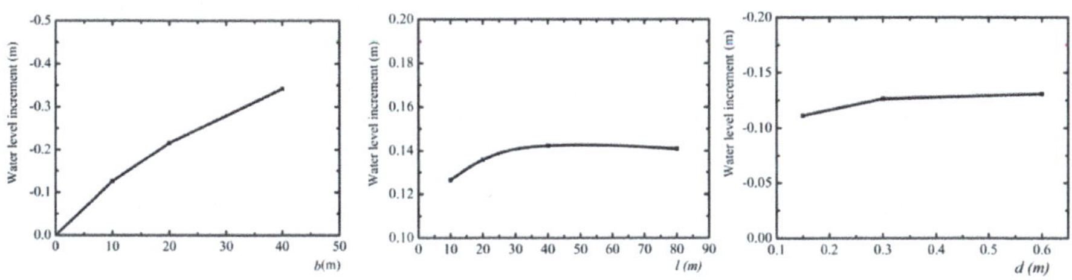

(a) influence of structure width (b) influence of structure length (c) influence of structure depth

Fig. 2. Water level increment influenced by the structure dimension

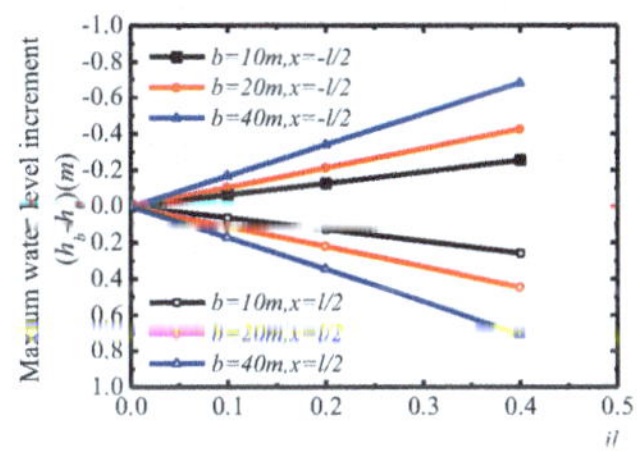

Fig. 3. Water level influenced by hydraulic gradient

2.3 Influence of Underground Structure on Underground Water Level in Layered Soil

To find the influence of the difference of the soil permeability between the soils on the effect of underground structures on the seepage in layered soils, a two-layer soil system is analyzed herein. The system is shown in Fig. 4. As shown in Table 1, six cases are calculated. In those cases, GK1-6 respectively represent the ratio of the permeability coefficient of the upper layer to that of the lower layer is set as $k_1/k_2 = 10^{-4}, 10^{-3}, 10^{-2}, 0.05, 0.1, 1$. And two structure depths are chosen for each case. The calculated maximum water level increase ratio is shown in Table 1. It can be seen, the soil permeability difference between the soils also influences the maximum water level increase ratio when considering the effect of the underground structure on seepage.

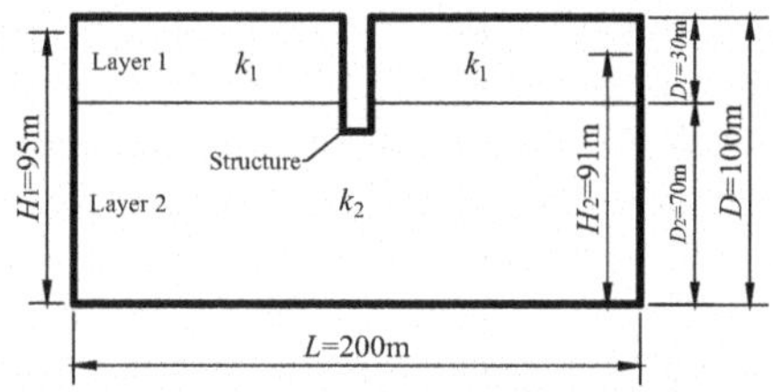

Fig. 4. Structure in a two-layer system

Table 1. Water level increment influence by the ratio of soil permeability coefficient

Structure dimension $b \times l \times d'$	Water level increment ratio $\Delta h'/il$					
	GK1	GK2	GK3	GK4	GK5	GK6
40 × 10 × 40	0	0	0.2%	0.9%	1.7%	11%
40 × 10 × 50	0	0	0.1%	0.6%	1.1%	6.5%

3 Simplified Method

Based on those calculations, an empirical method is proposed herein. As shown in Fig. 5, the distribution of the underground water level along the widthwise line at the upstream of the structure is similar to a normal distribution curve. The maximum water level increase occurs at the center of the upstream of the structure. Thus, the distribution of the underground water level along the widthwise line at the upstream of the structure can be calculated by Eq. (1).

$$\frac{h_y - h_{y=100}}{h_{y=0} - h_{y=100}} = e^{-[0.1\ln(b/d') + 0.83]\cdot(2\cdot y/b)^2} \tag{1}$$

Figure 6 shows the relationship between maximum water level increase ratio and the dimension of the structure. To fit all the curves in Fig. 6, the maximum water level increase ratio can be calculated as following.

$$(h_b - h_0)/(i \cdot l) = A\ln(d'/b) + B \tag{2}$$

where, h_b is the underground water level when there is structure at the center of the upstream of the structure, h_O is the underground water level when there is no structure, d' is the effective structure depth.

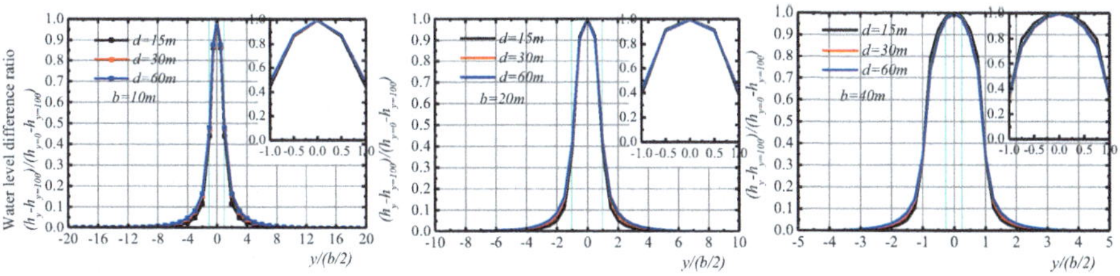

(a) b=10m, l=10m (b) b=20m, l=10m (c) b=40m, l=10m

Fig. 5. Water level along the widthwise at the upstream of the structure

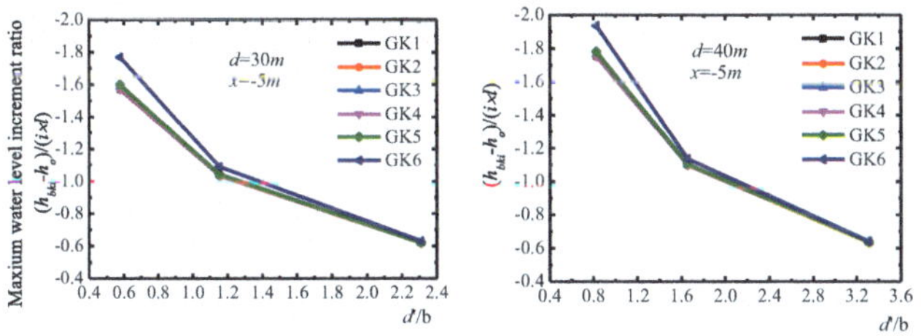

(a) d=30m (b) d=40m

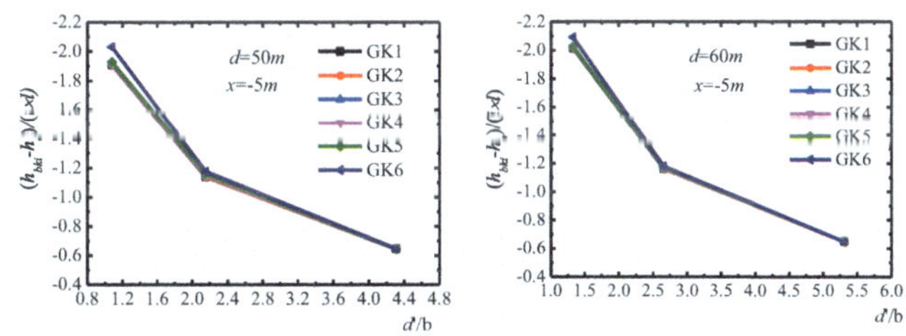

(c) d=50m (d) d=60m

Fig. 6. The relationship between maximum water level increment ratio and the structure depth and width ratio in layered soil

In order to make the empirical method suitable to the structures in layered soil, we assumed that the soil is divided into two layers separated by the bottom of the structure. All the layers above the bottom of the structure are assumed to be one equivalent layered. For example, the permeability of the equivalent layered in Fig. 4 can be calculated by the following equation.

$$k_1' = [k_1 D_1 + k_2(d - D_1)]/d \tag{3}$$

The coefficients A and B in Eq. (2) are related to the soil permeability coefficient. The relationship of those two coefficients and the soil permeability coefficient ratio is shown in Fig. 7. Fitting the curves in Fig. 7, Eq. (2) can be expressed as following.

$$\begin{aligned}(h_b - h_0)/(i \cdot l) = & [-0.025 \cdot \ln(k_1'/k_2) + 0.24] \cdot d'/l \cdot \ln(d'/b) \\ & + d'/l \cdot [-0.03 \cdot \ln(k_1'/k_2) - 0.47]\end{aligned} \tag{4}$$

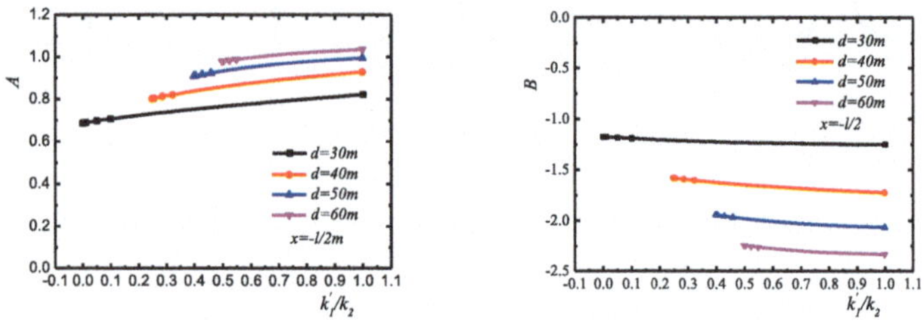

(a) Coefficient A (b) Coefficient B

Fig. 7. Coefficient related to the soil permeability coefficient

It is clear that the block effect of the underground structure on seepage increases the underground water level at the upstream of the structure comparing to the static underground water level. And it decrease the underground water level at the downstream of the structure. As shown in Fig. 8, we can calculate the buoyancy on the underground structures based on the static underground water level when we obtain the difference of the underground water level through Eq. (4).

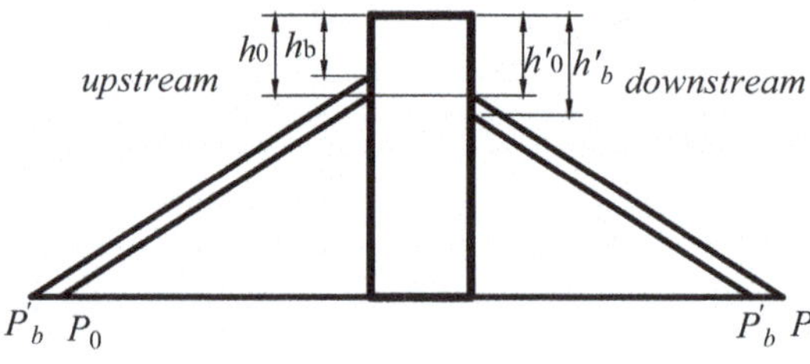

Fig. 8. Distribution of pore pressure around the structure

The pore pressure at the bottom of the structure, which is used to calculate the buoyancy of the structure, can be calculated by the following equation according to the calculated water level considering the seepage.

$$P_b' = \gamma_w h_b \tag{5}$$

4 Verification

The multi-layer soil system shown in Fig. 9 is analyzed by the proposed method. As seen in Fig. 9(a), a underground structure, the dimension of which is 40 m × 10 m (Depth × Length), is placed in the three-layer soil system. The depths of the three layers of the soils are 30 m(D_1), 30 m(D_2) and 40 m(D_3). The permeability coefficients of the soils are $k_1 = 5 \times 10^{-6}$ m/s, $k_2 = 1 \times 10^{-4}$ m/s and $k_3 = 1 \times 10^{-5}$ m/s. Firstly, make the three-layered soil system equivalent into two-layered soil system which is shown in Fig. 9(b). According to Eq. (3), the equivalent permeability coefficients of the two layers of the soil are $k_1' = 2.87 \times 10^{-5}$ m/s, $k_2' = 4 \times 10^{-5}$ m/s. The maximum underground water level increments at the upstream of the structure calculated by the FEM and the empirical method are shown in Table 2. It can be seen that the results calculated from the empirical method fit the results from the FEM well for all those cases. We can conclude that the proposed method is reasonable to evaluate the underground water level considering the effect of the underground structure on seepage.

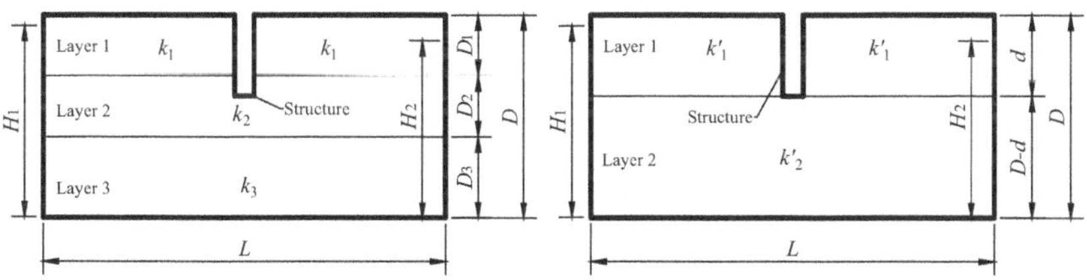

(a) Three-layer soil system (b) Equivalent two-layer soil system

Fig. 9. Multi-layer system

Table 2. Comparison of maximum water level increment calculated from FEM and proposed method in multi-layer soil

Structure dimension $b(m) \times l(m) \times d(m)$	Maximum water level increment(m)		Difference
	FEM	Proposed method	
10 × 10 × 40	−0.128	−0.108	15%
20 × 10 × 40	−0.237	−0.224	5.5%
40 × 10 × 40	−0.355	−0.335	5.6%

5 Conclusion

A serious of FEM analysis on underground water level considering the effect of underground structures on seepage were carried out herein. Through the FEM analysis, it is found that the boundary effect can be ignored when the structure length and the model length ratio is less than 0.1, the structure width and the model width ratio less than 0.4 and the structure depth and the model depth ratio is less than 0.6. The dimensions of the structure and the ratio of different soil permeability coefficients influence the effect of underground structure on seepage significantly. And an empirical method is proposed to calculate the underground water level around the structure considering the effect of structure on seepage. Through validation, it is found that the proposed method is reasonable to calculate the underground water level, which is used to calculate buoyancy on the structure, considering the seepage.

References

1. Xu, J., Wang, G., Li, X.: A preliminary research on potential effect of underground engineering on groundwater environment. J. Eng. Geol. **01**, 15–19 (1999)
2. Geotechnical Control Office (GCO): Mid-levels study: report on geology, hydrology and soil properties. Geotechnical Control Office, Hong Kong (1982)
3. Pope, R.G., Ho, C.S.: Effect of piles and caissons on groundwater flow. Hong Kong Eng. **10** (11), 25–27 (1982)
4. Attia, M.H.: Geotechnical behavior of diaphragm walls. University of Alberta, Canada (1989)
5. Brassington, R.: Field Hydrogeology, 2nd edn. Wiley, New York (1998)
6. Ding, G.P., Jiao, J.J., Zhang, D.: Modelling study on the impact of deep building foundations on the groundwater system. Hydrol. Process. **22**, 1857–1865 (2008)
7. Jiao, J.J., Wang, X.S., Nandy, S.: Preliminary assessment of the impacts of deep foundations and land reclamation on groundwater flow in a coastal area in Hong Kong, China. Hydrogeol. J. **14**(1–2), 100–114 (2006)
8. Jiao, J.J., Leung, C.M., Ding, G.P.: Changes to the groundwater system, from 1888 to present, in a highly-urbanized coastal area in Hong Kong, China. Hydrogeol. J. **16**(8), 1527–1539 (2008)
9. Cao, Y., Huang, R., Shen, S., Xu, Y., Ma, L.: Investigation of blocking effect on groundwater seepage of piles in aquifer. Rock Soil Mech. **06**, 1617–1622 (2014)
10. Xu, Y., Shen, S., Ma, L.: Cutoff effect of groundwater seepage due to existence of underground structure. J. Zhejiang Univ. (Eng. Sci.) 10, 1902–1906 (2010)
11. Xu, Y.: Evaluation of the behaviour of groundwater seepage and land subsidence via considering infrastructures penetrated into aquifers. Shanghai Jiaotong Univeristy, China (2010)
12. Xu, Y.S., Shen, S.L., Du, Y.J., Chai, J.C., Horpibulsuk, S.: Modelling the cutoff behavior of underground structure in multi-aquifer-aquitard groundwater system. Nat. Hazards **66**, 731–748 (2013)
13. Fan, Y.: Experimental Study and Analysis of Sand - tank Characteristics of Water - blocking Characteristics of Underground Structures. South China University of Technology, China (2006)
14. Zheng, P., Xu, Y., Shen, S.: Laboratory investigation on cutoff effect of underground structure to groundwater seepage of aquifer. Chin. J. Undergr. Space Eng. **02**, 253–256 (2011)

Numerical Characterization of Basalt Platform Slope Under Rainfall Condition

Zhen-ming Shi[1], Chen-xi Zhao[1], Qing-zhao Zhang[1(✉)], and Jing Song[2]

[1] Department of Geotechnical Engineering, Tongji University, Shanghai 200092, China
67515962@qq.com
[2] Shanghai Construction Group Co., Ltd., Shanghai 200086, China

Abstract. Basalt platform landslide triggered by rainfall is one of the most important landslides in China, which is highly dangerous. Based on the saturated-unsaturated seepage theory, this paper adopts finite element method to simulate the unsteady seepage of basalt platform slope under continuous rainfall, so as to analyze the distribution characteristics of seepage field inside the slope and the influence of matric suction on slope stability. The results show that: (1) Vertical infiltration is the main form for slope seepage filed above groundwater level and bedding flow is the main form below groundwater level; (2) Cracks are beneficial to rainfall infiltration and catchment areas are formed at the crack of the top; (3) Rainfall intensity and weak layer have an inverse relationship with the factor of safety of slope; (4) Matric suction has the effect to enhance slope stability.

Keywords: Basalt platform slope · Rainfall infiltration · Cracks Numerical simulation

1 Introduction

In recent years, with the increase in the number and scale of construction projects, landslide hazards occur frequently in basalt platform areas, resulting in the increase in losses and damages year by year. According to the past researches, rainfall infiltration has an important influence on slope stability [1]. Rainfall infiltration changes the water content in the unsaturated area of the slope; with the increase in water content, the matric suction in the unsaturated area of the slope decreases accordingly, which has great influence on soil strength [2]. Heavy rain will cause dramatic decrease of negative pore water pressure and lead to decrease of shear strength [3]. Research indicated that the factor of safety of slope decreases with the duration of rainfall [4]. In addition, the slope stability will be influenced by rainfall intensity [5].

Most soil slopes containing a weak layer are easily softened in the presence of water due to its poor physical and mechanical properties, which will easily reduce the shear strength of soil. The weak interlayer will cause the change of rainfall infiltration distribution, thus reduce the slope stability [6]. In addition, the cracks are beneficial to the infiltration of rainfall into the underlying soft layer, thus forming a sliding surface.

L. Hu et al. (Eds.): GSIC 2018, *Proceedings of GeoShanghai 2018 International Conference: Multi-physics Processes in Soil Mechanics and Advances in Geotechnical Testing*, pp. 307–314, 2018.
https://doi.org/10.1007/978-981-13-0095-0_35

For expansive soil slope containing cracks, rainwater goes deep into the soil along the cracks resulting in the reduction of suction and strength of the soil. Therefore it is necessary to take the influence of cracks into consideration [7]. After taking cracks into account, the factor of safety of expansive soil slope significantly reduces [8]. At present, there are few researches on the failure of basalt platform slope. For example, there is no systematic understanding of the influence of basalt vertical cracks and weak layer on the slope stability. Therefore, the failure mechanism of basalt platform slope needs further investigation.

This paper adopts Geo-studio to simulate the unsteady seepage of the basalt platform slope under continuous rainfall based on the saturated-unsaturated seepage theory, and draws the distribution of seepage field. It compares the seepage field and stability of slope under different conditions, analyzes the influences of rainfall intensity, rainfall duration and the thickness of weak layer on slope stability, and discusses the influence of rainfall infiltration on matric suction and shear strength of soil, which will provide reference for the analysis of deformation and failure of basalt slope.

2 Seepage Analysis of Basalt Platform Slope

2.1 Basic Equation

In unsaturated soil, the permeability coefficient is a variable and a function of saturation or moisture content. When the main direction of penetration is consistent with the coordinate axis, two-dimensional differential equation for unsaturated soil seepage is as shown in Eq. (1):

$$\frac{\partial}{\partial x}\left(k_x\frac{\partial h}{\partial x}\right)+\frac{\partial}{\partial y}\left(k_y\frac{\partial h}{\partial y}\right)=\gamma_w\frac{\partial\theta_w}{\partial\psi}\frac{\partial h}{\partial t} \tag{1}$$

k_x and k_y are the permeability coefficient of direction x and y. γ_w is the unit weight of water; θ_w is the volumetric moisture content; ψ is soil suction; h is the total hydraulic head and t is time.

2.2 Calculating Model

A numerical model is established by taking a basalt landslide in a highway section as an example. The prototype slopes can be divided into three layers: upper, middle and lower layer, respectively being clayey soil containing gravel, basalt weathered soil and clayey soil (weak layer). According to a contrastive analysis of the topographic slope of 114 landslides in the basalt platform region of Zhejiang, it is discovered that the landslides in the basalt platform region mainly occur to slopes with the gradient of 0°–30°, especially during the range of 10°–20° [9]. Therefore, the landslide is divided into three sections in modeling, the gradient of which is 30°, 15° and 10° respectively. The simulated model is as shown in Fig. 1.

The contact surface between the bedrock of the model and the bottom of the slope is impermeable boundary, the constant flow of which is zero. The rainfall process is

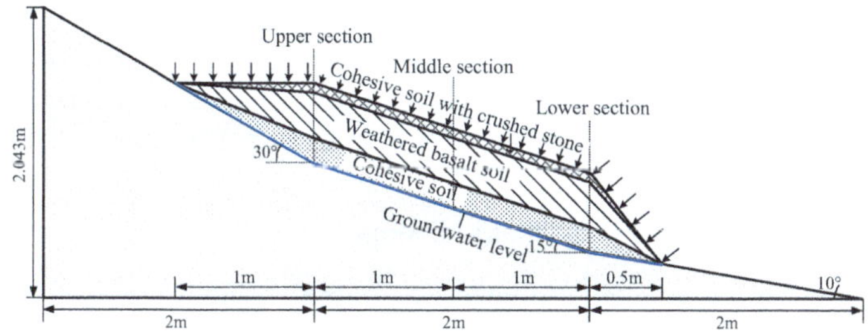

Fig. 1. Model diagram

Table 1. Model scheme

Model number	Upper layer: cohesive soil with crushed stone (cm)	Middle layer: weathered basalt soil (cm)	Lower layer: cohesive soil (cm)
1	8	32	20
2	10	40	10
3	12	48	0

simulated at the upper boundary of the model and the experimental conditions are set as continuous rainfall. It is assumed that the initial groundwater level is located at the bottom of the slope.

In order to study the influence of soft layer thickness on the seepage field of slope, 3 groups of model geometries are designed in this numerical simulation. Maintaining the total thickness of 60 cm, the thickness of weak layers is respectively 20 cm, 10 cm and 0 cm. The specific model scheme is as shown in Table 1.

2.3 Materials Properties

In the seepage analysis, the soil materials are clayey soil containing gravel, basalt weathered soil and clayey soil, the physical properties of which are shown in Table 2.

Considering that the soil skeleton is incompressible, the average value of measured data is used as the setting value of saturated moisture content of the soil. However, the permeability coefficient of the soil material will vary according to the variation of water

Table 2. Index properties of the three soil materials used in the seepage analysis

Soil materials	Saturated water content ($m^3 \cdot m^{-3}$)	Residual water content ($m^3 \cdot m^{-3}$)	Permeability coefficient ($m \cdot s^{-1}$)	Specific gravity ($kN \cdot m^{-3}$)	Cohesion (kPa)	Internal friction angle (°)
Cohesive soil with crushed stone	0.45	0.06	$1.53 * 10^{-5}$	19.1	15	9.5
Weathered basalt soil	0.42	0.05	$1.12 * 10^{-5}$	18.8	24	25
Cohesive soil	0.59	0.1	$8.00 * 10^{-7}$	16.2	20	24

content, the hydraulic characteristics of the soil material, i.e. the permeability coefficient curve and soil water characteristic curve, shall be calculated before seepage analysis is carried out. In this simulation, the soil and water characteristic curve is estimated according to empirical characteristic curve of the material.

2.4 Characteristics of Seepage Field of Slope

The soil slope is in unsaturated-saturated state during rainfall with complex seepage field condition. The control variables in simulation test include rainfall intensity (60 mm/h and 30 mm/h); thickness of the weak layer (20 cm, 10 cm and 0 cm); In addition, since the vertical cracks of basalt weathered soil is rather developed, a group of test without cracks is used as a control. The test conditions are as shown in Table 3.

Table 3. Simulated test groups and conditions

Simulated test groups	A	B	C	D	E
Rainfall intensity ($mm \cdot h^{-1}$)	60	60	30	60	60
Thickness of weak layer (cm)	20	20	20	10	0
Vertical fractures	Yes	No	Yes	Yes	Yes

The simulation results are as shown in Fig. 2. The red lines in the test slope represent vertical cracks. The black lines represent the dividing lines of different layer areas, which are, from top to bottom, clayey soil containing gravel, basalt weathered soil and clayey soil with weak layer. The dark blue line represents the water level.

Figure 2(a) shows the seepage characteristic of the test A. The seepage velocity is the fastest in the cracks and catchment areas are easily formed on the top of cracks. In the slope, the upper seepage velocity is larger, followed by the middle layer. The lower part is clayey soil, which has very small permeability coefficient and the smallest seepage velocity. In addition, at the bottom of the slope, there is large seepage velocity. The water level passes through the cracks and forms a smaller crest. In the upper area near the water level, the direction of seepage is relatively smooth.

From the seepage characteristic of the test B in Fig. 2(b), we can find that the seepage velocity decreases successively from top to bottom while the seepage velocity is rather large at the toe of the slope. Besides, near the interface of the basalt weathered soil and the clayey soil, the direction of seepage velocity changes greatly. Compared with test A, the water level within the slope containing no cracks is lower than that in the slope containing vertical cracks.

It can be seen from Fig. 2(a) and (c), the greater the rainfall intensity is, the greater the saturated area within the slope is, the water level rises rapidly and the water line is higher. The water line in the slope of test D in Fig. 2(d) is higher than that in the slope of test C, which means the influence of weak layer thickness decreasing from 20 cm to 10 cm on slope seepage is smaller than the influence of rainfall intensity decreasing from 60 mm/h to 30 mm/h. Figure 2(e) shows the pore water pressure distribution of slopes containing no weak layer in test E, the water level rises the highest and the pore water pressure line is substantially parallel to the substrate. The direction of seepage is basically parallel to the water level in the soil below the water level.

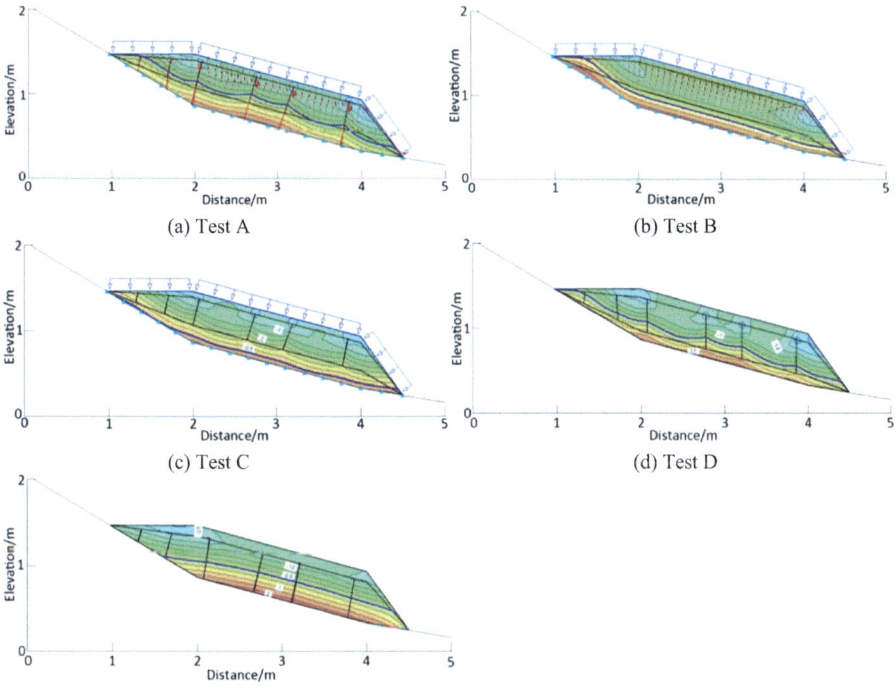

(a) Test A (b) Test B (c) Test C (d) Test D (e) Test E

Fig. 2. Characteristic of seepage field in the slope

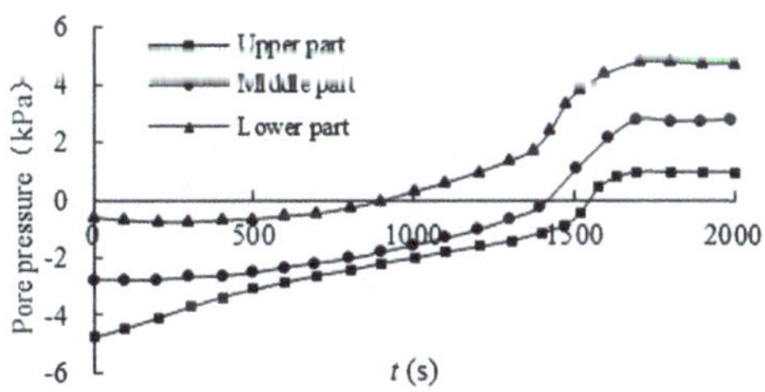

Fig. 3. Variation curves of pore pressure in different soil layers

Figure 3 shows the variation curves of pore water pressure in different soil layers under the condition of test A. Three measuring points, respectively the upper, middle and lower one, are laid at the level of 10 cm, 30 cm and 50 cm. The curves have similar variation law. The middle and upper pore pressure variation range is smaller than the middle and lower pore pressure variation range. It is because that the upper part is unsaturated area while the middle and lower part are saturated areas.

3 Calculation of Slope Stability During Rainfall

3.1 Effect of Rainfall Intensity and Thickness of Weak Layer on Slope Stability

Based on the pore pressure distribution results, the minimum factors of safety of slope at each moment are calculated based on the theoretical foundation of Mohr-Coulomb Criterion and Bishop Method. According to the simulation results, the factors of safety of slope under various conditions are shown in Fig. 4.

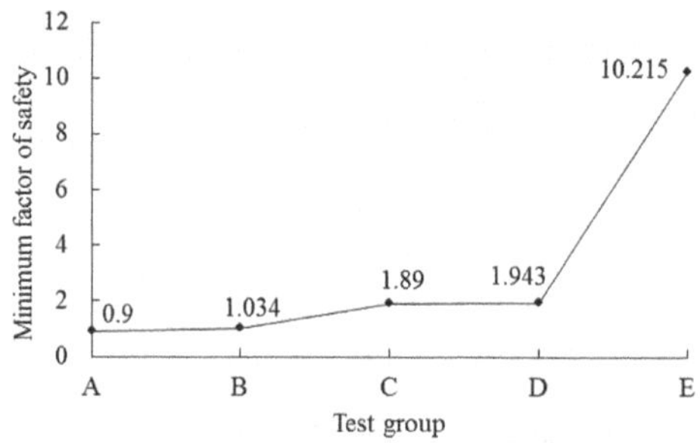

Fig. 4. Minimum factor of safety for each group

From the Fig. 4, we can find that the factor of safety of slope containing no cracks is higher than that of slope containing cracks through comparing the Test A and Test B. The factor of safety decreases by 12.96% when the cracks exist in the stratum. The cracks will increase rainfall infiltration, lead to decreases of soil strength and reduction of slope stability. The greater the rainfall intensity, the smaller the factor of safety is. As the rainfall intensity increases from 30 mm/h to 60 mm/h, the factor of safety decreases by 52.38%. It can be observed through comparing the Test A, Test D and Test E, remaining equal rainfall intensity, the greater the thickness of weak layer is, the smaller the factor of safety is. The comparison of test C and D illustrates again that there is greater influence on slope stability when the rainfall intensity changes 30 mm/h than when the thickness of weak layer changes 10 cm.

3.2 The Influence of Matrix Suction on Shear Strength

Figure 5 show the pore water pressure distribution, matric suction distribution and ϕ^b (ϕ^b refers to the internal friction angle related to matric suction) distribution on the sliding surface under the condition of test D.

The pore water pressure increases first and then decreases with the sliding slope, and reaches maximum 1 cm away from the initial section of the sliding surface. The matric suction is related to the saturated area in the slope. The matric suction decreases from 0.26 kPa to zero from the starting point of the sliding surface to 0.2 m. Combining Fig. 6, it can be seen that the sliding surface is below the water level of the slope at this time, and the area of sliding surface is completely saturated and the matric

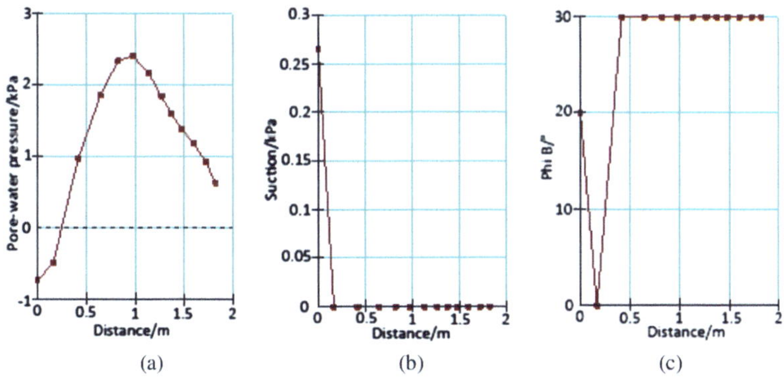

Fig. 5. Distribution of pore pressure (a), matrix suction (b) and ϕ^b on sliding surface (c)

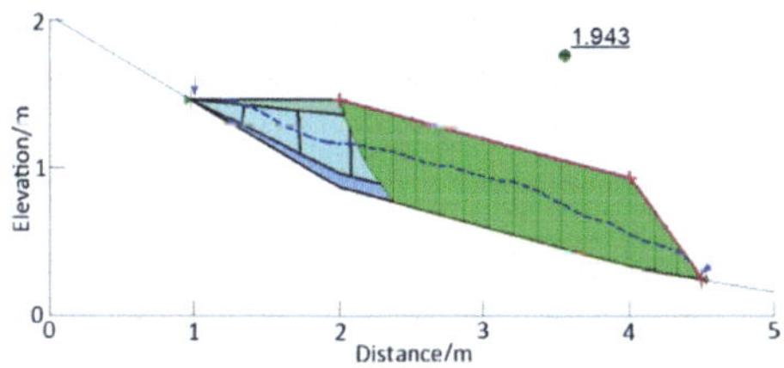

Fig. 6. Factor of safety of slope for test D

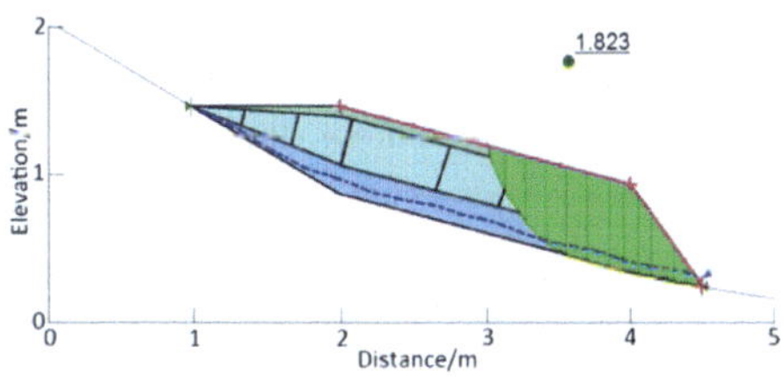

Fig. 7. Factor of safety of slope without matrix suction

suction is zero. According to ϕ^b curve, all the values of the sliding surface are positive, which is the inherent property of the sliding surface. Its matric suction is not only related to ϕ^b, but also related to the distribution of pore water pressure.

Figure 7 shows the factor of safety of slope without matrix suction under the condition of test D, namely setting ϕ^b as zero while keeping other parameters unchanged. The minimum factor of safety obtained through the numerical simulation is 1.823. Compared with 1.943 in test group D, it increases by 6.5% which is a bit small. In addition, considering that the area of the sliding slope in the unsaturated zone is small, namely, the area where the matric suction can exert effect is small, the result of the test shows that the matric suction has an important effect on the slope stability and the factor of safety of slope is larger when considering the matrix suction.

4 Conclusion

In this paper, the slope seepage field distribution characteristics and the impact of matric suction on slope stability during rainfall using numerical simulation are analyzed. The main conclusions are as follows:

(1) The seepage field in the slope has obvious differentiation characteristics with the groundwater level line as the boundary: vertical infiltration is the main form for slope seepage filed above groundwater level and bedding flow is the main form below groundwater level; there is a big turning point in the direction of seepage near the water level line.
(2) Vertical cracks have influence on the distribution of seepage field. Cracks are beneficial to rainfall infiltration and catchment area will be formed at the top of cracks. The factor of safety decreases by 12.96% when the cracks exist in the stratum.
(3) The increase in rainfall intensity and thickness of weak layer will reduce the factor of safety. The numerical simulation shows that there is greater influence on slope stability when the rainfall intensity changes 30 mm/h than when the thickness of weak layer changes 10 cm.
(4) The pore water pressure and the matric suction will change the stability of the slope. The increase of pore water pressure will reduce the factor of safety of slope, while matric suction has certain strengthening effect on slope stability, yet the strengthening effect is not obvious for basalt weathered soil. The factor of safety of slope with matrix suction increases only by 6.5%.

Acknowledgments. This work is supported by National Key R&D Program of China (NO. 2017YFC0806004), Fundamental Research Funds for the Central Universities, National Natural Science Foundation of China (NO. 41602287), Key Project of Natural Science Foundation of China (NO. 41731283) and National Natural Science Foundation of China (NO. 41372272).

References

1. Cho, S.E., Lee, S.R.: Evaluation of surficial stability homogeneous slope considering rainfall characteristics. J. Geotech. Geoenviron. Eng. **128**(9), 756–763 (2002)
2. Wu, J.J., Wang, C.H., Li, G.X.: Influence of matric suction in unsaturated soils on slope stability. Rock Soil Mech. **25**(5), 732–736 (2004). (in Chinese)
3. Anderson, M.G., Richards, K.S. (eds.): Slope Stability: Geotechnical Engineering and Geomorphology. Wiley, Chichester (1987). 620 pp
4. Huang, R.Q., Qi, G.Q.: The effect of unsaturated soil suction on slope stability. J. Eng. Geol. **10**(4), 343–348 (2002). (in Chinese)
5. Ng, C.W.W., Shi, Q.: Influence of rainfall intensity and duration on slope stability in unsaturated soils. Q. J. Eng. Geol. **31**(2), 105–113 (1998)
6. Wang, R., Zhang, G., Zhang, J.M.: Centrifuge modeling of rainfall-induced deformation of slopes with weak layers. Chin. J. Geotech. Eng. **32**(10), 1582–1587 (2010). (in Chinese)
7. Chen, T.L., Deng, G., Chen, S.S., et al.: Effects of fissures on stability of unsaturated soil slope. Chin. J. Geotech. Eng. **02**, 210–215 (2006). (in Chinese)
8. Chen, S.X., Dai, Z.J., Lu, D.J., et al.: Stability analysis considering fracture distribution and strength for expansive soil slope. J. Hydraul. Eng. **45**(12), 1442–1449 (2014). (in Chinese)
9. Tang, X.M., You, S.Y., Shang, Y.Q.: Geomorphic features of basalt platform and geological disaster in Zhejiang Province. J. Zhejiang Univ. Nat. Sci. **36**(2), 231–235 (2009). (in Chinese)

A Finite Element Analysis of Stress Change in Pavement Subjected to Freeze-Thaw

Bin Luo[1], Tatsuya Ishikawa[1(✉)], and Tetsuya Tokoro[2]

[1] Hokkaido University, Kita 13, Nishi 8, Kita-Ku, Sapporo, Japan
bin-luo@eis.hokudai.ac.jp, t-ishika@eng.hokudai.ac.jp
[2] National Institute of Technology, Tomakomai College, 443 Nishikioka, Tomakomai, Japan

Abstract. The prediction of stress change in pavement structure is quite essential for the design, construction, and maintenance of pavement in the seasonal cold region. Much attention is concentrated on the change of elastic modulus of soils due to freezing and thawing. To overcome the limitation of conventional design method, in which the elastic modulus of materials needs to be assigned manually to various thermal state layer by layer, a comprehensive analysis method is proposed in this study. This analysis can specify temperature-dependent modulus, and a typical pavement in the cold region was selected and simulated. It is revealed that snow cover on footpath can affect the thermal pattern in the pavement, which leads to differential frost penetration along the cross-section of the pavement. The stress change induced by frost action of soil shows the reverse trend between frozen and unfrozen (thawed) layers. The stress increment in subgrade during spring season should be accounted properly.

Keywords: Stress change · Freezing and thawing · Flexible pavement

1 Introduction

In the seasonal cold region, the pavement is exposed to climatic change which causes periodic temperature change and freeze-thaw of soils in the pavement. For porous material like soil, the phase transition of pore-water to ice results in changes of the physical and mechanical index including permeability, thermal properties, and elastic modulus, etc. These changes highly affect the performance of the pavement structure which is treated as a multilayer system in engineering design. A series of the study was conducted to investigate the relation between change of resilient modulus and freeze-thaw of soils by former researchers [1–3], along with thaw weakening of pavement structures [4]. The in-situ survey and laboratory experiments were carried out to investigate the effects of climatic change and material properties on mechanistic-empirical pavement design (MEPDG) in cold region; it is indicated that there is a need to consider elastic modulus related to freezing and thawing for the design based on multilayer elastic theory [5]. Recently from an investigation on the stress distribution in the ground beneath pavements, the change of stress distribution associated with freezing and thawing was pointed out [6]. For the conventional stress analysis method, the value

L. Hu et al. (Eds.): GSIC 2018, *Proceedings of GeoShanghai 2018 International Conference: Multi-physics Processes in Soil Mechanics and Advances in Geotechnical Testing*, pp. 315–324, 2018.
https://doi.org/10.1007/978-981-13-0095-0_36

of elastic modulus needs to be assigned manually for frozen and thawed stage; a time-depend stress change is hard to be achieved. The prerequisite for accurate stress analysis is the determination of temperature distribution in the pavement; then the proper elastic modulus values can be set. As the thermal analysis is a complex boundary value problem, which is also associated with seepage analysis, a comprehensive analysis method that can implement multi-field analysis is expected. In this study, the GSUPRA software is applied for analysis [7], and the governing equations, numerical model, material properties, boundary conditions and analysis procedures were presented first. Then freezing and thawing pattern of pavement is discussed. Finally, the stress change caused by the change of elastic modulus is revealed.

2 Governing Equations

A coupled thermo-hydro-mechanical model is applied to characterize the freeze-thaw behavior of soils in the pavement structure [8]. As pavement soil is unsaturated in most occasions, the unsaturated-saturated seepage flow should be properly considered and is described as below

$$\frac{\partial}{\partial x_i}\left[-k_{ij}\frac{\partial}{\partial x_j}\left(\frac{u_w}{\rho_w g}+z\right)\right]=\left(\frac{nS_e}{K_w}-n\frac{\partial S_e}{\partial u_w}\right)\frac{\partial u_w}{\partial t}+nS_e\alpha_T\frac{\partial T}{\partial t}-S_e\frac{\partial \varepsilon_v}{\partial t} \quad (1)$$

Where x_i, x_j = Cartesian coordinates; k_{ij} = hydraulic conductivity of soil; u_w = pore-water pressure; ρ_w = density of water; g = gravitational acceleration; z = elevation head; n = porosity; S_e = effective degree of saturation; K_w = bulk modulus of pore-water; t = time; α_T = thermal expansion coefficient of water; T = temperature; ε_v = volumetric strain.

The relation between negative pore-water pressure termed as matric suction and the effective degree of saturation for unsaturated soils are formulated by the widely used van Genuchten (VG) model.

$$\psi=\frac{1}{\alpha}\left[(S_e)^{\lambda/(1-\lambda)}-1\right]^{1/\lambda} \quad (2)$$

Where ψ = matric suction; α, λ = constants.

The climatic change drives temperature variation in pavement and results in freeze-thaw of pavement soil. The heat transfer is governed by Eq. 3, in which the first and second terms on the right side of the equation represent the conductive heat transfer of soil and convective heat transfer of water flow, respectively. Since the soil is a kind of porous material, phase change and consumption of latent heat of pore-water should be taken into consideration. For the sake of simplification, the latent heat is quantified as part of apparent heat capacity.

$$C\rho\frac{\partial T}{\partial t}=-\frac{\partial}{\partial x_i}\left(\lambda_T\frac{\partial T}{\partial x_j}\right)+C_w\rho_w\frac{\partial (v_i T)}{\partial x_i} \quad (3)$$

Where C = specific heat of wet soil; ρ = density of wet soil; λ_T = thermal conductivity of wet soil; C_w = specific heat of pore-water; v_i = average velocity of pore-water in coordinate direction.

An elastic strain-stress framework is adopted to reproduce the effect of freeze-thaw on stress change in the pavement soil, which is slightly modified from basic form and is able to consider the contribution of pore-water pressure, thermal expansion and frost heave.

$$\left(\frac{1}{2}D_{ij}\left(\frac{\partial u_{j,i}}{\partial t}+\frac{\partial u_{i,j}}{\partial t}\right)+\frac{\partial \sigma_0}{\partial t}\delta_{ij}\right)_{,j}+\rho g_i=0 \tag{4}$$

Where D_{ij} = rigidity matrix; u_i and u_j = displacements in the coordinate direction; σ_0 = stresses are attributable to pore-water pressure, thermal expansion and frost heave; δ_{ij} = Kronecker's delta; g_i = acceleration in the i direction.

The elastic modulus is a function of temperature and freeze-thaw cycle.

$$E=\begin{cases} E_u\ T > T_f\ \text{(before frozen)} \\ E_f\ T \le T_f \\ E_t\ T > T_f\ \text{(after thawed)} \end{cases} \tag{5}$$

Where E = elastic modulus; E_u, E_f and E_t = elastic modulus of unfrozen soil, frozen soil and thawed soil; and T_f = frozen temperature of the soil dependent on physical properties.

3 Numerical Model

A typical pavement structure in the cold snowy region, Hokkaido (Kitami city), is selected and simulated (Fig. 1). The pavement is composed of surfacing layer, base course, frost protection layer (subbase) and subgrade. In the simulation, only one-half of the pavement is considered for simplicity, and simulated domain extends 7 m in width and 5 m in depth. Subsurface layers, which are close to pavement surface and influenced by climatic change, are given fine meshing (0.1 m) to capture time-dependent stress change. Elements of larger size (0.25 m) are assigned to lower part of the model to reduce calculation scale and save computation time. There is no potential frost heave reported in investigation mentioned before [6], therefore the stress change induced by frost heave is not accounted. Thermal expansion and contraction are also ignored to examine the main concern of this study, that is the effects of frost-induced change of elastic modulus of subsurface soils on stress change in the pavement.

3.1 Materials Properties

The flexible pavement in this simulation consists of several types of materials. Surfacing layer is made of asphalt concrete, while graded gravel (C40) constitutes base course distributing wheel load to underneath layers. The footpath is constructed by

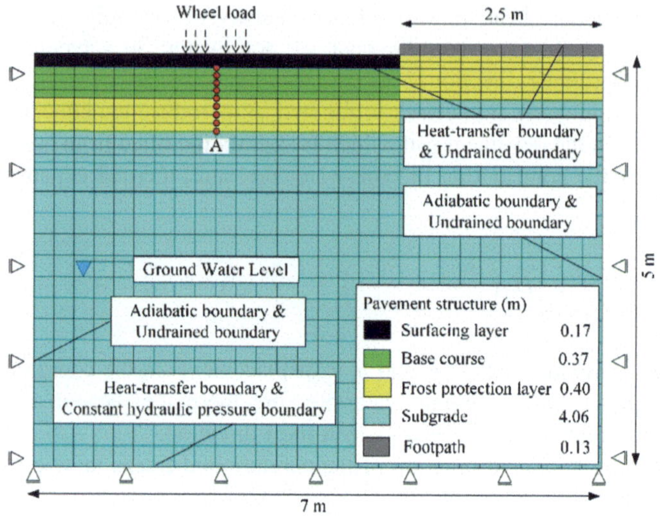

Fig. 1. Profile of simulated pavement.

using concrete material. Since detailed information on in-situ soils is not available, two types of volcanic soils used in pavement engineering in Hokkaido are specified for frost protection layer and subgrade, which may somewhat differ from the field condition. For frost protection layer, non-frost susceptible coarse-grained volcanic soil (Komaoka soil) can limit or prevent frost penetration in frost-susceptible subgrade soil and further load distribution. A representative fine-grained volcanic soil (Touryo soil) is chosen as subgrade soil.

Representative values for input parameters of these materials are selected for the sake of generality (Table 1). More specifically, the values of elastic modulus of unfrozen, frozen and thawed soils are referenced from previous study, in which the falling weight deflectometer test (FWD test) and resilient modulus test (MR test) are conducted to estimate the climatic effects on modulus of pavement materials [5]. Note that the temperature dependence of elastic modulus of asphalt concrete is roughly considered by specifying two different values for unfrozen and frozen asphalt. It is also should note that the constant thermal conductivity may reduce the frost penetration in this research.

3.2 Boundary Conditions

As the ground surface is hardened by impervious material like asphalt concrete and cement, no infiltration (rainfall and snowfall) is considered. Accordingly, the constant pore-water pressure is employed at the bottom boundary to achieve constant groundwater level (2.8 m below surface). A linear suction gradient (negative pore-water pressure) is set above the groundwater level, then water content is determined based on VG model. A sinusoidal function is applied to approximate and simplify periodic temperature variation (Eq. 6) [9], and the recent 5 years' meteorological data is used to

Table 1. Physical parameters of materials.

Parameters	Asphalt	C–40	Komaoka	Touryo	Concrete
Porosity (n)	0.030	0.230	0.670	0.450	0.050
Constant (α) 1/MPa	–	8125.14	121.69	89.63	–
Constant (λ)	–	1.42	2.09	1.46	–
Specific heat of dry soil (C_d) J/m^3 °C	1.8E + 6	1.8E + 6	1.9E + 6	9.9E + 5	2.0E + 6
Dry density of soil (ρ_d) kg/m^3	2100	2000	1400	1400	2200
Thermal conductivity (λ_T) W/m °C	1.45	1.66	2.16	1.61	0.94
Permeability of saturated soil (k_s) m/s	1.0E–10	1.0E–5	1.0E–5	1.0E–8	1.0E–10
Final freezing temperature (T_f) °C	0	–0.05	–0.10	–0.20	0
Elastic modulus of unfrozen soil (E_u) MPa	6000	300	200	40	21700
Elastic modulus of frozen soil (E_f) MPa	12000	2000	2000	2000	21700
Elastic modulus of thawed soil (E_t) MPa	6000	150	100	20	21700
Poisson's ratio (υ)	0.35	0.35	0.35	0.40	0.20

perform curve fitting and determine the coefficients in the equation. Likewise, the ground temperature at the bottom of the model is also described by a sine function which is weakened and delayed behind surface temperature (Eq. 7). It is well known that snow accumulation significantly reduces heat transfer between ground and atmosphere. The surface temperature of the footpath, therefore, is assumed to be 0 °C when the air temperature drops below 0 °C and snow fall accumulates on footpath, to roughly estimate thermal effects of the snowpack.

$$T_s = T_m + A \cdot \sin\left(\frac{2\pi t}{p} + \varphi\right) + \Delta T \tag{6}$$

Where T_s = surface temperature; T_m = mean annual temperature, 6.77 °C; A = surface temperature amplitude, 15.37 °C; t = time (day); p = corresponding period, 365 days; φ = initial phase, –2.98 rad; ΔT = increment of temperature for black surface pavement, 3 °C.

$$T_z = T_m + A \cdot \exp(-z\sqrt{\frac{\pi}{\alpha_u p}}) \cdot \sin\left(\frac{2\pi t}{p} + \varphi - z\sqrt{\frac{\pi}{\alpha_u p}}\right) \tag{7}$$

Where T_z = ground temperature; z = depth below the air temperature boundary, 5 m; α_u = thermal diffusivity, 5.17E–2 m^2/day.

The initial temperature condition is obtained through a 5 years' heat transfer till thermal equilibrium with the temperature boundaries described previously. Distributed loads are imposed at the surface layer for interested time steps to simulate wheel load (0.61 MPa), which is derived from standard traffic load (49 kN), and the change of stress increment induced by traffic load can be studied in this multi-layer system.

4 Results and Discussion

4.1 The Freeze-Thaw of Pavement

Frost line penetrates pavement gradually when the air temperature drops below 0 °C (Fig. 2). Surfacing layer and base course retain less water, accordingly 0 °C line goes through these two layers rapidly. However, it takes apparently longer time to freeze frost protection layer that has a larger water content and heat capacity. For this reason, the existence of frost protection layer contributes to resisting the advance of frost penetration into subgrade that always keeps unfrozen.

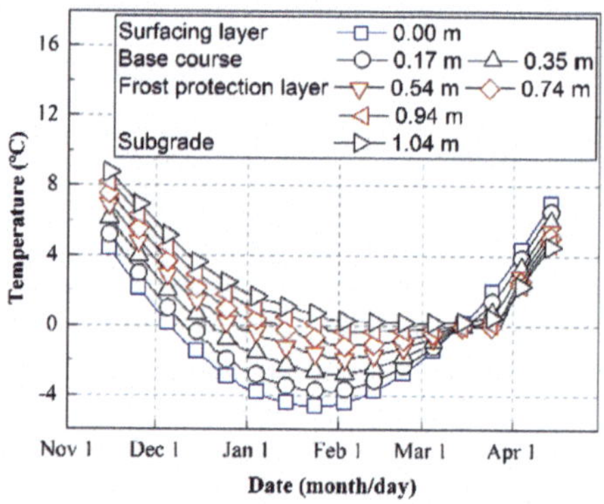

Fig. 2. Temperature history in the pavement.

The snowpack on footpath changes thermal condition in pavement during the winter season. Figure 3 presents the temperature distribution during the freezing process, the temperature under pavement is lower than that under footpath with snow cover of which thermal conductivity is quite small. This thermal pattern gives rise to a frost penetration that is not parallel with the ground surface. It is also believed that a differential cooling accumulation along the cross-section of pavement results in a special thawing pattern, namely the thawing of the central section of pavement comes later than that close to the footpath (Fig. 4).

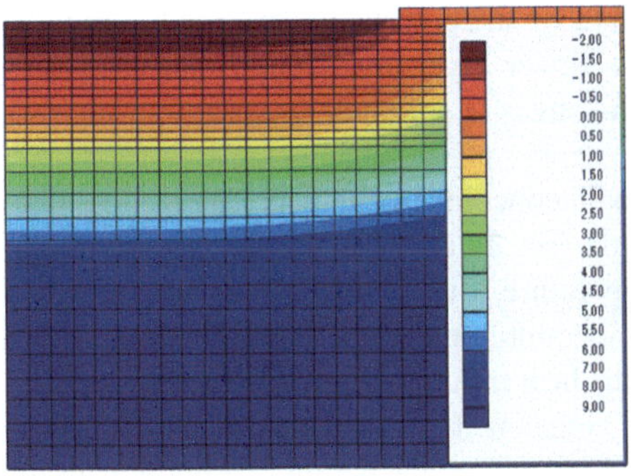

Fig. 3. Temperature distribution on March 1st.

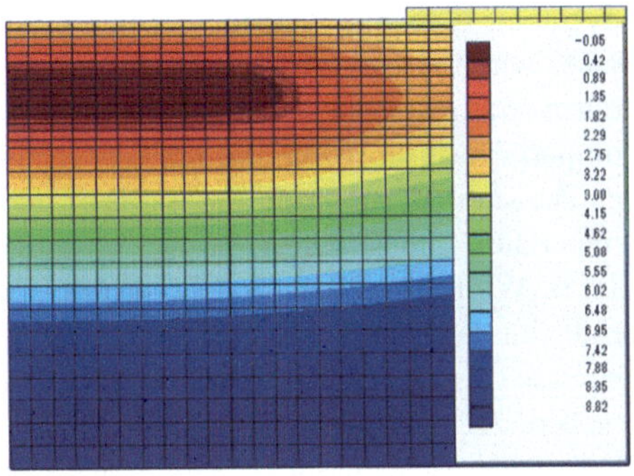

Fig. 4. Temperature distribution on April 1st.

4.2 Stress Change in Pavement

The layers below surface become frozen with the advance of frost line, and therefore the phase change of water bonds soil particles and get the soils stiffer. The mechanical meaning of this physical change is that elastic modulus significantly increases. A resembling phenomenon is the weakness of mechanical properties of thawed soil during springtime, namely the decreased elastic modulus is even smaller than the value before freezing, as freeze-thaw cycle tends to increase the net volume of soil and looser soil structure decrease resilient modulus [3].

Figure 5 presents stress variation in the pavement during winter and spring, and the locations of measuring points are represented by red circles in Fig. 1. For the freezing process, stress in upper frozen layers increases compared with the unfrozen state, while a decrease of stress is observed in the lower unfrozen layer. The stiffer base course and frost protection layer enhance load dispersion in a wider region, as a result, the stress in subgrade soil is reduced. In springtime, the thawing proceeds from both the top downward and bottom upward with the arrival of warm surface temperature, therefore, the stress change appears within one month corresponding to the thawing duration of

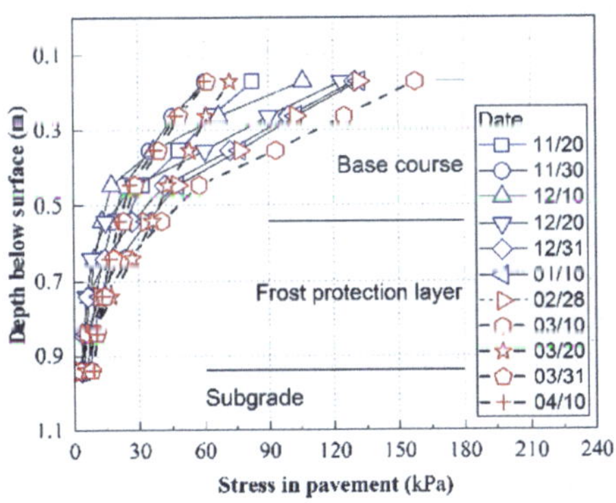

Fig. 5. Stress variation in the pavement.

pavement. A reverse tendency is that reduction of stress in thawed upper layers takes place, while stress concurrently rises in the lower layer. Eventually, stress in base course and frost protection layer decreases and stress in subgrade increases through the comparison of unfrozen and thawed values.

The stress in subgrade soil is one of the main concerns in engineering practice. Both temperature and stress at the top of subgrade (see Point A in Fig. 1) are plotted in Fig. 6. With the descending of surface temperature, overlying base course and frost protection layer gradually get frozen, the temperature at Point A drops together with a decrease of stress. After more than two months heat transfer, the thermal equilibrium is reached. As a result, temperature sustains a constant that is always slightly above 0, along with constant stress. The entire pavement is thawed within one month during subsequent spring season due to two sides warming described previously, and the final value of stress is larger than unfrozen one.

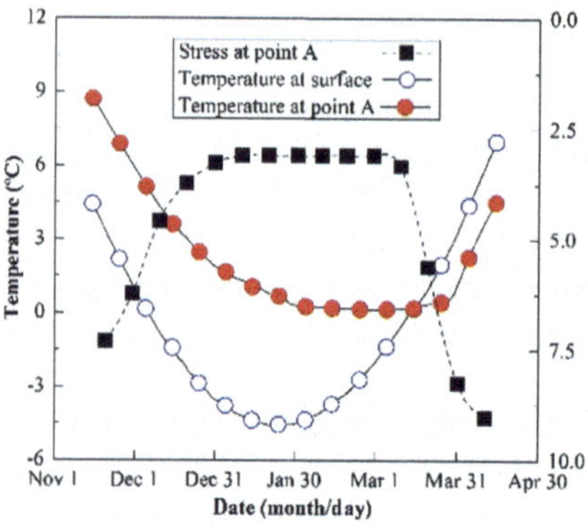

Fig. 6. Temperature and stress at the top of the subgrade.

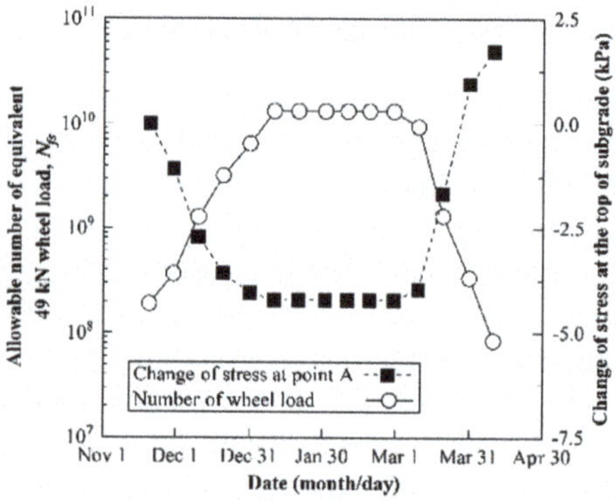

Fig. 7. An allowable number of equivalent wheel load and change of stress at the top of the subgrade.

As we know, the stress and strain level at the top of subgrade can be employed to estimate the possibility of permanent deformation of pavement indirectly. To quantify the effect of stress change on the performance of pavement structure, the allowable number of equivalent 49 kN wheel load (N_{fs}) is applied to evaluate lifespan of pavement [10]. A larger N_{fs} represents a longer service time which is calculated by the empirical equation below.

$$N_{fs} = \beta_{s1} \cdot \left(1.365 \times 10^{-9} \cdot \varepsilon_z^{-4.477\beta_{s2}}\right) \tag{8}$$

Where N_{fs} = required number of cyclic wheel loads to cause 15 mm permanent deformation in layers which results in rutting in flexible pavement; β_{s1} and β_{s2} = empirical correction coefficient; and ε_z = compressive strain at the top of the subgrade.

Figure 7 shows the relation between the change of stress at the top of subgrade and an allowable number of equivalent 49 kN wheel. The decrease of stress temporarily strengths the ability of the pavement to resist residual deformation during the winter season. However, the stress increment of 1.73 kPa can highly shorten the service time of pavement during the spring season compared with the unfrozen state. To alleviate this problem, load restriction is suggested to account.

5 Conclusions

The following conclusions can be drawn from this study:

1. The thermal pattern in pavement located in the cold region was reproduced. The snow cover on the footpath that affects temperature distribution in pavement should be properly considered. The presence of frost protection layer can effectively resist and limit frost penetration in the subgrade.
2. The stress change in pavement is highly associated with the variation of elastic modulus that is sensitive to temperature. The stress variation in base course and frost protection layer presents reverse trend compared with subgrade. The stress increment of subgrade in springtime may greatly deteriorate the performance of pavement and shorten its service life.

Acknowledgments. This research was supported in part by Grant-in-Aids for Scientific Research (A) (16H02360) from Japan Society for the Promotion of Science (JSPS) KAKENHI. The support from the China Scholarship Council (CSC) is also greatly acknowledged.

References

1. Cole, D.M., Irwin, L.H., Johnson, T.C.: Effect of freezing and thawing on resilient modulus of a granular soil exhibiting nonlinear behavior. Transp. Res. Rec. **809**, 19–26 (1981)
2. Simonsen, E., Isacsson, U.: Soil behavior during freezing and thawing using variable and constant confining pressure triaxial tests. Can. Geotech. J. **38**(4), 863–875 (2001)
3. Simonsen, E., Janoo, V.C., Isacsson, U.: Resilient properties of unbound road materials during seasonal frost conditions. J. Cold Reg. Eng. **16**(1), 28–50 (2002)
4. Janoo, V.C., Berg, R.L.: Thaw weakening of pavement structures in seasonal frost areas. Transp. Res. Rec. **1286**, 217–233 (1990)
5. Abe, R., Kumagai, M., Maruyama, K.: A study of materials and environmental conditions for mechanistic-empirical pavement design in cold snowy regions. J. Jpn. Soc. Civil Eng. **67**(3), 17–25 (2011). Ser. E1 (Pavement Engineering) (in Japanese)
6. Kishikawa, T., Otgonjargal, D., Kawaguchi, T., Nakamura, D., Yamashita, S.: Influence of freeze-thaw on stress propagation in the ground. In: Proceedings of Technical Report of the Annual Meeting of the JGS Hokkaido Branch, vol. 57, pp. 27–34 (2017). (in Japanese)
7. Ishikawa, T., Kijiya, I., Tokoro, T., Sato, M.: Numerical experiments on freeze-thaw of soils with coupled thermo-hydro-mechanical FE analysis. In: Computer Methods and Recent Advances in Geomechanics, pp. 415–420. Taylor & Francis Group, London (2015)

8. Luo, B., Ishikawa, T., Tokoro, T., Lai, H.: Coupled thermo-hydro-mechanical analysis of freeze-thaw behavior of pavement structure over a box culvert. Transp. Res. Rec. J. Transp. Res. Board **2656**, 12–22 (2017)
9. Andersland, O.B., Ladanyi, B.: Frozen Ground Engineering. John Wiley & Sons, New Jersey (2004)
10. Japan Road Association: Pavement Design Handbook. Japan Road Association, Tokyo (2006). (in Japanese)

Geotechnical In-Situ Testing and Monitoring

Large Diameter Soil Pressure Sensors Employed in Dynamic Shallow Foundation Testing

Anne Lemnitzer[1(✉)], Lisa Star[2], Lohrasb Keykhosropour[1], Antonio Marinucci[3], and Steve Keowen[4]

[1] Department of Civil and Environmental Engineering, University of California, Irvine, USA
alemnitz@uci.edu
[2] Department of Civil Engineering and Construction Engineering Management, California State University, Long Beach, USA
lisa.star@csulb.edu
[3] V2C Strategists, LLC, Brooklyn, NY, USA
antmarinucci@gmail.com
[4] AFB Engineered Test Systems, Simi Valley, CA, USA

Abstract. Measurements of distribution and magnitude of static and dynamic earth pressures resulting from self-weight and applied loading is essential to the design, behavior, and performance of many civil engineering-type structures involving soil-structure interaction. Advancements in testing and instrumentation have allowed researchers to improve upon the original, now classic analytical models to predict earth pressures acting on retaining walls, shallow foundations, and various deep foundation elements. Ideally, soil pressure instrumentation should be tailored towards the respective engineering application and comply with the stiffness of the structural system. In addition, it is desirable to provide sufficient sensing surface area to avoid pressure localization, and install the pressure surface flush with its surroundings to minimize arching effects. A new and simple load cell based pressure sensor with a diameter of 10 cm and a capacity of 144 kPa was developed, built, and deployed in a large-scale experimental investigation. Upon describing details pertaining to the design, fabrication, and calibration of the new sensor, results from the large-scale experimental investigation on dynamic shallow foundation performance is presented.

Keywords: Pressure sensors · Instrumentation · Dynamic loading Shallow foundations

1 Introduction

Seismic soil pressures are of critical importance for the design and construction of foundation elements in regions with load demands such as earthquake-induced ground shaking and dynamic loading (e.g., wave motions, impact pressures, or machine excitations). A wide range of geotechnical instrumentation is commercially available

L. Hu et al. (Eds.): GSIC 2018, *Proceedings of GeoShanghai 2018 International Conference: Multi-physics Processes in Soil Mechanics and Advances in Geotechnical Testing*, pp. 327–335, 2018.
https://doi.org/10.1007/978-981-13-0095-0_37

for implementation in short term foundation testing and long term monitoring efforts. Soil pressure sensors are commonly utilized in shallow foundation systems (wind turbine foundations, mat foundations) and retaining structures. Hereby it is desirable to capture soil-structure interaction effects across a large area to minimize localization effects. This manuscript will present a homemade pressure sensor that was employed in a series of experimental studies (e.g., [3, 4, 7]). Specifically, the paper will describe the components of the pressure sensor, its versatility, and its implementation in a dynamic testing program that investigates the pressure distribution beneath a rectangular shallow foundation excited through eccentric shakers.

2 Newly Developed Simple Soil Pressure Sensor

A schematic of the soil-pressure sensor's cross section and a photograph of the assembled pressure sensor are shown in Fig. 1. The sensing face consists of a steel plate (i.e., sensing plate) with a thickness of about 6.35 mm along the edges and about 9.5 mm in the center, where the plate is in contact with a high precision, commercially available load cell. This load cell is rigidly connected to the center of the housing. The load cell (i.e., capacity) is interchangeable, and can be selected to capture the expected range of pressures. The sensing plate rests on two stiff rubber O-rings located near the edge of the plate (minimizing potential tilting) and is in intimate and constant contact with the load cell. The sensing plate has a diameter of 9.5 cm and a sensing area of about 78.54 cm^2. The complete housing assembly has an outside diameter of approximately 10 cm (Fig. 1b). To reduce or eliminate arching effects, the stiffness of the material comprising the sensing plate ensures that no deflection in the plate would occur, which means that the calibration of the sensor is insensitive to the specific soil material/type transmitting the pressure to the sensing face. The sampling frequency of the data can be adjusted depending on the specific type of testing to be performed (i.e., higher frequency for dynamic testing, and lower frequency for static testing). The limitations of the sampling frequency are essentially dictated by the capabilities of the respective data acquisition system. The installation of the sensor can be altered to the needs of each specific application. For instance, the SPS can be embedded in the carrier structure

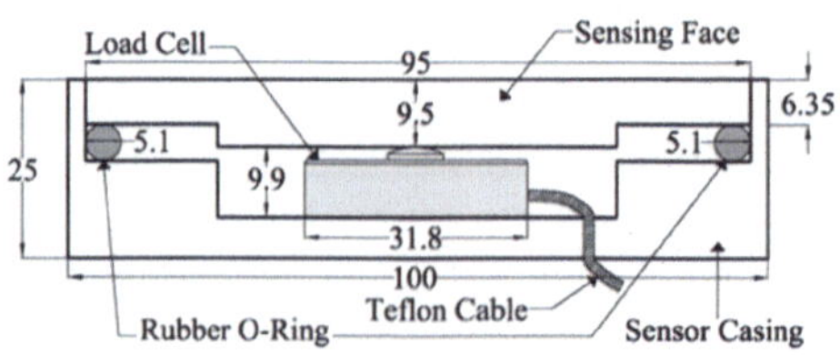

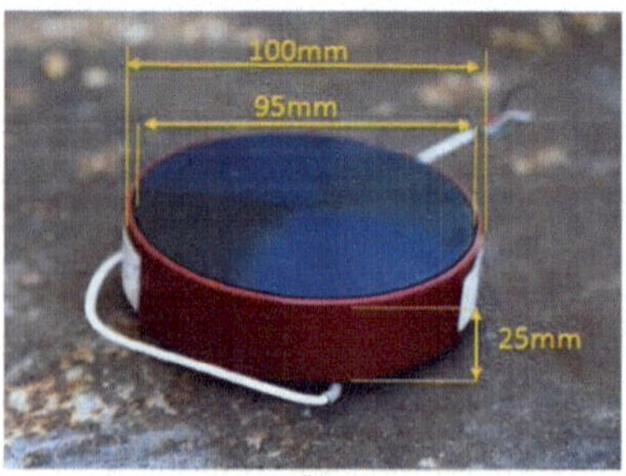

Fig. 1. Soil pressure sensor: (a) cross-section A-A schematic view (in mm), and (b) photograph of complete pressure cell

though a cut-out or perforation in the carrier element (e.g., concrete cutout). Similarly, the SPS can be installed with an adapter ring from the inside of a tin-walled hollow structure. Depending on the application and the stiffness of the material to which the SPS will be attached, different materials can be used for the sensor housing. For example, the application presented later in this manuscript, utilized a SPS composed of steel, as the sensors were embedded in a stiff foundation slab made of concrete. In another experiment by the authors, the sensor housing was made of aluminum [4].

3 Dynamic Calibration

Prior to implementing the sensor in the field, the sensor performance was tested under static and dynamic loading. A detailed description of the static sensor calibration is omitted for brevity. A static calibration can be easily performed with manual weights or universal testing machines if the applied load is well defined and calibrated laboratory equipment it used.

To evaluate the sensor performance under dynamic loading, particularly the behavior of the entire sensor assembly (e.g., load cell and housing), a two stage dynamic calibration setup was developed. First, dynamic loading up to 5 Hz was applied using a universal testing machine. A pressure sensor with a load capacity of 1110 N (i.e., 144 kPa pressure capacity) was exemplarily selected. Testing in the UTM consisted of applying a step sine motion with an amplitude of 50 lbs (Fig. 2a). The frequency of the sine wave was gradually increased from 0.5 to 5 Hz (Fig. 2b). This frequency range represents typical earthquake motions. Data shown in the plot records (Fig. 2b) of a reference load cell and the measured response of the pressure cell. Figure 2b indicates that data from the reference load cell and the newly developed pressure sensor match reasonably well and discrepancies were less than 10%.

In the second stage of dynamic calibration, the frequency-dependent response of the pressure sensor under a wider range of frequencies (1 to 16 Hz) was investigated using

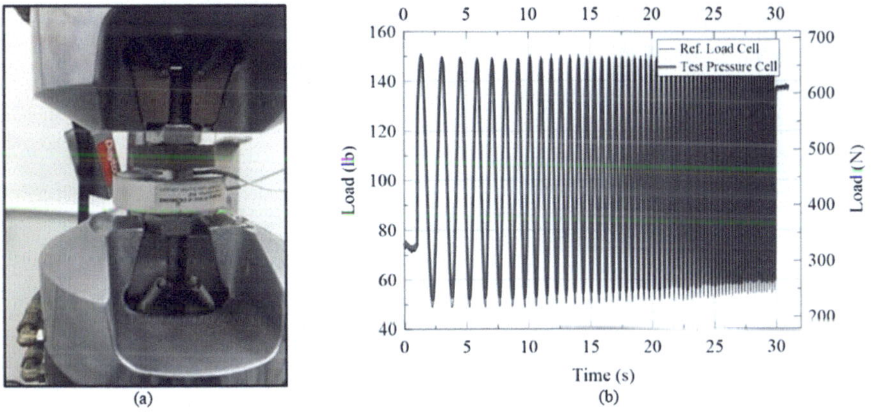

Fig. 2. Dynamic calibration using (a) the universal testing machine (UTM) and (b) recorded data

small shake table equipment. The shake table setup is shown in Fig. 3a. The shake table was placed vertically on the pressure sensor and buffered by a 2.5 cm steel plate to allow for uniform distribution of dynamic pressures on the cell. Another steel plate with a thickness of 5 cm was used between the pressure cell and the ground floor to ensure a level surface. Two accelerometers were employed: Accelerometer No. 1 was placed on the moving part of the shake table to measure the acceleration of the applied motion. Accelerometer No. 2 was used as reference sensor and placed on the frame of the shake table. The pressure sensor was subjected to step sinusoidal excitations and frequencies were increased from 1 to 16 Hz using 2 Hz increments. The net acceleration of the moving plate (i.e., recording of Accelerometer No. 1 minus Accelerometer No. 2) was multiplied by its mass (M_{moving} = 3.083 kg) to calculate the total applied force on the pressure cell. The unfiltered force-time histories and the measured force for the pressure cell are shown in Fig. 3b (top row). An amplitude of 40 mVpp was used for the 2 Hz loading, while an amplitude of 100 mVpp was selected for frequencies of 4 Hz and higher. The recorded data were filtered by first transforming the signal from the time domain to the frequency domain using the Fast Fourier Transformation (FFT), and then applying a low-pass, 8th order (n_f = 8) Butterworth filter to remove frequencies above 16 Hz. Hereafter, signals were transformed back into the time domain using the inverse

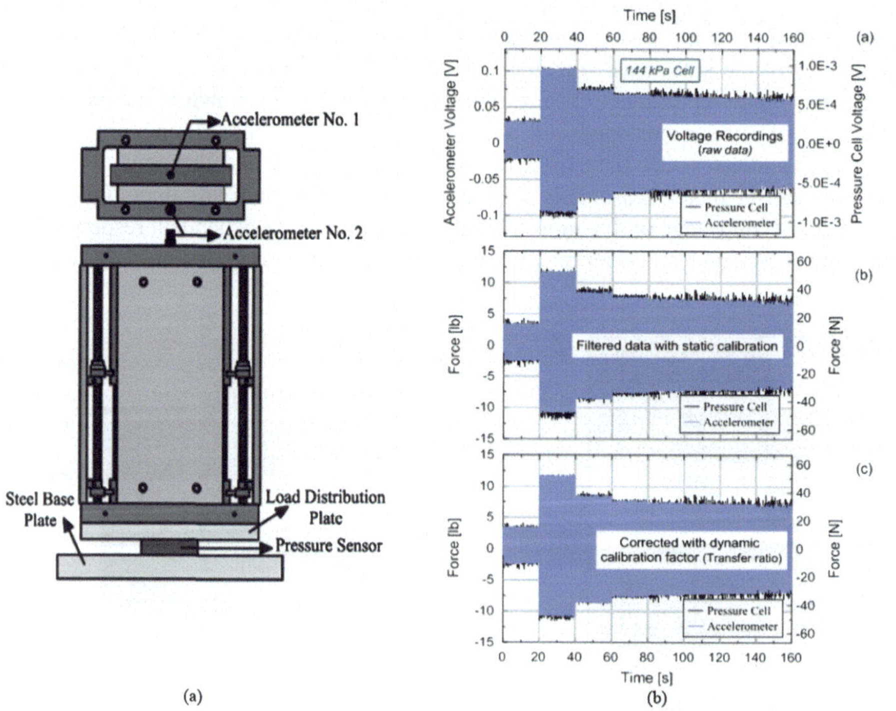

Fig. 3. Test setup for dynamic calibration using (a) a shake table and (b) pressure cell response to dynamic loading: unfiltered (top row), filtered (center row), and calibrated (bottom row) dynamic signals

FFT. The static calibration factor was applied to the voltage data, and the filtered force-time histories are shown in Fig. 3b (2nd row).

Following [2], a frequency dependent transfer function (i.e., a transfer ratio) between the accelerometer and the pressure cell was defined as the ratio of the Fourier force amplitude of the accelerometer, $P_{accelerometer}$, to the Fourier force amplitude of the pressure cells ($P_{pressure}$ cell) at each of the frequencies of the applied motion (i.e., 2 Hz, 4 Hz, 6 Hz, etc.), as expressed in Eq. (1).

$$TF(f) = \frac{|P_{accelerometer}(f)|}{|P_{Pressure\,cell}(f)|} \tag{1}$$

This transfer ratio is an index of the pressure cell accuracy in recording forces at different frequencies. The variation of the transfer ratio with the loading frequency is shown in Fig. 4b. The multiple points can be interpreted as the different response ratios for continuous cycles at the target frequency. As it can be seen from Fig. 4b, the value of the dynamic transfer ratio is very close to 1.0 throughout the entire frequency range. Differences between applied and measured pressures were small ($\leq 15\%$) and the sensor assembly has a little impact on the pressure recordings. In order to get more accurate results and to remove any minor frequency effect from the pressure cell response, recordings of the pressure cell (Fig. 3) were corrected with the transfer ratios of Fig. 4b. The corrected sensor response was obtained as the product of the transfer function and the corresponding pressure sensor recording at each frequency, as expressed in Eq. (2).

$$P_{pressure\,cell-corrected}(f) = TF(f) \times P_{pressure\,cell}(f) \tag{2}$$

The corrected Fourier force amplitude spectra obtained using Eq. (2) was transformed back into time domain using an inverse FFT. Figure 3b (bottom row) presents the dynamically calibrated force time history of the pressure sensor. The sensor performance can be considered frequency independent.

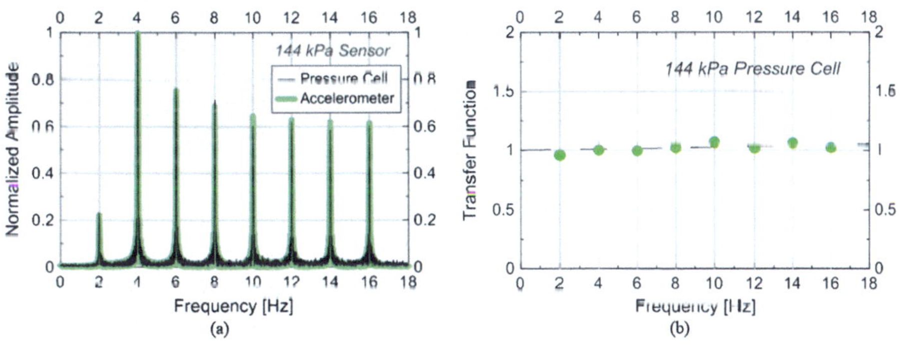

Fig. 4. (a) Normalized Fourier amplitudes and (b) Transfer function between applied and recorded acceleration measurements

4 Example Application: Dynamic Shallow Foundation Bearing Pressure

The pressure sensors were used in a series of field tests to measure the static and dynamic vertical contact pressures beneath a shallow foundation slab subjected to forced vibration and earthquake loading. The applied loading resulted in rocking and translation of the shallow foundation system. A test structure with steel columns and braces supporting a concrete top slab supported on a shallow foundation was purpose-built for this test series. A schematic of the test setup for the foundation and superstructure is shown in Fig. 5a–c. The foundation consisted of a simple, at-grade cast-in-place reinforced concrete mat foundation. Its dimensions were 4.3 m long by 2.14 m wide and 0.6 m thick. This thickness was sufficient to consider the slab nearly rigid. The mat foundation was composed of concrete with a nominal 28-day compressive strength of 27.6 MPa, and was reinforced with No. 5, Grade 60 steel rebar. The column-slab system with bracing installed had a high fixed-base natural frequency between of approximately 34 Hz and 22 Hz, in the X and Y directions respectively. The total structural mass was approximately 22,000 kg, which resulted in a uniform static bearing pressure of approximately 23.5 kPa. In addition a light lift-frame (not-shown) was located on the roof slab of the structure during this testing series.

The structure was located at the Wildlife Liquefaction Array (WLA), maintained by NEES@UCSB (nees.ucsb.edu/facilities/wla), near Brawley, CA. The site stratigraphy consists of about 2.0 m to 3.0 m of silty clay to clayey silt overlying silt, silty sand, and sandy silt that is about 3.0 m to 4.0 m in thickness. This relatively coarse layer is underlain by silty clay to clayey material. The average moist unit weight of the soil was approximately 17.0 kN/m^3. Within the upper 1.6 m to 2 m of soil, the shear wave velocity was measured using cross-hole tests and was determined to be about 106 m/s post-construction.

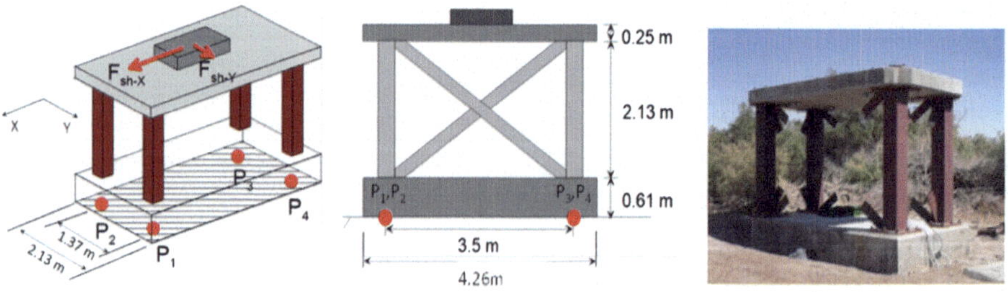

(a) (b) (c)

Fig. 5. Configuration of the UCLA portable SSI test structure with location of pressure sensors: (a) isometric view with directions of shaker force indicated, (b) plan view and (c) elevation view of foundation and pressure sensor locations, and (d) photograph showing the test structure and foundation without the bracing installed

Four high-capacity pressure sensors were arranged and installed at each corner on the bottom of the foundation, as shown in Fig. 5d. The sensors were installed after placement of the reinforcing steel but prior to casting the concrete for the foundation. The sensors were each fitted with a bolt and washer assembly and were tied directly to the reinforcing steel using steel wire ties. This assembly prevented any sensor movement during the concrete pour and restricted the sensors from displacing during testing. This cast-in place installation requires care, but reflects the only realistic installation option during real-world construction. A similar approach has been used in other applications including furnishing shallow foundations of on-shore wind turbine generators with pressure sensor [8]. The high-capacity pressure sensors had a maximum pressure capacity of 274 kPa. The instrument cables were encapsulated within and protected by plastic tubing to reduce potential physical damage and corrosion during foundation casting, and emerged at the edge of the foundation slab for convenient attachment to the data acquisition system. The pressure sensors were powered by a 10 V power source or a battery with the output voltage recorded. Additional instrumentation included eight Kinemetrics EpiSensor ES-T triaxial accelerometers, located at the four corners of the top and bottom slabs of the superstructure. All sensor output was recorded using a sampling frequency of 200 Hz.

Data from the pressure sensors was collected during dynamic excitation of the test structure. A small portable linear mass shaker, namely, the Acoustic Power System "ELECTROSEIS", long stroke shaker model 400 which can induce a maximum force of about 440 N, was used as excitation source. The linear mass shaker imposes low-amplitude loads on the structures, and soil behavior is expected to remain nearly linear. The shaker was located on the roof slab of the structure and could be configured to apply an excitation force in either longitudinal (X) or transverse (Y) direction. Waveforms applied to the specimen included continuous linear frequency sweeps with a nearly constant amplitude force and variable frequencies ranging from 4 Hz to 44 Hz. Complete details about the experimental program, including information about the test specimen, site, and instrumentation, are available in [7] and [8].

Sample data from a series of forced vibration tests is shown in Fig. 6. Figure 6(a) plots cycles of overturning moment versus base rotation for approximately two cycles of loading of the braced structure in the X (blue) and Y (green) directions. Overturning moment includes the shaker force plus inertial forces due to the acceleration of the structure and foundation masses. The shaker force amplitude is approximately the same for both X and Y directions of shaking for the loops shown (frequency of approximately 8.5 Hz and an amplitude of approximately 200 N). Vertical accelerometers located on the foundation are used to determine the rotational acceleration of the foundation. The rotational acceleration is double integrated to find the base rotation in radians. Despite the similar shaker loading level, the X-direction moment-rotation loops have a smaller amplitude than the Y-direction loops, indicating that the rotational response of the system is smaller in the X direction compared to the Y direction. Additionally, X-direction moment-rotation loops are steeper than the Y-direction loops, indicating that for the same level of applied moment the SSI system rotation is smaller in the X direction compared to the Y direction, and soil-foundation rotation in the Y direction has a stiffer response than in the X direction. This is consistent with analytical predictions of foundation impedances including [1, 5, 6], which indicate that rotation about the transverse axis of a foundation will have a stiffer foundation response than rotation about the longitudinal axis.

Because of the applied moment, the foundation rotation leads to dynamic changes in the vertical contact pressure at the four corners of the foundation slab. The response of the four exterior load sensors, P_1 to P_4, is shown in Fig. 6 (b). The pressure plotted is the dynamic or change in cell pressure, with the static load removed. For shaking in the X direction, load sensors P_1 and P_2 experience positive pressure when the system rotation is positive. The dynamic pressures on sensors P_3 and P_4 are negative when the rotation is positive. For shaking in the Y direction cells P_1 and P_4 are in phase and sensors P_2 and P_3 are out of phase, as expected. The pressure-rotation loops indicate a periodic reduction in the contact pressure during loading, but the structure does not experience full uplift, and at no time do the cells loose contact with the soil or experience pounding. The amplitudes of the pressure-rotation loops are similar for each pressure cell, as expected for a structure with symmetric loading and a low level of eccentricity. The plots also show that smaller rotations are required to generate the same level of soil pressures for loading in the X direction, compared to the Y, but that overall pressures generated are larger for Y.

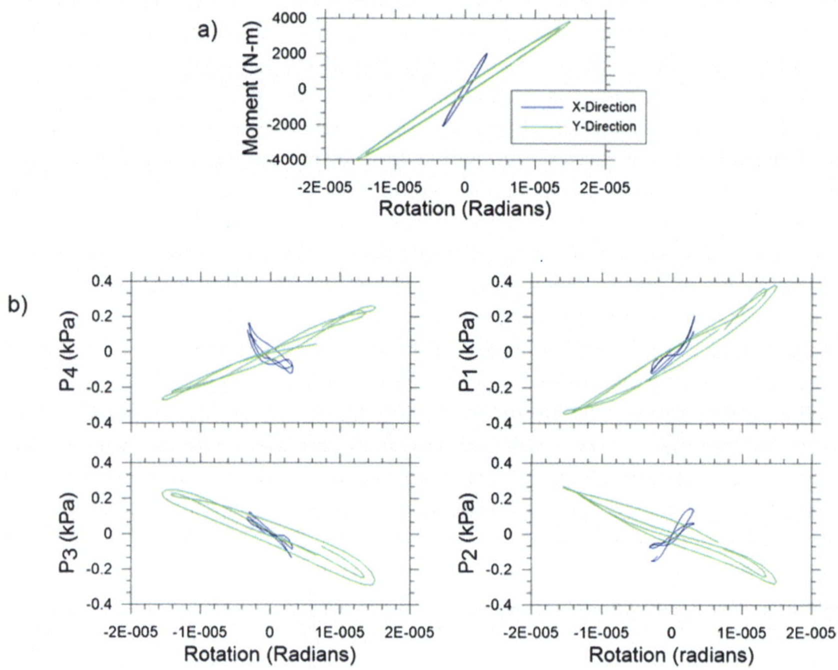

Fig. 6. Forced vibration loading of SSI system in the X (blue) and Y (green) directions: (a) overturning moment versus base rotation, (b) dynamic pressure of exterior pressure cells versus base rotation

5 Conclusions

The design, construction, calibration, and implementation of a simple, easily constructible, and robust soil pressure sensor are presented. One of the primary advantages of the sensor is its versatility, namely, the pressure sensor can be constructed with a housing material compliant with the stiffness of the structural element to which the sensors are being affixed. The internal load cell can be selected to accommodate the expected range of applied pressures during testing. The sensor performance was assessed during dynamic loading and load calibration factors in form of a transfer function were derived. Dynamic calibration showed that the sensors housing has very little impact on the sensors performance and the sensor response is largely frequency independent. Prototype sensors were implemented in dynamic research testing using forced vibration testing of shallow foundation components. The recorded data were found to be of high quality and match well with expected engineering solutions and pressure results.

References

1. Gazetas, G.: Foundation vibrations. In: Fang, H.-Y. (ed.) Foundation Engineering Handbook, 2nd edn., Chap. 15. Chapman and Hall, New York (1991)
2. Gillis, K., Dashti, S., Hashash, Y.: Dynamic calibration of tactile sensors for measurement of soil pressures in centrifuge. ASTM Geotech. Testing J. **38**(3), 261–274 (2015)
3. Lemnitzer, A., Hilson, C., Taciroglu, E., Wallace, J.W., Stewart, J.P.: Effect of backfill relative density on lateral response of a Bridge Abutment Wall. In: 15th World Conference on Earthquake Engineering, Lisbon, Portugal, 24–28 September 2012
4. Lemnitzer, A., Keykhosropour, L., Kawamata, Y., Towhata, I.: Dynamic response of underground structures in sand: experimental data. Earthquake Spectra **33**(1), 347–372 (2017)
5. Mylonakis, G., Nikolaou, S., Gazetas, G.: Footings under seismic loading: analysis and design issues with emphasis on bridge foundations. Soil Dyn. Earthquake Eng. **26**, 824–853 (2006)
6. Pais, A., Kausel, E.: Approximate formulas for dynamic stiffnesses of rigid foundations. Soil Dyn. Earthquake Eng. **7**(4), 213–227 (1988)
7. Star, L.M., Givens, M.J., Nigbor, R.L., Stewart, J.P.: Field testing of structure on shallow foundation to evaluate soil-structure interaction effects. Earthquake Spectra **31**, 2511–2534 (2015)
8. Yilmaz, M., Schubert, S., Tinjum, J., Fratta, D.: Foundation soil response to wind turbine generator loading. In: Geo-Congress 2014 Technical Papers, pp. 1493–1502 (2014)

Cavity Expansion Analysis for CPT in Granular Soils at Small Gravity Fields

Pin-Qiang Mo[1,2(✉)]

[1] State Key Laboratory for GeoMechanics and Deep Underground Engineering, China University of Mining and Technology, Xuzhou, China
pinqiang.mo@cumt.edu.cn

[2] Key Laboratory of Transportation Tunnel Engineering, Ministry of Education, Southwest Jiaotong University, Chengdu, China

Abstract. Cone penetration test (CPT) is a widely used and versatile in-situ tool for determination of soil properties and delineation of soil stratigraphy, which could also act as a useful device for the site characterization of surface soils of other planets during deep space exploration. An analytical solution of drained spherical/cylindrical cavity expansion is developed using a simple non-associated flow model for sand, with consideration of small gravitational environments. The soil model adopted a family of yield loci for triaxial compression based on the normality criterion and a simple stress-dilatancy rule, which showed the ability to model many behaviors of isotropically consolidated clays and sands. The large strains for soil in the plastic region are assumed in this solution, together with the approach of auxiliary variable to convert the Eulerian formulation to the Lagrangian description. The proposed solution is validated against the numerical results for the recovered Cam-clay model. The interpretation of CPT data is then analyzed by this proposed cavity expansion analysis, through the correlation between the cone tip resistance and the calculated cavity pressure. In order to investigate effects of gravity on planets like moon, a simplified approach of normalization is adopted for the analysis of penetration resistance. The predictions of normalised penetration resistance provide good comparisons with previous experimental and numerical results, which indicates the effectiveness of the proposed method for analysis of CPT data under various gravitational environments.

Keywords: Cavity expansion analysis · Cone penetration test
Granular soils · Small gravity fields

1 Introduction

Deep space exploration has gained more attention in recent years for the purposes of further scientific discovery and space resources exploration that have great potential value to humanity. Some countries and companies have launched their plans to land on the Moon and other planets with small gravitational environments. Cone penetration test (CPT) is arguably the most widely used and versatile in-situ tool for determination of soil properties and delineation of soil stratigraphy, with the reliable and repeatable readings including cone tip resistance and sleeve friction. Therefore, CPT could also

L. Hu et al. (Eds.): GSIC 2018, *Proceedings of GeoShanghai 2018 International Conference: Multi-physics Processes in Soil Mechanics and Advances in Geotechnical Testing*, pp. 336–344, 2018.
https://doi.org/10.1007/978-981-13-0095-0_38

serve as a useful device for the site characterization of surface soils of other planets. Space soils (e.g. Lunar soil, Martian soil), transformed from the space rock, are believed to be granular materials with no aqueous water existed in the pores. The understanding of mechanical properties and behaviors of these granular materials are essential for the development of future space bases, including the lunar outpost.

Cavity expansion analysis in geomaterials has been developed rapidly during the past three decades and has shown its ability for the analysis of geotechnical problems, including the in-situ testing, foundations, tunnelling and shaft stability. The prediction of penetration resistance using the cavity expansion method has led to the development of CPT data interpretation. However, the existing analytical solutions for granular geomaterials are still limited, especially for the space regolith with strong angularity, weak cohesion and small gravity fields.

This paper aims to propose an analytical solution of cavity expansion using a simple non-associated flow model for sand [1], with consideration of small gravitational environments. Following the framework of [2], a drained spherical/cylindrical cavity expansion solution is developed by assuming large strains for soil in the plastic region, and adopting the approach of auxiliary variable to convert the Eulerian formulation to the Lagrangian description. The proposed solution is validated against the numerical results, and is then applied for the interpretation of CPT data under small gravity fields. The predicted results of normalised penetration resistance are therefore compared with previous experimental and numerical results.

2 Cavity Expansion Analysis

The quasi-static analysis of cavity expansion has been taken as a simplified model for many geotechnical problems, while the initial spherical or cylindrical cavity with a radius of a_0 is embedded in geomaterials under an initial hydrostatic stress condition. After applying the compression-positive notation, the equilibrium equation is shown as:

$$\sigma_\theta - \sigma_r = \frac{r}{m}\frac{d\sigma_r}{dr} \tag{1}$$

where σ_r and σ_θ are the radial and tangential stresses, respectively; r is the radius of material element and m is used to combine cylindrical ($m = 1$) and spherical ($m = 2$) cavities; 'd' denotes the Eulerian derivative for every material particle at a specific moment. Note that the drained analysis is employed in this study, thus the excess pore pressure is not taken into account. Following the previous study [2, 3], the mean and deviatoric stresses for cavity expansion problems are defined as $p = (\sigma_r + m \cdot \sigma_\theta)/(1+m)$ and $q = \sigma_r - \sigma_\theta$, whereas the volumetric and shear strains are expressed as $\delta = \varepsilon_r + m \cdot \varepsilon_\theta$ and $\gamma = \varepsilon_r - \varepsilon_\theta$. To consider the elastic-plastic behaviour of soil, the soil surrounding the cavity wall performs with both elastic and plastic regions after expansion. The small strain assumption is used for soil in the elastic region ($\varepsilon_r = -du/dr; \varepsilon_\theta = -u/r$), while the logarithmic strains are adopted for the plastic region where large deformation occurs, i.e. $\varepsilon_r = -\ln(dr/dr_0); \varepsilon_\theta = -\ln(r/r_0)$.

3 Soil Model

McDowell [1] proposed a simple non-associated flow model for sand, with a family of yield loci based on micro mechanical considerations and the plastic work dissipated in friction and fracture. The model is therefore adopted in this study to simulate the granular soils. The expressions of the yield surface and non-associated flow rule are shown as:

$$\eta = N\left[(\alpha + 1)\ln\left(p_y/p\right)\right]^{\frac{1}{\alpha+1}} \tag{2a}$$

$$\frac{D\delta^p}{D\gamma^p} = \frac{D\varepsilon_v^p}{D\varepsilon_q^p} \times \frac{m}{m+1} = \frac{M^{\beta+1} - \eta^{\beta+1}}{\eta^\beta} \times \frac{m}{m+1} \tag{2b}$$

where η is the stress ratio and p_y is the preconsolidation pressure; N is the peak value of q on the yield locus, M is the critical state dissipation constant; α and β are the model constants for both yield surface and plastic potential; 'D' denotes the Lagrangian derivative for a given material particle. The conventional volumetric hardening rule is also applied as $D\gamma^p = (\lambda - \kappa)/\upsilon \cdot Dp_y/p_y$, where υ is the specific volume. While the original Cam-clay model can be recovered by making $N = M$ and $\alpha = \beta = 0$, the Modified Cam-clay can be assumed for $N = M$ and $\alpha = \beta = 0.4427$ (i.e. the reference state parameter $\xi_R = (\lambda - \kappa)/(\alpha + 1) \cdot (M/N)^{\alpha+2} = (\lambda - \kappa)\ln 2$), as shown in Fig. 1 (a). The schematic of yield locus and flow rule for a general case is then presented in Fig. 1(b). As reported by [1], the behaviour of both loose and dense sands in drained and undrained conditions can be modelled, and the soil model remains to be flexible, powerful, simple and easy to apply. As to the merits of this model, the introduced parameters could be easily determined from conventional triaxial compression tests, and the employed non-associated flow rule enables the model to capture more sand behaviour. Although non-coaxiality is not considered in this model and the soil model parameters have not been correlated to sand particle properties, this model with its simplicity serves as a useful and effective critical state-based model for sand with a non-associated flow model, which could also be taken as a framework for development of sand models.

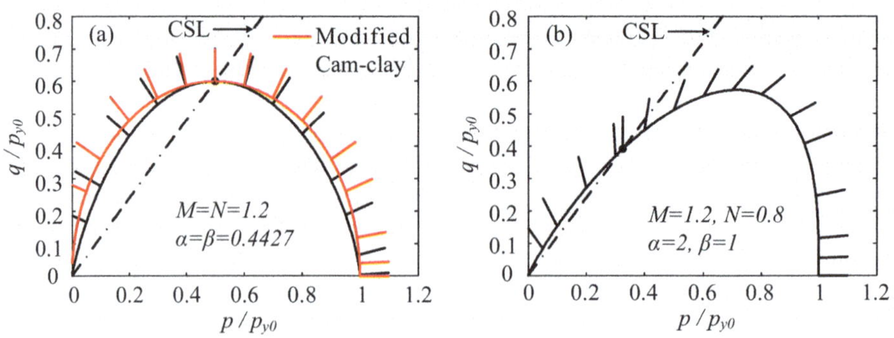

Fig. 1. Schematic of yield locus and flow rule: (a) comparison with the Modified Cam-clay; (b) a general case for non-associated flow model for sand.

4 Cavity Expansion Solution

When the elastic–plastic boundary is referred to as 'c', the elastic region is for soil at $r > c$. Thus the elastic strain rates are expressed as follows:

$$\begin{aligned} D\delta^e &= \frac{\kappa}{\upsilon p} Dp \\ D\gamma^p &= \frac{[1+(m-1)\mu]\kappa}{(1+m)(1-2\mu)\upsilon p} Dq \end{aligned} \tag{3}$$

Therefore, according to the solutions of [2, 4], the distributions of stresses and strains in the elastic region can be concluded as:

$$\begin{aligned} \sigma_r &= p_0 + \frac{B_1}{r^{1+m}} \\ \sigma_\theta &= p_0 - \frac{B_1}{m \cdot r^{1+m}} \\ \delta &= 0 \\ \gamma &= \frac{[1+(m-1)\mu]\kappa}{(1+m)(1-2\mu)} \cdot \frac{B_1(1+m)}{\upsilon_0 p_0 m \cdot r^{1+m}} \end{aligned} \tag{4}$$

where B_1 can be derived based on the boundary condition at $r = c$, giving that $B_1 = q_c m c^{1+m}/(1+m)$, $q_c = N \cdot [(\alpha+1)\ln R_0]^{1/(\alpha+1)} \cdot p_0$, and $R_0 = p_{y0}/p_0$, which represents the overconsolidation ratio.

For soil in the plastic region ($a<r<c$), the volumetric strain is related to the specific volume: $\delta = -\ln(\upsilon/\upsilon_0)$. An auxiliary variable is thus adopted to convert the Eulerian formulation to the Lagrangian description [2, 5], i.e. $\chi = (r - r_0)/r$. The expression of deviatoric strain can then be deduced as $\gamma = -\ln\left[\upsilon/\upsilon_0 \cdot (1-\chi)^{m+1}\right]$.

Following the use of auxiliary variable, the equilibrium equation can be rewritten as follows:

$$-q = \frac{1}{m}\left[1 - \chi - \frac{\upsilon_0}{\upsilon(1-\chi)^m}\right] \frac{D\left(p + \frac{m}{m+1} q\right)}{D\chi} \tag{5}$$

Based on the volumetric strain, the expression of specific volume is derived as:

$$\upsilon = [\upsilon_0 + \lambda \ln p_0 + (\lambda - \kappa) \ln R_0] - \lambda \ln p - \frac{\lambda - \kappa}{(\alpha+1)N^{\alpha+1}} \eta^{\alpha+1} \tag{6}$$

And the derivative form of Eq. (6) gives:

$$D\upsilon = -\lambda \frac{Dp}{p} - \frac{\lambda - \kappa}{(\alpha+1)N^{\alpha+1}} \eta^\alpha \left(\frac{Dq}{p} - \eta \frac{Dp}{p}\right) \tag{7}$$

In terms of the deviatoric strain, the derivative expression is deduced as:

$$D\gamma = -\frac{D\upsilon}{\upsilon} + \frac{m+1}{1-\chi} D\chi = \left[\frac{[1+(m-1)\mu]\kappa}{(1+m)(1-2\mu)\upsilon} + \frac{\lambda-\kappa}{\upsilon} \frac{\eta^{\beta}}{M^{\beta+1}-\eta^{\beta+1}} \frac{m+1}{m} \left(\frac{\eta}{N}\right)^{\alpha} \frac{1}{N} \right] \frac{Dq}{p} + \frac{\lambda-\kappa}{\upsilon} \frac{\eta^{\beta}}{M^{\beta+1}-\eta^{\beta+1}} \frac{m+1}{m} \left[1 - \left(\frac{\eta}{N}\right)^{\alpha+1} \right] \frac{Dp}{p} \tag{8}$$

Therefore, the numerical calculation from $r = c$ ($\chi_c = \frac{(c-c_0)}{c} = \frac{[1+(m-1)\mu]\kappa}{[(1+m)(1-2\mu)]} \cdot \frac{q_c}{[(1+m)\upsilon_0 p_0]}$, $p_c = p_0, \upsilon_c = \upsilon_0$) to $r = a$ ($\chi_a = \frac{(a-a_0)}{a}$) can be applied to obtain the distributions of stresses and strains in the plastic region. The location of soil element is then revived by integration from a to r:

$$\int_a^r \frac{dr}{r} = ln\frac{r}{a} = \int_{\chi_a}^{\chi} \frac{d\chi}{1-\chi-\upsilon_0/[\upsilon(1-\chi)^m]} \tag{9}$$

The elastic/plastic boundary c is thus calculated from the integration of Eq. (9). Note that the detailed derivations of this solution can be found in an unpublished report, and the readers may request a copy of this report from the author.

5 Results of Cavity Expansion and CPT at Small Gravity Fields

5.1 Solution Validation for the Modified Cam-Clay

The proposed solution is firstly verified in this section by comparisons with the results of numerical simulation of cylindrical cavity expansion using the Modified Cam-clay model (after [5]). It should be noted that the adopted non-associated flow model for sand can be adjusted for the prediction of the Modified Cam-clay model, having $N = M$ and $\alpha = \beta = 0.4427$. For the scenario of $R_0 = 3$ and expansion of $a/a_0 = 2$, the stress paths in both $p - q$ space and $\upsilon - p$ space are provided in Fig. 2. Despite of the differences on the yield surface and the flow rule (see Fig. 1a), the stress paths show a good agreement with the numerical results. After expansion, the distributions of stresses and the specific volume are presented in Fig. 3, together with the comparisons of numerical simulation from [5]. The general responses of cavity expansion are verified using the proposed solution, including the elastic-plastic boundary. However, the ability of the solution lies on the prediction of behaviour of granular soils with a non-associated flow rule, which is examined in the following sections.

5.2 CPT at Small Gravity Fields

The proposed analytical solution of cavity expansion is then applied for the analysis of cone penetration tests. The penetration is treated as a spherical cavity expanded from an initial size of sand particle ($d_{50} = 0.11$ mm, a typical size of lunar regolith simulant from [6]) to a final size of penetrometer diameter ($B = 35.7$ mm, a standard cone size).

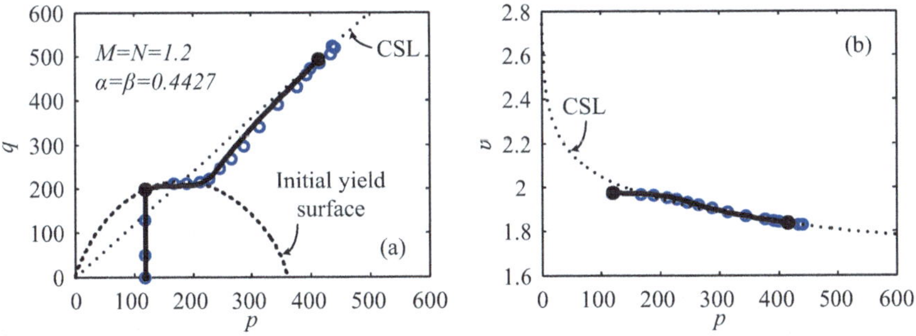

Fig. 2. Stress path of cylindrical cavity expansion: (a) in $p - q$ space; (b) in $\upsilon - p$ space

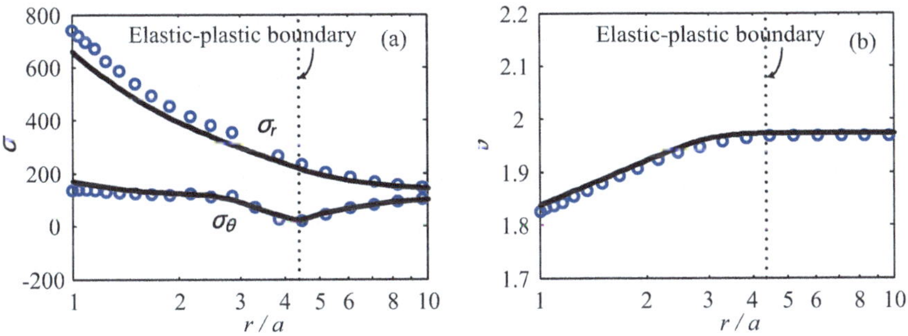

Fig. 3. Distributions of (a) stresses; and (b) specific volume

The average density of lunar regolith is assumed as $\rho = 1.8 \cdot 10^3\,\mathrm{kg/m^3}$ (based on NASA report [7]). According to the gravity of the Moon, the g-level is set to be $g_{simu} = 1/6\,\mathrm{g}$, where g is the earth gravity $\approx 9.81\,\mathrm{m/s^2}$, and the g-level $N_g = g_{simu}/g = 1/6$. The calculated cavity pressure is related to the penetration resistance, using the correlation: $q_c = \sigma_{r,wall} \cdot \left(1 + \sqrt{3}\tan\phi\right)$, where $\sin\phi = 3M/(6+M)$. The predicted penetration resistance and its initial stress condition are shown in Fig. 4(a), showing a linear increase of q_c with depth, as also reported by [8, 9]. Using the normalization of $Q = (q_c - \sigma_{v0})/\sigma_{v0}$, a relative constant value of $Q \approx 9.8$ is observed, while the slightly decrease with depth is confirmed by the centrifuge tests of CPT [10]. When the g-level varies from 0 to 5, the normalised penetration resistance decreases with N_g, as shown in Fig. 4(b), showing a similar trend with experimental data [11] and numerical simulation [12].

When the soil sample is assumed to be uniform with a given density ($\upsilon_0 = 2.325$, based on the experimental tests of [11]), the decrease of normalised penetration resistance with g-level is more significant and comparable to the experimental and numerical results. As shown in Fig. 5, when the critical state dissipation constant $M = 1.2$, the prediction of Q is underestimated for $N_g < 1$ and overestimated for $N_g > 1$, compared with the experimental data [11]. However, the triaxial compression tests of a granular material under various small gravity fields indicate that the critical

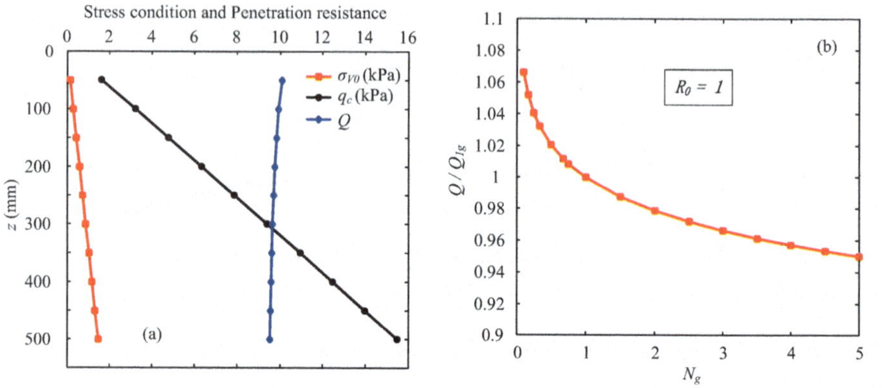

Fig. 4. Prediction of penetration resistance of CPT (a) penetration resistance against depth for $N_g = 1/6$; (b) normalised penetration resistance against N_g.

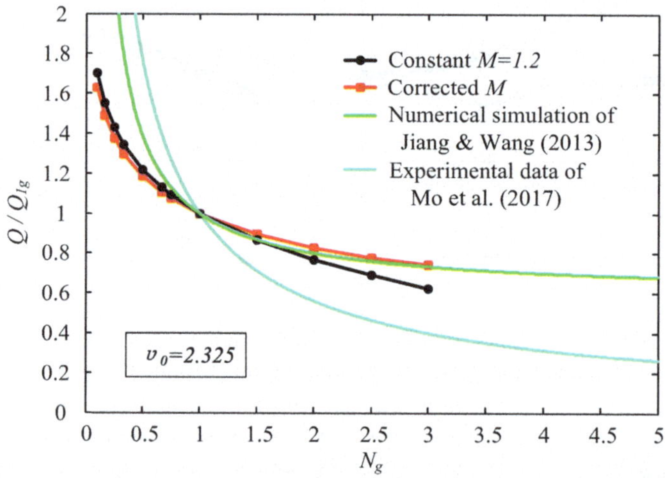

Fig. 5. Prediction of normalised penetration resistance of CPT against N_g.

state friction angle decreases with N_g at a rate of 1.24 [13]. The value of M is thus corrected based on the relationship: $M = 6 \sin \phi/(3 - \sin \phi)$, and the friction angle is assumed as $\phi = 30 - 1.24(N_g - 1)$. Although the underestimation of Q remains for $N_g < 1$, the prediction for $N_g > 1$ agrees well with the DEM simulation of CPT by [12]. It should be noted that the other soil model parameters are not changed during the calculation, which may also be influenced by the gravity; further study is therefore needed to focus on the investigation of soil parameters effects, and their relationships between g-level and the resultant penetration resistance. With the back analysis, the obtained CPT measurements on a planet could be used reversely for the estimation of the mechanical properties and behaviour of the relevant space soils.

6 Conclusions

An analytical solution of both cylindrical and spherical cavity expansion is proposed in this paper using a simple non-associated flow model for sand, which serves as a critical state-based soil model for granular materials. The solution provides the stress paths and distributions of stress/strain during and after expansion of a cavity, which is then verified by the numerical simulation using the Modified Cam-clay model. Relating the cavity pressure to the cone penetration resistance, the effect of small gravity fields is examined by varying the initial stress condition. For a normally-consolidated soil, the normalised penetration resistance decreases with N_g within the range from 0 to 5. When the soil sample is assumed to be uniform with a given density, the decrease of normalised penetration resistance with g-level is more significant and comparable to the experimental and numerical results. By correcting the critical state friction angle based on triaxial compression tests of a granular material under various small gravity fields, the prediction for $N_g > 1$ agrees well with the DEM simulation of CPT. Following this preliminary study, further study is required for the investigation of soil parameters effects and their relationships with g-level.

Acknowledgments. The author would like to acknowledge financial support from the Natural Science Foundation of Jiangsu Province (no. BK20170279), the Foundation of Key Laboratory of Transportation Tunnel Engineering (Southwest Jiaotong University), Ministry of Education (no. TTE2017-04), the Fundamental Research Funds for the Central Universities (no. 2017QNB10), China Postdoctoral Science Foundation (no. 2017M621866), and Jiangsu Planned Projects for Postdoctoral Research Funds (no. 1701196B).

References

1. McDowell, G.R.: A simple non-associated flow model for sand. Soils Found. **4**(2), 65–69 (2002)
2. Mo, P.Q., Yu, H.S.: Drained cavity expansion analysis with a unified state parameter model for clay and sand. Can. Geotech. J. (2018). https://doi.org/10.1139/cgj-2016-0695
3. Mo, P.Q., Yu, H.S.: Undrained cavity expansion analysis with a unified state parameter model for clay and sand. Géotechnique **67**(6), 503–515 (2017)
4. Yu, H.S., Houlsby, G.T.: Finite cavity expansion in dilatant soils: loading analysis. Géotechnique **41**(2), 173–183 (1991)
5. Chen, S.L., Abousleiman, Y.N.: Exact drained solution for cylindrical cavity expansion in modified Cam Clay soil. Géotechnique **63**(6), 510–517 (2013)
6. Arslan, H.: JSC-1a Geotechnical Properties Experiments, Document No. 105525, University of Colorado (2007)
7. Sibille, L., Carpenter, P., Schlagheck, R., French, R. A.: Lunar Regolith Simulant Materials: Recommendations for Standardization, Production, and Usage. NASA/TP-2006-214605 (2006)
8. Bolton, M.D., Gui, M.W., Garnier, J., Corte, J.F., Bagge, G., Laue, J., Renzi, R.: Centrifuge cone penetration tests in sand. Géotechnique **49**(4), 543–552 (1999)
9. Mo, P.Q., Marshall, A.M., Yu, H.S.: Centrifuge modelling of cone penetration tests in layered soils. Géotechnique **65**(6), 468–481 (2015)

10. Mo, P.Q.: Centrifuge Modelling and Analytical Solutions for the Cone Penetration Test in Layered Soils. Ph.D. thesis, The University of Nottingham (2014)
11. Mo, P.Q., Gao, F., Zhou, G.Q., Li, R.L.: Magneto-gravity simulation of cone penetration test in cohesionless soil under small gravity fields. Accepted by GeoChina 2018 (2017)
12. Jiang, M.J., Wang, X.X.: Numerical analysis of cone penetration tests under different gravity fields by distinct element method. Rock Soil Mech. **34**(3), 863–873 (2013)
13. Zhou, G.Q., Gao, F., Mo, P.Q.: An experimental study on strength and deformation characteristics of cohesionless soil under small gravity field. Accepted by GeoChina 2018 (2017)

A Full-Scale Experimental Study on the Performance of Jacking Prestressed Concrete Cylinder Pipe with Misalignment Angle

Yao Zhang[1,2(✉)], Zhi-guo Yan[1,2], and He-hua Zhu[1,2]

[1] State Key Laboratory of Disaster Reduction in Civil Engineering, Tongji University, 1239 Siping Road, Shanghai 200092, China
2011zhangyao@tongji.edu.cn
[2] Department of Geotechnical Engineering, Tongji University, 1239 Siping Road, Shanghai 200092, China

Abstract. The full-scale test of Jacking Prestressed Concrete Cylinder Pipe (JPCCP), a novel jacking pipe to be used in practical engineering, is conducted in this paper. The purpose of this work is to investigate the structural performance of JPCCP under axial jacking force without misalignment angle and eccentric jacking force with different misalignment angles, respectively. The experimental results indicate that misalignment of the pipeline is the major factor causing tensile strain which usually appears on the exterior reinforced concrete surface. In addition, both the maximum tensile and compressive axial strain increase with the misalignment angle. When the misalignment angle reaches 0.461°, the axial maximum strain is 106με which exceeds the tensile strain limit (100με), which indicates that the increase of section stiffness is required to resist the generation and propagation of the possible crack and protect the prestressed wire from potential erosion. Moreover, larger misalignment angle can not only reduce contact area between pipes ends, but also lead to a more serious stress concentration. As a result, the prestress near the concentrated region increases. The experimental results can provide a reliable reference for pipe design and construction.

Keywords: Full-scale test · Misalignment angle · Eccentric jacking force Prestressed wire · JPCCP

1 Introduction

Prestressed Concrete Cylinder Pipe (PCCP), a popular type of composite open-cut pipe consisting of high strength steel wire, steel and concrete cylinder, and a mortar coating applied to the exterior of the pipe, has been widely used in the world during last decades due to its evident advantages [1–5], such as high pressure resistance, and seepage resistance. PCCP, however, as a kind of open-cut pipe, may bring extremely damage to the environment and inconvenience to citizens in the built-up area.

Compared with the open-cut construction, pipe jacking is a kind of trenchless construction technology for the installation of underground pipelines, and offers

L. Hu et al. (Eds.): GSIC 2018, *Proceedings of GeoShanghai 2018 International Conference: Multi-physics Processes in Soil Mechanics and Advances in Geotechnical Testing*, pp. 345–354, 2018.
https://doi.org/10.1007/978-981-13-0095-0_39

advantages of non-disruptive construction coupled with a unique one without need for temporary ground support or secondary lining [6–9]. In addition, Pipe jacking can not only avoid the unnecessary disruption to the public life, but also reduce the damage to the nearby structures [10–19]. Because of the advantages of both PCCP and pipe jacking, utilizing PCCP as a jacking pipe to satisfy the requirement of civilized construction is becoming attractive [20, 21]. Most attempts have been devoted to studying PCCP as a buried pipe. To investigate the effects of prestress on PCCP, some models were developed by Zarghamee [2–4] to predict its mechanical behaviors under combined load (internal pressure and external load). These results shed light on the performance of PCCP pipe and facilitate its application greatly [4, 22, 23]. However, few studies have been reported about PCCP as a jacking pipe during last decades.

Jacking prestressed concrete cylinder pipe (JPCCP), an improved PCCP, has a reinforced concrete coating applied to the exterior of the pipe instead of a mortar coating applied by PCCP. As a novel type of jacking pipe, its mechanical behaviors under construction with misalignment, which plays an important role in governing the safety of pipe structure, should be analyzed. Prototype test is an attractive procedure because it can illustrate the mechanical performance in real condition, which is also evident to verify analytical solutions and numerical studies. For these reasons, full-scale test is presented in the paper to study the structural behaviors of JPCCP pipe under axial even jacking force with and without misalignment angle.

2 Test Program

2.1 JPCCP Specimen

As shown in Fig. 1, JPCCP consists of inner concrete core, steel cylinder, intermediate concrete core, prestressed wire, and exterior concrete. The detailed geometry parameter of JPCCP pipe is depicted in Fig. 1. To facilitate the assembly of pipe in the construction, bell and socket connection is adopted. According to national standard Code

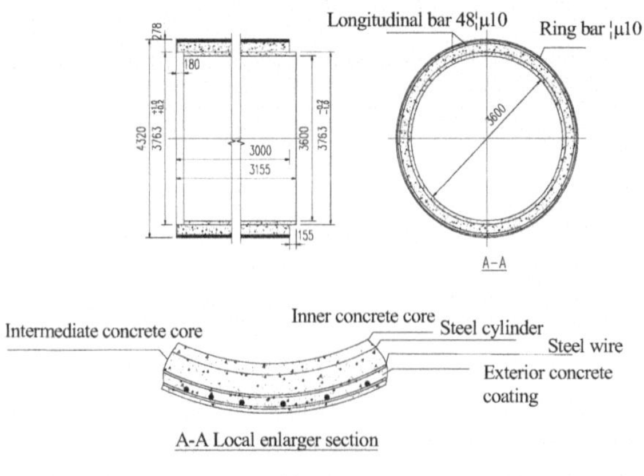

Fig. 1. The structure diagram of JPCCP (mm)

for Design of Concrete Structures (GB50010-2002) [26], the basic material parameters are listed in Table 1.

Table 1. JPCCP pipe material parameter

Item	Value
Modulus of concrete(GPa)	35.5
Modulus of Steel cylinder(GPa)	206
Modulus of Steel wire(GPa)	206
Tension control stress of steel wire(GPa)	1.04
Ultimate strain of concrete($\mu\varepsilon$)	3300(100)[a]

[a]The value in the bracket is tensile ultimate strain.

2.2 Test Arrangement

Generally, jacking pipes will be under two conditions during the construction: edge loading condition and diagonal loading condition, as shown in Fig. 2. This is mainly caused by the misalignment angle of the pipeline. In this study, the edge loading condition is simulated and studied. Two JPCCP pipes are utilized in each test (see Fig. 3a), whose length are 3 m. In addition, six main hydraulic jacks (see Fig. 3b) are used to provide the designed experimental jacking force. Besides, six lateral hydraulic jacks are installed to maintain the presupposed misalignment angle of pipeline. In addition, a simulative starting shaft is built to sustain the reaction force caused by hydraulic jacks.

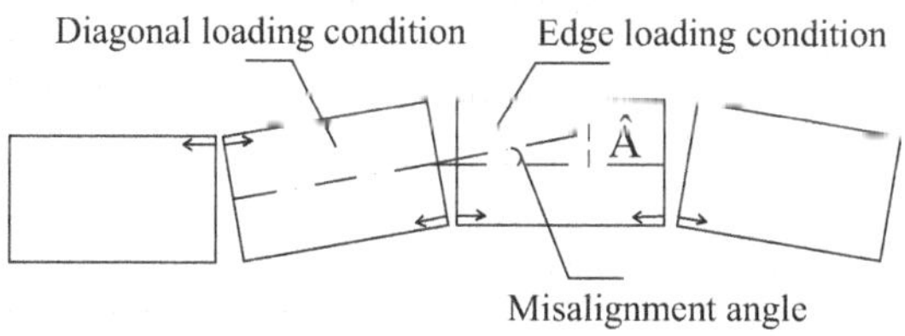

Fig. 2. Two typical loading conditions

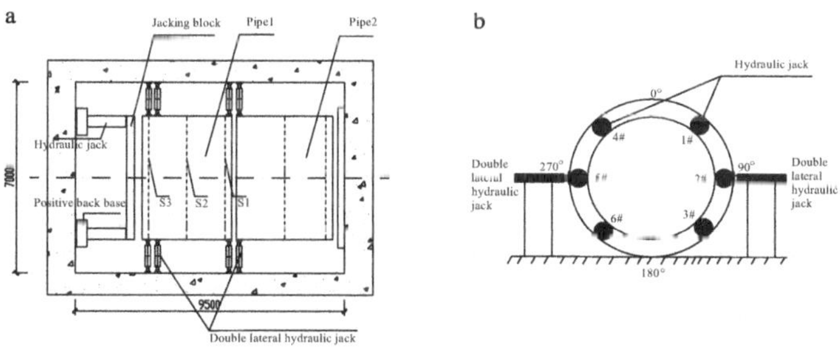

Fig. 3. Experimental sketches (a) experimental plane sketch of pipe arrangement, (b) distribution of hydraulic jack (mm).

Three testing sections are arranged for each JPCCP pipe. The first and third measuring sections are 300 mm away from pipe end. While the second testing section is located at the middle of the pipe (see Fig. 3a). The axial strain measuring points are fixed every 45° in the circular direction on each measuring section, which contains the axial strain of inner concrete core, intermediate concrete core, and exterior concrete coating. Hoop strain gauges are attached on each measuring section every 90° in the circular direction.

In addition, several pressure sensors are fixed every 30° on the contact area between pipe ends. The misalignment angle of pipeline can be obtained through joint's relative opening displacement according to formula (1) [24].

$$\tan\beta = \frac{l_r}{4.32} \tag{1}$$

where l_r is the relative opening displacement of measuring points fixed at pipe ends. According to the value measured by the two displacement sensors fixed at $\theta = 90°$ (l_{90}) and 270° (l_{270}) of the two testing pipe contact end, the relative opening displacement l_r can be calculated by: $l_r = |l_{90} - l_{270}|$.

2.3 Test Series

Table 2 shows the specific test series in this study. As listed in this table, series 1 is to simulate straight jacking without misalignment of pipeline. Series 2 to 5 are to simulate jacking conditions with misalignment angle 0.38° and 0.5° in the practical engineering under edge loading condition as depicted in Fig. 2, respectively. Obviously, the presupposed misalignment angle will change under jacking force. So it is inevitable to adjust misalignment angle many times to fit and approach the presupposed misalignment angle.

Table 2. The contents of each test series in the test

Test series	Jacking load distribution	Total jacking load(kN)	Presupposed misalignment angle	Ultimate misalignment angle β^u
1	1#, 2#, 3#, 4#, 5#, 6#	21000	0°	$\beta_1^u = 0.013°$
2	1#, 2#, 3#	11250	0.38°	$\beta_2^u = 0.44°$
3				$\beta_3^u = 0.266°$
4			0.5°	$\beta_4^u = 0.422°$
5				$\beta_5^u = 0.46°$

Firstly, adjust two pipes to an align position and reset the displacement sensors. And then adjust the misalignment angle of the pipeline to the presupposed value listed in Table 2, according to Eq. (1). According to Standard Methods for Testing Concrete Structures (GB50152-2012) [26] and Technical Specification for Pipe Jacking of Water Supply and Sewerage Engineering (CECS246-2008) [25], the maximum experimental jacking force is 21000 kN, one half the design values. The exterior reinforced concrete coating and intermediate concrete core at the pipe ends are the main loading area. Stepwise loading method is adopted in the test, and it will be hold for 2 min after each loading step until the total jacking force reaches the presupposed value listed in Table 2. And after the last loading step, it will be hold for 15 min. In the process of loading, the state of pipe should be observed carefully to make sure whether it can continue sustaining or not. At the meantime, crack distribution on the pipe's external concrete coating surface is recorded, specifically. All of the aforementioned data are collected and stored automatically by a data acquisition system (Datataker) with a suitable sampling frequency.

3 Test Results and Discussion

The mechanical behavior of JPCCP is very complicated and its accurate response is difficult to calculate and evaluate through a theoretical method due to the interaction among prestressed steel wire, concrete layers, steel cylinder, and steel bar under the jacking force. This study is concentrated on the influences of the misalignment angle on the mechanical properties of pipe. And the experimental results can be divided into five cases: β_1 to β_5 according to ultimate misalignment angle, as listed in Table 3. In the whole process of this test, the state before loading is taken as a reference, so the measured value of each point is an increment.

Table 3. The pressure distribution on the contact area between pipe ends under maximum jacking force (MPa)

β	Circular position θ					
	0°	60°	90°	270°	330°	Open position
$\beta_1 = 0.013°$	12.5	6.8	10.5	12.5	8.3	270°
$\beta_2 = 0.266°$	2.2	6.5	16.1	0	1.1	270°
$\beta_3 = 0.422°$	0.6	8.6	8.3	0	0	270°
$\beta_4 = 0.44°$	0	3.2	16.5	0	0	270°
$\beta_5 = 0.461°$	0	2.8	17.5	0	0	270°

3.1 Characteristics of Strain Distribution

The strain in JPCCP is compressive in the case β_1. While the tensile area emerges and extends with β gradually. Even though the total jacking force loading on JPCCP pipe is largest in the case $\beta_1 = 0.013°$, the axial maximum compressive strain is not the largest, even smaller than some other cases. Meanwhile, larger compressive strains always appear at $\theta = 90°$ where jacking force is applied while tensile strains emerge on another side ($\theta = 270°$). Figure 4 shows the maximum (Fig. 4a) and minimum (Fig. 4b) strain on each measuring section (S1, S2 and S3). In comparison with the strain distribution under axial force without misalignment angle, it can be conclude that the misalignment angle is a critical factor to the initiation of tensile strain. Moreover,

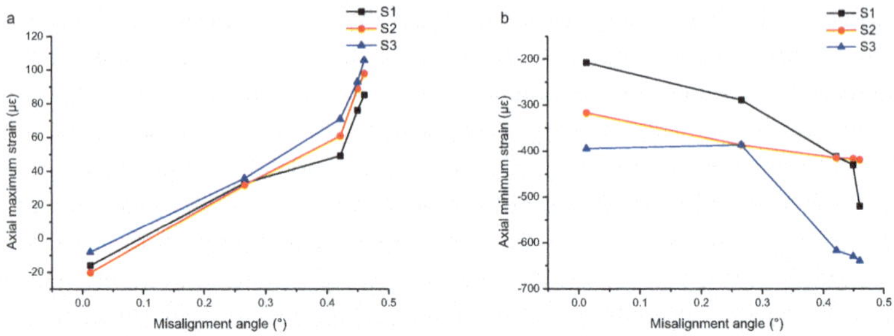

Fig. 4. Axial strain distribution on each measuring section. (a) maximum axial stain, (b) minimum axial strain.

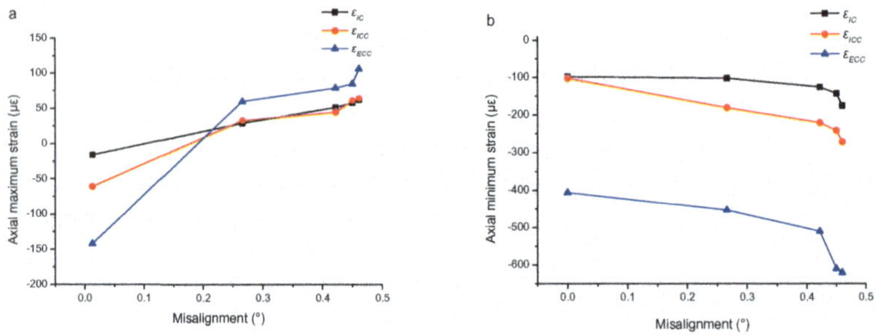

Fig. 5. The variation of maximum (minimum) axial strain upon the misalignment angle. (a) axial maximum strain, (b) axial minimum strain. (ε_{IC}, ε_{ICC} and ε_{ECC} correspond to strain of inner concrete core, intermediate concrete core and exterior reinforced concrete coating, respectively).

from this figure, it's clear that the magnitude of tensile strains on the measuring Sects. 2 and 3 are higher than Sect. 1. Compared to the strain on Sect. 2, the variation of axial minimum strain distribution on the measuring Sects. 1 and 3 is more scattered. The maximum compressive strains on the measuring Sect. 1 are smaller than the measuring Sect. 3 in all cases. This is mainly because the wood packing material has a better ability to redistribute stress than the steel jacking block.

Both the maximum and minimum strains in the inner concrete core, intermediate concrete core and exterior concrete coating of JPCCP pipe with different misalignment angles are given in Fig. 5. As is shown, both tensile and compressive strains increase with the misalignment angle. Since the maximum compressive strain of JPCCP pipe is 636 με (much less than the compressive strain limit 3300 με), and the maximum tensile strain is 106 με (exceeds the tensile strain limit), tensile cracking prior to compressive crush may occur, which indicates that cracking will occur firstly on JPCCP pipe's surface. When the crack extends the thickness of exterior concrete coating completely, a serious erosion of steel wire may be induced. Consequently, the pipe's performance and durability will deteriorate. Seriously, it may lead to a burst failure. Hence, it's conceivable that a more strict limit for JPCCP design than traditional reinforce concrete jacking pipe should be made.

From the discussion above, it can be seen that the occurrence of misalignment angle of the pipeline is a key factor leading to the damage of jacking pipe. In addition, tensile strain appears on the external concrete coating of JPCCP, which plays an essential role in governing its durability. Therefore, several corresponding measures, such as increasing the reinforcement ratio and enlarging the area of the section, should be adopted to enhance the stiffness of JPCCP.

3.2 The Pressure Distribution on Pipe End

Based on several laboratory test results about small scaled model pipes [13], pipe under jacking force tends to come to an aligned position instead of remaining in the presupposed misalignment position. As shown in Table 2, however, the ultimate misalignment angle is larger than the presupposed in case β1 and β2, while an opposite change appears in other cases. This reveals that the misalignment angle of the pipeline may not always come to an aligned position. As the aforementioned above, this phenomenon may be dominated by two factors: (1) inconsistent loading in these six hydraulic jacks; (2) lateral constraint.

Because it is an insurmountable task to measure the pressure distribution continuously in the circular direction, the contact region is quantified through 12 points at pipe ends in this test. The pressure distribution on the interface between two pipe ends with different misalignment angles is summarized in Table 3 (the damaged sensors are not listed out). In the case β_1, the open position is in the region of $\theta = 270°$ where the pressure is 12.5 MPa, even a little larger than that of $\theta = 90°$ (10.5 MPa). So it is credible that the initial geometric roughness exists at two pipe ends. Further, more serious stress concentration will appear, and pose a threat to the safety of JPCCP pipe.

In addition, when the misalignment angle becomes larger, more serious stress concentration can be induced. So protective measures are needed to control the deviation of pipeline in the construction, such as installing automatic correction system.

3.3 Behavior of Prestressed Steel Wire

JPCCP pipe is a novel composite structure which consists of inner concrete core, steel cylinder, intermediate concrete core and exterior concrete coating. Concrete core will expand under axial compressive load, which is caused by Poisson effects, while prestressed steel wire will constrain this kind of expansion. The interaction between concrete core and prestressed steel wire makes the mechanical response of JPCCP pipe more complicated. Figure 6 exhibits the variation of prestress with the load during the test. The positive values mean prestress increment and the negative means prestress reduction.

The prestress increases almost linearly with the jacking force in the case β_1 (see Fig. 6a). However, the decrease of prestress occurs when the misalignment angle becomes larger. From Fig. 6, it is clear that larger prestress increments always appears on the side (θ = 90°) where jacking force loads, while the prestress decrease appears on the opposite side (θ = 270°). Compared with the strain distribution on each measuring section (see Fig. 4), it can be found that prestress increase occur in the compressive stress concentrated region. This indicates the prestressed wire to some extent can

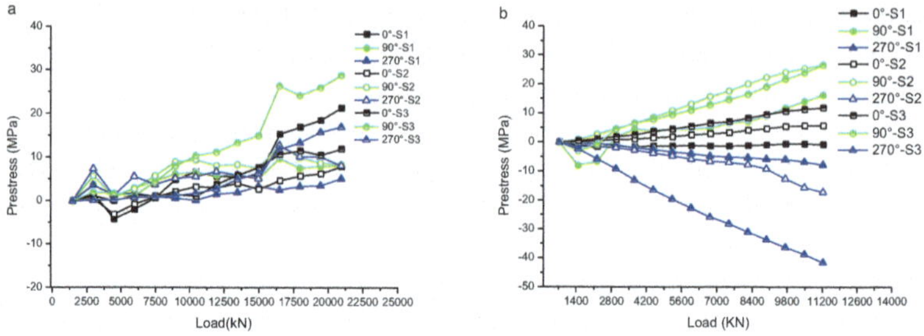

Fig. 6. The prestress increment upon load. (a) β_1, (b) β_5.

constrain the hoop deformation and improve the bearing capacity of the pipe. According to the test results, the prestress decrease almost reaches 40 MPa, and the maximum prestressed increment approaches almost 30 MPa, which illustrates that debonding between the steel wire and concrete may occur though it is intact and undamaged observed from the surface. Based on the analysis by Gong et al., these extreme state may also be caused by the lateral forces provided by lateral jacks [20]. Nevertheless, the damage of prestressed wire should not occur during the construction because the lateral force offered by the surrounding soil rather than concentred force in this test.

4 Conclusions

In this paper, the experiment upon the mechanical behavior of large diameter prototype JPCCP pipe has been conducted and carefully analyzed considering the effects of eccentric jacking force and possible misalignment angle of the pipeline under edge loading condition. Based on experimental test results, the following conclusions can be drawn.

(1) The maximum axial compressive strain in this test is far less than the ultimate strain. Based on the test results, it can be found that the misalignment angle plays an important role in tensile strain. The test results also indicate that tensile strain appears in the reinforced concrete coating rather than other parts of JPCCP pipe, which suggests that the stiffness of pipe section should be improved to bear the effects caused by misalignment angle, and constrain the propagation of possible crack.
(2) Larger misalignment angle can not only cause serious axial strain concentration, but also cause considerable incremental prestress. Meanwhile, it reveals that the prestressed wire can play a role in constraining concrete hoop tensile strain.
(3) The contact area at the pipe ends is an important factor to the strain state. If the size of contact region approaches a point, a serious stress concentration may occur. Hence, it's critical to control the misalignment angle of pipeline and correct timely during the construction. And proper soft packing material with a certain thickness is also anticipated to enlarge the contact area between pipes ends and

reduce the severity of stress concentration caused by the misalignment of the pipeline.

(4) In the practical engineering, the combined interaction of soil and water pressure, and internal water pressure pose a challenge to the pipe structural performance assessment. To acquire more real mechanical behavior of JPCCP, further study on site test is necessary.

References

1. Zarghamee, M.S., Ojdrovic, R.P., Nardini, P.: Prestressed concrete cylinder pipe condition assessment-what works, what doesn't, what's next. In: Proceedings of Pipelines, ASCE, Reston, VA, pp. 182–194 (2011)
2. Zarghamee, M.S., Heger, F.J., Dana, W.R.: Experimental evaluation of design methods for prestressed concrete pipe. J. Transp. Eng. **114**, 635–655 (1988)
3. Zarghamee, M.S., Ojdrovic, R.P., Dana, W.R.: Coating delamination by radial tension in prestressed concrete pipe I: experiments. J. Struct. Eng. **119**, 2701–2719 (1993)
4. Zarghamee, M.S., Fok, K.: Analysis of prestressed concrete pipe under combined loads. J. Struct. Eng. **116**, 2022–2039 (1990)
5. Xiong, H., Li, P., Li, Q.: FE model for simulating wire-wrapping during prestressing of an embedded prestressed concrete cylinder pipe. Simul. Model. Pract. Theory **18**(5), 624–636 (2010)
6. Milligan, G.W.E.: Pipe jacking research. World Tunnelling (1993)
7. Senda, T., Maeda, Y., Shimada, H., Sasaoka, T., Matsui, K.: Studies on surrounding soil during construction using the deep pipe jacking method in the deep strata. Procedia Earth Planet. Sci. **6**, 396–402 (2013)
8. Senda, T., Shimada, H., Sasaoka, T., Matsui, K.: Behavior of surrounding soil during construction and its countermeasures using pipe jacking method in deep strata. Open J. Geol. **03**(02), 44–48 (2013)
9. Milligan, G.W.E., Norris, P.: Site-based research in pipe jacking—objectives, procedures and a case history. Tunn. Undergr. Space Technol. **11**(95), 3–24 (1996)
10. Zhou, S., Wang, Y., Huang, X.: Experimental study on the effect of injecting slurry inside a jacking pipe tunnel in silt stratum. Tunn. Undergr. Space Technol. Inc. Trenchless Technol. Res. **244**, 466–471 (2008)
11. Haslem, R.F.: Structural interaction at joints in pipe-jacked tunnels. Struct. Eng. **74**(10), 165–171 (1996)
12. Li, Z., Chen, J., Zhou, Q.: Larger diameter GRP pressure jacking pipe design and on-site monitoring. In: International Conference on Pipelines and Trenchless Technology, ICPTT 2009, Shanghai, pp. 989–995 (2009)
13. Palmeira, E.M.: The study of soil-reinforcement interaction by means of large scale laboratory tests. Ph.D. thesis, University of Oxford (1987)
14. Khazaei, S., Shimada, H., Kawai, T., Yotsumoto, J., Matsui, K.: Monitoring of over cutting area and lubrication distribution in a large slurry pipe jacking operation. Geotech. Geol. Eng. **243**, 735–755 (2006)
15. Xavier, B.: Lubrication and soil conditioning in pipe jacking and tunnelling, D. Phil. thesis, Cambridge University, (2006)
16. Marshall, M.A.: Pipe-jacked tunnelling: jacking loads and ground movement, D. Phil. thesis, University of Oxford (1999)

17. Norris, P.: The behavior of jacked concrete pipes during site installation. D. Phil. thesis, University of Oxford (1992)
18. Ripley, K.J.: The performance of jacked pipes. D. Phil. thesis, University of Oxford (1989)
19. Shou, K., Yen, J., Liu, M.: On the frictional property of lubricants and its impact on jacking force and soil–pipe interaction of pipe-jacking. Tunn. Undergr. Space Technol. **25**(4), 469–477 (2010)
20. Gong, Q., Zhu, H., Yan, Z., Huang, B., Zhang, Y., Dong, Z.: Fracture and delamination assessment of prestressed composite concrete for use with pipe jacking method. Math. Probl. Eng. **2015**, 1–11 (2015)
21. Zhang, Y., Yan, Z., Zhu, H.: Experimental study on the mechanical behaviors of jacking prestressed concrete cylinder pipe. Tunn. Undergr. Space Technol. **73**, 60–70 (2018)
22. Baumert, M., Allouche, E.: Real-time monitoring for quality delivery of directional drilling installations. J. Infrastruct. Syst. **9**(1), 35–43 (2003)
23. Dai, J., Ueda, T., Sato, Y.: Development of the nonlinear bond stress-slip model of fiber reinforced plastics sheet-concrete interfaces with a simple method. American Society of Civil Engineers (2005)
24. Zhou, J.Q.: Numerical analysis and laboratory test of concrete jacking pipes, D. Phil. thesis, University of Oxford (1998)
25. CESCE 246-2008 Technical Specification for Pipe Jacking of Water Supply and Sewerage Engineering. China Planning Press, Beijing (2008)
26. GB 50010-2010 Code for Design of Concrete Structures. Architecture and Building Press, Beijing (2010)

Evaluation of Liquefaction Potential of Saturated Sands Based on Resistivity Piezocone Penetration Testing

Guojun Cai[1(✉)], Haifeng Zou[1], Yan Yang[1], Songyu Liu[1], and Anand J. Puppala[2]

[1] Southeast University, Nanjing, Jiangsu, China
focuscai@163.com
[2] The University of Texas at Arlington, Arlington, TX, USA

Abstract. The liquefaction of saturated soils is a major concern for the earthquake damage of foundation. Due to the difficulty and cost constraint in obtaining high-quality undisturbed samples, in-situ testing is commonly applied to evaluating the potential of soil liquefaction. In high risk projects, a comprehensive evaluation based on various methods is usually adopted, and thus the research on new method to evaluate the liquefaction is still necessary. In this research the feasibility of using resistivity piezocone penetration test (RCPTU) for predicting liquefaction resistance is investigated. The resistivity of saturated silts and sands is measured using RCPTU at a test section of Suqian-Xinyi expressway. First, the relationship between normalized cone tip resistance and resistivity is analyzed and can contribute to the evaluation of soil liquefaction based on resistivity. Second, the study on combination of resistivity and soil behavior type index to directly calculate the cycle resistance ratio (CRR) is conducted with the CRR from Robertson modified liquefaction evaluation model as reference. The influence of thin cohesive layers and transition zones is also analyzed. It is shown that the resistivity and soil behavior type index can be used for effective evaluation of liquefaction potential of saturated soils.

Keywords: Resistivity · Piezocone penetration test · Liquefaction Soil behavior type index · Cycle resistance ratio

1 Introduction

Liquefaction of saturated cohesionless soils is a major cause of damage to a series of geotechnical issues. Therefore, the liquefaction of saturated soils should be evaluated properly. The advanced piezocone penetration test (CPTU) has been widely used in geotechnical site characterization due to its high accuracy and repeatability [1–3]. It has been recognized that the liquefaction characteristics of sands are influenced not only by the relative density of deposit, but also by factors including the structural orientation of the particles, gradation of sands, and in-situ stress conditions [4, 5]. Resistivity is one of the fundamental parameters of soil, which is highly related to some soil parameters such as porosity, saturation, shape and size, gradation, and grain orientation [6, 7].

L. Hu et al. (Eds.): GSIC 2018, *Proceedings of GeoShanghai 2018 International Conference: Multi-physics Processes in Soil Mechanics and Advances in Geotechnical Testing*, pp. 355–363, 2018.
https://doi.org/10.1007/978-981-13-0095-0_40

Hence it is supposed that the resistivity may contribute to assessing the liquefaction resistance of soil.

Arulmoli *et al.* (1985) suggested that the porosity is a useful intermediate parameter to establish the correlation between liquefaction potential and bulk resistivity of soil [8]. Based on laboratory cyclic triaxial test, it had proposed an electrical resistivity based approach for liquefaction assessment by accounting for anisotropy and grain size. Subsequently, Arulanandan *et al.* (1988) applied numerical simulation to verify the evaluation method of liquefaction proposed by Arulmoli *et al.*, and demonstrated that the electrical resistivity model provides a non-destructive method to predict the liquefaction resistance and avoids the difficulties in obtaining undisturbed samples [9]. However, it is supposed that the resistivity model developed by Arumoli is more suitable for relatively homogeneous soil. In sophisticated site conditions this assumption is perhaps over-simplified as the spatial variability in grain size of soil can be very significant.

Researches revealed that the soil behavior type index (I_c) obtained from CPTU can be directly proposed for soil classification [10–12]. Enhanced resistivity piezocone penetration test (RCPTU) provides continuous profiles of both I_c and resistivity, and hence is especially applicable to complex geological conditions. In this research the feasibility of evaluating liquefaction potential based on RCPTU data is investigated.

2 Experimental Investigation

2.1 Site Description

The testing site is located in Suqian-Xinyi expressway in Jiangsu province. Rivers are widely distributed in this region. The testing section belongs to an alluvial plain of the abandoned Yellow River, and is in an active seismic zone with a design magnitude (M_w) of 8. The soils in the testing section are mainly partially unconsolidated Quaternary sediments. The ground water table (GWT) is around 3.8 m below the ground surface. The main physical and mechanical indexes of the test site soil are shown in Table 1. In practice the soils ranging from 0 to 15 m in depth are investigated for evaluating the possibility of soil liquefaction.

Table 1. Main physic-mechanical index of site soil

Layer	Soil type	Thickness (m)	Specific gravity (G_s)	Clay content (%)	Moisture content (%)	SPT-$N_{63.5}$ value
1	Plain fill	1.5	2.71	14.9	22.4	9
2	Silt	3.8	2.70	3.7	24.1	4
3	Silty sand	9.4	2.69	5.4	33.6	10
4	Silt	Not exposed	2.68	5.6	23.6	24

2.2 Testing Equipment

The multi-functional, digital, vehicle-mounted RCPTU system was produced by the Vertek–Hogentogler company in USA and introduced by the Geotechnical Engineering Research Institute of Southeast University. The structure chart of the probe is shown in Fig. 1.

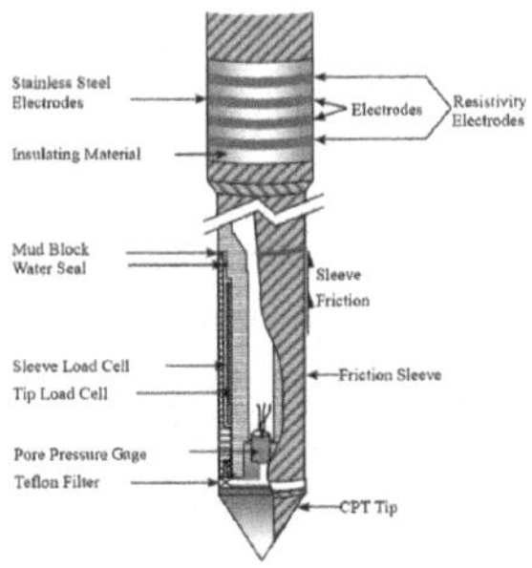

Fig. 1. Schematic of Hogentogler resistivity CPTU probe

The resistivity testing part of the RCPTU, the hardcore of this equipment, is mainly composed of four copper electrodes and an internal circuit system, using insulating plastic between the electrodes to form an O-type ring seal system. A voltage difference is applied to the two outer electrodes and the resistivity of the soil is determined with the aid of the two inner electrodes. The two adjacent copper electrodes are spaced 5 cm apart and the operating frequency is 1 kHz.

2.3 Testing Results

This paper presented the database of nine RCPTU trial holes in the construction site. As shown in Fig. 2, the cone tip resistance is generally small in silt and silty sand. Aggregation and dissipation of excess pore water pressure are not observed in the CPTU profiles, indicating that the soils are in a relatively loose state. The resistivity of soil is large above ground water table, but it decreases remarkably when the probe

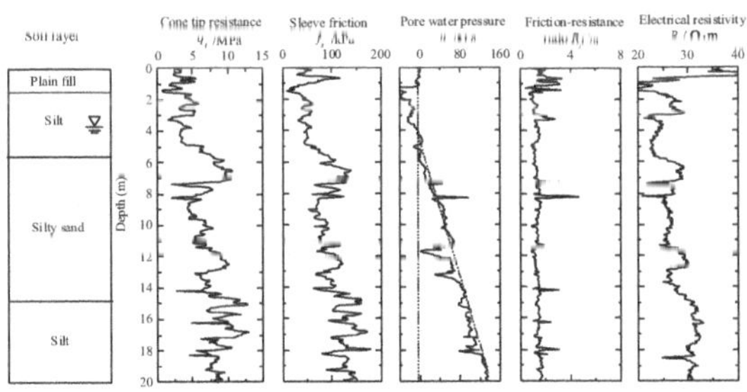

Fig. 2. Typical results of resistivity piezocone tests

penetrates under the ground water table. For saturated soil, the resistivity value of soil increases with cone tip resistance. However, it is expected that the porosity and void ratio of saturated cohesionless soil both decrease as resistivity increases according to the Archie (1942) equation [13]. As a result, the cone tip resistance should decrease with the porosity and void ratio. Since the silty sand layer is completely below the groundwater table, the liquefaction potential of this unit is the major concern in this study.

3 Resistivity Index of Liquefaction Assessment

According to the liquefaction mechanism, both cyclic and static liquefaction can occur in the saturated, loose silty sand or clean-sand. In this paper the unified approach proposed by Robertson (2009) for predicting liquefaction potential of both cohesionless and cohesive soils is referred as Robertson (2009) modified method (RMM) [14]. Since laboratory tests are limited, the CRR values estimated from RMM are taken as reference values. Robertson mentioned that this approximation is not reliable in high risk projects and should be used in low to moderately risk projects. Besides, due to the limited database in geotechnical practice, the reliable assessment of CRR in cohesive soil is seldom available in liquefaction analysis. Hence, this research only investigates the liquefaction resistance of cohesionless soil (silty sand). Based on cone tip resistance and soil behavior type index, the CPTU profiles of silty sand were subdivided into different segments. Besides, data corresponding to $I_c > 2.5$ were removed to ensure the accuracy of analysis.

3.1 Correlation Between ρ and Normalized Cone Tip Resistance Q_{tn}

Noted that in all the CPTU-based simplified methods for liquefaction evaluation, the CRR is correlated to Q_{tn}. The accuracies of those simplified methods mainly depend on the boundary's uncertainty of their CRR-Q_{tn} relationship. Thus, in the following section, the correlation between CRR and Q_{tn} is analyzed first. Then feasibility of applying the electrical resistivity (ρ) rather than Q_{tn} in evaluating the CRR of saturated cohesionless soil is discussed.

The relationship of normalized electrical resistivity ρ/ρ_w, normalized cone tip resistance Q_{tn} and soil behavior type index I_c is shown as Fig. 4. It is quite evident that Q_{tn} is well correlated to the derived parameters of both ρ and I_c. As cone tip resistance and resistivity is affected by soil types simultaneously, this correlation also indicated that I_c is one of the useful indices to connect mechanical and electrical characteristics of saturated cohesionless soils. Moreover, as shown in Fig. 3, the resistivity (ρ) tends to increase with Q_{tn} for silty sand. This is consistent with the research results conducted by Weemees [15]. However, Q_{tn} tends to decrease for normally consolidated soil when I_c increases. This phenomenon is also conformity with the application of I_c in soil classification, which illustrates that the Q_{tn} of cohesionless soil generally decreases with the fines content.

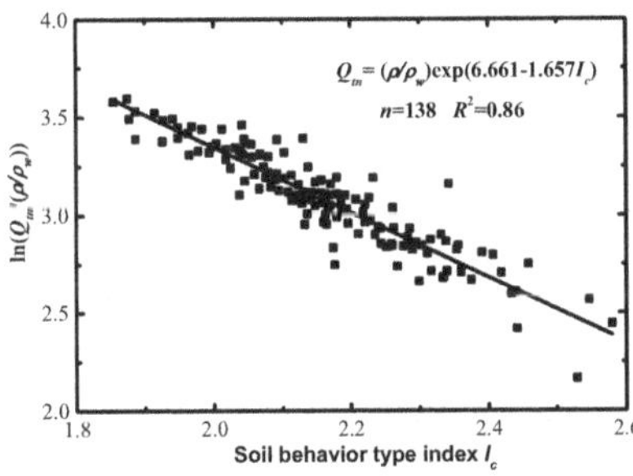

Fig. 3. Relationship among Q_{tn}, ρ and I_c

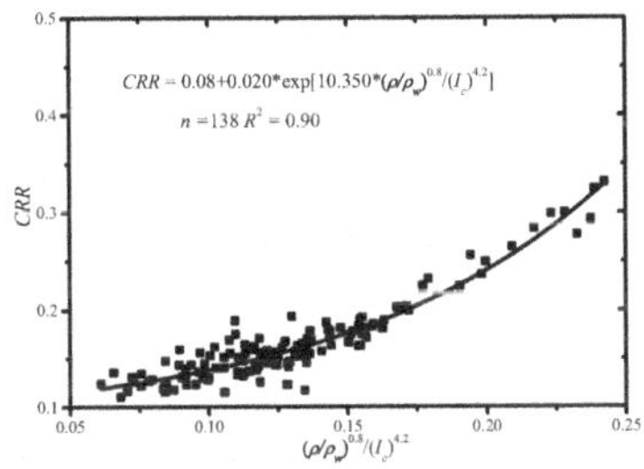

Fig. 4. Relationship among CRR, ρ and I_c

According to Fig. 4, the following equation for correlation among ρ/ρ_w, Q_{tn} and I_c is determined:

$$Q_{tn} = \begin{cases} (\rho/\rho_w)\exp(6.661 - 1.657I_c) \\ R^2 = 0.86 \end{cases} \quad (1)$$

Where ρ = soil electrical resistivity; ρ_w = resistivity of pore water. The coefficient of determination (R^2) is 0.86. Equation (1) is valid for silt and silty sand with I_c ranging from 1.8 to 2.6. When I_c is beyond this range, the above function still needs further validation.

As most literatures revealed that the CRR is mainly reflected by Q_{tn}, it is reasonable for the application of ρ rather than Q_{tn} in developing new correlation for evaluation of liquefaction resistance. However, the uncertainties associated with Eq. (1) might be transferred and magnified. To reduce the potential model errors, the correlation between CRR and ρ will be investigated directly using the experimental database.

3.2 Relation Between ρ and CRR

To develop a direct correlation between CRR and resistivity, database of the 138 segments of silty sand is investigated. Previous analysis demonstrated that the CRR tend to increase with the resistivity of soil, but decreases with I_c. Hence, a correlation between CRR and the derived variable, $(\rho/\rho_w)^n/(I_c)^m$, is analyzed. Using the multivariate regression analysis the coefficients (n and m) were determined as $n \approx 0.8$ and $m = 4.2$. The fitted curve along with the scatter plot for the correlation between CRR and $(\rho/\rho_w)^n/(I_c)^m$ is shown in Fig. 4. The following expression can be obtained according to Fig. 4:

$$CRR = 0.08 + 0.020\exp[10.350 \times (\rho/\rho_w)^{0.8}/(I_c)^{4.2}] \quad (2)$$

It is supposed that Eq. (2) is mainly suitable for normally consolidated cohesionless soils with I_c varying from 1.8 to 2.5. Application of above function in other soils should be validated using more testing data.

3.3 Thin Clay Layer Correction

The liquefaction mechanism of cohesionless soil is different from that of cohesive soil, indicating that the calculation procedure for estimating CRR should also be adjusted [16, 17]. Figure 5 compares CRR values calculated from the proposed resistivity-based method with those obtained from RMM. The database for this comparison consists of 138 sets of silty sand segments and other 230 clayey seams in the nine CPTU soundings. For silty sand ($1.8 < I_c < 2.5$), as shown in Fig. 5, the values of CRR calculated using the proposed resistivity-based method in this research are consistent with those obtained from RMM. The CRR values are located in a low level of CRR, ranging from 0.1 to 0.35. In this range the RMM has been demonstrated as one of the most effective tool for estimating CRR values. Therefore, it is expected that the resistivity-based method developed in this research is also highly reliable.

It is expected that the CRR values of cohesive soils are relatively higher than those of cohesionless soils. Due to high cohesion of clayey soils, the strength of cohesive soils would not reduce to zero even if the excess pore water pressure reaches the effective overburden stress during seismic activities. Hence the liquefaction mechanism of cohesive soils is generally identified as the static liquefaction rather than cyclic liquefaction.

The reliability of CRR estimated by the proposed resistivity method in this research can also be affected by the effect of transition zones. The influence zone ahead of and behind a cone during penetration will cause an impact on both the cone resistance and electrical resistivity of surrounding soils near the interface between two different soil types. The small resistivity of the upper clay may lead to underestimate the resistivity

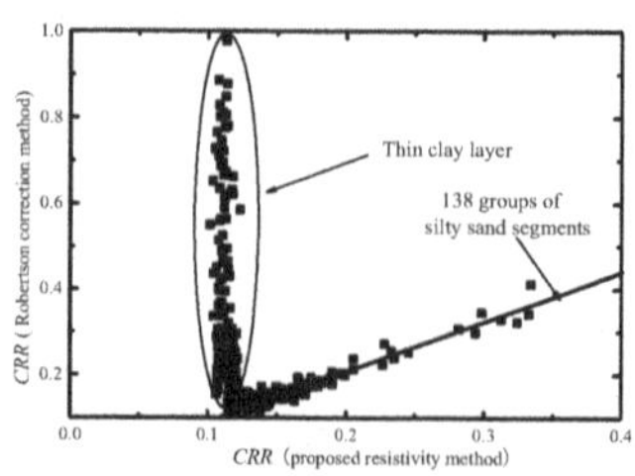

Fig. 5. Comparison of CRR from proposed method and RMM

of cohesionless soil near the interface when the cone penetrates from a clay unit to a sand unit. As a result, the predicted CRR values would also be smaller. This paper proposed that the transitional zone should be limited to 15 cm which is the maximum spacing of the electrodes.

3.4 Validation of the Proposed Method

In this section the CRR values estimated from the proposed resistivity-based model are compared to those obtained using the standard penetration test (SPT) data. The SPT-based model for estimating CRR of cohesionless soil was developed by Seed et al. (1971) [18]. Comparison between the estimates of CRR based on the proposed resistivity-based method and those obtained from RMM is also conducted. The factor of safety (FS) is used to provide more intuitive inspection on the evaluation of liquefaction potential of the cohesionless deposits.

The estimates of CRR for two adjacent RCPTU test bores and SPT are shown in Fig. 6, which indicates that the CRR values estimated from the resistivity-based method in this research are consistent with those obtained from RMM. Besides, similar evaluation of liquefaction potential is achieved and confirmed for the SPT-based approach and the resistivity-based method developed in this study. The main difference between the estimates of SPT-based method and those of resistivity-based method occurred when $I_c > 2.5$, as shown in Fig. 6, e.g., ranging from 6.3 to 6.5 m in depth. Based on RMM the liquefaction mechanism of soils when $I_c > 2.5$ is identified as static liquefaction. However, in this research the data of $I_c > 2.5$ are insufficient for developing a correlation between the CRR and electrical resistivity, which still requires further research.

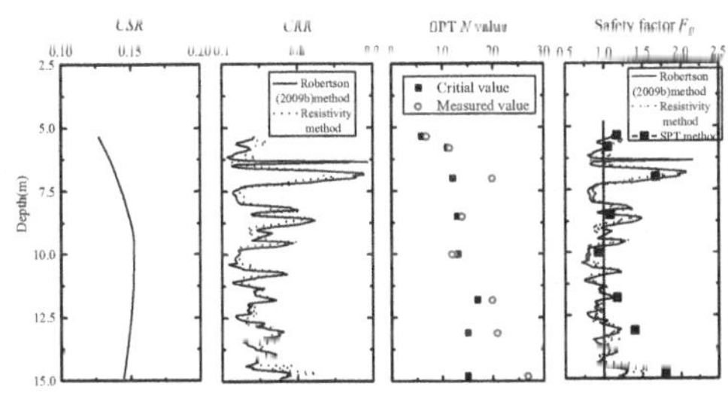

Fig. 6. Representative results of model validation

4 Conclusions

(1) For normally consolidated soil ($1.8 < I_c < 2.5$), correlation between normalized cone tip resistance (Q_{tn}) and electrical resistivity (ρ) is not specific. It is observed that Q_{tn} is a function of both ρ and I_c. Hence the soil behavior type index and resistivity can be used as effective parameters for evaluating the liquefaction potential of saturated cohesionless soil.

(2) For normally consolidated silty sand, correlation for predicting cyclic resistance ratio (CRR) using ρ and I_c is developed in this research. The estimates of CRR obtained from the proposed resistivity-based method are consistent with those from RMM. However, the resistivity-based method provides a novel and more convenient way for estimating the CRR.
(3) Due to the different liquefaction mechanisms between cohesive and cohesionless soils, the CRR values of cohesive soils would be underestimated when adopting the resistivity-based method to estimate liquefaction resistance.

Acknowledgements. Majority of the work presented in this paper was funded by the National Key Research and Development Program of China (Grant No. 2016YFC0800201) and the National Natural Science Foundation of China (Grant No. 41672294).

References

1. Lunne, T., Robertson, P.K., Powell, J.J.M.: Cone Penetration Testing in Geotechnical Practice. Chapman & Hall, London (1997)
2. Cai, G.J., Liu, S.Y., Anand, J.P.: Liquefaction assessments using seismic piezocone penetration (SCPTU) test investigations in Tangshan region in China. Soil Dyn. Earthq. Eng. **41**, 141–150 (2012)
3. Cai, G.J., Liu, S.Y., Tong, L.Y., et al.: Evaluation of liquefaction of sandy soils based on cone Penetration test. Chin. J. Rock Mech. Eng. **27**(5), 1019–1027 (2008)
4. Seed, H.B., Peacock, W.H.: Test procedures for measuring soil liquefaction characteristics. J. Soil Mech. Found. Div. **97**(SM8), 1099–1119 (1971)
5. Sharp, M.K., Dobry, R., Philips, R.: CPT-based evaluation of liquefaction and lateral spreading in centrifuge. J. Geotech. Geoenviron. Eng. **136**(10), 1334–1346 (2010)
6. Rinaldi, V.A., Cuestas, G.A.: Ohmic conductivity of compacted silty clay. J. Geotech. Geoenviron. Eng. **128**(10), 824–835 (2002)
7. Cai, G.J., Chu, Y., Liu, S.Y., et al.: Evaluation of subsurface spatial variability in site characterization based on RCPTU data. Bull. Eng. Geol. Environ. **75**, 401–412 (2015)
8. Arulmoli, K., Arulanandan, K., Seed, H.B.: New method for evaluating liquefaction potential. J. Geotech. Eng. **111**(1), 95–114 (1985)
9. Arulanandan, K., Muraleetharan, K.K.: Level ground soil-liquefaction analysis using in situ properties. J. Geotech. Eng. **114**(7), 753–770 (1988)
10. Jefferies, M.G., Davies, M.P.: Use of CPTU to estimate equivalent SPT N60. Geotech. Test. J. **16**(4), 458–468 (1993)
11. Robertson, P.K.: Interpretation of cone penetration tests - a unified approach. Can. Geotech. J. **46**(11), 1337–1355 (2009)
12. Li, D.K., Juang, C.H., Andrus, R.D., et al.: Index properties-based criteria for liquefaction susceptibility of clayey soils: a critical assessment. J. Geotech. Geoenviron. Eng. **133**(1), 110–115 (2007)
13. Archie, G.: The electrical resistivity log as an aid in determining some reservoir characteristics. Trans. Am. Inst. Min. Metall. Pet. Eng. **146**, 54–61 (1942)
14. Robertson, P.K.: Performance based earthquake design using the CPT. Keynote Lecture, IS-Tokyo (2009)
15. Weemees, I.A.: Development of an electrical resistivity cone for groundwater contamination studies. The University of British Columbia, Vancouver (1990)

16. Moss, R.E.S., Seed, R.B., Kayen, R.E., et al.: CPT-based probabilistic and deterministic assessment of in situ seismic soil liquefaction potential. J. Geotech. Geoenviron. Eng. **132** (8), 1032–1051 (2006)
17. Juang, C.H., Fang, S.Y., Khor, E.H.: First-order reliability method for probabilistic liquefaction triggering analysis using CPT. J. Geotech. Geoenviron. Eng. **132**(3), 337–350 (2006)
18. Seed, H.B., Idriss, I.M.: Simplified procedure for evaluating soil liquefaction potential. J. Soil Mech. Found. Div. **97**(9), 1249–1273 (1971)

Multivariate Correlations Among SCPTU Parameters of Jiangsu Cohesionless Soils

Haifeng Zou[1], Songyu Liu[1], Guojun Cai[1(✉)], and Anand J. Puppala[2]

[1] Southeast University, Nanjing, China
focuscai@163.com
[2] Department of Civil Engineering,
University of Texas at Arlington, Arlington, TX, USA

Abstract. Estimation of V_s from other testing indices has gained increasing interest over the decades. The enhanced seismic piezocone penetration testing (SCPTU) provides a convenient method to measure V_s simultaneously with other three indices including the cone tip resistance, sleeve frictional resistance and pore water pressure in a close proximity. In this study, a database containing 612 sets of data points is compiled for the stress-normalized SCPTU parameters of the Jiangsu cohesionless soils. An approach based on the multivariate distribution method is utilized to develop the multivariate correlations among the normalized shear wave velocity (V_{s1}) and other three indices. It is shown that the correlations are reliable to describe the dependence of the V_{s1} on other three indices for the Jiangsu cohesionless soils. However, caution shall be taken when these correlations are applied in other soils.

Keywords: Multivariate correlation · Shear wave velocity
Piezocone penetration test · Cohesionless soil

1 Introduction

The shear wave velocity, V_s, is a crucial parameter of cohesionless soils in a variety of geotechnical issues such as the seismic hazard analyses. Several geophysical methods provide reliable V_s measurements for different soils [1, 2]. It is of interest to evaluate V_s from other testing indices when the direct V_s measurements are not available, and these correlations have been widely studied in the literature [1, 2]. However, it shall be cautious to establish a correlation between V_s and other indices because V_s highly depends on the soil density, void ratio, effective stress, soil strength, stress history and soil type [1, 2]. These impacting factors contribute to the difficulty of developing a reliable V_s correlation for the cohesionless deposits as obtaining the high quality samples can be a challenging and expensive task.

The seismic piezocone penetration (SCPTU) provides a convenient way to simultaneously measure the V_s, cone tip resistance (q_t), sleeve frictional resistance (f_s), and pore water pressure (u_2) in a close proximity. Correlations among these four parameters have been systematically studied for cohesive soils over the decades [1, 2]. However, it seems that the correlations for the cohesionless soils have achieved insufficient analyses in the literature.

L. Hu et al. (Eds.): GSIC 2018, *Proceedings of GeoShanghai 2018 International Conference: Multi-physics Processes in Soil Mechanics and Advances in Geotechnical Testing*, pp. 364–372, 2018.
https://doi.org/10.1007/978-981-13-0095-0_41

In this study, a database containing four stress-normalized SCPTU parameters is compiled and analyzed for the Jiangsu cohesionless soils using a multivariate distribution model. This multivariate method has been shown to be a sounding approach to capture the multi-dependencies among different soil properties [3, 4]. The correlations for the normalized shear wave velocity conditional on the other three SCPTU parameters are developed and validated. The feasibility of the multivariate distribution model in describing the correlations of the SPTU parameters is therefore demonstrated.

2 Jiangsu SCPTU Database for Cohesionless Soils

In this section, the site conditions and seismic piezocone penetration tests for the Jiangsu cohesionless soils are introduced. A SCPTU database containing the four testing indices is presented and analyzed.

2.1 Site Conditions and SCPTU Tests

A comprehensive geotechnical investigation has been conducted on the Quaternary soils of 32 sites in the Jiangsu province, China. The geologic formations and conditions of most testing sites are available in the literature [4]. A total of 225 SCPTU soundings were performed over these sites. Only the cohesionless soils are investigated in this study. The SCPTU probe is in accordance with the international standard [2]. Figure 1 presents a representative SCPTU profile for the Jiangsu cohesionless soils.

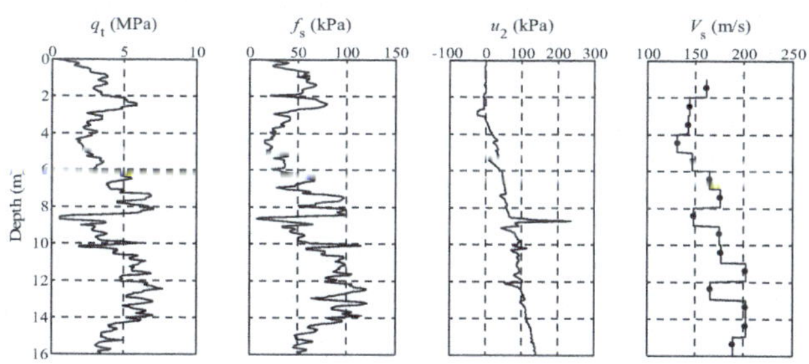

Fig. 1. Representative SCPTU profile for the Jiangsu cohesionless soils

It is evident in Fig. 1 that the four SCPTU measurements tend to increase with the increase of depth in a homogeneous soil unit. This is probably due to the impact of the in-situ stress on the soil strength and stiffness. It is therefore more rational to establish the correlations for the stress-normalized parameters, instead of the direct measurements. The following four normalized parameters are determined as:

$$Y_1 = q_{t1N} = q_t \cdot C_N / p_a \quad where \quad C_N = \left(p_a/\sigma'_{v0}\right)^{\alpha} \leq 1.70 \quad \alpha = 1.338 - 0.249(q_{t1N})^{0.264}$$

$$Y_2 = I_c = \sqrt{(3.47 - \lg q_{t1N})^2 + (1.22 + \lg F_r)^2} \quad where \quad F_r = f_s/\left(q_t - \sigma'_{v0}\right) \times 100\%$$

$$Y_3 = 1 + B_q = 1 + (u_2 - u_0)/(q_t - \sigma_{v0})$$

$$Y_4 = V_{s1} = V_s\left(p_a/\sigma'_{v0}\right)^{0.25}$$

where q_{t1N} is the normalized cone tip resistance, I_c is the soil behavior type index, F_r is the normalized friction ratio in %, B_q is the pore pressure parameter ratio, V_{s1} is the normalized shear wave velocity in m/s, C_N is the overburden correction factor, σ_{v0} is the vertical total stress in kPa, σ'_{v0} is the vertical effective stress in kPa, u_0 is the hydrostatic pressure in kPa, and p_a is the reference pressure of 100 kPa.

The definitions of q_{t1N} and I_c follow those given by Juang et al. (2006) [5]. Previous studies have demonstrated that the q_{t1N} is strongly affected by the soil strength, density and compressibility and thus can be correlated to V_{s1}. The I_c, instead of the direct f_s measurement, is used because the former is considered an effective representation of the fines content within the soils [6] and thus may be better correlated to V_{s1}. The B_q is deemed a potentially useful description of the soil density and void ratio which impact V_{s1}. Since u_2 can be less than u_0 in the cohesionless soils and thus B_q can be negative, the term $(1 + B_q)$ is used for Y_3 to ensure that it is consistently positive in the analysis. This positivity is an essential requirement of the subsequent data transformation used to develop the correlations. Besides, the usage of $(1 + B_q)$ is also more consistent with the previous studies [1, 2].

2.2 Data Processing

The processing of $\{Y_1, Y_2, Y_3, Y_4\}$ data is crucial to the developing correlations for V_{s1}. It is fundamental that these data points are representative for the cohesionless soils. To achieve this goal, two main procedures are applied in this study. Firstly, the V_{s1} evaluates the averaged seismic behavior every one meter. Therefore, the q_{t1N}, I_c, and B_q profiles are also averaged every one meter to ensure the consistency of the compiled database. These averaged indices are then used in the subsequent analysis.

Secondly, it shall be important to exclude the data points corresponding to the soil mixtures and hence the compiled dataset can represent the behavior of typical cohesionless soils. To address this issue, the criteria of identifying physically homogeneous soil units from a soil profile suggested by Uzielli et al. (2005) [7] are applied. In these criteria, the coefficients of variation (COVs) of $\lg q_{t1N}$ and $\lg I_c$ of a homogeneous soil unit shall be less than 0.10. Therefore, the data points with COVs of $\lg q_{t1N}$ and $\lg I_c$ larger than 0.10 within the one-meter averaging interval are removed.

Using the above screening processes, six hundred and twelve (612) sets of $\{Y_1, Y_2, Y_3, Y_4\}$ data points are compiled and portioned into a calibration dataset for developing the multivariate distribution model and a validation dataset for demonstrating the performance of the model. The validation dataset is formed by extracting two or more SCPTU soundings from each site. The sample sizes of the calibration and validation datasets are 480 and 132, respectively.

2.3 Soil Behavior Types of SCPTU Parameters

The two SCPTU datasets are plotted in the Robertson soil classification chart [6] to illustrate the soil types, as shown in Fig. 2. Most data points are identified as sandy silt to sands with $B_q < 0.1$, indicating the drained to partially drained conditions during the cone penetration. The I_c varies from 1.7 to 2.6, corresponding to the common range of the coarse-grained soils [6]. Therefore, these data are deemed representative for the

Jiangsu cohesionless soils. Besides, among the four parameters, $Y_1 = q_{t1N}$ shows the highest variability with COV = 0.47, perhaps due to the diversity and complexity of soil strength, density and compressibility. $Y_3 = 1 + B_q$ is associated with the lowest variability with COV = 0.009. This is reasonable since most u_2 measurements in the loose cohesionless soils are concentrated around the hydrostatic pressures and therefore most B_q values are within the narrow range of −0.04 to 0.03.

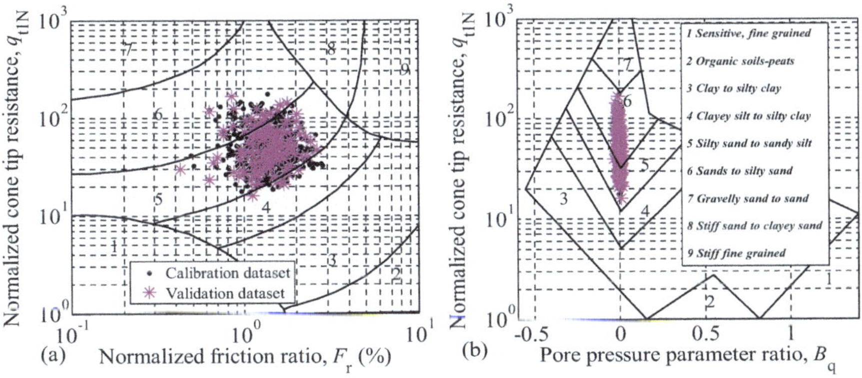

Fig. 2. Soil behavior types identified using the Robertson soil behavior type chart.

3 Multivariate Correlation Analyses

The construction of a multivariate distribution model includes the data transformation, correlation analysis, Bayesian updating, and back transformation [4]. The data transformation converts a non-Gaussian Y parameter to a normal X variable, the correlation analysis estimates the linear (Pearson) correlation coefficients between any pair of the transformed variables, the Bayesian updating derives the predicted statistics such as percentiles of a transformed variable conditional on other given variables, and the back transformation provides the predicted statistics of the physical soil parameters in the raw measurement scale [4]. These procedures are implemented in this section.

3.1 Transformation of Soil Parameters

In this study, the Box-Cox method is used to achieve the data transformation. Three transformation parameters including a transformation power (λ) and two standardization parameters (a and b) are used in the Box-Cox transformation, and more details are discussed in the literature [4]. Figure 3 shows the histograms and fitted normal distribution functions (smooth curves) of the transformed X variables. The estimated transformation parameters for the four variables are also presented in Fig. 3.

A visual inspection on Fig. 3 indicates that the fitted normal curves match the histograms well, implying that the each X variable may individually follow a normal distribution. A more rigorous check of the normality is conducted using the Kolmogorov-Smirnov (KS) test. The p values of the KS test are 0.65, 0.73, 0.90, and

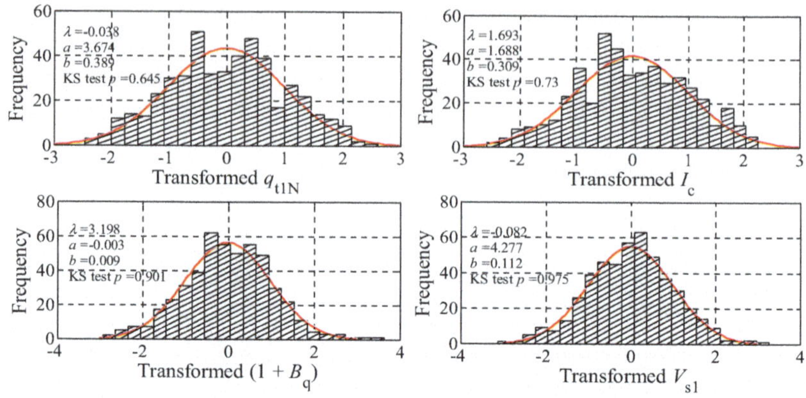

Fig. 3. Transformation parameters and histograms of the transformed variables.

0.98 for the transformed q_{t1N}, I_c, B_q, and V_{s1}, respectively. All these four p values are larger than 0.05, indicating that each X variable can be viewed to follow a normal distribution reasonably.

It shall be realized that even if each soil variable individually follows a normal distribution, the four transformed variables do not have to jointly follow a multivariate normal distribution as the joint distribution cannot be accessed by the marginal distributions. Nevertheless, past studies revealed that meaningful results can still be derived in the case that the transformed variables do not strictly follow a multivariate normal distribution [3]. Therefore, a rigorous test on the multivariate normality of the four transformed variables is not conducted here. It is simply assumed that after transformation these four X variables follow a multivariate normal distribution and thus multivariate correlations can be derived analytically. Instead of verifying this hypothesis, this study is focused on developing the multivariate correlations for V_{s1} since the latter is of main interest for geotechnical engineers.

3.2 Correlations Among Transformed Variables

The next step is to evaluate the linear correlation between any pair of the X variables. The estimated correlation matrix for the calibration dataset of transformed $\{q_{t1N}, I_c, 1 + B_q, V_{s1}\}$ is given in Table 1. The positive correlation between q_{t1N} and V_{s1} is in accordance with the studies conducted in the literature [1, 2]. The I_c is found to be negatively correlated to q_{t1N} and V_{s1}. This can be explained by the fact that the increase of the fines content may decrease the soil strength and stiffness. The correlation between $(1 + B_q)$ and V_{s1} is also found to be negative, perhaps due to the fact that a negative observation of B_q indicates a high density and low void ratio of the cohesionless soils and thus V_{s1} shall be high. Besides, the correlation between X_1 and X_2 is highly strong. This is not surprising since q_{t1N} is directly impacted by the soil type (I_c).

Based on the above discussion, the correlation matrix shown in Table 1 is deemed reasonable. Therefore, it will be used in the subsequent Bayesian updating and back transformation to develop the correlations for V_{s1}. It shall be noted that the correlation

Table 1. Estimated Pearson correlation coefficients among X variables

Variable	X_1	X_2	X_3	X_4
X_1	1.000	−0.917	−0.283	0.554
X_2	−0.917	1.000	0.286	−0.273
X_3	−0.283	0.286	1.000	−0.465
X_4	0.554	−0.273	−0.465	1.000

coefficients among V_{s1} and $\{q_{t1N}, I_c, 1 + B_q\}$ are moderate, implying that bivariate correlation for V_{s1} conditional on any one of $\{q_{t1N}, I_c, 1 + B_q\}$ is not advisable. Multivariate correlations that involve multiple indices to reduce the uncertainties within soil parameters shall be better to improve the accuracy of the predictions.

3.3 Correlations for V_{s1} Conditional on Other Parameters

For a set of X variables following a multivariate normal distribution, the mean value and variance of one variable conditional on other variables can be derived using a Bayesian updating technique [3]. Based on these two statistics, the percentiles and moments of the corresponding Y variable can be obtained using a back transformation method [4]. More details on the Bayesian updating and back transformation procedure can be found in the literature [4]. In this section, only the final mathematical expressions for the predicted median of V_{s1} are presented.

The bivariate $V_{s1} - q_{t1N}$, $V_{s1} - I_c$, and $V_{s1} - B_q$ correlations derived from the developed multivariate distribution model are given in terms of median as follows:

$$V_{s1} = \left(0.344Q_t^{-0.038} + 0.354\right)^{-12.240} \tag{1a}$$

$$V_{s1} = \left(0.0048I_c^{1.693} + 0.632\right)^{-12.240} \tag{1b}$$

$$V_{s1} = \left[0.155\left(1 + B_q\right)^{3.198} + 0.498\right]^{-12.240} \tag{1c}$$

Figure 4 presents the median values (smooth curves) of V_{s1} predicted using Eqs. (1a), (1b) and (1c). The calibration dataset, validation dataset and the predicted 95% confidence intervals that describe the 95% ranges in which the actual measurements are located are also shown as the dash curves in Fig. 4. It is evident that the predicted median values of V_{s1} agree with the calibration and validation data well, and the predicted 95% CIs are also in accordance with the scatter of these two datasets. Therefore, the constructed multivariate distribution model is deemed reasonable to describe the V_{s1}-related bivariate correlations for the Jiangsu cohesionless soils.

It is also shown in Fig. 4 that the predictions are notably scattered around the measurements. Thus, the performance of the above bivariate correlations may not be satisfactory. Enhanced accuracy of the predicted V_{s1} can be achieved by developing the multivariate correlations. The correlations for the median of V_{s1} conditional on $\{q_{t1N}, I_c\}$, $\{q_{t1N}, B_q\}$, and $\{q_{t1N}, I_c, B_q\}$ are obtained as follows:

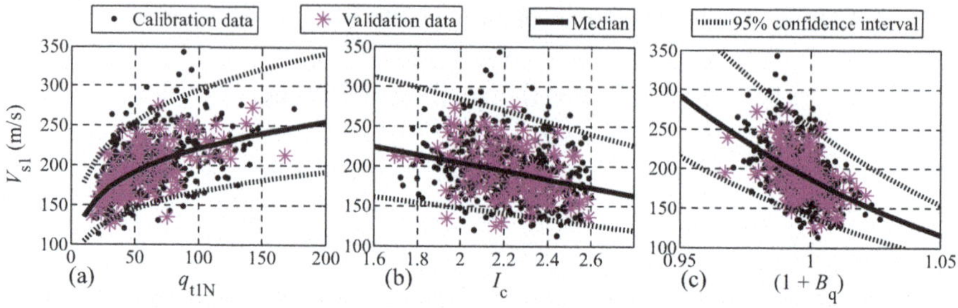

Fig. 4. Predicted V_{s1} conditional on: (a) q_{t1N}; (b) I_c; (c) $1 + B_q$

$$V_{s1} = \left(1.184 q_{t1N}^{-0.038} - 0.026 I_c^{1.693} - 0.269\right)^{-12.240} \tag{2a}$$

$$V_{s1} = \left[0.285 q_{t1N}^{-0.038} + 0.111 (1 + B_q)^{3.198} + 0.295\right]^{-12.240} \tag{2b}$$

$$V_{s1} = \left[1.153 q_{t1N}^{-0.038} - 0.027 I_c^{1.693} + 0.126 (1 + B_q)^{3.198} - 0.363\right]^{-12.240} \tag{2c}$$

The medians of V_{s1} predicted using Eqs. (1a), (1b) and (1c) are plotted against the actual measurements for the calibration and validation datasets in Fig. 5. The coefficients of determination (R^2) of these three equations are also presented in Fig. 5. The R^2 values of Eqs. (2a), (2b) and (2c) are estimated to be 0.65, 0.40, and 0.78 for the calibration dataset, and 0.66, 0.49, and 0.79 for the validation dataset, respectively. The advantage of the multivariate distribution model is evident here, as the accuracy of the predicted V_{s1} can be improved by including more texting indices in the correlations. It is also noted in Fig. 5 that the V_{s1} may be underestimated by Eqs. (2a), (2b) and (2c) when $V_{s1} > 300$ m/s. This bias is most significant in the $V_{s1} - \{q_{t1N}, B_q\}$ correlation. This is perhaps because the sample size of the high V_{s1} data is too limited. More data are still necessary to evaluate the performance of these correlations for the stiff cohesionless soils.

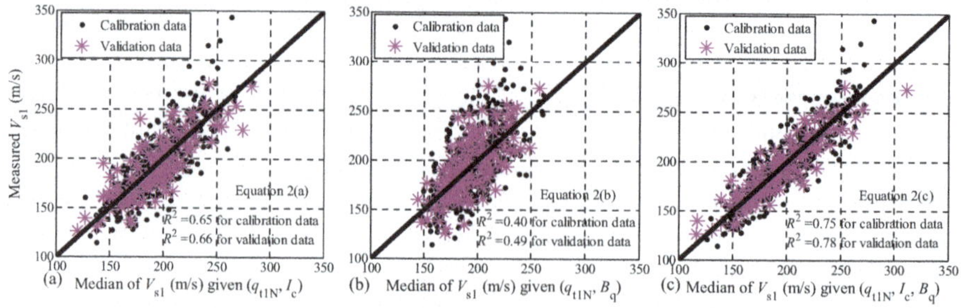

Fig. 5. Predicted median of V_{s1} conditional on: (a) $\{q_{t1N}, I_c\}$; (b) $\{q_{t1N}, B_q\}$; (c) $\{q_{t1N}, I_c, B_q\}$

The above three multivariate correlations are considered suitable for the Jiangsu cohesionless deposits. More work shall be done to investigate the potential application of these correlations in other cohesionless soils at other locations. This type of universal validation requires sufficient data points for the cohesionless soils reported in the literature. However, it seems that these data points have gained insufficient publication and thus the universal validation is not conducted here. It is always worth exercising due caution when the correlations developed in this study are applied for other soils as site-specific effect may play an important role in the soil behavior [3, 4].

Moreover, extra caution shall be exercised when the extrapolating the correlations developed in this study to the cohesive soils. For example, Long and Donohue [1] reported that $V_s = 1.961 q_t^{0.579}(1 + B_q)^{1.202}$ for the Norwegian clays. This correlation indicates a positive correlation between V_s and B_q, which is completely different from the negative correlation achieved in this study. The detailed reasons for this difference are still unclear and deserve further investigation. It is always advisable to check the suitability of the correlations reported in this study in other soils before application.

4 Conclusions

In this study, a multivariate distribution model is used to describe the multi-dependency within four SCPTU parameters of Jiangsu cohesionless soils. Based on the constructed model, multivariate correlations for V_{s1} are developed for the Jiangsu cohesionless soils. It is shown that the model is capable of the multivariate correlations among the four SCPTU parameters, and the uncertainties associated with the V_{s1} can be consistently reduced by incorporating more than one testing index in the prediction. However, it seems that the V_{s1} may be underestimated by the multivariate correlations when $V_{s1} > 300$ m/s. Besides, more work shall be done to illustrate the performance and universal applicability of these correlations in other soils.

Acknowledgements. Majority of the work presented in this paper was funded by the National Key Research and Development Program of China (Grant No. 2016YFC0800201), the National Natural Science Foundation of China (Grant No. 41672294) and the Foundation of Jiangsu Province Outstanding Youth (Grant No. BK20140027). These financial supports are gratefully acknowledged.

References

1. Long, M., Donohue, S.: Characterization of Norwegian marine clays with combined shear wave velocity and piezocone cone penetration test (CPTU) data. Can. Geotech. J. **47**(7), 709–718 (2010)
2. Cai, G., Puppala, A.J., Liu, S.: Characterization on the correlation between shear wave velocity and piezocone tip resistance of Jiangsu clays. Eng. Geol. **171**, 96–103 (2014)
3. Ching, J.Y., Arroyo, M., Chen, J.R., Jorge, C., Lansivaara, T., Li, D.Q., Mayne, P.W., Phoon, K.K., Prakoso, W., Uzielli, M.: Transformation models and multivariate soil databases. Final Report of Joint TC205/TC304 Working Group on "Discussion of statistical/reliability methods for Eurocodes", International Society for Soil Mechanics and Geotechnical Engineering (ISSMGE) (2017). Chap. 1

4. Zou, H., Liu, S., Cai, G., Puppala, A.J., Bheemasetti, T.: Multivariate correlation analysis of seismic piezocone penetration (SCPTU) parameters and design properties of Jiangsu quaternary cohesive soils. Eng. Geol. **228**, 11–38 (2017)
5. Juang, C.H., Fang, S.Y., Khor, E.H.: First order reliability method for probabilistic liquefaction triggering analysis using CPT. J. Geotech. Geoenvironmental Eng. ASCE **132**(3), 337–350 (2006)
6. Robertson, P.K.: Interpretation of cone penetration tests - a unified approach. Can. Geotech. J. **46**(11), 1337–1355 (2009)
7. Uzielli, M., Vannucchi, G., Phoon, K.K.: Random field characterization of stress-normalized cone penetration testing parameters. Géotechnique **55**(1), 3–20 (2005)

Visual Measurement Device and Experiment of Ground Water Level in Vacuum Preloading

Jianying Lai[1,2], Ping Li[1,2(✉)], Wei Liu[2], and Jianhui Tang[2]

[1] Key Laboratory of Geomechanics and Embankment Engineering of Ministry of Education, Hohai University, Nanjing 210098, China
hero2008lai@163.com

[2] College of Civil and Transportation Engineering, Hohai University, Nanjing 201198, China

Abstract. Due to the inaccuracy and incapability of the conventional methods of monitoring the groundwater level in vacuum preloading, a visual apparatus of the measuring groundwater level has been designed. The proposed visual apparatus was efficiently developed based on the video camera technology. Utilizing a video camera system, it could be transmitted to the location of the groundwater level in real time to the image receiver. This paper proposes a novel approach using the visual measurement device and assess the conventional methods for performing ground improvement in an area located in Lianyungang, China. The groundwater level, vacuum pressure and the shear strength of soil were recorded during the vacuum consolidation process. Since groundwater level varied in vacuum consolidation, it would be necessary to analyze and assess the variation of shear strength of soil caused by changing the groundwater level. The results showed that the proposed approach is more reliable and intuitive than conventional methods in order to measure the groundwater level under vacuum preloading. Eventually, it is worth mentioning that, the shear strength of soft soil was different at above and below the groundwater level.

Keywords: Groundwater level · Shear strength · Vacuum preloading

1 Introduction

Vacuum preloading method contains number of advantages, including high efficiency, low greenhouse gases and low noise [1–3], in which it is widely used in the treatment of the newly developed soft soil with high water content, high-compressibility, low shear strength and low permeability [4]. The variation of groundwater level caused by this soft foundation reinforcement method contains the characteristics of importance, complexity and nonuniformity [5]. A traditional groundwater level test, which named open type test, uses an ordinary water pipe, while it requires to open the pipe nozzle during measurement. If the measurement could be immediately carried out after opening, the position of the groundwater level rapidly varies (the drop speed can reach 3.45 mm/s [6]) in a short period, indicating that the conventional test methods cannot truly measure the groundwater level under negative pressure. Meanwhile, based on the previous theories developed by a number of scholars [5–7] pointed out that utilizing the

L. Hu et al. (Eds.): GSIC 2018, *Proceedings of GeoShanghai 2018 International Conference: Multi-physics Processes in Soil Mechanics and Advances in Geotechnical Testing*, pp. 373–380, 2018.
https://doi.org/10.1007/978-981-13-0095-0_42

conventional methods in order to assess the groundwater level under negative pressure conditions would be inappropriate as well.

Under the negative pressure conditions, the groundwater level decreases, unchanging and rising were reported by several scholars [8]. This originally backs to lack of reliable and practical underground water level testing techniques to be used under vacuum preloading. To overcome the mentioned challenges, in this study, a novel visual apparatus method has been proposed, and it was employed for field to investigate the effects of variations of groundwater level on the shear strength of soil.

2 Visual Measurement Device

A visual new groundwater level test device [9, 11] mainly includes a groundwater level pipe (GLP) with scale, floating block and image receiver (mobile phone, laptop computer) and other components. This method is using one camera, which mounted in a floating block in GLP, to pass the water level corresponding to the scale to the image display port. The specific structure is shown in Fig. 1(a), the physical diagram of the new water level measuring device is shown in Fig. 1. The new visual apparatus described in Li et al. [9] has three active components, as follow:

(1) A new groundwater level pipe (NGLP) is shown in Fig. 1(b). The most important feature of this NGLP is the attachment of one stainless steel scale inside the NGLP. The corresponding inner diameter, is about 110 mm, which is enough large to facilitate the floating blocks in the pipe to be smoothly floated up and down with the variation of water level. Generally, a NGLP is divided into permeable and impermeable sections. The upper part is impermeable section, which is buried under the soil with the depth of within 2 m to ensure that the pipe is impermeable and airtight [7]. The permeable section is composed of a NGLP with a filter hole, which the diameter is 8 mm on the pipe wall, and a geotextile is wrapped outside the pipe to avoid forming mud and other soil particles clogging the filter hole. During the installation of the pipe, it is vital to consider verticality in order to avoid sticking the floating blocks into the pipe.

(2) A floating block (Fig. 1(b)), which is composed of a lightweight material, is a cylinder and its outer diameter is less than the inner diameter of NGLP. The upper part of the floating block is equipped with a mirror hole to be used for placing an endoscope camera. The endoscope camera, as shown in Fig. 1(b), has the diameter of 7 mm, facilitating easy installation. The endoscope passes through the top hole of the floating block and it is fixed by screws. Then, leads the endoscope camera's data line to the pipe and seal the upper nozzle.

(3) An image receiver, is composed of a mobile phone, or a laptop or other electronic devices, which is connected to a camera in order to receive the captured images, indicating the position of floating block corresponding to the scale position. The groundwater level obtained from an image receiver is shown in Fig. 1(c).

3 Field Testing

3.1 Test Location and Soil Profile

The test site was located at a land reclamation project in the northeastern of Lianyungang, China. The test area (see Fig. 2) was 209 × 68 m^2. The soil, which investigated at the test, consisted of various layers, including filling sand (1.0 m), silt clay (4.5 m), soft silty (6 m), and clay (5.5 m). The groundwater level was about 0.1 m below the ground surface before applying the vacuum consolidation. Prefabricated vertical drains (PVDs), with a cross-section of 100 mm × 4 mm, were placed in the depth of 15 m in a square pattern with the spacing of 1.0 m. Vacuum preloading operates for 75 days and the average vacuum pressure under the geomembrane was 90 kPa.

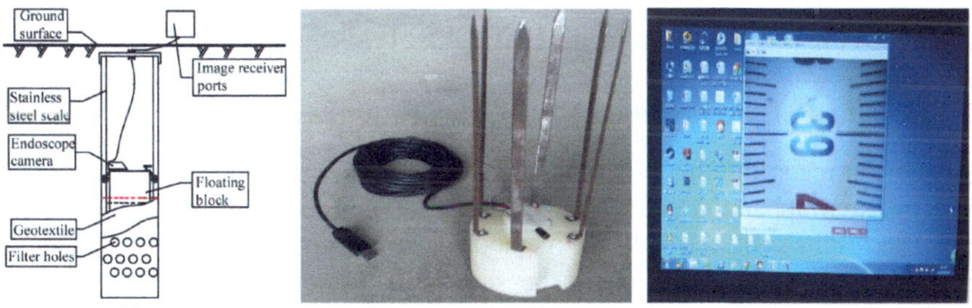

(a) (b) (c)

Fig. 1. Illustration of measuring for groundwater level under vacuum pressure: (a) Schematic diagram for the new apparatus; (b) Floating block; (c) Image receiver ports

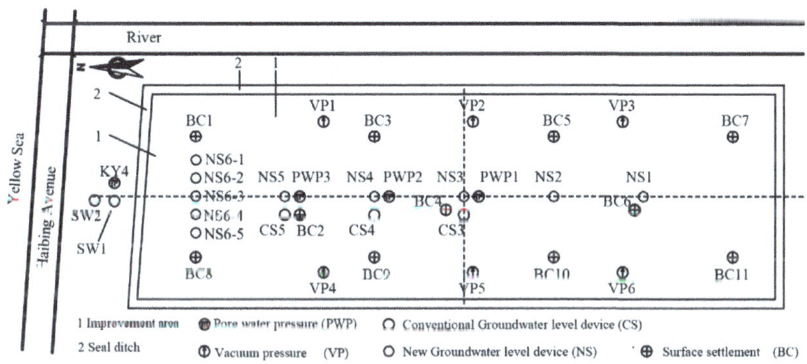

Fig. 2. Field test plan

3.2 Instrumentations

The following methods were employed to estimate the groundwater level: an open type test and the proposed visual apparatus. As shown Fig. 2, the GLP has two types: NS was applied by the proposed visual apparatus and CS was utilized by an open type test. In the improved zone, three NGLPs locations close to the PVDs, from NS3 to NS5, were designed in one section with spacing of 30 m, which the NS3 was in the center of the test zone. Near the NGLPs, CS3 to CS5, were placed at a distance of 1.0 m with NS3 to NS5, respectively. Outside the improved area, two open GLPs (SW1 and SW2) were installed. Additionally, all GLPs and NGLPS were installed in the level of 8 m (groundwater level drops by less than 6 m [6]). During the test, two periods, 26th to 29th October, and 25th to 28th November, which the disappearance of vacuum preloading was caused by power outage.

The in situ testing was undertaken to evaluate the improvement of soft grounds where pore water pressure (PWP), vacuum pressure (VP), surface settlement (BC) were considered as well, as shown in Fig. 2. In addition, the field vane shear test before and after the vacuum preloading was conducted.

4 Measurement Results and Data Analysis

4.1 Validation of the Proposed Visual Apparatus

A comparison regarding the groundwater level, which obtained by the methods of the proposed visual apparatus and open type test was done, and the result is shown in Fig. 3. The results achieved by these methods demonstrated that the groundwater level dropped and the variations showed the same trend during the vacuum preloading. However, results of the groundwater level achieved by the open type method have deeper than that the proposed visual apparatus. The maximum differences of the groundwater level, at the end of the vacuum preloading operation, were approximately found 1.1, 1.1, and 0.6 m between CS3 and NS3, CS4 and NS4, and CS5 and NS5, respectively. Because the open type method must open the GLP then can be tested, accordingly, the atmospheric pressure immediately goes into the pipe, which caused that the groundwater level to be quickly decreased.

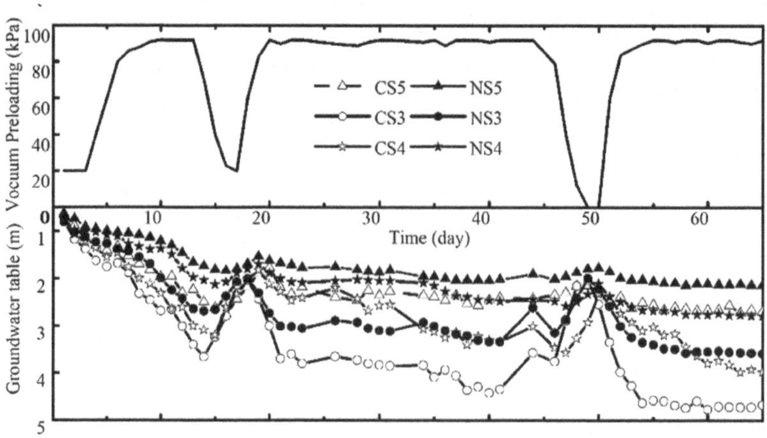

Fig. 3. Comparison curves of the two measuring methods

In addition, under the same environmental conditions, the results of the groundwater level achieved by the proposed visual apparatus showed to be more stable than the open type test. This can be justified that by employing the proposed visual apparatus, rather than using the open type test should open GPL nozzle, the negative pressure variations are applied to the GPL. The speed of the operates will different made the results differently, since the slow operation will take more time for changing the groundwater level in GPL, while less time is available for changing the groundwater level in the GPL.

In summary, it can be concluded that the proposed visual apparatus represents the following advantages over the open type test. (a) The whole process of testing the groundwater level was conducted under the negative pressure conditions, and the conventional methods had failed to be appropriately measured; (b) It could be found that the proposed visual apparatus can directly detect the groundwater level by an image receiver; (c) The expense of one float and one camera is about 200RMB, meaning that the cost of the proposed visual apparatus is slightly higher than the open type test, while this apparatus can be reused several many times, (d) Required less manpower, only one person is enough to handle this test, while at least two people is required to perform the open type test.

4.2 The Material of the Groundwater Pipe

NGLP is such an important material which is used in groundwater level test. The particular feature of this method is that a ruler is set in the wall side of the groundwater level. For choosing the appropriate ruler material, the authors have conducted the experiment contrast. As displayed in Fig. 4, the surface of the rulers of both materials is complete before the pipe is laid. After passing 90 days of vacuum reinforcement, these two rulers are depicted in Fig. 5. As can be seen from Fig. 5, a conventional steel ruler has been fully corroded by seawater, while the stainless steel ruler was kept intact. Therefore, when the water used in the reinforcement zone is seawater, it is highly recommended to utilize a stainless steel ruler.

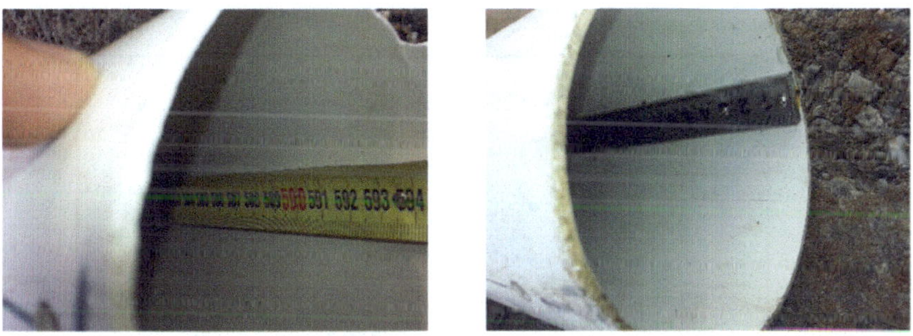

(a) (b)

Fig. 4. The ruler inside of the groundwater pipe before test, (a) The ruler material is a common Steel, (b) the material of the ruler is the stainless

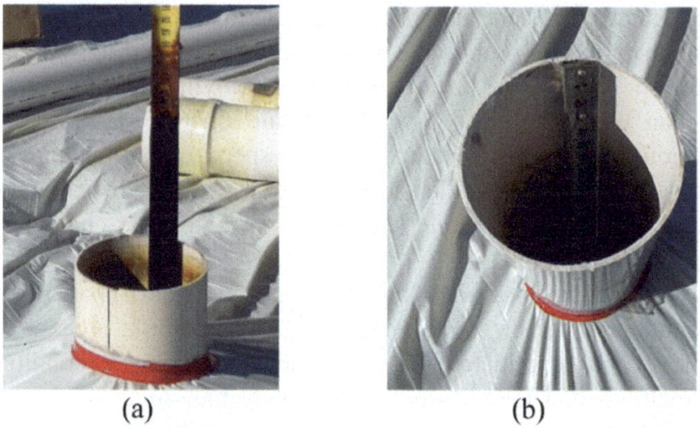

Fig. 5. The ruler inside of the groundwater pipe after test, (a) The ruler material is a common Steel, (b) the material of the ruler is the stainless

4.3 Comparing the Variation of Groundwater Level Inside and Outside the Improved Area

In order to study the variation of groundwater level outside the reinforcement zone under negative pressure, NS5 point in the studied reinforcement area with SW1 and SW2 points, outside the region, were compared which the results are shown in Fig. 6. As illustrated in Fig. 6, the variation of groundwater level inside and outside of the reinforcement region is basically the same. During the vacuum strengthening, the groundwater level at the SW2 point was the highest, and was 0.4 m higher than at SW1 point. Additionally, the ground water level of NS5 was within 0.8 m lower than that of SW1. This shows that under the negative pressure, the groundwater level in the reinforcement zone quickly decreased, and the speed of seepage consolidation was maximum as well. The farthest point from the boundary of the reinforcement zone outside

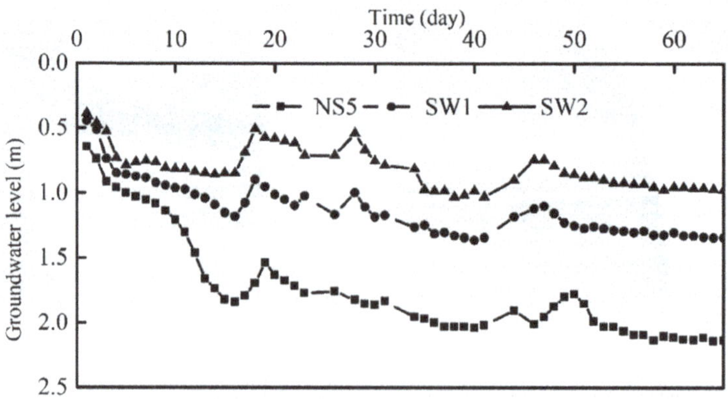

Fig. 6. Comparison of the groundwater level changes for inside and outside of the improvement zone

the SW2 point indicated that the negative pressure was also affected at the distance of 10 m away from the boundary of the reinforcement zone. Simultaneously, the groundwater level in the reinforcement area inside and outside of SW1 and SW2 points, after passing 18 and 50 days, could suddenly rise, and when the vacuum negative pressure reached to 90 kPa, the variation trend of the groundwater level gradually decreased in which this reduction rate was stable.

4.4 Undrained Shear Strength

To investigate the undrained shear strength behavior connected to vacuum preloading operation, Bao [10] pointed that the shear strength of soil was different at below and above the groundwater level. Considering the effect of groundwater level, an analytical solution was presented by Bao [10] as follows:

$$\Delta S_u = \begin{cases} (q+\eta P_v - \gamma_m h)\frac{\sin \varphi'}{1+\sin \varphi'} & (h \le h_0) \\ (q+\eta P_v - \gamma_m h_0)\frac{\sin \varphi'}{1+\sin \varphi'} & (h > h_0) \end{cases} \tag{1}$$

where, q is the surcharge preloading, P_v is the vacuum pressure, h denotes the groundwater level at any time, h_0 is the initial groundwater level, and φ' represents the friction angle.

The groundwater level was around 1.0 m over the membrane as surcharge preloading. The field samples used to test the van shear strength were located at the NS3, where the average groundwater level was about 3.0 m. The results of the computations and field tests are summarized in Table 1. The results demonstrated that the shear strength of soft soil increased faster at above the groundwater level than the below one, and the computation was in a good agreement with the field test data in which the rate of error was 9%.

Table 1. Comparing the measured and predicted undrained shear strength values.

Depth (m)	P_v (kPa)	$(q+\eta P_v - \gamma_m h)$ & $(q+\eta P_v - \gamma_m h_0)$ (kPa)	φ (°)	s_{u0} (kPa)	s_u (kPa)	Field data	Δs_u (kPa)	Error	Δs_u ($h_0 = 0$) (kPa)	Error
0.5–0.7	90	94.12	30	43.6	13	30.6	31.4	2.5	33.3	8.9
2.5–2.7	90	74.52	29.7	38.4	12	26.4	24.7	6.5	33.1	25.5
4.5–4 7	90	70.6	24	47.3	26	21.3	20.4	−4.2	28.9	35.7
6.7–6.9	90	70.6	26.4	41.5	21	20.5	21.7	6.0	30.8	50.1
8.7–8.9	90	70.6	23.6	35.1	13	22.1	20.2	−8.7	28.6	29.4
11.7–11.9	90	70.6	23.1	39.3	18	21.3	19.9	−6.6	28.2	32.3
13.7 13.9	90	76.48	23.5	42.5	21	21.5	21.8	1.4	28.5	32.6

5 Conclusions

The main objective of this paper is to present a new apparatus in order to measure the groundwater level in the vacuum preloading. The reliability of the proposed measurement device was verified by monitoring data derived from field testing. Under the negative pressure, the groundwater level decreased and the impact ranges of negative pressure on outside the reinforcement area could reach at the distance of 10 m. For GLP, considering high-salinity of seawater, the durability of rule material using stainless steel ruler is better than an ordinary steel. In addition, the results achieved from the field test also demonstrated that it would be vital to consider groundwater level variation in order to predict the shear strength of soft soil.

Acknowledgments. This work was supported by the national natural science foundation of China (Grant No. U1134207 and 51408187) and the program for new century excellent talents in university (Grant No. NCET-12-0843).

References

1. Rujikiatkamjorn, C., Indraratna, B.: Environmental sustainability of soft soil improvement via vacuum and surcharge preloading. In: Proceedings of the 2014 Geo-Congress, pp. 3658–3665. ASCE, Atlanta (2014)
2. Yan, S.W., Chu, J.: Soil improvement for a road using the vacuum preloading method. Ground Improv. **7**(4), 165–172 (2003)
3. Griffin, H., O'Kelly, B.C.: Sustainability of combined vacuum and surcharge preloading. In: Proceedings of the 2014 Geo-Congress, pp. 3826–3835. ASCE, Atlanta (2014)
4. Yan, L.: Soft Soil Reinforced by Vacuum Method of Preloading, 2nd edn. China Communications Press, Beijing (2013)
5. Gao, Z.W., Hou, J.F., Liang, A.H.: Analysis and test method for groundwater level by vacuum preloading. Chin. J. Geotech. Eng. **35**(s2), 684–688 (2013). (in Chinese)
6. Liu, H.L., Zhou, Q., Gu, C.C.: New method for measuring groundwater under vacuum preloading. Chin. J. Geotech. Eng. **31**(1), 48–51 (2009). (in Chinese)
7. Zhang, G.X., Dong, Z.L., et al.: Discussion and improvement of underground water level measurement technology in vacuum preloading. Rock Soil Mech. **28**(9), 1899–1903 (2007). (in Chinese)
8. Li, P., Jin, Y.T., et al.: Review of research on characteristics of seepage-induced consolidation of soil under negative-pressure reinforcement conditions. J. Hohai Univ. (Nat. Sci.) **44**(02), 115–121 (2016). (in Chinese)
9. Li, P., Liu, W., et al.: Visual measurement device for underground water level under negative pressure condition and measurement method: China, CN104316144A (2015). (in Chinese)
10. Bao, S.F., Mo, H., Dong, Z.L., Chen, P.S.: Increment calculation of soil shear strength for sand drain foundations consolidation groundwater level. Chin. J. Rock Mech. Eng. **33**(6), 1269–1277 (2014). (in Chinese)
11. Lai, J.Y., Li, P., Tang, J.H.: An apparatus for measuring groundwater level under negative pressure and experimental investigation. China Harb. Eng. **37**(11), 35–39 (2017). (in Chinese)

CPT-SPT Correlation Analysis Based on BP Artificial Neural Network Associated with Partial Least Square Regression

Xiaocong Liang[1,2](✉), Zhiguang Qin[1,2], Sheng Chen[1,2], and Deyong Wang[1,2]

[1] CCCC Fourth Harbor Engineering Institute Co., Ltd., Guangzhou 51000, China
mariobeatrice@sina.com

[2] Key Laboratory of Environment Protection and Safety of Transportation Foundation Engineer of CCCC, Guangzhou 51000, China

Abstract. Most of the correlations of CPT-SPT have been widely investigated based on a statistical method without considering the multicollinearity among the influenced factors. Therefore, one model combining back propagation neural network (BP ANN) with partial least square regression (PLS), which could consider the multicollinearity influence, is proposed for building CPT-SPT correlation. The sensitivity analysis based on the ANN model and the PLS model is conducted, and the four most sensitive factors are obtained, i.e., cone resistance (q_c), soil behavior type (SBT), friction resistance (fs) and soil behavior index (Ic). Further, these four most sensitive factors plus fine content (Fc%) and effective stress (σ'_o) are adopted as the input factors of the combined model (BP ANN associated PLS). And 362 group data are collected from New Doha Port for building the combined model. The result shows the combined model has a correlation index R^2 of 0.83311 and further demonstrates the combined model is effective. Finally, additional 50 group data are applied to verify the combined model. The result indicates that it can improve the correlation coefficient between the predicted (q_c/pa)/N value and the measured one from 0.6429 to 0.7523 and reduce the relative error from 28% down to 21% compared with the solely PLS model. Given the advantage of the combined model, it can provide a more effective and reliable method for CPT-SPT correlation analysis in practice.

Keywords: Multicollinearity · Partial least square regression BP artificial neural network · CPT-SPT correlation

1 Introduction

The in situ standard penetration test (SPT) has been widely used because of its simplicity and operability; however, it has disadvantages of poor repeatability and low reliability because the N value can be influenced by many factors. Alternatively, the cone penetration test (CPT), as one of the continuous in situ tests with higher repeatability and reliability, is increasingly applied in design aspect [1–4]. Therefore, to

L. Hu et al. (Eds.): GSIC 2018, *Proceedings of GeoShanghai 2018 International Conference: Multi-physics Processes in Soil Mechanics and Advances in Geotechnical Testing*, pp. 381–390, 2018.
https://doi.org/10.1007/978-981-13-0095-0_43

fully use the existing SPT empirical experience, an in-depth study of the correlation between SPT and CPT is required. Based on a detailed referenced review of the literature [5–13], most of the correlation was mainly based on the mathematical statistics correlated with single influence parameters, such as mean grain size (D_{50}), fine content (FC (%)), soil behavior type (SBT) and soil behavior index (Ic). However, these correlations are normally built on multiple factors without considering the multicollinearity of influence factors.

Back propagation artificial neural network (BP ANN) is an artificial intelligence algorithm that can be applied to build the complicated nonlinear relationship through a reasonably accurate model. However, because of the high dimension of independent variables and multicollinearity among variables, the BP ANN model is more unstable compared with the partial lease square (PLS) approach [14–19]. The PLS approach, as a regression analysis method, with its advantage of eliminating the mulitiсollinearity between variables, can be used to perform regression analysis. Thus, to take advantage of the BP method, the PLS methodology was proposed to be combined with the BP model to analyze the correlation between CPT-SPT.

2 Method of Partial Least Squares

The partial least squares method, a multivariate statistics method that can extract the most effective interpretative information for independent variables with a limited sample to achieve higher predicted accuracy and reduce the mulitiсollinearity of influence factors, has become increasingly common in actual engineering [20–22]. When partial least squares analysis is being conducted, the first principal components t1 and u1 are extracted from the dependent and independent variables and are the most represented component of the variable nature and have the greatest relevance. Next, the second or third principal component is continuously extracted from the residual information of variables until the precision of multivariate regression analysis for all of the extracted principal components can meet the requirement.

When performing partial least squares fitting analysis, the variable importance projection (VIP) can be used to measure the explanatory capacity of each independent variable to dependent variable. The VIP value of the jth independent variable can be calculated as Eq. (1) [21, 22].

$$VIP_J = \sqrt{\frac{P}{Rd(y; t_1, \cdots, t_m)} \sum\nolimits_{h=1}^{m} R_d(y; t_h) w_{hj}^3} \quad (1)$$

Where y is the dependent variable, $R_d(y; t_h)$ is the explanatory capacity of principal component t_h to dependent variables y, $Rd(y; t_1, \cdots, t_m)$ is the cumulative explanatory capacity of each principal component to dependent variable y, and w_{hj} is used to calculate the marginal contribution of independent variables x_i on the principal components t_h.

According to the above formula, the importance of the independent variable is mainly transferred through the principal component. If the explanatory capacity of the principal components for the dependent variable is stronger with higher fitting precision, then it can verify the greater importance of the independent variable for the dependent variable. Therefore, the VIP value can be used to evaluate the sensitivity between different independent variables, thereby screening out factors that have a major influence on the dependent variable.

3 Principle of BP ANN Associated with PLS

The BP ANN is a complex nonlinear system. It can establish a high accuracy of the mapping relationship between the existing input and output parameters; this relationship can be used for output variable prediction if there are enough samples for network training [19, 23]. When adopting BP ANN to perform an analysis, the input and output variables correlation degree will affect the stability of the model structure. Thus, the mean impact value index (MIV) was introduced to quantitatively evaluate the importance of input variables to the output variable [24]. By comparing the absolute value of MIV, the importance of the input variables can be judged. Hence, the input variable evaluation and selection can be achieved.

Since most of the BP ANN input variables are selected based on researcher professional knowledge and experience, no strict theoretical system is available regarding the screening and analysis of input variables. However, because the validity of the input variables and independence will affect the fitting precision of the network structure, the analysis of the effectiveness and independence of input variables shall be carried out first based on the VIP value using the BP ANN model and the MIV value using the partial least square method. Next, the reasonable and effective value is screened based on sensitivity analysis of each independent variable. Second, on the basis of the partial least square method, a number of valid principal components are extracted as input variables of the BP ANN for building a reasonable network model [25, 26].

4 CPT-SPT Correlation Study for the New Doha Port Project

The New Doha port in Qatar was using the dry excavation method with a coastline with a total length of 8050 m, excavation basin area of 3.385 million m^2 and total excavation amount of approximately 65 million m^3. The ground treatment area was approximately 6.3 million m^2, with an average thickness of approximately 6 to 8 m. The backfilling material is sand with fine content less than 15% with thickness of nearly 2 to 4 m. The natural subsoil was sedimentary soil. Most of the soil was silty sand, sandy silt, and fine or medium sand with a silty pocket (maximum fine content of up to 70% 80%).

Dynamic compaction was adopted for ground improvement, and CPT was selected as the verification test. SPT was also selected for comparison with the CPT result within 1 m distance. Because the SPT N value was counted through 30 cm, the average qc correlated with same depth will be used for comparison. A total of 362 comparing groups are collected for model building, and an additional 50 groups' data were adopted for model verification.

4.1 Sensitivity Analysis of the CPT-SPT Correlation Influence Factors

According to previous research data, the main factors affecting the $(q_c/pa)/N$ were mainly related to cone tip resistance (q_c), friction resistance (fs), penetration depth (D), the effective stress for overlying strata (σ'_o), fine content (Fc%) (<0.063 mm) particles, friction ratio (Fr%), mean particle diameter (D_{50}), soil behavior index (Ic) [27], and soil behavior type SBT. The sensitivity of influence factors will be determined to analyze the importance of an impact factor based on the BP ANN MIV value and PLS VIP value. Thus, more sensitivity values can be screened out for the BP network input factor based on sensitivity analysis.

The MIV values based on BP network analysis of different impact factors are shown in Fig. 1. It shows that the most influential factors were q_c, SBT, Ic, fs, σ'_o, D, Fr(%), Fc(%) and D_{50}. Three of them, i.e., Ic, Fr(%), and Fc(%), show negative correlation, and the remaining impact factors show positive correlation in order.

Based on the PLS method, the VIP values obtained are shown in Fig. 2 below. It reveals that the most explanation capacity of each independent variable to dependent variable was in the following order: q_c, SBT, fs, Ic, F_C(%), D, σ'_o, Fr(%) and D_{50}.

The two sensitivity analysis methods introduced above show that the analysis results are close to each other. Based on the analysis result, four of these factors, i.e., q_c, SBT, fs and Ic, are the most important impact factors. The strongest sensitivity of these four factors will have a great impact on the dependent variable $(q_c/pa)/N$, while the rest of the five impact factors obviously have less impact. Moreover, because the penetration depth D was linearly correlated with σ'_o and Fr(%) can be obtained from q_c and fs, these two factors will be removed when performing further correlation analysis to avoid multicollinearity. Further, D_{50}, with the lowest MIV value and VIP value, was excluded. Thus, taking into account the above analysis, the six impact factors qc, SBT, fs, Ic, F_C% and σ'_o will be selected for correlation analysis.

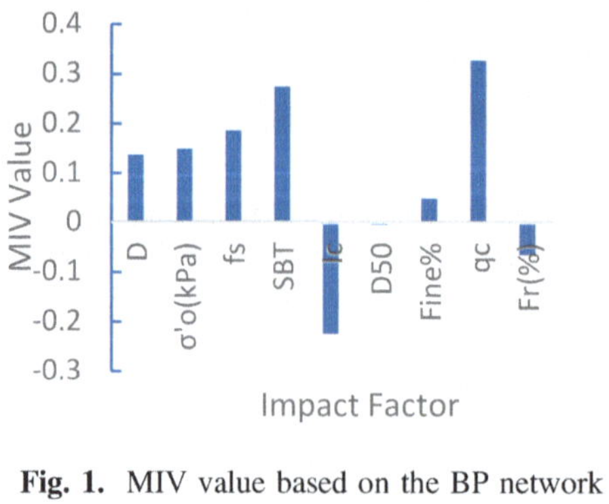

Fig. 1. MIV value based on the BP network

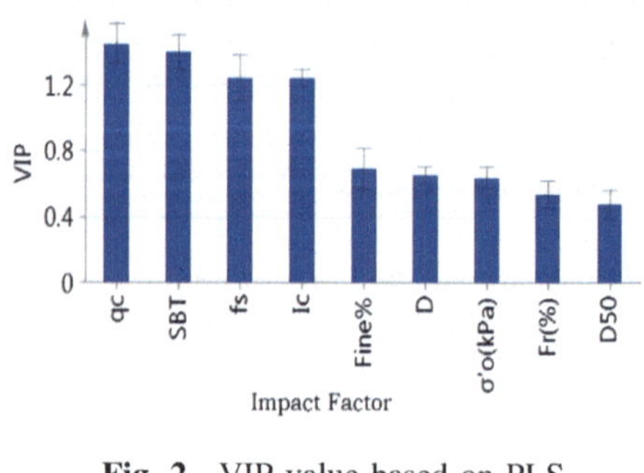

Fig. 2. VIP value based on PLS

4.2 CPT-SPT Correlation Analysis Based on the PLS Method

According to the sensitivity analysis of the independent variable above, six variables are selected for further analysis. The correlation coefficients between the independent variables are shown in Table 1 below.

Table 1. Variables correlation matrix

Variables	Σo (kPa)	qc	fs	SBT	Ic	Fine%	qc/pa/N
σo(kPa)	1.000	–0.161	–0.135	–0.302	0.675	0.454	–0.221
qc		1.000	0.849	0.776	–0.654	–0.292	0.765
fs			1.000	0.606	–0.448	–0.249	0.650
SBT				1.000	–0.828	–0.497	0.752
Ic					1.000	0.575	–0.668
Fine%						1.000	–0.275
qc/pa/N							1.000

Based on the above-mentioned correlation matrix calculation result, a strong correlation exists between variables. To reduce multicollinearity between variables, the PSL method should be applied to extract the principal component that corresponds to cross validation. Set Q_h^2 as the cross validation test result of t_1 and $Q_h^2(\text{cum})$ as the cumulative cross validation value. The principal component and related cross validation test results are shown in Table 2 below.

Table 2. Overall cross validation for components

Principal component	Q_h^2	$Q_h^2(\text{cum})$
t1	0.615	0.615
t2	0.105	0.655
t3	0.00953	0.659

According to Table 2, the cross validation value for the third component was 0.00953, which was less than 0.0975. Thus, the extraction of principal components will be terminated according to the introduction of PLS method in Sect. 2. The first two principal components with cross validation values larger than 0.0975 will be selected for fitting analysis. To further verify the correlation between the independent variable and dependent variable, the extracted principal t1 from the independent variables and u1 extracted from the dependent variables are used to conduct a correlation analysis.

Figure 3 shows that the independent variables and dependent variables have a strong linear relation with a correlation coefficient of 0.62, which indicates that the applied model was reasonable. Further, the regression analysis is based on the original impact factors transferred from the selected two principal components. The coefficient of regression equation is shown in Fig. 4. It reveals that the four impact factors, cone resistance q_c, soil behavior type SBT, friction resistance fs and soil behavior index Ic,

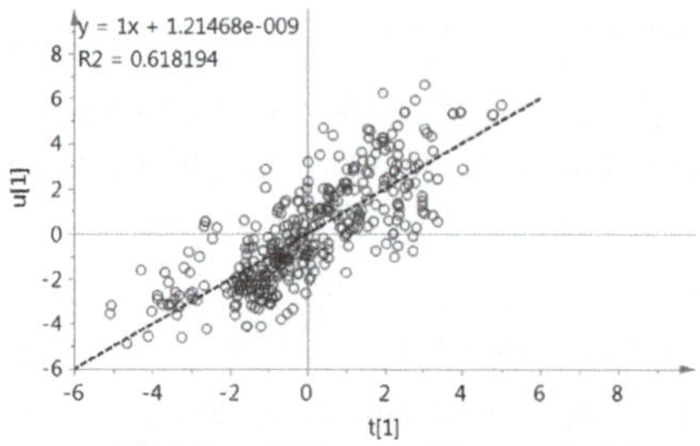

Fig. 3. t1/u1 regression analysis

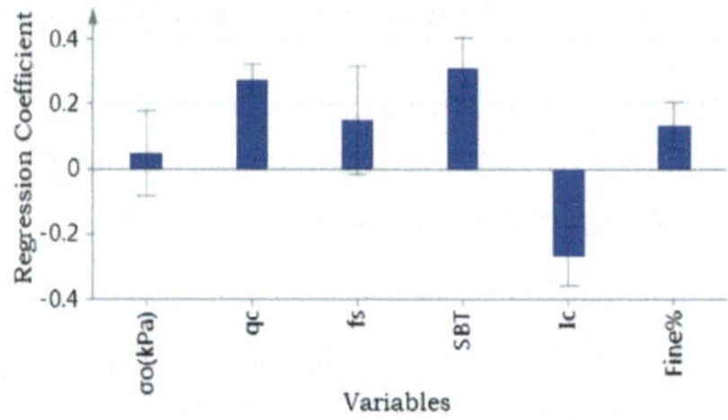

Fig. 4. Coefficient of regression plot

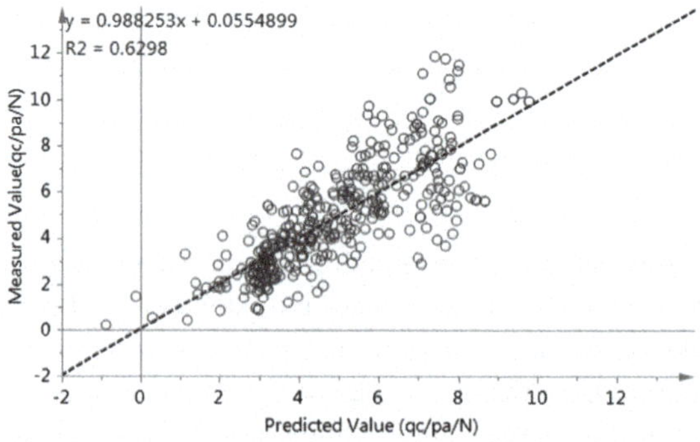

Fig. 5. Correlation analysis of predicted and measured values

with larger coefficient values indicated that these four impact factors have a great impact on the dependent variables.

To evaluate the prediction accuracy of the regression equation, the independent variable selected from the additional verification groups will be adopted as the input of the regression equation in Fig. 4. Afterwards, the output value calculated from the regression equation was compacted with actual results. The scatter distribution of actual values and predicted values is shown in Fig. 5. The predicted values are found to have a good fitting degree of regression, with a correlation coefficient of 0.6298. Based on the analysis above, the PLS model using the selected impact factors clearly has a good prediction result and can reduce multicollinearity.

4.3 Correlation Analysis Based on the Combined Method

(1) Building of the combined method

Based on the theory of Sect. 3, the two principal components extracted from the PLS method will be taken as the input variables of the optimized BP neural network model, and the ratio $(q_c/pa)/N$ will be taken as the output variable. Based on the MATLAB neural network toolbox, three layers of the BP ANN model are adopted. For information regarding the determination of the number of hidden layers, please refer to the literature [17].

The collected 362 group data, which were automatically separated into three groups with a ratio 2:1:1, will be used for BP network model training, validation and prediction. The correlation analysis result of three stages can be obtained after model analysis running is completed. The correlation coefficients for the three stages based on the three groups are 0.85535, 0.8381, 0.71662, as shown in Fig. 6.

The correlation coefficient based on all of the data is 0.83311 and was found to be higher compared with the correlation coefficient of 0.6298 obtained from the partial least squares model. As a result, the optimized BP neural network model associated with the PLS method can achieve higher fitting precision and obtained a better prediction.

(2) Applicability of the combined model

To further verify the model BP neural network associated with PLS reliability and feasibility, an additional independent 50 sets of data were collected from another area that was not incorporated in model building and training. The principal components based on selected impacted factors of 50 sets of data are input in the established combined model. After model building and training, the predictive output dependent variable was obtained, i.e., $(q_c/pa)/N$. Next, correlation analysis is conducted for the predictive and actual dependent variable. Moreover, the correlation analysis for the predicted variable using the single PLS model was also conducted. The correlation results are shown separately in Fig. 7 (a) and (b). The fitting index and average error for these two models are calculated in Table 3.

Based on the figure and the table referenced above, the combined model has less root mean square error RMSE, and the relative error can be reduced from 28% for the

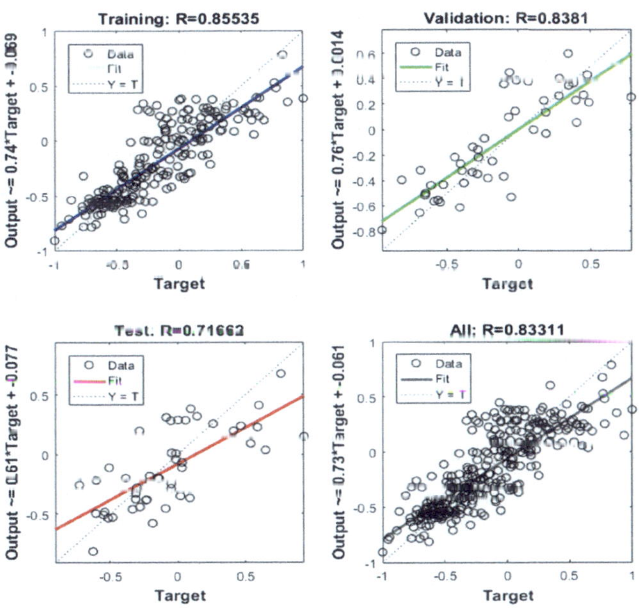

Fig. 6. Correlation analysis of the predicted and measured values for different groups

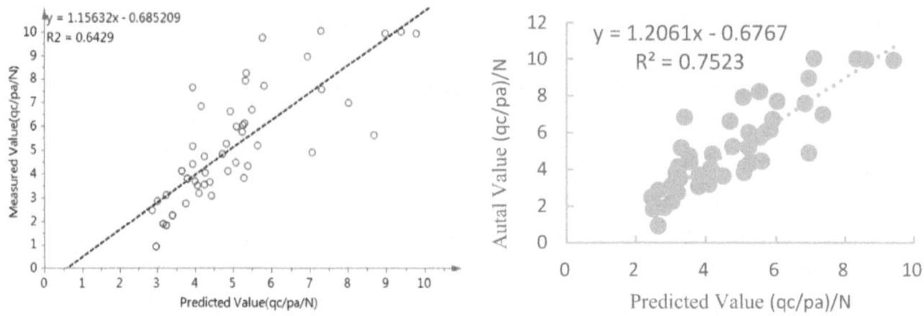

(a) PLS model (b) Combined model

Fig. 7. Correlation analysis of the predicted and measured values

Table 3. Statistical table of mode's precision parameter

Model	RMSE	Relative error(%)	Correlation index-R2
PLS	1.34	28	0.64287
BP associated with PLS	1.22	21	0.7523

single PLS model to 21%. Further, the correlation index was improved from 0.6429 to 0.7523. As a result, the combined model, i.e., the BP network associated with PLS, had a higher prediction precision and a better fitting correlation compared with the model that only applied PLS.

5 Conclusion

Based on the above analysis regarding the optimized BP neural network associated with PLS for CPT-SPT correlation, the following conclusions can be drawn:

(1) The four most important impact factors regarding the correlation of CPT-SPT were cone resistance (q_c), soil behavior type SBT, friction resistance (fs) and soil behavior index (Ic) based on the sensitivity analysis using the MIP and VIP values.
(2) The four most important impact factors plus FC% and effective stress σ_o' are taken as the input variables for building the combined model. The result shows that the combined model provides a great performance with correlation index R^2 of 0.83311 and further demonstrates the combined model is effective.
(3) Fifty additional sets of data were collected to verify the applicability and reliability of the combined model. The verification results show that the combined model can improve the correlation index R^2 from 0.6429 to 0.7523 and reduce the relative error from 28% down to 21% compared with the solely PLS model.
(4) The PLS model could reduce the effect of multicollinearity among variables, and the BP ANN model could build a non-linear relationship with high precision. Therefore, the BP network associated with the PLS method, which makes full use of these two advantages, provides a new method for CPT-SPT correlation development.

References

1. Clayton, C.R.I.: The Standard Penetration Test (SPT): Methods and Use, pp. 30–50. Construction Industry Research Information Association, London (1995)
2. Lunne, T., Robertson, P.K., Powell, J.J.M.: Cone Penetration Testing in Geotechnical Practice. Blackie Academic and Professional, London (1997)
3. IRTP: International Reference Test Procedure for Cone Penetration Test (CPT). Report of the ISSMFE Technical Committee on Penetration Testing of Soils, TC 16, vol. 7, pp. 6–16. Swedish Geotechnical Institute, Linkoping, Information, Swedish (1999)
4. Robertson, P.K., Cabal, K.L.: Guide to Cone Penetration Testing for Geotechnical Engineering-guide to Cone Penetration Testing, 5th edn. Gregg Drilling & Testing Inc, California (2015)
5. Salehzade, H., Hozouri, A., et al.: Correlation between cone and standard penetration test. In: 5th Symposium on Advances in Science &Technology (2011)
6. Chin, C., Duann, S., Kao, T.: SPT–CPT correlations for granular soils. In: Proceedings of the 1st International Symposium on Penetration Testing (ISOPT-1), pp. 335–339 (1988)
7. Kara, O., Gunduz, Z.: Correlation between CPT and SPT in Adapazari, Turkey. In: International symposium on cone penetration testing, California, USA, pp. 9–11 (2010)
8. Shahri, A.A., Juhlin, C., Malemir, A.: A reliable correlation of SPT-CPT data for Southwest of Sweden. Electron. J. Geotechn. Eng. **19**, 1013–1032 (2014)
9. Kulhawy, F., Mayne, P.: Manual on estimating soil properties for foundation design. Report no. EPRI-EL-6800, Electric Power Research Institute, EPRI (1990)
10. Schmertmann, J.H.: Guidelines for Cone Penetration Test, Performance and Design. Report No.FHWA-TS-78-209. U.S. Department of Transportation, Washington, D.C. (1978)
11. Meyerhof, G.G.: Shallow foundations. J. Soil Mech. Found. Div. ASCE **91**(SM2), 21–31 (1965)
12. Elbanna, M., Quinn, J., Martens, S.: SPT – CPT correlations for oil sands tailings sand. In: Proceedings of Tailings and Mine Waste, Vancouver, BC, Canada (2011)
13. Kasim, A.G., Chu, M.Y., Jensen, C.N.: Field correlation of cone and standard penetration tests. J. Geotech. Eng. **112**(3), 368–372 (1986)
14. Hubick, K.T.: Artificial neural networks in Australia. Department of Industry, Technology and Commerce, Commonwealth of Australia, Canberra (1992)
15. Waszczyszyn, Z.: Fundamentals of artificial neural networks. In: Waszczyszyn, Z. (ed.) Neural Networks in the Analysis and Design of Structures, CISM International Centre for Mechanical Sciences (Courses and Lectures), vol. 404 (1999)
16. Koprinkova-Hristova, P.D., Mladenov, V., Kasabov, N.: Artificial Neural Networks Methods and Applications in Bio /Neuroinformatics (2015)
17. Yan Sheng, T.: The Theory of Neural Network and its Application in Engineer. The Mapping Publishing Company, Beijing (2006)
18. Liu, X.: Discussion on neural network application in civil engineer. Chin. J. Geotech. Eng. **25**(4), 514–515 (2003)
19. Xiaochuang Wang, T.: MATLAB Neural Network 43 Case Analysis. Beihang University Presss, Beijing (2013)
20. Vinzi, V.E., Chin, W.W., Henseler, J., et al.: Handbook of Partial Least Squares. Springer, Heidelberg (2010)
21. Wang, H., Zaibin, W., Meng, H.: Partial Least Square Regression-Linear and Non-Linear Methods. National Defence of Industry Press, Beijing (2006)
22. Huiwen Wang, T.: Partial Least Squares Regression Method and Application. National Defence of Industry Press, Beijing (1999)

23. Zeng, W., Wei, R., Chen, H.: Research and application of BP neural network based on improved PSP algorithm. Comput. Technol. Dev. **18**(4), 50–52 (2008)
24. Dombi, G.W., Nandi, P., Saxe, J.M., et al.: Prediction of rib fracture injury outcome by an artificial neural network. J. Trauma **39**(5), 915 (1995)
25. Zhang, G., Weiya, X.: Determination of multicorrelation for property parameters of geotechnical engineering materials. Chin. J. Rock Mech. Eng. **23**(7), 1109–1113 (2004)
26. Huang, S., Feng, X., et al.: Study of method of comprehensive elevation for parameters of constitutive model of rock mass. Chin. J. Rock Mech. Eng. **27**(s1), 2624–2630 (2008)
27. Robertson, P.K., Wride, C.E.: Evaluating cyclic liquefaction potential using the cone penetration test. Can. Geotech. J. **37**(3), 270–273 (1998)

A Two-Fold Empirical Approach for Estimating the Preconsolidation Stress in Clay Deposits

Karim Kootahi[1(✉)] and Paul W. Mayne[2]

[1] Department of Civil Engineering, Saghez Branch, Islamic Azad University, Saghez, Iran
kootahi@iausaghez.ac.ir

[2] Geosystems Engineering Group, School of Civil and Environmental Engineering, Georgia Institute of Technology, Atlanta, USA

Abstract. The conventional method for determining preconsolidation stress of clay is via one-dimensional consolidation tests on undisturbed samples but a first-order approximate value can be estimated from simple empirical models that relate preconsolidation stress to easily measured soil properties, such as natural water content, liquid limit, plasticity index, etc. In this paper, a two-fold simple empirical model for predicting preconsolidation stress that bifurcates at an overconsolidation ratio (OCR) of 3 is developed. Two independent data sets which consist of about 2000 samples of fine-grained soils are used as model-building and model-validation data sets. The results of applying the new and existing simple empirical models to validation data set indicate that (1) the proposed model provides quite acceptable estimates for soils with different stress histories and sensitivities, (2) the predictive ability of the new model is quite superior to existing models, and (3) the performance of prior existing models can be classified as unacceptable.

Keywords: Preconsolidation stress · Index properties · Empirical models

1 Introduction

Among important engineering properties of clay, the preconsolidation stress (σ'_p) is the most important one, as this key parameter affects shear strength and compressibility of fine-grained soils. Specifically, the stress history of the soil is represented by the preconsolidation stress (or "yield stress"), often reported in a dimensionless form as the overconsolidation ratio ($OCR = \sigma'_p/\sigma'_{vo}$).

Conventionally, σ'_p is determined by performing one-dimensional consolidation tests on undisturbed soil samples and test results are interpreted by the method of Casagrande (1936). A variety of problematic issues may arise during either or both processes (i.e., during sampling and interpreting the results) which makes the determined values of preconsolidation stresses somewhat doubtful. As such, several different graphical or numerical methods have been developed to either better define σ'_p or correct for the effects of sample disturbance. Considering these difficulties, empirical

L. Hu et al. (Eds.): GSIC 2018, *Proceedings of GeoShanghai 2018 International Conference: Multi-physics Processes in Soil Mechanics and Advances in Geotechnical Testing*, pp. 391–399, 2018.
https://doi.org/10.1007/978-981-13-0095-0_44

statistics-based models that use various physical and/or index properties as entry information appear to be very beneficial. The virtue of using simple index properties as predictor variables is that (1) these indices are almost insensitive to sample disturbance, (2) they are familiar to all geoengineers, and (3) they can be easily and economically determined. Such empirical models can also be useful in cross-checking and validating laboratory-determined values of σ'_p (Kootahi and Mayne (2017). Simple correlations for σ'_p have been linked to few index properties (e.g., Stas and Kulhawy 1984; Nagaraj and Murthy 1986). Table 1 summarizes the currently available simple empirical models for estimating σ'_p.

Table 1. Available simple empirical models for predicting the preconsolidation stress.

Correlation	Applicability	Reference
$\sigma'_p/p_a = 10^{(1.11-1.62LI)}$	Clays with $S_t < 10$	Stas and Kulhawy (1984)
$\sigma'_p\ (\mathrm{kPa}) = 10^{[5.97-5.32(wn/LL)-0.25\log\sigma'_{vo}]}$	Overconsolidated uncemented soils	Nagaraj and Murthy (1986)
$\sigma'_p\ (\mathrm{kPa}) = 10^{(2.9-0.96\,LI)}$	Onshore and offshore clays	DeGroot et al. (1999)
$\sigma'_p/p_a = 1.070\,LI^{-0.295}$	Sensitive to quick clays	Ching and Phoon (2012)

Notes: w_n = natural water content; LL = liquid limit; LI = liquidity index; S_t = sensitivity; σ'_{vo} = effective overburden stress; p_a = atmospheric pressure (=1 atm $\approx$ 100 kPa)

Accordingly, this study aimed at developing a robust simple empirical model for estimating the preconsolidation stress of clays. For this purpose, the methodology of this study used a two-phase strategy, consisting of an initial model-building process, followed by a model-validation procedure. That is, a good quality model-building data set is collected first to derive the empirical expression for predicting σ'_p. Later, a separate model-validation data set is compiled and it is used for assessment purposes.

2 Model-Building and Model-Validation Data Sets

Two independent data sets of experimental data which consisted of about 2000 samples of fine-grained soils were compiled from different sources. The data set for developing new model (model-building data set) consists of 120 data points from 59 different sites all over the world. An independent model-validation data set which contains 1850 data points from 194 different sites worldwide was compiled for assessing the performance of both newly proposed and existing empirical models. However, after careful screening, data were selected for the model-building and model-validation data sets only if they were obtained from high-quality undisturbed sampling techniques. Due to space limitations, a detailed description of the two data sets is not given here and only summaries of their characteristics are given. Detailed description and discussion of the both model-building and model-validation data sets (with a full list of sources utilized) can be found in Kootahi and Mayne (2016, 2017).

The model-building data set consists of index properties of different clay soils and their corresponding consolidation curves that were taken from different sources and processed consistently. That is, most of the consolidation curves were digitized from the original sources and redrawn again, and then preconsolidation stresses were determined using the method of Casagrande (1936). The observed measurements include natural void ratio e_n, natural water content w_n, liquid limit *LL*, plastic limit *PL*, plasticity index *PI*, liquidity index *LI*, preconsolidation stress σ'_p, and effective overburden stress σ'_{vo}. This database contains clay soils of different physical and mechanical properties. The model-validation data set consists of index properties and preconsolidation values of different fine-grained soils collected from the literature. In order to ensure a robust and effective validation process, the records in the model-validation data set completely differ from those used in developing new and existing models. The model-validation data set covers a wide variety of soil types. The statistics for main geotechnical characteristics of the both model-building and model-validation data sets are summarized in Table 2. It is seen from Table 2 that the parameter ranges for validation data set are wider than those for model-building data set and thus a conclusive validation test is expected. Moreover, model-validation data set covers most conditions encountered in practice.

Table 2. Statistics for main geotechnical characteristics of the both model-building and model-validation data sets.

Parameter	Model-building data set				Model-validation data set			
	Mean	Standard deviation	Min	Max	Mean	Standard deviation	Min	Max
σ'_p/p_a	4.50	5.7	0.2	29.3	2.2	4.0	0.1	62.5
σ'_{vo}/p_a	2.3	4.2	0.1	23.2	1.2	3.1	0.1	56.8
OCR	4.0	3.8	1.0	19.2	2.6	3.5	0.8	57.5
PL	23.6	9.1	10.0	74.0	28.9	11.5	4.0	151.0
LL	49.6	24.7	18.0	131.0	68.7	31.7	15.1	232.0
w_n	39.1	22.3	9.5	153.0	63.3	30.2	14.8	215.0
LI	0.6	0.6	−0.6	3.5	1.0	0.7	−1.2	7.2

3 Statistical Analysis and Discussion

3.1 New Model and Its Performance on Model-Building Data Set

The new model for predicting σ'_p was developed by statistically-fitting equations to the model-building data set. More specifically, multiple linear regression analyses have been performed to express the effective preconsolidation (σ'_p) in terms of the properties influencing σ'_p. which were identified through Pearson's correlation matrix. However, following the finding of Nagaraj and Murthy (1986), which states that estimates of log (σ'_p) can be obtained based on knowledge of log (σ'_{vo}) and ratio of w_n/LL, logarithmic transformations of variables (i.e., log (σ'_p), log (σ'_{vo}), log (w_n), and so forth) were considered. Moreover, all stresses have been made dimensionless by use of a reference stress equal to atmospheric pressure: p_a = 1 atm ≈ 100 kPa.

The correlation matrix, calculated for logarithmic values of variables (Table 3), indicated a very strong positive correlation between log (σ'_p) and log (σ'_{vo}) and a significant correlation between log (σ'_p) and log (*LI*). The correlation matrix also indicated mild correlation between log (σ'_p) and log (w_n) or log (*LL*), but no significant correlation at all between log (σ'_p) and log (*PL*) or log (*PI*).

Table 3. Correlation matrix for logarithmic values of different soil properties.

Soil property	σ'_{vo}/p_a	w_n	*LL*	*PL*	*PI*	*LI*
σ'_p/p_a	0.732	−0.340	0.252	−0.070	0.008	−0.572
σ'_{vo}/p_a	–	0.016	0.171	0.160	0.143	−0.308
w_n	–	–	0.821	0.804	0.674	0.418
LL	–	–	–	0.828	0.931	−0.160
PL	–	–	–	–	0.588	0.033
PI	–	–	–	–	–	−0.281

The model-building data set was split into two subsets and regression equations were fitted to each subset of data. An overconsolidation ratio (OCR) of 3 (OCR < 3 and OCR ≥ 3) was chosen as distinguishing parameter between the two subsets because these two ranges of OCR (i.e., OCR < 3 and OCR ≥ 3) represent two different classes of soil behavior (i.e., contractive vs. dilative behavior). Indeed, soils are usually grouped by OCR as follows: normally- to lightly-overconsolidated (NC-LOC) clays and moderately- to heavily-overconsolidated (MOC-HOC) clays. The results of the applied statistical analyses are expressed in the following two-fold model:

$$\log(\sigma'_p/p_a) = 0.21 + 0.89\log(\sigma'_{vo}/p_a) + 0.12\log(LL) - 0.14\log(w_n) \quad \text{for OCR}_\text{s} < 3 \tag{1a}$$

$$\log(\sigma'_p/p_a) = 0.90 + 0.71\log(\sigma'_{vo}/p_a) + 0.53\log(LL) - 0.71\log(w_n) \quad \text{for OCRs} \geq 3 \tag{1b}$$

Based on the results of analysis of variances, both of these regression equations are significant to a confidence level of 99.99%. The signs of regression coefficients in Eqs. 1a and 1b are exactly what are expected. In fact, the larger the preconsolidation stress, the smaller the natural void ratio (or equivalently the smaller the natural water content). The liquid limit is the limiting value for natural water content and thus it should be positively correlated with preconsolidation stress. A positive correlation between σ'_p and σ'_{vo} is expected for most overconsolidated soils, in which overconsolidation is due to changes in overburden pressure (not due to cementation).

For ease of use, the proposed relationships (Eqs. 1a and 1b) can be retransformed to the original units, and therefore the direct expressions are:

$$\sigma'_p/p_a = 1.62(\sigma'_{vo}/p_a)^{0.89}(LL)^{0.12}(w_n)^{-0.14} \quad \text{for} \quad \text{OCRs} < 3 \tag{2a}$$

$$\sigma'_p/p_a = 7.94(\sigma'_{vo}/p_a)^{0.71}(LL)^{0.53}(w_n)^{-0.71} \quad \text{for} \quad \text{OCRs} \geq 3 \tag{2b}$$

Note that these equations are consistent with the model developed by Nagaraj and Murthy (1986). Specifically, the σ'_{vo} and ratio of w_n/LL exist in both models. However, the signs of coefficients in Nagaraj and Murthy's model are not consistent with what are expected for overconsolidated soils (e.g., the sign of σ'_{vo} should in fact be positive). Equation (2a) has a coefficient of determination (R^2) equal to 0.95 and Eq. (2b) has an R^2 of about 0.75. The calculated overall R^2 for the complete model ($1 \leq \text{OCRs} \leq 19$) is $R^2 = 0.89$. The performance of retransformed model on model-building data set is presented graphically and quantitatively, as shown in Fig. 1. The quantitative measures include coefficient of determination (R^2) and standard error of estimate (SEE). It can be seen from Fig. 1 that the proposed model provides a very good fit to the model-building data set. Note that since OCR depends on σ'_p, some engineering judgment is needed to decide whether equation (Eq. 2a or 2b) is applicable to a specific depth in a specific deposit. In the next section, an efficient classification scheme for discriminating between NC-LOC clays and MOC-HOC clays is presented.

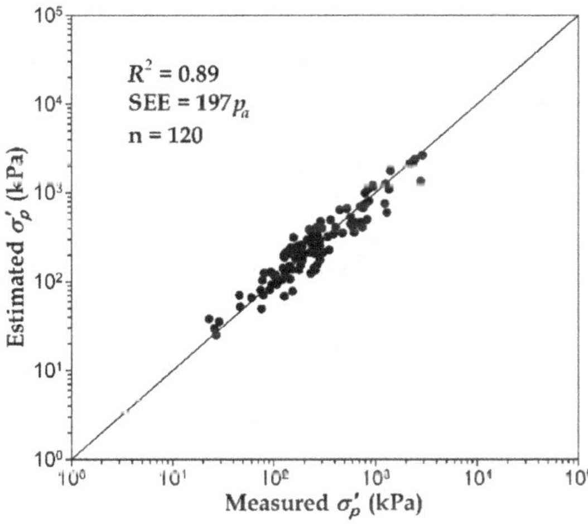

Fig. 1. Performance of proposed model on model-building data set.

3.2 Discriminant Function for Differentiating NC-LOC from MOC-HOC Clays

In order to find a mathematical expression for predicting which class (NC-LOC vs. MOC-HOC) a particular sample with known index properties (w_n, LL, etc.) and effective overburden stress (σ'_{vo}) belongs, a linear discriminant analysis (LDA) was performed. LDA finds a discriminant function (DF) that is linear combination of the variables and best discriminates between predefined groups. The coefficients of the DF are estimated so that the distance between the means of populations is maximized. The populations (or predefined groups), which were intended to make their distance as large as possible, are as follows: NC-LOC clay samples and MOC-HOC clay samples. Discriminant functions with different predictor variables were tried, yielding the following significant function, in which DS is the discriminant score:

$$DS = 5.152 \log \left(\sigma'_{vo}/p_a\right) - 0.061\, LL - 0.093\, PL + 6.219 e_n \tag{3}$$

This DF is significant to a confidence level of 99.99%. The mean discriminant score (cut-off value) for this discriminant function is 1.123 and NC-LOC clay samples have $DS < 1.123$ and MOC-HOC clay samples have $DS > 1.123$. The performance of DF on model-building and model-validation data sets is as follows: In the model-building data set, DF correctly classified 107 of the 120 samples, for 89% correct classification rate. In the model-validation data set, which is of particular interest, DF correctly classified 1668 of the 1850 samples, for 90% correct classification rate.

The fact that the discriminant function presented here works very well over a wide range of inputs suggests that it may provide a good scheme for guessing whether equation (Eq. 2a or 2b) is applicable to a specific depth in a specific deposit. Thus, Eqs. 2a–3 is combined into a single model.

3.3 External Model Validation and Comparison with Existing Models

In order to assess the generality of the new model and to assess its relative performance in comparison to existing simple empirical approaches, the newly proposed and existing models were applied to the compiled model-validation data set. In the present study, the following existing empirical models are examined: Stas and Kulhawy (1984), Nagaraj and Murthy (1986), DeGroot et al. (1999), and Ching and Phoon (2012). In the application of the proposed model to model-validation data set, it was assumed that the OCR was not available for each case. Therefore, in order to predict σ'_p for each case, combination of Eqs. 2a–3 was used. The performance of the newly proposed model on validation data set is presented graphically and quantitatively, as shown in Fig. 2. Moreover, Fig. 3 shows the performance of the existing empirical models on validation data set. The quantitative measures of model performance, which are used to compare the overall accuracies of models, include coefficient of determination (R^2), coefficient of efficiency (E), mean absolute error (MAE), and mean and coefficient of variation of K (μ_K, COV_K), where K is the ratio of predicted σ'_p over the corresponding measured σ'_p.

From Figs. 2 and 3, it is evident that the predicted values of σ'_p using the proposed model are closer to the perfect prediction line compared to the existing models. In the case of the proposed model, the major source of disagreement between predicted and measured preconsolidation stresses is attributable to the misclassified cases for which the wrong equation was used for predicting σ'_p and it is not due to the inherent defects

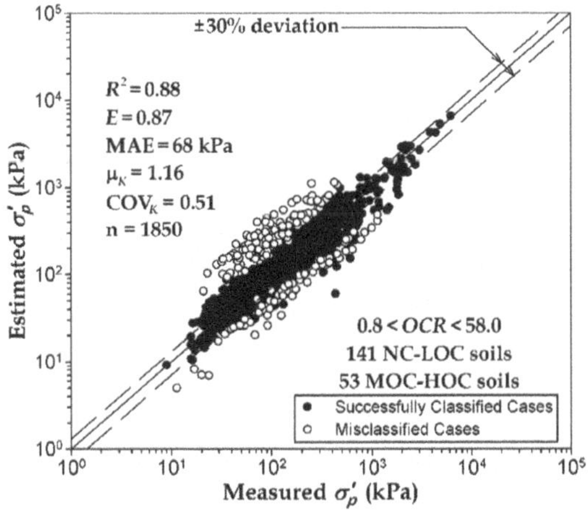

Fig. 2. Performance of proposed model on model-validation data set.

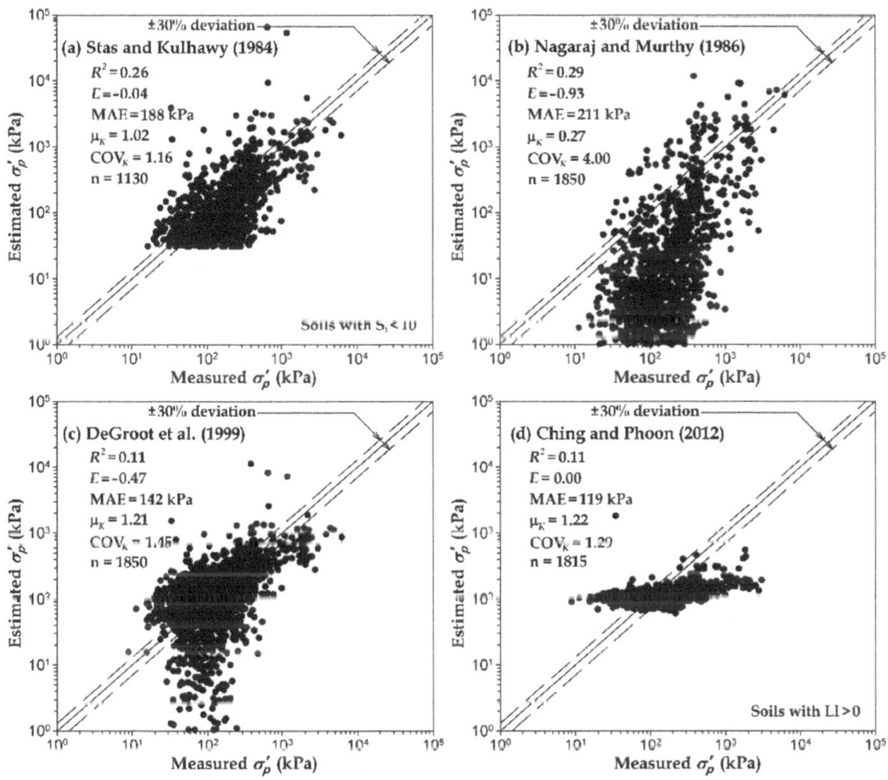

Fig. 3. Performance of existing empirical models on model-validation data set.

of the proposed expressions (Eqs. 2a and 2b). Fortunately, there are only 182 misclassifications out of the 194 soil deposits in the validation data set (i.e., only one misclassification in each soil deposit) and it should not be a worrisome issue. Moreover, reduction in the number of misclassified cases and consequently improvement in the performance of the proposed model can be obtained by using newer data mining approaches such as artificial neural networks instead of using Eq. 3.

The quantitative measures of model performance in Figs. 2 and 3 demonstrate good performance of the proposed model and its superiority over existing models. It is noted that coefficient of efficiency (E), with values ranging from $-\infty$ to 1, has several advantages over R^2 in assessing model performances (See e.g., Kootahi and Mayne 2016). An E-value of 1 indicates perfect agreement between estimated and measured values but values of $E < 0$ indicate unacceptable model performances. Based on the calculated E values, predictive performances of all existing empirical models can be evaluated as unacceptable, as their E-values range from –0.93 to 0.00. The proposed model also has the smallest COV and COVs for existing models are quite large. Indeed, it may be seen from Fig. 3 that existing models either severely underestimate or severely overestimate the preconsolidation stress.

4 Summary and Conclusions

Using a two-phase strategy, which allows for developing and validating statistical-empirical models, a new approach has been developed for predicting the preconsolidation stress of clays (σ'_p) from simple index properties (w_n, LL, PL) and the effective overburden stress (σ'_{vo}). The compiled data were processed consistently to form a good quality model-building data set and then multiple linear regression techniques were applied, which yield a highly significant algorithm that bifurcates at an OCR = 3. Discriminant analysis was also applied to find a discriminant function for discriminating between soils with OCR < 3 and soils with OCR $\geq$ 3. The performance of the new model, along with existing simple empirical models, was evaluated using an external model-validation data set and four quantitative measures of model performance (namely, R^2, E, MAE, and COV) were used to compare the overall accuracies of models. Based on the results of external validation, the performance of the proposed model was an improvement (R^2 = 0.88, E = 0.87, MAE = 68 kPa, COV = 0.51) over any of the existing models (R^2 range 0.11–0.29; E range −0.93–0.00, MAE range 119–211 kPa, COV range 1.16–4.00). Furthermore, according to the calculated coefficient of efficiencies, all of the prior existing models have unacceptable performance.

References

Casagrande, A.: The determination of preconsolidation load and its practical significance, Discussion D-34. In: 1st International Conference on Soil Mechanics and Foundation Engineering, pp. 60–64. Harvard University Press, Cambridge (1936)

Ching, J., Phoon, K.K.: Modeling parameters of structured clays as a multivariate normal distribution. Can. Geotech. J. **49**(5), 522–545 (2012)

DeGroot, D.J., Knudsen, S., Lunne, T.: Correlations among p'_c, s_u, and index properties for offshore clays. In: Singh, S. K., Lacasse, S. (eds.) International Conference on Offshore and Nearshore Geotechnical Engineering, pp. 173–178. Balkema, Rotterdam (1999)

Kootahi, K., Mayne, P.W.: Index test method for estimating the effective preconsolidation stress in clay deposits. J. Geotech. Geoenviron. Eng. **142**(10) (2016)

Kootahi, K., Mayne, P.W.: Closure to "Index test method for estimating the effective preconsolidation stress in clay deposits" by Karim Kootahi and Paul W. Mayne. J. Geotech. Geoenviron. Eng. **143**(10) (2017)

Nagaraj, T.S., Murthy, B.R.S.: Prediction of compressibility of overconsolidated uncemented soils. J. Geotech. Geoenviron. Eng. **112**(4), 484–488 (1986)

Stas, C.V., Kulhawy, F.H.: Critical evaluation of design methods for foundations under axial uplift and compression loading. Report No. EL-3771, Electric Power Research Institute, Palo Alto, California (1984)

Influence of Pore Water Salinity Effect on Identification of Soft Clay Based on CPTu in Lianyungang

Hao-Chen Xue[1], Zi-Long Wu[1], Yong-Feng Deng[1(✉)], and Yong-Hong Miao[2]

[1] School of Transportation, Institute of Geotechnical Engineering, Southeast University, Nanjing, China
nodon@seu.edu.cn

[2] School of Civil Engineering and Mechanics, Jiangsu University, Zhenjiang, China

Abstract. In-situ tests classify soils by their macro mechanical behaviors, while many laboratory investigations suggested that the chemical composition of pore water has significant influence on both physical and mechanical properties of clay. In this paper, Cone penetration test (CPTu) was conducted in Lianyungang (LYG) marine-deposit plain in order to investigate the influence of pore water salinity effect on in-situ test. Hereafter, the influence on the classification parameters of the Soil Behavior Type chart and the Eslami-Fellenius chart were discussed respectively. The results suggested that the Eslami-Fellenius chart was more suitable to identify LYG marine soft clay. Theoretical solutions of cavity expansion in modified Cam clay were proposed to interpret CPTu. It was inferred that pore water salinity effect could influence the classification parameters and lead to the deviations of projections in classification chart. Despite the parameters used in Eslami-Fellenius chart were also affected, soft clay was generally identified for the low sensitivity of their parameters to the change of pore water salinity. As a conclusion, pore water salinity effect should be taken into consideration while interpreting CPTu data with respect to marine clay.

Keywords: CPTu · Marine clay · Soil classification · Salinity effect

1 Introduction

CPTu has been widely used in geotechnical investigation. A continuous and rapid soil profiling is one of the advantages it has over conventional investigation approaches. Many criterions have been proposed based on different database collected by CPTu [1–3]. Essentially, in-situ tests categorize soils by the difference of their macro mechanical behaviors. Previous researches have paid attention to the change of mechanical properties with different stress history [4]. Nevertheless, in addition to physical interactions between the particles, there is the coupling of the pore fluid flow with the solid matrix deformation that should be taken into consideration.

L. Hu et al. (Eds.): GSIC 2018, *Proceedings of GeoShanghai 2018 International Conference: Multi-physics Processes in Soil Mechanics and Advances in Geotechnical Testing*, pp. 400–409, 2018.
https://doi.org/10.1007/978-981-13-0095-0_45

Pore water salinity effect refers to the effect of pore water salinity on the physical and mechanical properties of soft marine clay. The interaction between montmorillonite and cations has proved to be a key factor of this effect [5]. The presence of soluble salts in pore water can affect engineering characteristics via cation exchange, flocculation, and dispersion of clay particles. Thickness of the diffuse double layer changes with the variation of cation concentrations, along with plasticity properties. On the other hand, with the increase of electrolyte concentration, the floc formation among the platelets of the clays gets accelerated due to the dominance of attractive forces among particles in the suspensions, leading to the formation of structure [6], along with the change of macro mechanical behavior. Meanwhile, the consolidation behavior of clays is also influenced by the presence of soluble salts [7].

However, with the leaching of salinity, the strength of the clay would decline dramatically [8]. In most situations, the clay with a higher saline concentration tends to be stiffer than that without salinity [9]. As a result, the in-situ tests only reveal the current mechanical behaviors which has been affected by salinity effect and would change with the leaching of salinity.

The aim of this study is to investigate the influence of pore water salinity effect on CPTu interpreting. LYG marine deposit was taken as an example and the influences on the classification parameters of the Soil Behavior Type classification, the Eslami-Fellenius chart were discussed respectively.

2 Test Sites and Methods

2.1 Test Sites

The test sites were located in LYG marine deposit plain, where soft clay was widely deposited. Note that the clays in this area were deposited under a sea regression background [7]. The clays in this plain have high water contents, compressibility, sensitivity, low permeability and strength [10], leading to many problems in engineering. Twelve sites were selected for this study. Figure 1 showed the location of each site.

2.2 Test Methods

In-situ Tests
CPTu was conducted by the Vertek Hogentogler standard cone penetration device. The data were recorded every 5 cm until the penetration of Holocene sedimentary.

Laboratory Tests
High quality samples were taken at each site. The basic properties of the sampled soils were investigated following JTG E40-2007 [11] and the results were listed in Table 1. The plasticity properties of each sample were demonstrated in Fig. 2.

The tested pore water was centrifuged from the sampled soils and passed through a 22 μm syringe filter. The chemical composition of pore water was determined by Ion Chromatograph (IC). As an exception, the concentration of SO_4^{2-} was measured by EDTA titrimetric method since the results measured by IC was interfered by the high

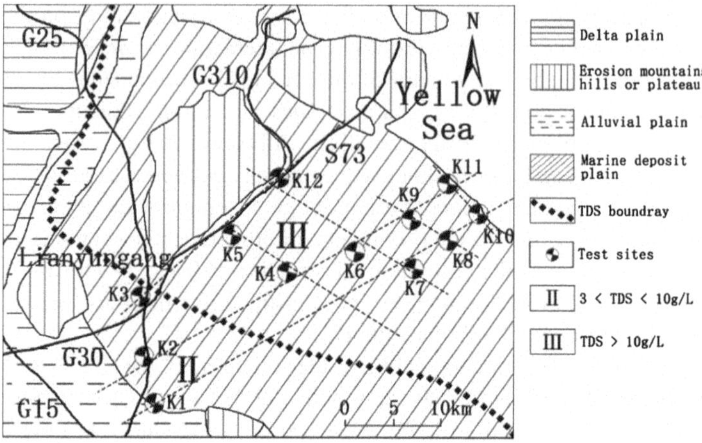

Fig. 1. Coastal landforms of LYG and distribution of test sites

Table 1. Statistics of physical and mechanical characteristics of tested soft clay

	Unit weight ($kN \cdot m^{-3}$)	Water content $\omega(\%)$	Void ratio	Plasticity index PI	Liquid index LI
Maximum	17.4	77.8	2.25	44.6	1.46
Minimum	15.4	37.9	1.27	21.2	0.56
Average	16.5	59.3	1.66	32.5	1.00
Standard error	0.6	10.4	0.29	7.0	0.20

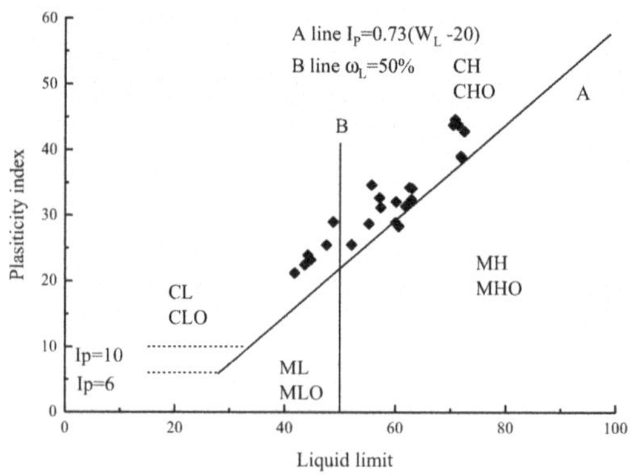

Fig. 2. Plasticity chart

concentration of Cl^-. The total saline concentration of the pore water was assumed to be approximate to the sum of all measured ion concentrations (including Cl^-, Na^+, K^+, Mg^{2+}, Ca^{2+}, $Fe^{2+/3+}$ and SO_4^{2-}). Furthermore, mineral compositions of the samples were measured by X-ray diffraction (XRD) following SY/T5163-2010 [12]. The results are presented in Table 2.

Table 2. Mineral composition of each site

Site		K2	K3	K4	K5	K6	K7	K8	K9	K10	K11
Non-clay mineral		77.0	55.6	70.5	51.8	83.2	54.0	56.6	48.0	59.6	75.5
Clay mineral	Illite/smectite	12.4	16.4	12.7	25.1	6.4	21.7	16.5	27.6	13.7	9.3
	Other clay mineral	10.6	28.0	16.8	23.1	10.4	24.3	26.9	24.4	26.7	15.2

3 Results and Discussion

3.1 Salinity Distribution

Due to the presence of sandy or silty interlayers, the test data were not intact at some positions. As a result, four boreholes with intact data, K4, K7, K8 and K10, were chosen to illustrate the salinity distribution and soil classification to demonstrate the relation between saline concentration and in-situ test results. The tip resistance curve and the salinity distribution are shown in Fig. 3 and Fig. 4, respectively. The tip resistance of K4 is generally higher than K7, followed by K8 and K10 in order. This sequence is in correspondence with the sedimentary age.

K4 is the farthest borehole away from modern coastal line and the earliest site to be deposited among the 4 sites. The long-time wash by runoff leads the saline concentrations to be almost identical at different depth, and they are all lower than the average

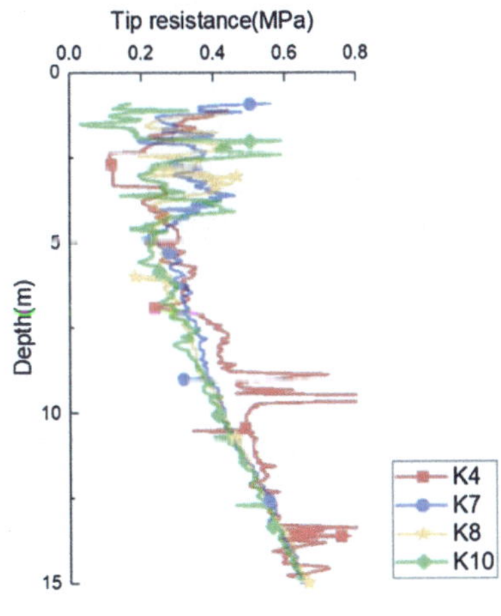

Fig. 3. Tip resistance of each site

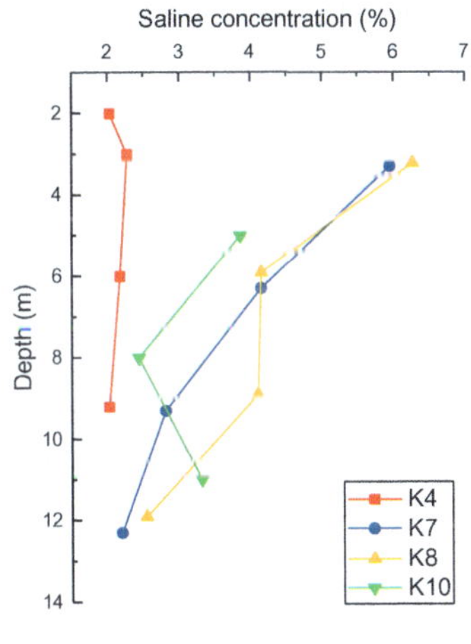

Fig. 4. Distribution of major ion concentration

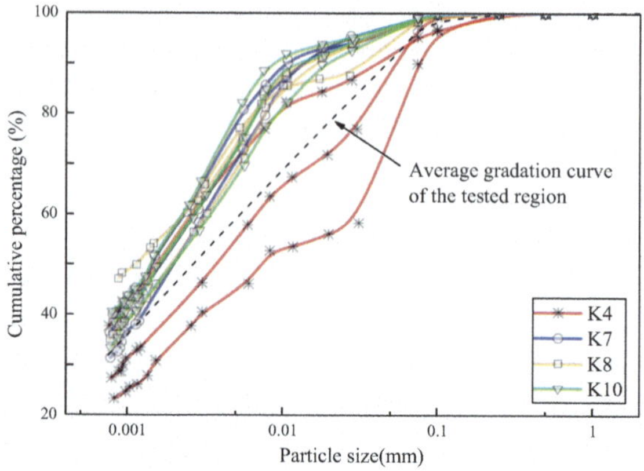

Fig. 5. Particle size distribution

concentration of Yellow Sea. Meanwhile, the particle size distribution curve shows a distinct discrepancy among the sediments at different depths (Fig. 5), suggesting the runoff intensity changes with the variation of sea level.

The gradation curve shows little difference among the tested samples, implying a relatively stable deposit environment. The lower layers of K7, K8 and K10 has lower concentrations than Yellow Sea, while the concentration of the upper layers declines with the increasing depth. The high concentration in the surface layer is mainly caused by surface water evaporation. Downwards the surface layer, the concentration depends mainly on the time of the sediment being washed by underground runoff, leading to a notable vertical saline distribution.

3.2 Soil Classification

A rapid and continuous soil classification is one of the advantages that in-situ tests surpass conventional investigation approaches. There are plenty of classification methods proposed by researchers based on tests conducted in different regions and by different means. Among them, many parameters have been proposed to classify soil type. Some parameters originate from theoretical solutions and others from empirical practices. Some variables in the theoretical solutions would be affected by the change of pore water salinity, resulting in the deviation and misjudgment of classification.

Soil Behavior Type Classification

Soil behavior type classification, proposed by Robertson [13], is one of the most widely used classification chart. Z1 and Z2 in Fig. 6 were merged as one zone representing sludge in correspondence to Chinese soil classification standard [3]. Two parameters are required to identify soils based on CPTu data, including normalized cone resistance (Q_t) and normalized friction ratio (F_R):

$$Q_t = (q_t - \sigma_{v0})/\sigma'_{v0} \quad (1)$$

$$F_r = f_s/(q_t - \sigma_{v0}) \times 100\% \quad (2)$$

Where σ_{v0} is vertical stress; σ'_{v0} is effective vertical stress.

The test results of soft clays in the 4 sites are listed in Fig. 6. Most sludge recorded in drill profile is identified as clay in this chart. The data points of K4 distributed in a wide range from Z1 to Z4 in Fig. 6 (note that with the increasing of opacity, the density of data points becomes greater). K10, located closest to latest coastline and deposited the most recently, shows the closest soil behavior to sludge. Generally, the clays deposited earlier located further from Z2.

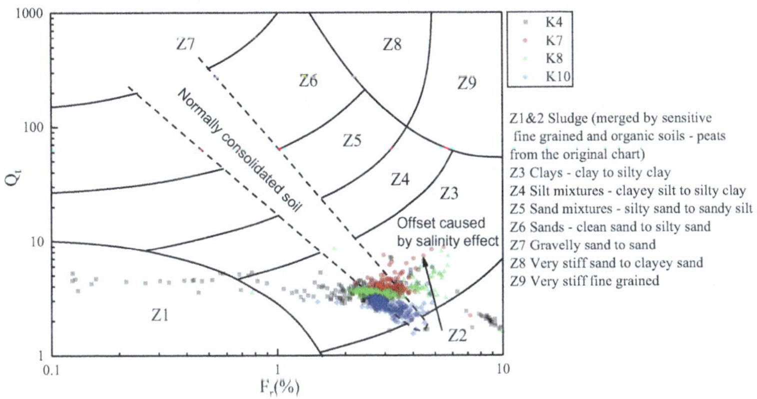

Fig. 6. CPTu results on the Robertson classification of soil behavior type

This result is similar to the classification of Norwegian marine clay [14]. The deviation was ascribed to the fact that Robertson diagram was based mainly on Canadian soils. Nevertheless, considering the similar result of LYG marine clay, it is speculated that pore water salinity effect is one of the factors resulting in this deviation.

Deng *et al.* [7] reported that pore water salinity effect had significant influence on the engineering properties of LYG clay, and a clay with a higher salinity behaved more like a silt. Zhang [15] investigated the influence of this effect on artificial clay mixed by montmorillonite and kaolinite and discovered that with the increasing saline concentration, the critical stress ratio (M), rigidity index (I_R) increased, plasticity index (I_p) decreased and the slope of the normal consolidation line in the e-ln σ'_v plane (λ) were slightly changed. With the interaction between montmorillonite and ions, it is presumed the influence mechanisms of natural clays are similar to that of artificial clays.

Mayne [16, 17] proposed a simplified theoretical solution of cavity expansion in modified Cam clay to evaluate the tip resistance of CPTu, expressed as:

$$q_t = \sigma_{v0} + [\frac{4}{3}(\ln I_R + 1) + \frac{\pi}{2} + 1] \cdot \frac{M}{2} \cdot (\frac{OCR}{2})^{\Lambda} \cdot \sigma'_{v0} \quad (3)$$

Where I_R is the rigidity index of the clay. I_R can be calculated by:

$$I_R = \frac{\exp(\frac{137 - I_p}{23})}{1 + \ln[1 + (\frac{OCR - 1}{26})^{3.2}]^{0.8}} \quad (4)$$

Thus, based on the above expressions, it can be inferred that tip resistance would be bigger with the increasing of saline concentration at a certain depth. To avoid the trouble of discussing the influence of depth, normalized tip resistance is given by:

$$Q_t = \frac{qt - \sigma_{v0}}{\sigma'_{v0}} = [\frac{4}{3}(\ln I_R + 1) + \frac{\pi}{2} + 1] \cdot \frac{M}{2} \cdot (\frac{OCR}{2})^{\Lambda} \quad (5)$$

Equation 5 indicates that when M and I_R increase under the influence of the interaction between montmorillonite and ions, Q_t would appear bigger. The saline concentration of K7 and K8 at different depth shows the regularity of vertical distribution. Meanwhile, the particle distribution of both sites is similar. Figure 7 demonstrates that the trend of saline solution and Q_t remains similar.

Tan *et al.* [18] proposed that the sleeve friction could be taken similar to Beta method for calculating side friction of piles. The sleeve friction of CPTu is given by:

$$f_s = K_0 \tan\delta' \sigma'_{v0} \quad (6)$$

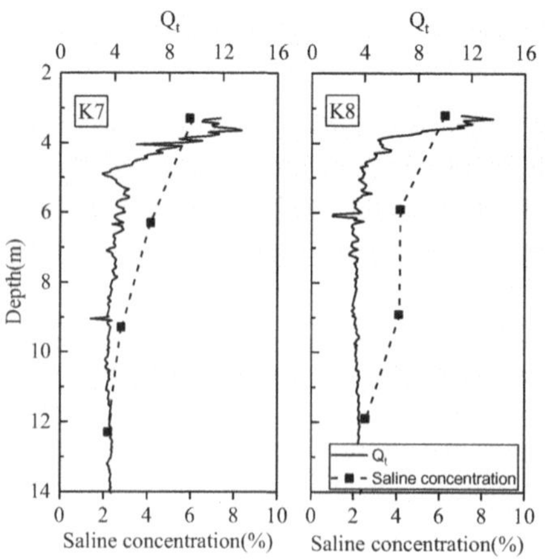

Fig. 7. Correlation between Qt and saline concentration

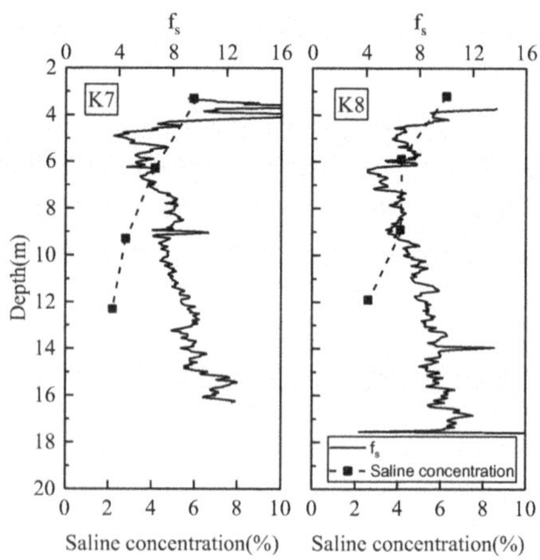

Fig. 8. Correlation between f_s and saline concentration

Where $tan\delta'$ is the interface friction between penetrometer and the soil.
K_0 can be calculated by:

$$K_0 = (1 - \sin\phi')OCR^{\sin\phi'} \tag{7}$$

Where ϕ' is the effective friction angle.
The normalized sleeve friction could be expressed by:

$$F_r = f_s/(q_t - \sigma_{v0}) \times 100\% = f_s/(Q_t \cdot \sigma'_{v0}) \tag{8}$$

The effective friction angle increases with the increase of saline concentration, leading to the decrease of K_0 and f_s [17]. Figure 8 gives the relation between f_s and saline concentration.

In SBTn chart, the changes of Q_t and F_r caused an offset of the data projection. In this case, under the effect of salinity effect, the increase of Q_t and decrease of F_r led the projection to move towards top-left, away from Z2. Therefore, the sludge was mis-classified as normal clays, which would cause serious accidents in practices.

Eslami-Fellenius Classification

Eslami and Fellenius noted that the assumption of independence between Q_t and F_r would lead to a distortion in SBTn chart. Furthermore, the parameters used in SBTn chart depend on overburden stress, which is hard to obtain directly from CPTu [1]. Therefore, a classification based on net tip resistance (q_E) and sleeve friction was proposed.

Figure 9 shows the database CPTu records plotted in the Eslami-Fellenius classification chart. It presents that LYG marine clays were accurately identified to be fine grains/sensitive/collapsible clay. Eslami found that the smaller the soil particle, the lower the location of data point in this chart. The data points of K4 are located higher

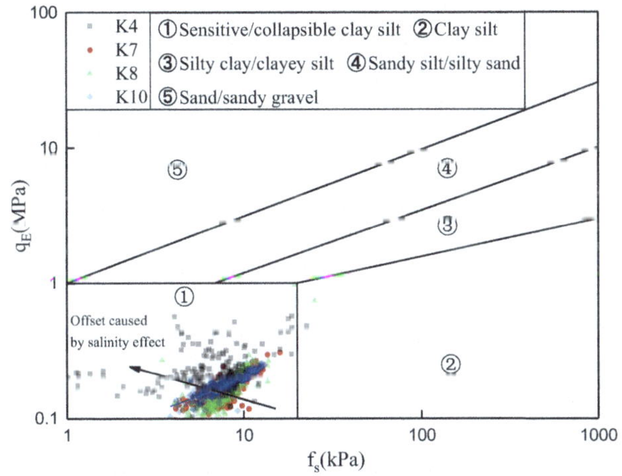

Fig. 9. Eslami-Fellenius classification chart

than others', indicating the particle sizes of K4 were bigger than others, which is consistent with the gradation distribution in Fig. 5.

Similar to SBTn classification, the parameters of Eslami-Fellenius classification chart were also influenced by salinity effect. Mayne proposed that the theoretical solution of the penetration pore water pressure at the cone shoulder could be expressed by:

$$u_2 = u_0 + \frac{4}{3}\ln I_R \cdot \frac{M}{2} \cdot \left(\frac{OCR}{2}\right)^{\Lambda} \cdot \sigma'_{v0} + \left[1 - \left(\frac{OCR}{2}\right)^{\Lambda}\right] \cdot \sigma'_{v0} \tag{9}$$

According to Eqs. (3) and (9), the net tip resistance can be calculated using:

$$q_E = q_t - u_2 = \sigma_{v0} - u_0 + \left[\left(\frac{OCR}{2}\right)^{\Lambda} - 1\right] \cdot \sigma'_{v0} + \left(\frac{7}{3} + \frac{\pi}{2}\right) \cdot \frac{M}{2} \cdot \left(\frac{OCR}{2}\right)^{\Lambda} \cdot \sigma'_{v0} \tag{10}$$

q_E increases with the increasing of M. Thus, the changes lead the projection to move towards top-left in the chart. On the other hand, experiments have proved that clays with montmorillonite flocculate in salty environment [6], driving the particles to grow bigger. As has been discussed above, larger particles locate higher in the chart. Therefore, the theoretical and experimental results are in good accord.

Zhang [17] reported that M almost doubled when the saline concentration raised from 0% to 5% in artificial clay composed by 20% montmorillonite and 80% kaolinite. After comparing Eqs. (5), (8) and (10), it can be inferred that theoretically, with the increasing of M led by salinity effect, Q_t is linearly multiplied with M, while just one part linearly magnified in q_E. This discrepancy results in the fact that parameters used in SBTn classification are much more sensitive to salinity variation than that in Eslami-Fellenius classification.

4 Conclusions

In this work, CPTu was conducted in LYG marine deposit plain in order to study the influence of pore water salinity effect on in-situ tests. Two CPTu-based soil identification methods were discussed in this study. The SBTn chart only identified a handful of soft clays, while the Eslami-Fellenius chart succeeded in identifying most of them. The theoretical solutions of the classification parameters were deduced based on cavity expansion theory in modified Cam clay. The results indicated that when M was increased by the interaction between montmorillonite and ions, Q_t, F_r in SBTn classification was remarkably influenced, leading to an offset away from sludge zone on the classification chart. However, the parameters used in Eslami-Fellenius chart was slightly affected so that the classification was generally accurate. Thus, pore water salinity effect of marine clay should be taken into consideration while processing in-situ test results in order to get a proper understanding of the clay properties.

Acknowledgements. This study is supported by the National Natural Science Foundation of China (Grant No. 51378117, 41572280), China Scholarship Council and Fundamental Research Funds for the Central Universities (Grant No. KYLX_0148), and Guangxi and Nanning Science and Technology Project Grant (No.14124004-4-12, 20143113).

References

1. Eslami, A., Fellenius, B.H.: CPT and CPTu data for soil profile interpretation: review of methods and a proposed new approach. Iran. J. Sci. Technol. **28**, 69–86 (2004)
2. Robertson, P.K.: CPT-based Soil Behaviour Type(SBT) classification system - an update. Can. Geotech. J. **53**, 1910–1927 (2016)
3. Liu, S.Y., Cai, G.J., Zou, H.F.: Practical soil classification methods in China based on piezocone penetration tests. Chin. J. Geotech. Eng. **35**, 1765–1776 (2013)
4. Józsa, V.: Soil classification and determination of over-consolidation from CPTu test in deep excavation. Pollack Periodica **8**, 53–63 (2013)
5. Yukselen-Aksoy, Y., Kaya, A., Ören, A.H.: Seawater effect on consistency limits and compressibility characteristics of clays. Eng. Geol. **102**, 54–61 (2008)
6. Gumaste, S.D., Iyer, K.R., Sharma, S., Channabasavaraj, W., Singh, D.N.: Simulation of fabric in sedimented clays. Appl. Clay Sci. **91**, 117–126 (2014)
7. Deng, Y.F., Yue, X.B., Cui, Y., Shao, G.H., Liu, S.Y., Zhang, D.W.: Effect of pore water chemistry on the hydro-mechanical behaviour of Lianyungang soft marine clay. Appl. Clay Sci. **95**, 167–175 (2014)
8. Bjerrum, L.: Geotechnical properties of Norwegian marine clays. Geotechnique **4**, 49–69 (1954)
9. Shinde, S.T., Juneja, A.: Engineering behavior of soils in electrolyte solutions. In: Geo-Chicago 2016, pp. 959–968 (2016)
10. Yi, M., Zhang, D.W.: Research into engineering characteristics and soil improvement method of marine clay in Lianyungang. J. Highw. Transp. Res. Dev. **05**, 52–55 (2005)
11. JTG E40-2007: Test methods of soils for highway engineering (2007)
12. SY/T5163-2010: Analysis method for clay minerals and ordinary non-clay minerals in sedimentary rocks by the X-ray diffraction (2010)
13. Robertson, P.K.: Soil classification using the cone penetration test. Can. Geotech. J. **27**, 151–158 (1990)
14. Thakur, V., L'Heureux, J., Locat, A.: Landslides in Sensitive Clays. Springer Nature, Cham (2017)
15. Zhang, T.W.: Pore water salinity effect on the physical, mechanical behavior and constitutive model of artificial soft clay, Ph.D dissertation, Southeast University (2017)
16. Mayne, P.W.: In situ determination of clay stress history by piezocone model. In: Predictive Soil Mechanics. Proceedings of the Wroth Memorial Symposium, St Catherine's College, Oxford (1992)
17. Mayne, P.W.: Determination of OCR in clays by piezocone tests using cavity expansion and critical state concepts. Soils Found. **31**, 65–76 (1991)
18. Tan, T.S., Phoon, K., Hight, D.W., Leroueil, S.: Characterisation and engineering properties of natural soils. In: Two Volume Set. Proceedings of the Second International Workshop on Characterisation and Engineering Properties of Natural Soils, Singapore, 29 November–1 December 2006, vol. 3. CRC Press (2006)

Analysis and Interpretation of Inclinometer and Pressure Cell Data on a Soil-Geofoam Embankment

Ali Shafikhani, Tejo V. Bheemasetti, Anand J. Puppala(✉), and Aritra Banerjee

The University of Texas at Arlington, Box 19308, Arlington 76019, USA
anand@uta.edu

Abstract. This paper presents analysis and interpretation of monitored vertical settlements and pressures of a rehabilitated bridge approach slab located in Johnson County, Texas. Four horizontal inclinometer casings and pressure cells (equipped with thermometers) were installed at the test site during the rehabilitation process. In this paper, collected data from the inclinometers, pressure cells, and thermometers were employed in an attempt, to understand the effect of climate changes on the vertical settlements and pressures of the approach slab. It has been observed that with an increase in temperature, the bridge structure enforced the movements of the geofoam blocks causing vertical settlement. Whereas, the bridge structure movements after a temperature decrease, induced vertical swelling at the top of the geofoam blocks. The pressure cells that were installed at the top and bottom of the geofoam embankment revealed that the stresses observed at the top of the geofoam were significantly reduced. Also, the pressure cells installed at the sides of the bridge and geofoam structure to evaluate the lateral pressure response lost contact and provided negative results. The loss of contact can be a response to the movement of the structure with respect to thermal changes in the structure. This research highlights the important observations of a bridge structure and its approach soil-geofoam embankment movements with respect to temperature and precipitation variations.

Keywords: Climate changes · Vertical movements · Swelling
Settlement

1 Introduction

The differential settlement between bridge approach and bridge deck, also known as 'bump phenomenon' is a common problem faced by approximately 25% of the 600,000 bridges across the United States [1, 2]. Millions of dollars are spent annually in an attempt to repair these bridges [1]. Recent studies on bridge infrastructure demonstrated that more than 28 states in the United States use an approach slab as an interface to the pavement and bridge deck. In a survey conducted in Texas, it was estimated that it costs annually more than 100 million dollars for the Texas Department of Transportation (TxDOT) to repair the bridges with bump problem in Texas [3–5].

L. Hu et al. (Eds.): GSIC 2018, *Proceedings of GeoShanghai 2018 International Conference: Multi-physics Processes in Soil Mechanics and Advances in Geotechnical Testing*, pp. 410–418, 2018.
https://doi.org/10.1007/978-981-13-0095-0_46

Several researchers have tried to determine the probable causes of bump phenomenon at bridge approaches [3, 6–19]. Based on the previous literatures, the primary factors causing the approach bump problem include consolidation settlement of foundation soil, poor embankment soil compaction close to the bridge approach, poor drainage system and soil erosion due to the age of the bridge, types of bridge abutment, and traffic volume.

In the past decade, extensive studies were performed to identify the causes of the bump phenomenon and several techniques have been proposed to mitigate the problem [8, 9, 12, 14, 15, 20–22]. Some of the widely-accepted techniques to resolve differential settlement problem include excavation and replacement, deep soil mixing (DSM) column, geosynthetic reinforcement, mechanically stabilized earth (MSE) wall, effective drainage, and erosion control method and use of lightweight materials to decrease the effective overburden pressure.

Thermally induced displacements generally occur at integral abutment bridges due to seasonal temperature changes [23–29]. These long-term deformations mainly take place because of the continuity of superstructure and substructure of bridges with integral abutment systems and can make a progressive contribution to accelerating the development of bump phenomenon. As per the seasonal temperature variations, lateral interaction at the studied bridge structure and the integrated hybrid soil-geofoam approach embankment system was monitored and presented.

In this research study, lightweight geofoam blocks were used in place of the top 1.83 m (6 ft) of the soil in order to reduce the approach slab-bridge deck differential settlement. The geofoam blocks despite being 100 times lighter than the soil, can provide comparable strength and stiffness properties as soils. Effect of temperature and precipitation was studied to monitor the performance of a Geofoam-soil hybrid system installed at a site which was facing the menace of the "bump phenomenon". The following sections provide the details of the instrumentation and analysis.

2 Site Description

The US 67 bridge over SH 174, located at Johnson County, Cleburne, Texas was constructed in 1995. The approach slab was constructed on the 12.2 m (40 ft) moderately high plastic clay embankment which was laterally supported by concrete retaining walls. More than 406.4 mm (17 in.) settlement was experienced by the approach slab over a period of 16 years. Excessive settlement happened primarily due to the consolidation settlement of the embankment soil, insufficient compaction of soils near the abutment and erosion [11]. During this period, several soil improvement techniques include soil nailing, grout injections, and hot mix overlays were tried but none of the applied techniques were found to be effective to mitigate the differential settlements at the bridge approach slab.

In January 2012, the Texas Department of Transportation (TxDOT) replaced the top 1.83 m (6 ft) of the embankment soil by the lightweight EPS 22 Geofoam. In order to monitor the long-term vertical deformations of the Geofoam layer under diverse environment and loading conditions, four horizontal inclinometer casings (US67 -1, US67 -2, US67 -3, US67 -4) were installed on the top of the Geofoam layer during the construction. Data monitoring in two years depicted a differential settlement of less than 38.1 mm (1.5 in.). This was considered as the allowable design settlement to evaluate long-term performance of the rehabilitation technique [11]. In addition, four pressure cells were installed at the top and bottom of the Geofoam layer to assess the lateral and vertical pressures transferred to and from the EPS Geofoam layer respectively (see Fig. 1). Figure 1(a) and (b) presents the pressure cells installed on top and bottom of the geofoam blocks for monitoring vertical pressures that were transferred from pavement layers. Whereas, Fig. 1(c) and (d) depicts the pressure cells installed laterally on the retaining wall and attached to the Geofoam blocks to monitor the lateral pressure that is subjected onto geofoam blocks. The laterally installed pressure cell at the top side of the bridge structure and geofoam interface as shown in Fig. 1(c) showed occasional negative values. This led us to analyze the role of thermal expansion and contraction of the bridge structure on the Geofoam material.

(a) (b) (c) (d)

Fig. 1. Four pressure cells installed on the top and bottom of the geofoam layer

3 Data Analysis and Interpretation

The data of two pressure cells installed vertically at the top and bottom of the Geofoam blocks were used to analyze the performance of the Geofoam layer under dynamic traffic loadings, temperature variations, and precipitation. Figure 2 shows a typical monthly (a) and daily (b) pressure diagram at the top and the bottom of the Geofoam layer for a given month and day respectively.

The recorded pressures include the dynamic traffic loading in addition to the static overburden pressure of 0.61 m (2 ft) rigid pavement. A significant decrease of the average pressure from the top (35.8 kPa) to the bottom (12.4 kPa) of the Geofoam layer is evident from both plots (Fig. 2(a) and (b)). However, no time-dependent viscous behavior can be observed under the combination of static and dynamic loadings in the either plots. Although Geofoam material has good damping characteristics, it was observed that the amplitude of recorded pressure by the bottom pressure cell was higher than that recorded by the ones at the top (Fig. 2(a)). This can be attributed to the stronger pressure cell rigid soil base at the bottom as opposed to the Geofoam blocks at the top.

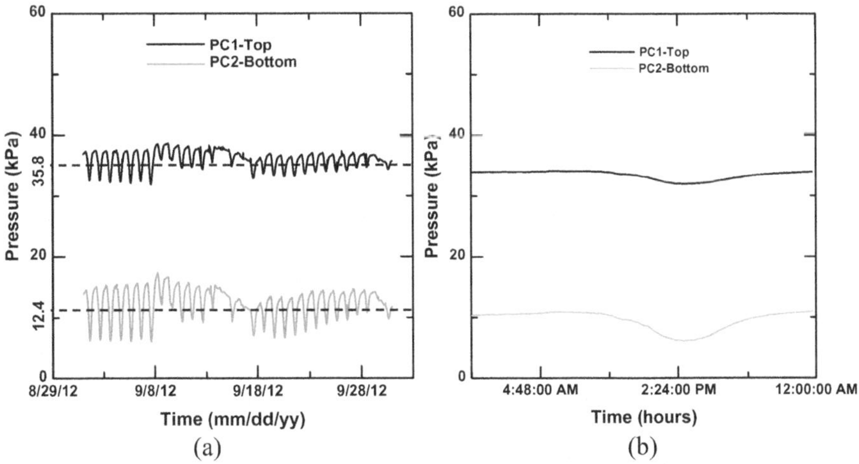

Fig. 2. (a) Typical monthly and (b) daily vertical pressures

3.1 Influence of Temperature Variation on Lateral Pressures

The temperature and pressure data were collected from the installed pressure cell (equipped with the thermometer) at the top side of the geofoam layer and the bridge structure interface (retaining wall). Because of the integral bridge-approach embankment system of the infrastructure, thermally induced lateral pressure interaction was recorded at the bridge and the approach embankment interface. Figure 3 shows the variation of temperature and pressure versus time at the top of the Geofoam layer. In the following plots, the left vertical axis represents the temperature variations and the right vertical axis depicts the lateral pressure changes. It should be noticed that both parameters (temperature and lateral pressure) were recorded simultaneously each 15 min' time intervals. It can be observed from the Fig. 3 that a consistent trend of lateral pressure variations was observed with change in temperature for the years 2012 and 2013. From January 2012, the lateral pressures increased gradually with the rise in temperature till the end of June 2012 and reached their peak magnitudes (see Fig. 3(a)).

As long as the retaining wall and the Geofoam layer were in contact, the lateral dynamic response to the traffic loading was amplified. Whereas, with a slight decrease in the temperature in July 2012, a rapid drop in lateral pressures was recorded which finally reached negative values at a temperature of 28.8 °C. In other words, with a decrease in temperature, bridge started to move back to its initial position. However, the difference in thermal induced stress-strain behavior of the concrete and Geofoam materials caused a quick decline on the lateral pressures in July 2012. Figure 3(b), shows a similar trend of the lateral pressure fluctuations with the seasonal temperature variations for 2013. This pattern reveals important information about the physical interaction between the bridge structure and the geofoam layer due to the seasonal temperature changes.

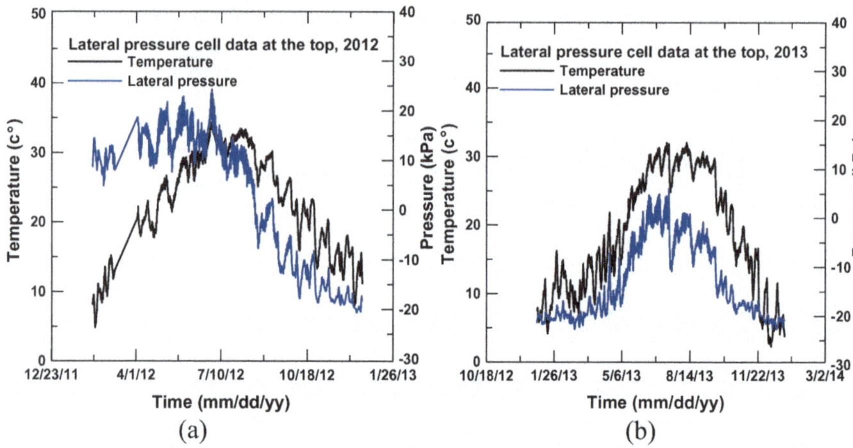

Fig. 3. Lateral pressure changes with temperature variations (a) 2012 (b) 2013.

3.2 Influence of Temperature Variation on Vertical Pressures

This section presents the influence of temperature variations on the vertical pressure at the top and bottom of the Geofoam layer. The temperature variations recorded by the top pressure cell was negligible since the vertically installed pressure cell was wrapped in geomembrane and embedded in the Geofoam, a material with negligible thermal conductivity (see Fig. 1a). Therefore, the variation in the recorded pressure data was due to the dynamic traffic loading rather than changes in temperature. However, the pressure cell at the bottom of the Geofoam layer (see Fig. 1b), has recorded reasonable temperature variations.

Figure 4(a) and (b) depicts vertical pressure changes and deformations corresponding with the thermal fluctuations at the bottom of the Geofoam layer. The behavior of Geofoam under the seasonal temperature changes can be better assessed at the soil-Geofoam interface due to the thermal insensitivity of the soil. Hence, a similar trend was observed for both vertical pressure change and temperature variation at the bottom of the Geofoam layer (see Fig. 4a and b). The effects of seasonal temperature variations were also studied on the collected vertical deformations at the top of the Geofoam blocks (Fig. 4(b)). It was observed that by increasing the temperature, settlements were raising up and by decreasing the temperature, swelling behavior was observed in the average vertical deformations.

3.3 Precipitation Effect

This section presents the effect of precipitation on the vertical settlements obtained from the inclinometer data. Precipitation data were obtained from database recorded from U.S. Climate Data. Figure 5 presents the precipitation data for 2012 and 2013.

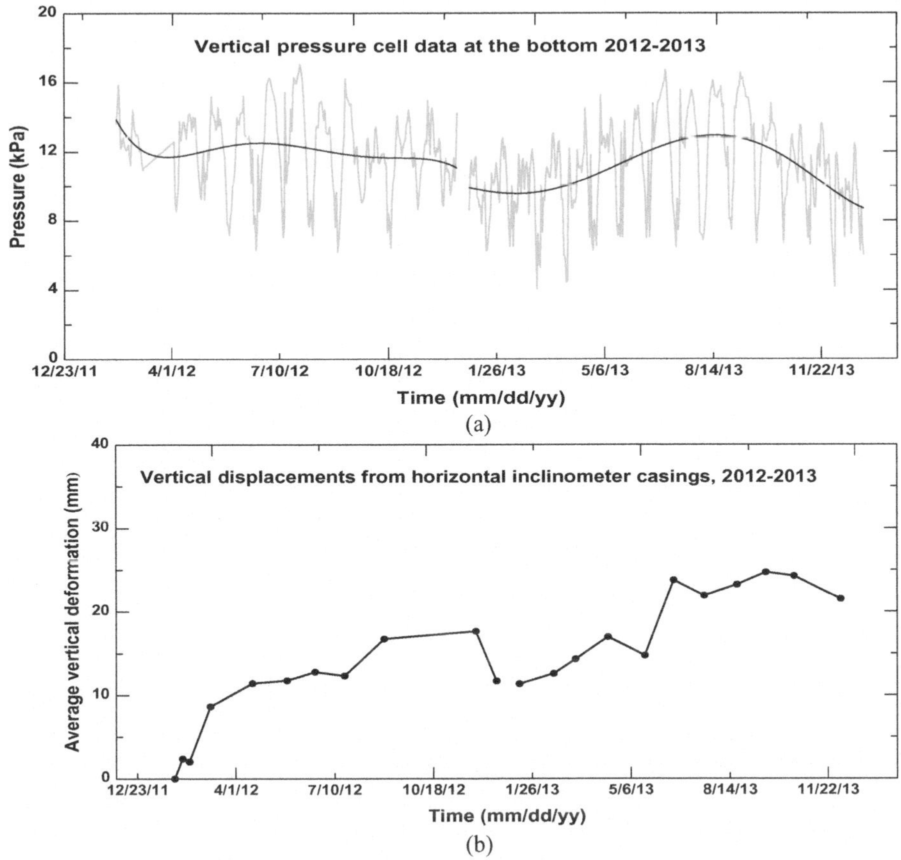

Fig. 4. (a) vertical pressure cell data at the bottom, (b) vertical deformation

Preliminary analysis of the data showed no immediate effect of precipitation on the vertical deformations in two years. This could be due to the very low permeability of the high plasticity clayey soil present in the embankment. A close study on the plots depicted the swell and shrink cycles after heavy rainfall and dry periods respectively [30]. However, a more detailed analysis is required before attributing the effects of precipitation on vertical deformation.

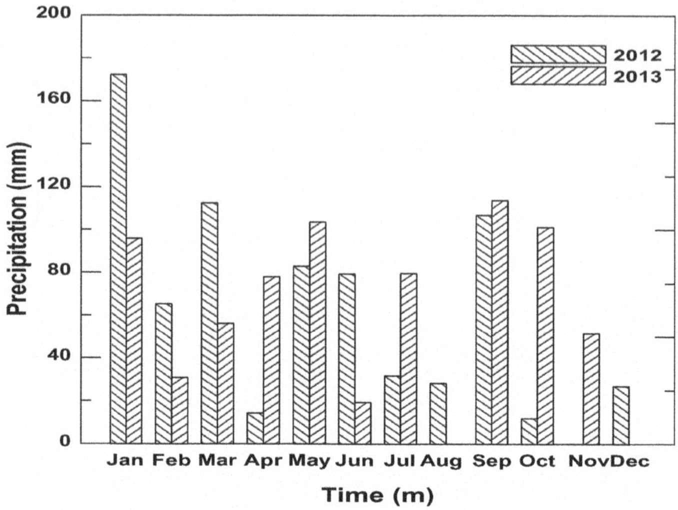

Fig. 5. Precipitation data (2012–2013)

4 Conclusion

- Vertical pressure cell data at the bottom of the Geofoam shows a considerable decrease in the average vertical pressures. This is due to the stress absorption of the geofoam material. The dynamic response from the traffic loading was studied, where the stress amplitude was increased due to the stronger rigid soil base at the bottom as opposed to the Geofoam blocks at the top.
- Analysis of the seasonal temperature variation on vertical deformations and lateral pressure showed that the Geofoam blocks were laterally pushed by the thermal expansion of the bridge structure. Also, it has been observed that the variation in the vertical deformation was consistent with the temperature variation.
- The effect of the precipitation data on the vertical settlement was studied for two years and no immediate effect was observed. This could be due to the low permeability of the embankment soils present in the considered test section and also the geomembrane wrapped around the geofoam block which provides an impermeable interface.

References

1. Islam, A.A.: On reducing bumps at pavement-bridge interface. Final report submitted to: YSU Center for Transportation and Materials Engineering (CTME). Youngstown State University, Youngstown, OH 44555 (2010)
2. Thiagarajan, G., Gopalaratnam, V., Halmen, C., Ajgaonkar, S., Ma, S., Gudimetla, B., Chamarthi, R.: Bridge Approach Slabs for Missouri DOT: Looking at Alternative and Cost-Efficient Approaches (No. OR 11.009) (2010)

3. Briaud, J.L., James, R.W., Hoffman, S.B.: Settlement of Bridge Approaches: (The Bump at the End of the Bridge), vol. 234. Transportation Research Board (1997)
4. Nassif, H., Abu-Amra, T., Shah, N.: Finite element modeling of bridge approach and transition slabs (No. FHWA-NJ-2002–007) (2002)
5. Saride, S., Puppala, A.J., Archeewa, E.: Bridge Approach Settlements—An Issue Due to Design or Construction Practices, pp. 210–214. The University of Texas at Arlington (2009)
6. Hopkins, T.C., Deen, R.C.: The Bump at the End of the Bridge (1969)
7. Yasrobi, S.Y., Ng, K.W., Edgar, T.V., Menghini, M.: Investigation of approach slab settlement for highway infrastructure. Transp. Geotech. **6**, 1–15 (2016)
8. Seo, J.B.: The bump at the end of the bridge: an investigation (Doctoral dissertation, Texas A&M University) (2005)
9. White, D.J., Mekkawy, M.M., Sritharan, S., Suleiman, M.T.: "Underlying" causes for settlement of bridge approach pavement systems. J. Perform. Constructed Facil. **21**(4), 273–282 (2007)
10. Stewart, C.F.: Highway structure approaches. FHWA/CA/SD-85-05, Office of Applied Research, Division of Structures, California Department of Transportation, Sacramento, California (1985)
11. Ruttanaporamakul, P., Puppala, A.J., Pedarla, A., Bheemasetti, T.V., Williammee, R.S.: Settlement Mitigation of a Distressed Embankment in Texas by Utilization of Lightweight EPS Geofoam Material. In: Transportation Research Board 95th Annual Meeting (No. 16-4179) (2016)
12. Puppala, A.J., Saride, S., Archeewa, E., Nazarian, S., Williammee Jr., R.: Bridge approach settlements: lessons learned from present case studies and ground improvement solutions. In: Ground Improvement and Geosynthetics, pp. 228–238 (2010)
13. Bhaskar, C.S., Saride, S., Puppala, A.J.: Superstructure Design, vol. 647 (2010)
14. Tadros, M.K., Benak, J.V.: Bridge abutment and approach slab settlement: Phase 1. Nebraska Department of Roads (1989)
15. Wahls, H.E.: Design and construction of bridge approaches, vol. 159. Transportation Research Board (1990)
16. Mahmood, I.U.: Evaluation of causes of bridge approach settlement and development of settlement prediction models (Doctoral dissertation, University of Oklahoma) (1990)
17. Kramer, S.L., Sajer, P.: Bridge Approach Slab Effectiveness. Final Report (No. WA-RD 227.1) (1991)
18. Puppala, A.J., Saride, S., Archeewa, E., Hoyos, L.R., Nazarian, S.: Recommendations for design, construction, and maintenance of bridge approach slabs: Synthesis report. Report No. FHWA/TX-09/6022, 1 (2009)
19. Chen, Y.T., Chai, Y.H.: Experimental study on the performance of approach slabs under deteriorating soil washout conditions. J. Bridge Eng. **16**(5), 624–632 (2010)
20. Dupont, B., Allen, D.: Movements and Settlements of Highway Bridge Approaches (No. KTC 02 18/SPR-220-00-1F) (2002)
21. Abu-Hejleh, N., Hanneman, D., White, D.J., Wang, T., Ksouri, I.: Flowfill and MSE Bridge Approaches: Performance, Coast, and Recommendations for Improvements (No. CDOT-DTD-R-2006-2). Colorado Department of Transportation, Research Branch (2006)
22. Hsi, J.: Bridge approach embankments supported on concrete injected columns. In: Geo-Congress 2008: Geo-sustainability and Geohazard Mitigation, pp. 612–619 (2008)
23. Arsoy, S., Barker, R.M., Duncan, J.M.: The behavior of integral abutment bridges, vol. 3, p. 13. Virginia Transportation Research Council, Charlottesville, VA (1999)

24. Puppala, A.J., Archeewa, E., Saride, S., Nazarian, S., Hoyos, L.: Recommendations for design, construction, and maintenance of bridge approach slabs (No. FHWA/TX-11/0-6022-2) (2012)
25. Seo, J., Ha, H., Briaud, J.L.: Investigation of settlement at bridge approach slab expansion joint: Numerical simulations and model tests (No. FHWA/TX-03/0-4147-2) (2002)
26. Rodriguez, L.E.: Temperature effects on integral abutment bridges for the long-term bridge performance program. Utah State University (2012)
27. White, H.: Integral abutment bridges: Comparison of current practice between European countries and the United States of America. Transportation Research and Development Bureau, New York State Department of Transportation (2007)
28. Onsa, E.H., Ahmed, A.A.: Effect of temperature variation and type of embankment soil on integral abutment bridges in Sudan. J. Civ. Environ. Eng. (2015)
29. Shafikhani, A., Bheemasetti, T.V., Puppala, A.J.: Effect of seasonal changes on a hybrid soil-geofoam embankment system. Int. J. Geosynthetics Ground Eng. **3**(4), 39 (2017)

Impact of Strike Energy on the Resolution of Dispersion Image in Active MASW Survey

Jumrik Taipodia[1] and Arindam Dey[2](✉)

[1] Department of Civil Engineering, NIT Arunachal Pradesh, Yupia, India
[2] Department of Civil Engineering, IIT Guwahati, Guwahati, Assam, India
arindamdeyiitg16@gmail.com

Abstract. Active MASW survey is a promising method of the recent days to evaluate the stiffness of the subsurface strata. Active MASW is largely dependent upon the energy generated from an active source. This paper describes the influence of source characteristics on the resolution of dispersion images and the depth of investigation. Higher is the applied energy, larger is the wavelength generated, and hence, larger is the depth of investigation. The efficacy of application of a low-weight sledgehammer in obtaining larger investigation depths with the aid of dispersion image stacking is reported. It is observed that larger depths of investigation can be achieved by 10 kg sledgehammer records stacked for 3-times, equivalent to single shot from 40 kg PEG.

Keywords: Active MASW · Strike energy · Resolution · Dispersion image

1 Introduction

Multichannel Analysis of Surface Waves (MASW) is a trending non-invasive seismic method which utilizes a multi-receiver approach for identifying the subsurface stratification in terms of the variation of shear wave velocity with depth. The technique operates on the dispersive characteristics of the propagating waves, where each frequency component (*f*) of the propagating waves possess different propagation velocity (or, the phase velocity, *C*). The dispersive nature of the propagating waves is used to infer the elastic characteristics of the substrata. One of the most prominent factors that govern the dispersive nature of the surface waves is the type of input energy source. Since the evolution of MASW [1], sources of seismic waves were classified as active and passive [1, 2]. It is shown that for the impulsive active sources like sledgehammer and weight drop, more than 2/3rd of total seismic energy is imparted into Rayleigh type surface waves [3]. The use of 8 kg sledgehammer to obtain the raw record was reported earlier by Park et al. [1]; since then, the sledgehammer has been used as an active source by several practitioners and researchers [4–6]. Apart from the sledgehammer, various other types of sources, such as heavier weight drops, has been reported to be used [7, 8], which led to larger depth of investigation. However, in many cases, the use of heavy weight drops failed to be economical or practical, owing to the shortcomings of its portability and ease of operation. As MASW survey became popular, new type of sources have evolved such as the vibrators [9, 10] and rubber-aided weight drops [11]. The impact of use of different weight of hammers such as 5 kg, 10 kg and Rubber

L. Hu et al. (Eds.): GSIC 2018, *Proceedings of GeoShanghai 2018 International Conference: Multi-physics Processes in Soil Mechanics and Advances in Geotechnical Testing*, pp. 419–427, 2018.
https://doi.org/10.1007/978-981-13-0095-0_47

aided weight drop (RAWD) on the resolution dispersion image obtained from active MASW survey has been explained by Park et al. [11]. The uncertainty and bias associated with MASW dispersion estimates obtained from both impact and harmonic sources, has been reported by Wood and Cox [12]. With the passage of time, explosive dynamites and gunfiring have also been used as active sources [13, 14]. A special SR-II (Kangaroo) source has also been referred by Neducza [15], having a weight of 80 kg, which is operated by a blank 12-bore shotgun cartridge, fired by an electric starter device. Though there are lot many different type of sources introduced over the time, there are little details available regarding the influence of power spectrum of the source type on the resolution of obtained dispersion image. Stacking is the process of summing up the wavefield or dispersion images, resulting in the enhancement in the signal-to-noise (SNR) ratio by the square root of the number of repetitions [16]. Stacking of dispersion images has also proven to have significant impact on the resolution of the dispersion images [2, 16, 17].

Active MASW survey with various combinations of data acquisition parameters, considering 1 m inter-receiver spacing and 4 m offset distance, was conducted at the IIT Guwahati campus, to check the influence of source type (10 kg sledgehammer and 40 kg PEG) on the resolution of dispersion image. The aim of this article is to recommend some guidelines regarding the application of low-weight source types for generating good resolution dispersion images from an active MASW survey.

2 Experimentation Program

The present article is based on active MASW tests conducted at the Cricket ground near Siang Hostel at the IIT Guwahati campus (Fig. 1). As per the already conducted Standard Penetration Tests (SPT), the site is characterized with soft soil up to a depth of nearly 15 m. Based on the SPT-N values recorded at the site, the approximate shear wave velocity along the depth was observed in the range of 80–200 m/s.

Fig. 1. Location of the test site

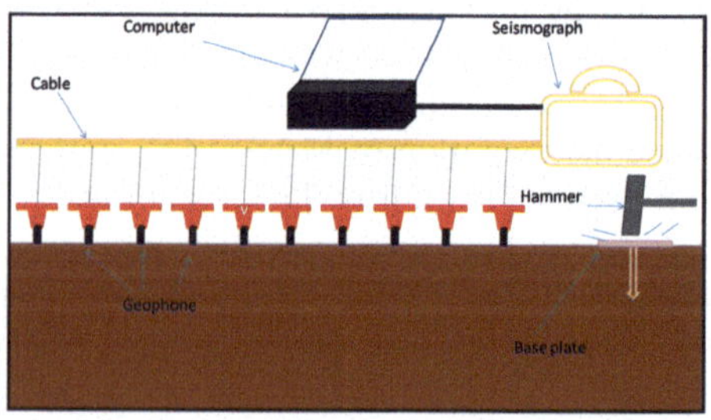

Fig. 2. Schematic of Active MASW Survey

Figure 2 highlights the schematic diagram of Active MASW survey conducted in the field. In general, seismic waves generated from impulse hammer strike travels through soil substrata, and are recorded by a linear array of geophone receivers, which are connected to a Data Acquisition System (DAQ). In the present study, 24 channels of 4.5 Hz geophones were used to record the seismic signals generated by 10 kg sledge-hammer or 40 kg Propelled Energy Generator (PEG). The equipment used in the present study is shown in Fig. 3.

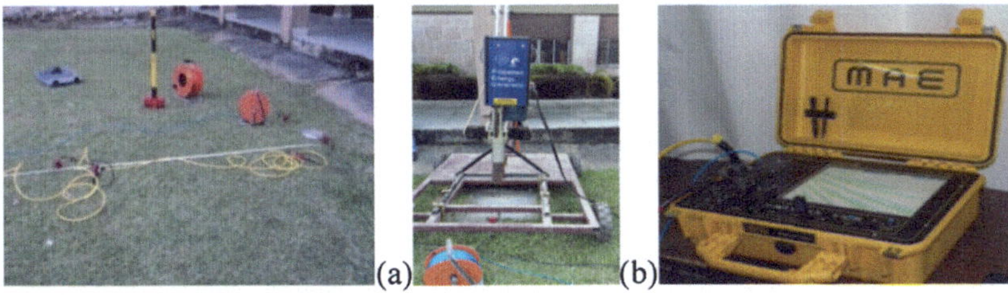

(a) (b) (c)

Fig. 3. Equipments of active MASW survey: (a) Sledgehammer, geophones, connection cables and base plate (b) Propelled Energy Generator (c) DAQ using MAE seismograph

3 Dispersion Analysis, Dispersion Image and Its Resolution

Dispersion analysis is the first step of data processing in MASW method. The aim is to obtain a dispersion curve, based on which the inversion process is carried out to obtain a shear wave velocity profile. The dispersion image is defined as the plot between the phase velocity and the frequencies, having a dominating fundamental mode. In MASW approach, first, an image space is constructed accounting for the multimodal influence, from where the dispersion characteristics are identified as the patterns of energy accumulation in the space. The necessary dispersion curve is extracted from imaging trends. In this imaging technique, a multichannel record in time (t)-space (x) domain is converted into either the domains of frequency (f)-wavenumber (K_x), or the frequency (f)-phase velocity (C_f). The *f-k* method generally results in the lowest resolution in imaging, while the phase-shift method provides images of highest resolution [18]. The resolution of dispersion images can be quantified based on the bandwidth of the dispersion curve. The resolution of a dispersion image is defined as the extent of resolving the image along both the velocity and frequency axes [19]. The resolution of surface wave dispersion depends on the field data acquisition parameters [7]. Optimized field parameters have been proposed to obtain the high resolution dispersion image. A background literature survey reveals that good resolution dispersion image indicates a distinguishable dispersion curve, with clearly recognizable fundamental and/or higher modes encompassing a sufficiently wide frequency band. Based on the highest energy accumulation at fundamental mode, the image should exhibit narrow bandwidth of the energy, so that dispersion curve extraction becomes simpler. There are various parameters that affect the resolution of dispersion image such as data processing parameters (offset, receiver spacing and source used) and data pre-processing parameters (filtering and muting) [20–23]. This article describes the effect of active source on the resolution of the dispersion image, and subsequently set up a guideline to obtain the best possible resolution from a qualitative perspective.

4 Results and Discussion

4.1 Influence of Energy Source

A seismic source generates both surface and body waves upon its impact on the ground surface. The impact energy is directly related to the generated range of wavelengths of surface waves, which determines the largest depth of investigation (Z_{max}). A powerful source is inadvertently required to increase the maximum investigation depth (Z_{max}). A sledge hammer, having weight greater than 10 kg, is commonly used impact source for achieving $Z_{max} \leq 30$ m. An accelerated weight-drop source can increase Z_{max} by 30% under the most favorable conditions. A projectile source, such as Buffalo Gun, increases Z_{max} by generating long wavelengths (low-frequency) containing more energy. The selection of such impact sources used in MASW survey is also governed by its convenience in use and regulation issues. Table 1 lists different sources commonly used for the conduct of an active MASW.

Table 1. Details of the various sources of MASW

Source	Maximum depth (m)	Type of active energy
Sledgehammer (10 kg)	20–30	Impulsive
PEG (40 kg)	50	Impulsive
Electromechanical shaker	Varying depth	Both harmonic and random
Bulldozer (in tons)	100	Semi-impulsive
Blast	Varying depth	Strong impulsive

The energy of impact sources can be calculated based on the delivered kinetic energy upon impact. The kinetic energy is estimated as 0.5 mv^2 (in J) (m - mass of the source in kg, v - velocity of fall of the impact weight in m/s). As an example, if the velocity of fall of impact hammers is 10 m/s, the typical kinetic energy for a 10 kg sledgehammer is 250 J, while that of 40 kg PEG would be 2000 J.

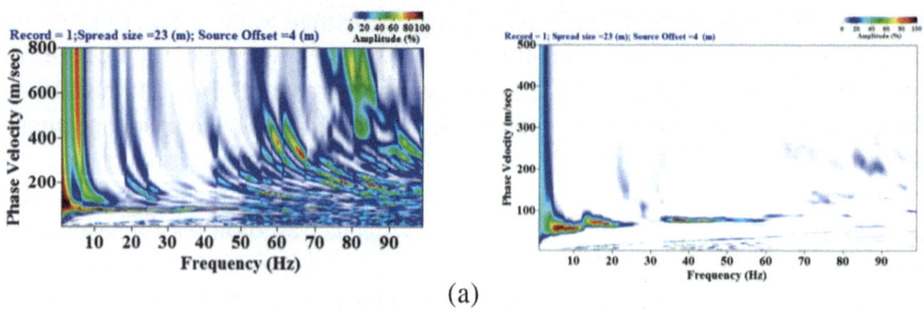

Fig. 4. Typical dispersion images from a single impact (a) 40 kg PEG (b) 10 kg sledge hammer

In active MASW survey, the source has to produce surface waves with a high signal-to-noise ratio (SNR) in a wide frequency band. This task of obtaining the same in the low frequency ranges of the significant frequency band is difficult, thus limiting the resolution of the larger investigation depths. Inadvertent to the type of active source used, the spectral power distribution over the frequency band is primarily site dependent. The spectrum of the source is related to the frequency content of the signal; however, the site of investigation plays a dominant role in altering the spectrum of the propagating signals. Different sites exhibit different frequency bands with spectral energy concentrations.

The use of an impact source is depends upon the desired depth of investigation. It is highly necessary that the signal contains enough low frequency content so that larger penetration depth can be achieved. Figure 4 depicts typical dispersion images obtained from a single shot of impact hammers. It can be observed that in comparison to the dispersion image produced from a single shot of 10 kg sledgehammer, the same generated by a single shot of 40 kg PEG comprises of significant energy in the low frequency ranges. The weight of the hammer conditions the frequency content of the generated pulse. Thus, a relatively lighter hammer mostly generates high-frequency waves, thus revealing only shallow depth characteristics.

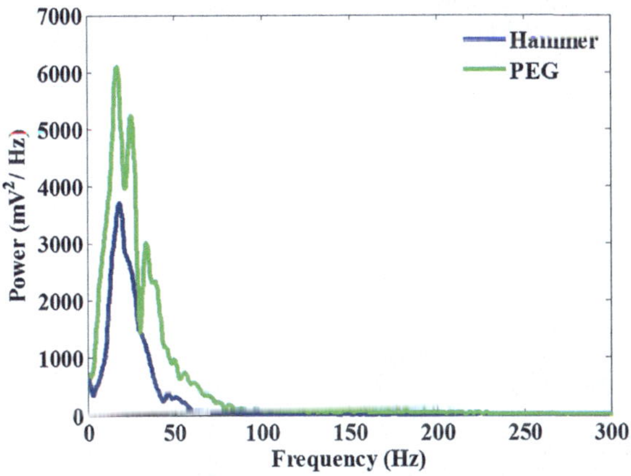

Fig. 5. Power spectrum of the signals obtained from a 10 kg sledgehammer and 40 kg PEG

Figure 5 exhibits typical power spectrum obtained from a single shot of impact hammers. A single shot of 40 kg PEG has recognizably more energy than that generated by a single 10 kg shot. The presence of high energy at lower frequencies makes the 40 kg PEG suitable enough to provide deeper sub-strata information.

The technique of stacking is used to combine the dispersion images developed from different strikes to obtain a resulting dispersion image having higher energy at different frequencies. In this process, even the energy associated with lower frequencies gets radically increased, thus providing shear wave velocity profiles for larger depths. Figure 6 exhibits the results of dispersion image stacking obtained from the 10 kg sledgehammer shots. The dispersion image exhibits more distinctness with increasing stackings.

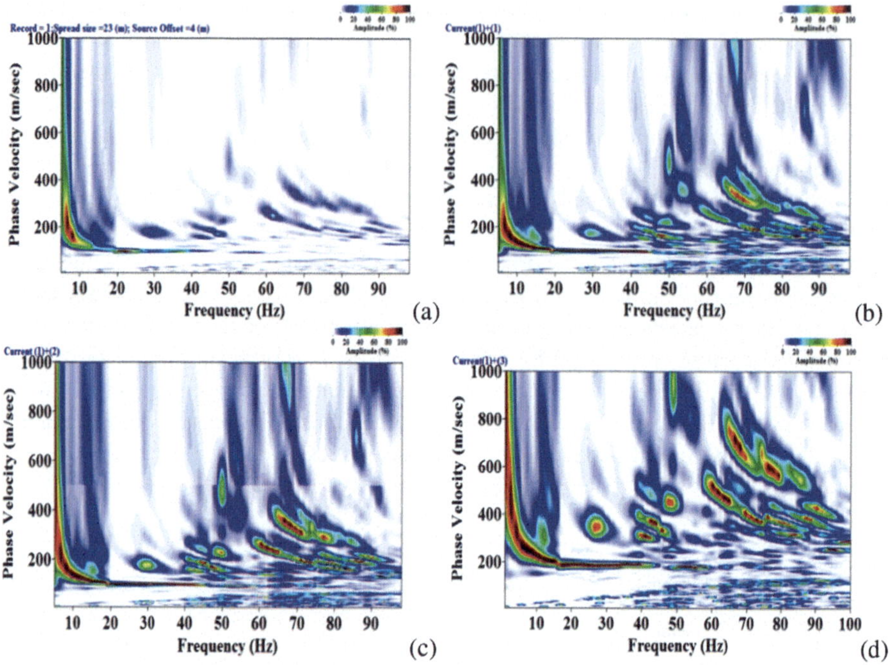

Fig. 6. Dispersion images obtained from stacking of 10 kg sledge hammer records for (a) No stack (b) Single stack (c) Two stacks (d) Three stacks

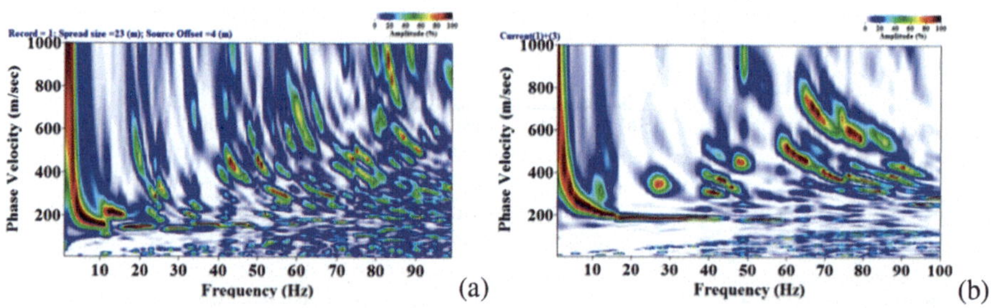

Fig. 7. Comparison of typical dispersion image obtained from (a) Single strike of 40 kg PEG (b) 3-stacked 10 kg sledge hammer record

In general, if it is assumed that the velocity of fall of hammer is identical for both the impulsive sources in consideration, the potential energy of a single shot of PEG exceeds the same obtained for 10 kg sledgehammer by nearly 3–4 times. Thus, stacking of dispersion images from multiple shot gathers of a comparatively lighter sledgehammer will lead to dispersion images with higher energy. Figure 7 shows near similar dispersion images (in terms of energy distribution along significant frequencies) obtained from a single shot of PEG and 3-stacked shots of the sledgehammer. Similarly, Fig. 8 shows that the V_s profile obtained from the above-stated considerations are

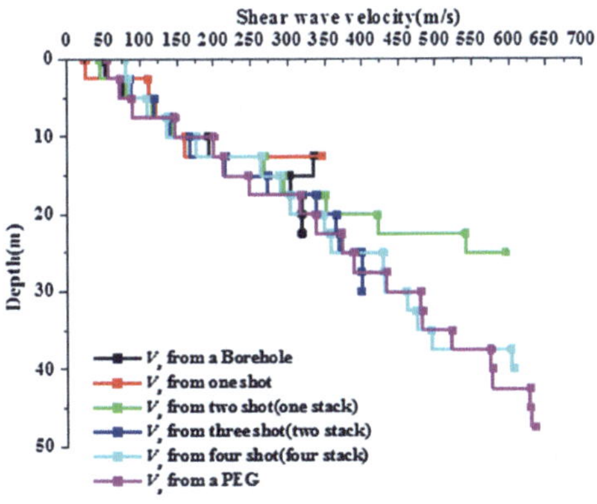

Fig. 8. Comparison of V_s profiles obtained from a single strike of 40 kg PEG and varying stacking of 10 kg sledge hammer records

in appreciable agreement. Thus, it can be stated that the energy content of the dispersion image generated from a single shot of 40 kg PEG is appreciably equivalent to the same obtained from 3–4 stacks of sledgehammer records. This ensures the applicability of a comparatively lighter sledgehammer in harnessing the information of deeper substrata using multiple shots by using dispersion image stacking. It can be also seen from V_s profile that the depth of investigation obtained from one single PEG ($\sim$48 m) record is more than the depth obtained from borehole ($\sim$22 m) and from one single sledgehammer record ($\sim$12 m).

5 Conclusions

An attempt was made to understand the influence of source energy on the resolution of dispersion image obtained from active MASW surveys conducted on site condition having V_s in the range of 80–200 m/s. The following conclusions are drawn based on the present study:

- A heavier weight source (e.g. 40 kg PEG) leads to larger depths of investigation than a 10 kg sledgehammer.
- Impact energy from one 40 kg PEG record equals 3–4 stacked 10 kg sledgehammer records.
- Stacking increases the resolution of the dispersion images and the depth of investigation.

References

1. Park, C.B., Miller, R.D., Xia, J.: Multichannel analysis of surface waves. Geophysics **64**, 800–808 (1999)
2. Park, C.B., Miller, R.D., Xia, J., Ivanov, J.: Multichannel analysis of surface waves (MASW)-active and passive methods. Lead. Edge **26**, 60–64 (2007)

3. Richart, F.E., Hall, J.R., Woods, R.D.: Vibrations of Soils and Foundations, p. 414. Prentice-Hall Inc., New Jersey (1970)
4. Shtivelman, V.: Using surface waves for studying the shallow subsurface. Bollettino di Geofisica Teorica ed Applicata **44**, 223–236 (2003)
5. Zhang, S.X., Chan, L.S., Xia, J.: The selection of field acquisition parameters for dispersion images from multichannel surface wave data. Pure. appl. Geophys. **161**, 185–201 (2004)
6. Gosar, A., Stopar, R., Roser, J.: Comparative test of active and passive multichannel analysis of surface waves (MASW) methods and microtremor HVSR method. RMZ Mater. Geoenvironment **55**, 41–66 (2008)
7. Stephenson, W.J., Louie, J.N., Pullammanappallil, S., Williams, R.A., Odum, J.K.: Blind shear-wave velocity comparison of ReMi and MASW results with boreholes to 200 m in Santa Clara Valley: implications for earthquake ground-motion assessment. Bull. Seismol. Soc. Am. **95**, 2506–2516 (2005)
8. Park, C.B., Miller, R.D., Ryden, N., Xia, J., Ivanov, J.: Combined use of active and passive surface waves. J. Environ. Eng. Geophys. **10**, 323–334 (2005)
9. Park, C.B., Miller, R.D., Xia, J.: Offset and resolution of dispersion curve in Multichannel analysis of surface waves (MASW). In: Proceedings of the SAGEEP SSM4, pp. 1–6 (2001)
10. Xia, J., Miller, R.D., Park, C.B., Ivanov, J.: Construction of 2-D vertical shear-wave velocity field by the multichannel analysis of surface wave technique. In: Proceedings of the Symposium on the Application of Geophysics to Engineering and Environmental Problems, pp. 1197–1206 (2001)
11. Park, C.B., Miller, R.D., Miura, H.: Optimum field parameters of an MASW survey. In: Expanded Abstract: Japanese Society of Exploration Geophysics (2002)
12. Wood, C.M., Cox, B.R.: A comparison of MASW dispersion uncertainty and bias for impact and harmonic sources. In: Geocongress, pp. 2756–2765. ASCE (2012)
13. Xu, Y., Xia, J., Miller, R.D.: Quantitative estimation of minimum offset for multichannel surface-wave survey with actively exciting source. J. Appl. Geophys. **59**, 117–125 (2006)
14. Kaufmann, R.D., Xia, J., Benson, R.C., Yuhr, L.B., Casto, D.W., Park, C.B.: Evaluation of MASW data acquired with a hydrophone streamer in a shallow marine environment. J. Environ. Eng. Geophys. **10**, 87–98 (2005)
15. Neducza, B.: Stacking of surface waves. Geophysics **72**, 51–58 (2007)
16. Foti, S., Lai, C.G., Rix, G.J., Strobbia, C.: Surface wave methods for near-surface site characterization. CRC Press, United States (2015)
17. Kanli, A.I., Tildy, P., Pronay, Z., Pinar, A., Hermann, L.: V_{s30} mapping and soil classification for seismic site effect evaluation in Dinar region, SW Turkey. Geophys. J. Int. **165**, 223–235 (2006)
18. Moro, D.G., Pipan, M., Forte, E., Finetti, I.: Determination of Rayleigh wave dispersion curves for near surface applications in unconsolidated sediments. In: Expanded Abstracts: Society of Exploration Geophysicists, pp. 1247–1250 (2003)
19. Park, C.B., Miller, R.D., Xia, J.: Imaging dispersion curves of surface waves on multi-channel record. In: Expanded Abstract: Society of Exploration Geophysics, pp. 1377–1380 (1998)
20. Dikmen, U., Arisoy, M., Akkaya, I.: Offset and linear spread geometry in the MASW method. J. Geophys. Eng. **7**, 211–222 (2010)
21. Luo, Y., Xia, J., Miller, R.D., Xu, Y., Liu, J., Liu, Q.: Rayleigh-wave dispersive energy imaging using a high-resolution linear radon transforms. Pure. appl. Geophys. **903**, 903–922 (2008)

22. Taipodia, J., Baglari, D., Dey, A.: Resolution of dispersion image obtained from active MASW survey. Disaster Adv. **10**, 34–45 (2017)
23. Taipodia, J., Baglari, D., Dey, A.: Recommendations for generating dispersion images of optimal resolution from Active MASW survey. Innovative Infrastruct. Sol. **3**(14), 1–19 (2017)

Application of Geotechnical Monitoring (Slope Monitoring and Early Warning System) for Risk Reduction in Philippine Infrastructure

John Erickson B. Delos Santos(✉), Roy Anthony C. Luna, Ramon D. Quebral, Rodgie Ello B. Cabungcal, and Jenna Carmela C. Pallarca

AMH Philippines, Inc., University of the Philippines, Diliman, 1101 Quezon City, Philippines
erickson.delossantos@amhphil.com

Abstract. The Philippines, an archipelago of more than 7,100 islands, and with a population of more than 100 million, is one of the counties most exposed to seismic hazards, being located in the Circum-Pacific Ring of Fire. It is also identified as one of the top 10 countries worldwide which are most vulnerable to the effects of climate change.

With a climate generally characterized by predominantly rainy season, many areas of high altitudes and characterized by thick residual soils and highly weathered rocks are highly susceptible to rainfall-induced or earthquake-induced slope failures. These failures may occur in populated areas without warning; thus, increasing the risks to the communities and essential utilities.

This paper presents the long list of the various geohazards in the Philippines, current efforts in the development of GIS-based regional hazard maps, and the current practice of conducting engineering geologic and geohazard assessment (EGGA) for critical projects – as part of disaster mitigation and risk reduction strategies. In few critical projects, geotechnical monitoring has been employed.

A case study involving the development of a slope monitoring program and early warning system for a power generation facility in the Philippines shall be presented. Monitoring equipment were installed, aimed at detecting slope movements and increase in pore water pressures – which became the basis for work suspension and evacuation of personnel working on or near slopes. The monitoring program proved to be effective as impending failures were detected, and the projects were completed without serious incident.

Keywords: Instrumentation · Geotechnical monitoring
Landslide mitigation · Risk reduction

1 Background

Roughly 90% of all seismic activities occur along a 40,000 km long horseshoe-shaped stretch called the Pacific Ring of Fire. This area consists of 452 volcanoes running along the southern tip of South America, Alaska, Japan, Philippines, and New Zealand.

L. Hu et al. (Eds.): GSIC 2018, *Proceedings of GeoShanghai 2018 International Conference: Multi-physics Processes in Soil Mechanics and Advances in Geotechnical Testing*, pp. 428–435, 2018.
https://doi.org/10.1007/978-981-13-0095-0_48

The Pacific Ring of Fire contains about three-fourths of dormant and active volcanoes in the world.

The Philippines, an archipelago of more than 7,100 islands, is one of the countries most exposed to seismic hazards. The country has 23 active volcanoes, 21 of which have historical eruptions according to the Philippine Institute of Volcanology and Seismology (PHIVOLCS). As a result, the country frequently experiences catastrophic events due to seismic and volcanic activities such as liquefaction, landslide and debris flow.

Aside from seismic and volcanic activities, other factors influence the susceptibility of the Philippines to such catastrophic events. With a climate generally characterized by predominantly rainy season, high amount of precipitation is experienced by the country year round, with strong typhoons occurring frequently, with an average of 20 typhoons per year passing through the Philippine Area of Responsibility (PAR). The highly-altered geologic formation of the Philippines also influences the hazards.

2 Hazard Mitigation and Risk Reduction

2.1 Risk Assessment

Risk is a function of three parameters: hazard, exposure and vulnerability. Hazard is a threat that can cause loss of life or damage to property. In the case of mountainous regions of the Philippines, the most common type of hazard is landslide.

Exposure refers to elements which have to confront a hazard. Exposure may be in terms of infrastructures and facilities - such as access roads, bridges, pipelines or buildings - and people, in the case of motorists travelling along access roads or occupants of buildings. Vulnerability refers to the capacity of a structure or facility, for example, to survive a hazard. Essential facilities located far from a slope are less vulnerable than those found along the base of a slope. Risk may be classified as low, moderate or high after taking into consideration the consequence of failure.

2.2 Risk Rating

Considering the three parameters of risk, the following ratings were generated, generally applicable for infrastructures exposed to landslide risk:

Low Risk is defined as an inconvenience that is easily corrected, not directly endangering lives or; Moderate Risk is defined as a more severe inconvenience, corrected with some effort, but not usually directed endangering lives or structures when it occurs; High Risk is defined as complete loss of roadways, important structures or complete closure of the way for some period of time. Lives are endangered during failure.

2.3 Disaster Risk

Since hazards are only threats that can cause loss of life or damage to property, it becomes a risk if two other parameters are present: exposure and vulnerability (Fig. 1).

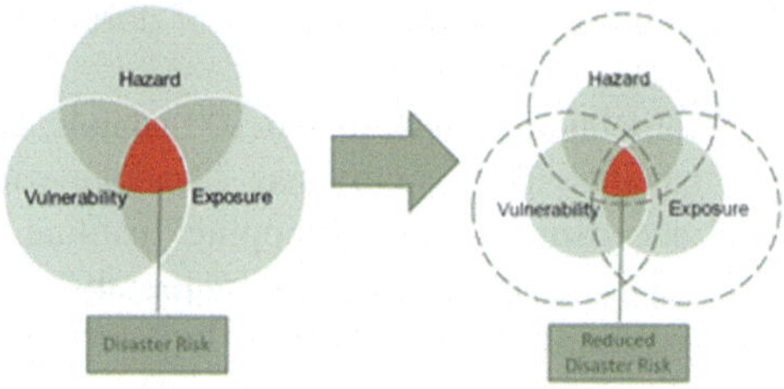

Fig. 1. Disaster risk parameters

2.4 Risk Objects

The three objects of risk are property, way of life and life. Their elements are subjective depending on the value system of the person, party or entity. For this risk assessment, the element of life pertains to the people; way of life pertains to the health status, public safety, livelihood, service, social and economic order and environment; property involves assets, physical structures, technical facility and system and land.

2.5 Disaster Risk Mitigation

Figure 2 presents the Disaster Management Steps that are taken to ensure the safety of life, way of life, and properties. Response refers to the emergency services and assistance rendered during or immediately after disasters. This is done to assess the disaster, to plan and pursue further actions. Rehabilitation, on the other hand, is to restore and improve the affected facilities, structures and/or livelihood of affected communities.

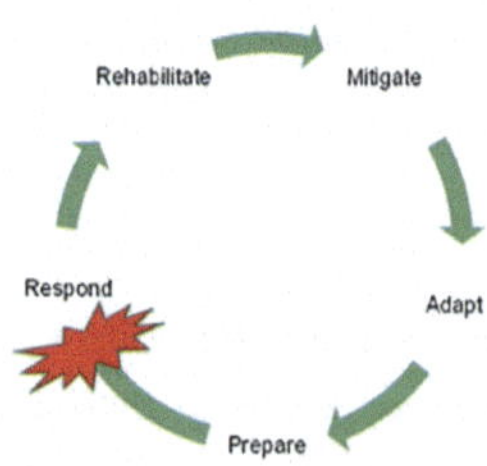

Fig. 2. Disaster management steps

Adapting to the natural environment such as site planning to anticipate potential problems is a proactive approach. An important part of the Disaster Management is mitigation, which can either be structural or non-structural in nature. These engineering measures reduce and/or prevent future damage related to hazards and disasters.

3 Hazard Mitigation and Risk Reduction

Instrumentation system for early detection of slope failure and monitoring of movements is essential in areas near essential facilities, and places where communities are exposed to slope-related geotechnical hazards. Slope monitoring through various instruments will be able to provide quantitative data on in order to constantly measure slope movements and thereby assess the stability of a slope. Lastly, it could help detect early signs of impending slope failure and can even send warning when the threshold parameters are exceeded.

The instruments used for slope monitoring measure several parameters that may be used to calculate the stability of the slope. The commonly measured parameters in slope monitoring are porewater pressure, deformation (directional and rotational), soils and rock stresses, temperature, and vibration.

Slope monitoring instrumentation systems may be classified into two groups, namely: contact and remote monitoring. The more traditional approach is the contact monitoring. It requires the instruments to be installed directly in the ground, either on the surface and/or at a certain depth below the surface.

In contrast, remote monitoring can collect data based on sensors installed far away from the site monitored. Some instruments may be partially remote such as sensors still required to be installed on the ground, but with considerable distance from the site monitored. Accordingly, fully remote instruments are available wherein no instruments are installed on site. Examples of different instruments used for remote monitoring are: Terrestrial Interferometric Synthetic Aperture Radar (TInSAR), and Light Detection and Ranging (LiDAR).

Remote monitoring will have no interaction with the ground or structure. While remote monitoring would have the advantage to cover a more extensive area of study and capability of increasing of the spatial information density, it would reduce the local precision of the monitoring point.

Usually, for sites with relatively small area/s of concern, it would be more practical to use the traditional contact monitoring. The case study presented in the next section utilized contact monitoring instrumentation system as early detection and monitoring system.

4 Case Study: Development of Monitoring Program as Early Warning System

4.1 Instrumentation

The monitoring instrumentation system used in the case study at a facility in the Philippines is composed primarily of data logger, MEMS tilt meters, soil moisture sensors, rain gauge, and an alarm.

Data logger is a device that records and stores data at a period of time. These data come from the external instruments and sensors attached to the logger. A computer connected to the data logger can view and process the collected data using a software.

MEMS tilt meters are instruments used to measure small and sudden movements or changes (in degrees) on the slope in terms of vertical and horizontal level. Tilt meters can measure up to ±15° tilt from vertical.

Soil moisture sensors measure the volumetric water content (in percentage) of the soil. These sensors monitor the saturation level of the slope at an approximate radius of 20 mm around the sensors.

Rain gauge (tipping bucket) is an instrument used to measure the amount of rainfall (in millimeters) in a period of time. Each tip on the rain gauge is equivalent to 0.254 mm rainfall.

An alarm is a device that activates when the data transmitted by the tilt meter exceeded the threshold values set for the said instrument.

4.2 Methodology

Installation of Monitoring System. One (1) data logger, two (2) tilt meters, three (3) soil moisture sensors, a rain gauge, a solar panel and an alarm are installed at the site of the study.

Data logger is placed in a temporary shelter, distant enough from the slope, to protect the logger from falling off and getting wet due to rainfall. Tilt meters and soil moisture sensors are installed at different locations (top, middle, and bottom portions) of the slope.

Tilt meters are attached to an angle steel bar which is embedded at the ground for stability. The orientation (x-axis, y-axis, and z-axis) of the tilt meters are properly set.

Soil moisture sensors are implanted completely deep enough (~0.5 to 1.0 m) from the ground surface. The rain gauge, solar panel, and the alarm are installed near the temporary shelter and the data logger. Rain gauge is elevated from the ground to collect rainfall properly.

All the instruments and sensors are then connected to the data logger by wires/cables and were calibrated to set the values to zero (0). This is to ensure the accuracy of the data to be recorded and collected (Fig. 3).

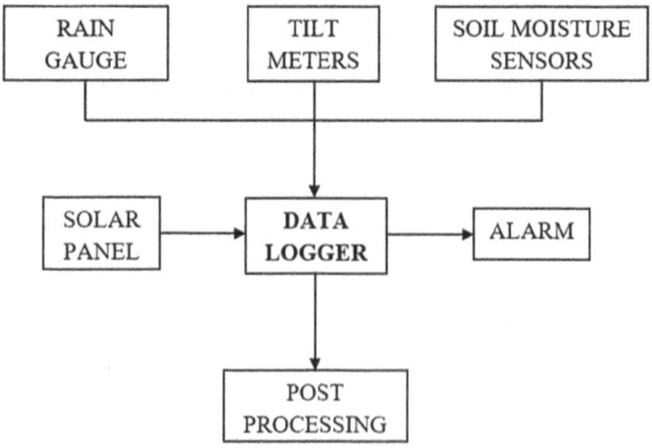

Fig. 3. Schematic diagram of the monitoring system

Monitoring Works. Monitoring of the slope is conducted on a daily basis. Prior to the daily monitoring, an initial 14-day monitoring (first phase) is conducted to derive the threshold values stated in the following subsection. Data are collected and analysed twice a day – one in the morning and one in the afternoon.

Data are presented graphically with corresponding interpretations. These are the basis for any work suspension and evacuation of staff working on the slope.

Threshold Values. Threshold values are set at the data logger for each of the instruments (Table 1). These are the basis of warning that a slope failure may potentially occur.

Table 1. Threshold values for each instrument

Instrument		Threshold
Rain gauge		15 mm/2 h
Soil moisture sensors	SM 1	14.00%
	SM 2	33.00%
	SM 3	23.50%
Tilt meters		>0.25% or <–0.25%

The threshold value for the rain gauge are based on the rainfall advisories of Philippine Atmospheric, Geophysical and Astronomical Services Administration (PAGASA). Three categories are included in the rainfall advisories; yellow warning (7.5–15 mm/2 h), orange warning (15–30 mm/2 h), and red warning (>30 mm/2 h). The monitoring system adopted the orange warning in which intense rain is observed and flooding is threatening.

For the soil moisture sensors and tilt meters, the threshold values are established based on the initial trend of the data on the 14-day monitoring. In case the threshold values for the soil moisture sensors were exceeded, resumption values are established to ensure that the area is generally safe before any work resumes.

4.3 Results and Findings

Based from the monitoring and analyzing of data, there are the instances where collected data exceeded the threshold values for any of the instruments.

One good example is the rainfall, and soil moisture data collected for four (4) consecutive days, from January 15 to 19, 2017, as shown in Fig. 4. During the first 12 h (January 15, 2017, 12:00 PM to January 16, 2017, 12:00 AM), accumulated amount of rainfall for any 2-h interval within the time period did not exceed the threshold value of 15 mm/2 h. Moreover, the soil moisture contents recorded for the three sensors did not exceed their corresponding threshold values.

However, during the succeeding hours (January 16, 2017, 12:00 AM onwards), the accumulated amount of rainfall per 2-h interval exceeded the threshold value of 15 mm/2 h. These increase of amount of rainfall can also be validated by the increase of values of the moisture sensors, which also exceeded their corresponding threshold

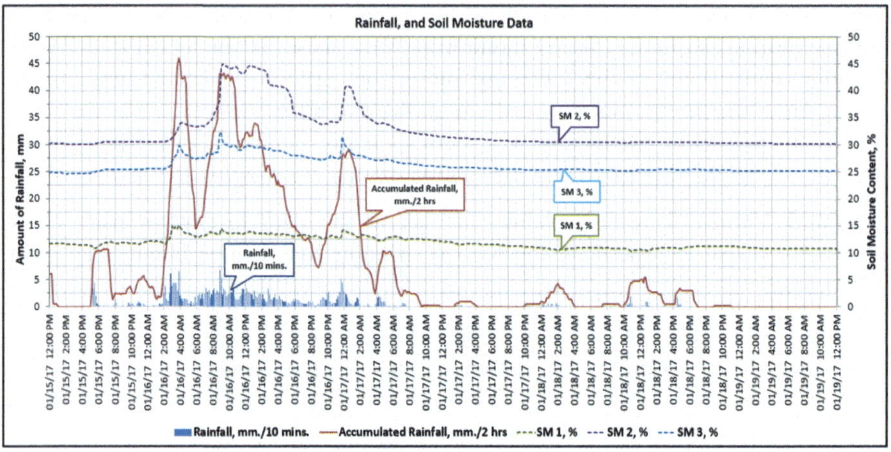

Fig. 4. Rainfall, and soil moisture content data for four (4) consecutive days

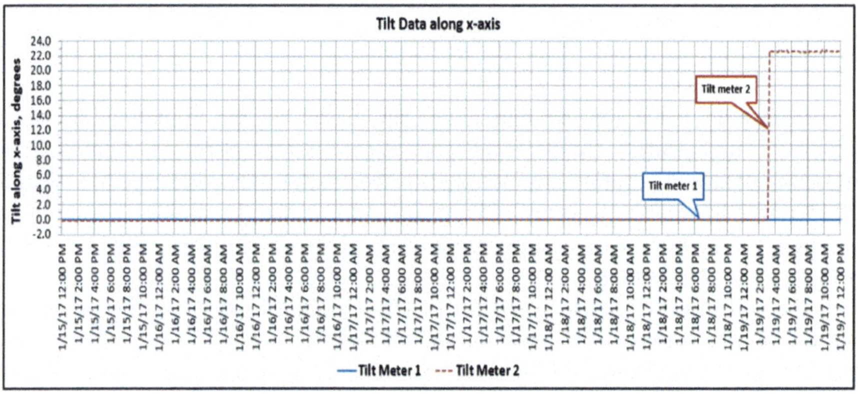

Fig. 5. Tilt data along x-axis for the two (2) tilt meters

values. The sudden increase of the values is largely due to the continuous rainfall caused by a tropical depression.

As for the tilt data along both x and y axis (Figs. 5 and 6), the threshold values of >0.25 or <–0.25 for the tilt, are also exceeded two (2) days after the recorded exceedance for the rainfall, and moisture content values. The data showed that there are sudden increase in the tilt values, which is an indication of a slope movement.

Prior to the sudden increase in tilt values, the graphs show that there are no (or minimal) changes/variations on the tilt data indicating that the slope is stable at that period of time.

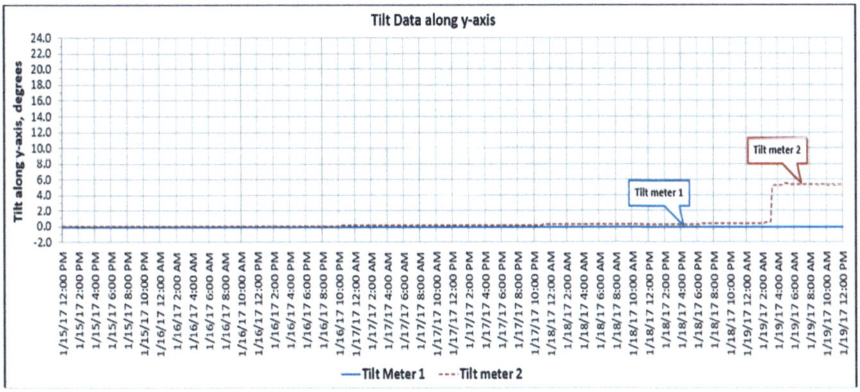

Fig. 6. Tilt data along y-axis for the two (2) tilt meters

True enough, a landslide occurred at the site which is a good validation of the data recorded. Slope works are recommended for suspension during the day threshold values exceeded and three (3) days following the exceedance.

Generally, the results and findings are used as the basis for any work suspension on the slope.

5 Conclusion

The paper presents the case study involving a slope monitoring system which was successfully implemented as an early warning system on an area with critical slopes. A 14-day monitoring program (first phase) was conducted to establish the threshold values. It also aimed at ensuring the general safety of personnel working on or near the slopes.

The threshold values initially established was deemed appropriate since landslides occurred when the threshold values were exceeded. Moreover, based on the data gathered, further analysis and evaluation is warranted to further optimize the threshold values, while ensuring safety along the slope.

An Automatic Data Processing Method for Deformation Analysis and Visualization of Tunnel Cross Sections Using Laser Scanning Data

Zixin Zhang[1,2(✉)], Tong Yin[1,2], Xin Huang[1,2], and Fan Zhang[1,2]

[1] Department of Geotechnical Engineering, College of Civil Engineering, Tongji University, Shanghai 200092, People's Republic of China
zxzhang@tongji.edu.cn

[2] Key Laboratory of Geotechnical Engineering, Ministry of Education, Tongji University, Shanghai 200092, People's Republic of China

Abstract. Based on Laser Scanning Technology (LST), a data processing method is proposed to automatically visualize the tunnel cross sections and collect deformation information. In this method, Cartesian coordinates of each data point are transferred to polar coordinates, and two-dimensional tunnel cross sections can be extracted from three-dimensional point cloud data obtained by LST. The tunnel deformation is calculated by subtracting radius vectors of deformed cross section from design cross section and visualized through adjustable deformation factors. The deformation data at arbitrary azimuthal angles could be gained with an accuracy of 0.1 mm. To achieve the ideal calculation accuracy and ensure visualization performance, a novel algorithm adopting point discretization and regeneration technique is developed to automatically detect and treat the noise and discontinuity points from the point cloud data. While the traditional data processing methods could only handle a specific type of tunnel cross section, the proposed method by us is able to deal with arbitrary shapes of tunnel sections as long as three-dimensional cloud data are available. The results show that the proposed method could provide accurate deformation data with high data processing efficiency for monitoring the deformation of tunnel cross sections in laboratory, and potentially for real-life tunnels.

Keywords: Laser scanning technology · Data processing
Tunnel cross section

1 Introduction

Measuring tunnel segment deformation is essential for engineers to evaluate the safety of tunnel structure so as to adjust design and construction parameters as well as for researchers to acquire tunnel mechanical behaviors. Traditionally, total station is used to measure tunnel segment deformation for accredited precision. However, this method is particularly time-consuming especially when it comes to large-scale projects or experiments. Total station measurement, which is also known as 'single point measurement', indicates it delivers only a set of points that can be monitored in the limited

L. Hu et al. (Eds.): GSIC 2018, *Proceedings of GeoShanghai 2018 International Conference: Multi-physics Processes in Soil Mechanics and Advances in Geotechnical Testing*, pp. 436–448, 2018.
https://doi.org/10.1007/978-981-13-0095-0_49

frame available for the measurements [1]. Therefore, if there are a large amount of points to be measured, the labour and time cost would be unbearable and the measurement would almost be unrepeatable for big data. Digital photogrammetry is another widely applied technique, but this technique requires sufficient lighting conditions and texture on the objects surface [2, 3], requirements that cannot be guaranteed in tunnel construction projects. Due to these shortcomings of total station and digital photogrammetry, in recent years, terrestrial laser scanning (TLS) technique is drawing increasing interests for its advantages of non-contact, rapidity, high accuracy and large scale [4]. As described in Ref. [5], static or mobile scanning systems have been used to compare the measured construction to a fitted surface. TLS is a method of recording large quantity (up to million) of three-dimensional points' positions with a high speed (up to 1 million points per second). Using TLS, instead of the deformation of sparse sample points on the surface, engineers and researchers could obtain the overall deformation of object to be measured by analyzing the recorded point information (also called point cloud data).

Since the points cloud data could be extremely large, it is quite important to develop algorithms to process and visualize them efficiently and automatically. Recent improvements regarding the speed, accuracy, software algorithms and the fall in price have introduced a high potential for large scale applications of this technology in highly demanding engineering environments such as tunnels [6]. The use of TLS in drilling and blasting tunnels and corresponding algorithm have been proposed [7]. It shows that the TLS system makes it possible to remotely acquire detailed rock mass and excavation information with a high speed without disrupting construction. Another method is an automatic method for extraction of tunnel cross sections using TLS [8]. The algorithm converts three-dimensional point cloud data acquired by TLS to two-dimensional planar images and achieves ideal accuracy at the same time.

Although the approaches mentioned above have shown satisfactory performance in practice, they can only handle one specific type of tunnel cross section, which severely limits their application. The aim of this study is to develop a novel method to process three-dimensional cloud data of arbitrary cross section automatically with high efficiency. Firstly, the methodology to extract tunnel deformation contour from three-dimensional cloud data as well as to treat noise and discontinuous points is elucidated. Then a designed Graphical User Interface (GUI) which could display the deformation data at any polar angles is developed. Finally, the validity of the proposed method, is confirmed by applying it to capture the cross-section changes in full-scale ring tests of a circular shield tunnel and a sub-rectangular shield tunnel. The proposed method is designed for measuring the deformation contour of tunnel segment during a load test but it could also be used for in-situ measurement with a simple improvement to the data processing algorithm.

2 Methodology

The proposed method contains four major operations: (1) raw cloud data pre-processing, (2) extraction of tunnel cross section, (3) treatment of discontinuous points, (4) determination of tunnel center. Firstly, the pre-processing is conducted, during which the raw three-dimensional cloud data are filtered in order to exclude obviously unreliable noise caused by sensors or wires as shown in Figs. 1 and 2. Then the filtered cloud data are projected onto a two-dimensional plane which is characterized by a pre-designed polar coordinates system. The polar angle and radius vector define the points on the tunnel outline (Fig. 3). Because of the introduced polar coordinates system, this method is cross-section-independent, and can handle tunnel cross section of arbitrary shapes.

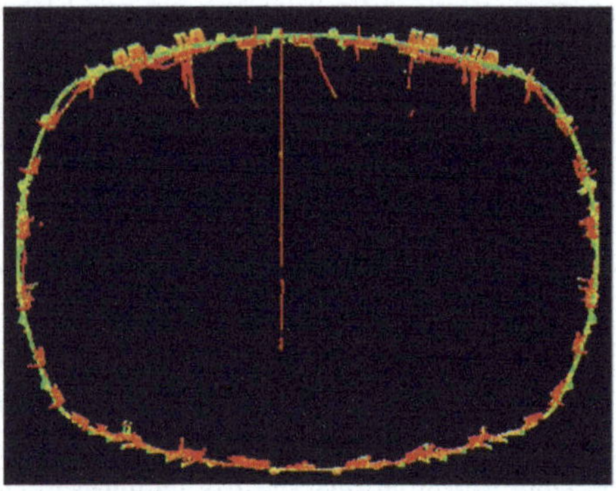

Fig. 1. Raw laser scanning data of a quasi-rectangular shield tunnel

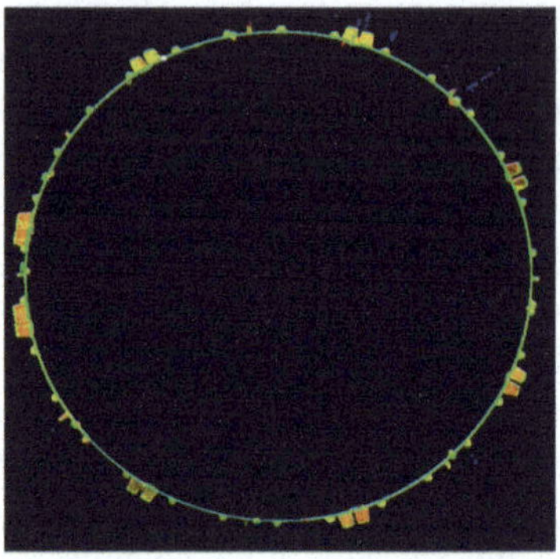

Fig. 2. Raw laser scanning data of a circular shield tunnel

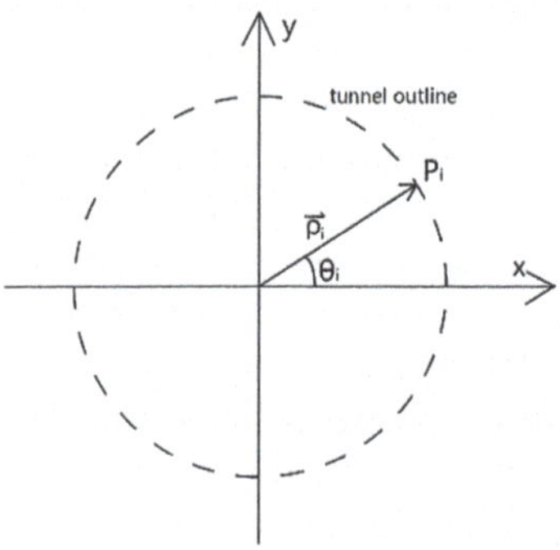

Fig. 3. Pre-defined polar coordinate system

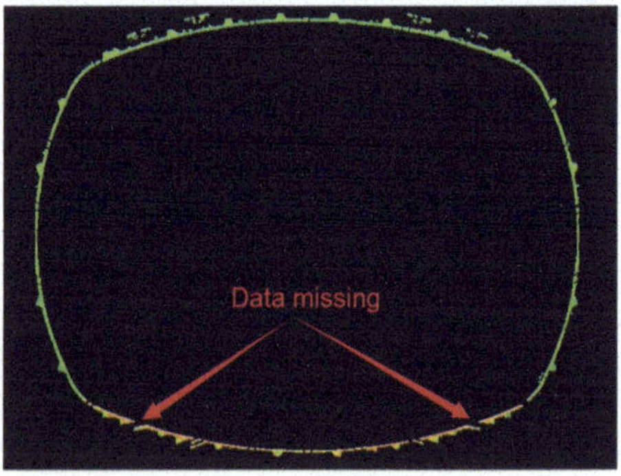

Fig. 4. Data missing caused by raw data processing

However, due to the existence of hand holes or other grooves on the segment, data missing is inevitable after operation (1) as shown in Fig. 4. Consequently, point jump occurs on the tunnel deformation curve (Fig. 5) because the deformation is calculated by the difference of the deformed outline and original outline. Operation (3) adopts point discretization and regeneration techniques to smooth the curve and acquire reasonable deformation values.

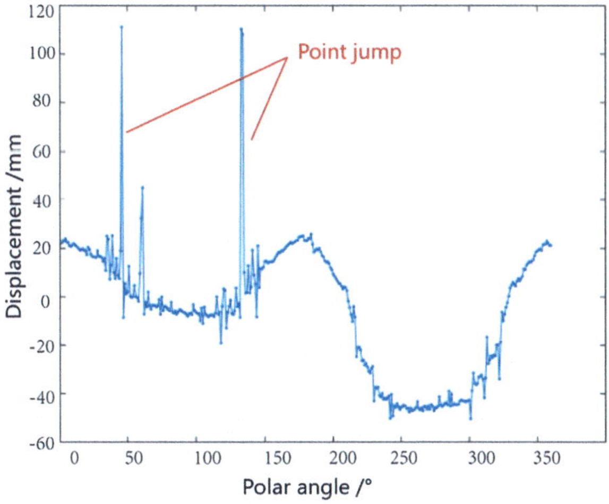

Fig. 5. Point jump on tunnel deformation curve

2.1 Raw Cloud Data Pre-processing

Before extracting the tunnel cross section, obvious noise needs to be treated during pre-process because noisy points would severely impair the efficiency and accuracy of data processing. This procedure is conducted in commercial software Cyclone for it provides efficient tools to treat this kind of noise.

2.2 Extraction of Tunnel Cross Section

A polar coordinates system is firstly established with polar angle ranging from 1° to 360°. In order to get deformation information, the design outline is also pre-set on the polar coordinates system. Then, the program formalizes the coordinates of points from the point cloud data by ergodic process. For arbitrary point P_i on the tunnel periphery, the polar angle θ_i and length of radius vector ρ_i are defined as:

$$\theta_i = \arctan\left(\frac{Y_i}{X_i}\right) \tag{1}$$

$$\rho_i = \sqrt{X_i^2 + Y_i^2} \tag{2}$$

where Xi and Yi are the x and y coordinates of Pi on the projection plane. Since the cloud data are so large that thousands of points exist at a certain polar angle θ_i, there are different values ρ_i for the same θ_i. In order to obtain accurate results, the algorithm adopts statistical approach. The maximal occurrence of radius vectors length is counted and recorded. Then those points whose occurrence of ρ_i is less than half of the highest occurrence would be excluded. Finally, the weighted mean value of the ρ_i of the remained points is calculated as the representative value at θ_i. After the above procedures, a points group is acquired which contains points with θ_i and weighted mean values of length of radius vectors at θ_i. By iterating θ_i from 1° to 360°, the representing value at each polar angle is gained.

After curve fitting using least square method based on the points group, two-dimensional tunnel centerline is extracted from three-dimensional cloud data. Then the deformation values at each point could be obtained by subtracting the length of radius vectors on extracted outline from design outline.

2.3 Treatment of Discontinuous Points

To deal with data missing (Fig. 4) and value jump (Fig. 5) caused by pre-processing, a point discretization and regeneration scheme has been designed to smooth the tunnel outline and acquire reasonable deformation values. The deformation of the adjacent two points is nearly continuous so the difference between the deformation of these two points would not be unreasonably large. Based on this fact, the point discretization and regeneration technique uses differential calculation in which 359 differential vectors di are obtained by differential calculating 360 radius vectors from 1° to 360°:

$$d_i = |\rho(i+1)| - |\rho(i)| \tag{3}$$

In accordance with the accuracy requirement, set corresponding error limit Δ and then select those differential vectors whose $|d_i| > \Delta$. Then the selected differential vectors are separated into different groups with difference of the subscript $i < \Delta_i$. After several calibrations, it is found that $\Delta_i = 5$ is accurate enough. If the absolute value of sum of differential vectors in one group $|\sum di| < \Delta$, the mean values of differential vectors are calculated. For example, if Δ = 10 mm, $d_1 = -80$ mm, $d_2 = 100$ mm and $d_5 = -26$ mm, so $|d_1 + d_3 + d_5| = 6$ mm < 10 mm and we set $d_1 = d_3 = d_5 = (d_1 + d_3 + d_5)/3 = -2$ mm. Finally, the treated values of differential vectors will be added to the original radius vectors and the smoothing process is completed. Those missing points are regenerated by assigning their radius vectors to equal the sum of design values and treated values of differential vectors. The smoothing result is shown in Fig. 6.

2.4 Determination of Tunnel Center

The proposed method requires that the origin of the spatial coordinate system must be located at the center of tunnel outline. Hence, determining the tunnel center is essential for data processing. For the circular cross section, Cyclone provides a "cylinder fit" tool to find the tunnel center. However, for tunnel cross sections of other types, there are no available tools to adopt. Consequently, the proposed method uses the concept of static moment to determine the tunnel center which is simple but effective for arbitrary cross section.

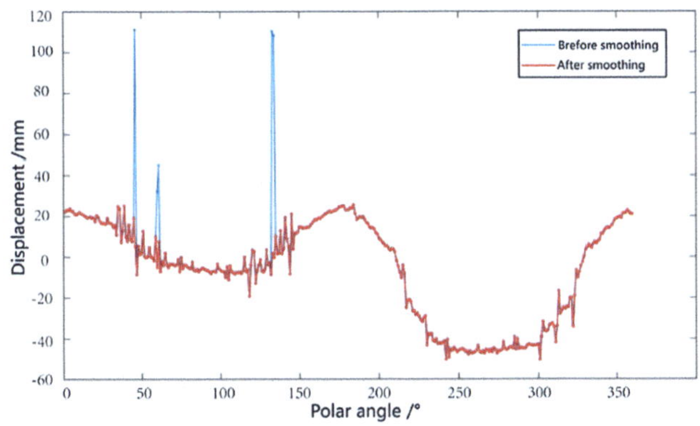

Fig. 6. The deformation curves before smoothing and after smoothing

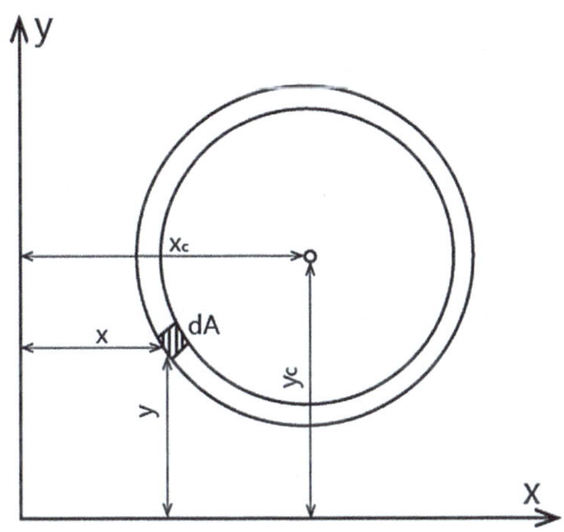

Fig. 7. Concept of static moment

As Fig. 7 shows, the static moment of the shaded area dA to x-axis and y-axis is defined respectively as:

$$S_x = \int_A ydA \tag{4}$$

$$S_y = \int_A xdA \tag{5}$$

And the coordinates of center could be calculated by:

$$y_c = \frac{\int_A y dA}{A} = \frac{S_x}{A} \quad (6)$$

$$x_c = \frac{\int_A x dA}{A} = \frac{S_y}{A} \quad (7)$$

As Figs. 1 and 2 indicate, the image of laser scanning data of tunnel segment is actually a ring because of its thickness which could be abstracted into the model depicted in Fig. 7. Consequently, once the noisy points and missing points have been well treated, using this method could easily determine the tunnel center.

3 Tunnel Deformation Visualization

In order to display the tunnel deformation in a direct way, we develop a Graphical User Interface named Cloud Viewer through which the deformed tunnel outline is shown on the display panel. The x component, y component of the deformation values and the resultant radius deformation values of arbitrary points on the tunnel outline could be obtained once the user imports treated laser scanning data (Fig. 8).

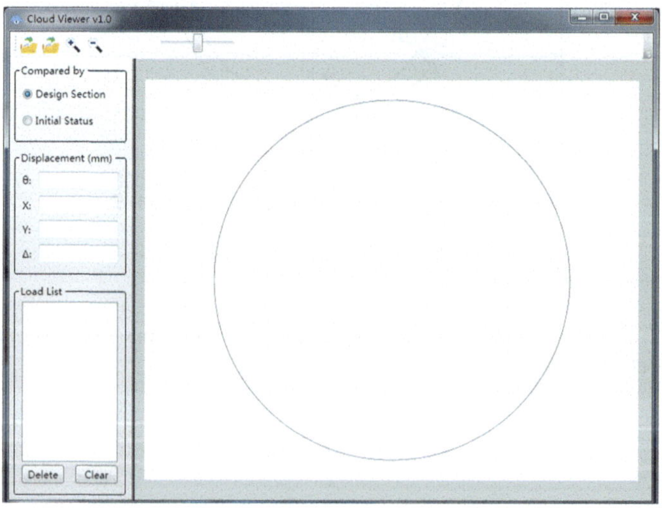

Fig. 8. The GUI of Cloud Viewer

The Cloud Viewer provides 2 calculation modes: a design section mode and an initial status mode. In the design section mode, the deformation values are calculated by subtracting the deformed tunnel outline from design outline, while in the initial status mode, the deformation values are the difference between current deformed status and the initial loading status. As Fig. 9 shows, to optimize the display performance, Cloud Viewer allows users to display the deformation by multiplying the real deformation values by different display factors. By assigning corresponding polar angles in the text box, the deformation values of at any locations can be obtained and presented in the text box.

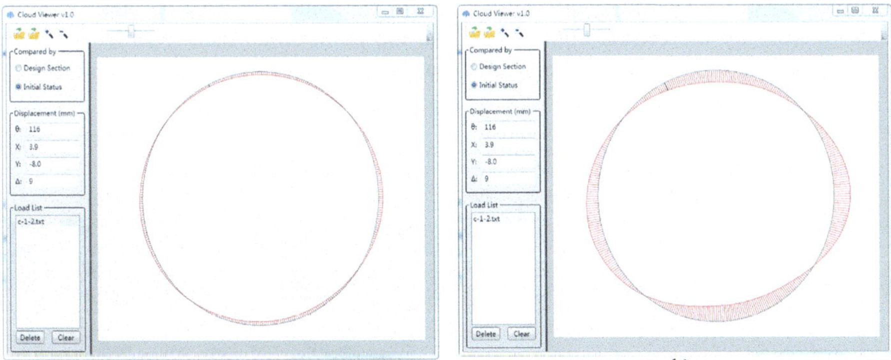

a) b)

Fig. 9. Deformation visualized by multiplying different display factors: (a) display factor = 1.0; (b) display factor = 4.0

4 Application

This section introduces two application examples of the proposed deformation analysis and visualization method: full-scale model tests on a circular shield tunnel and a quasi-rectangular shield tunnel. The results show that the proposed method is capable of processing different shapes of cross section with favourable accuracy.

4.1 Full-Scale Model Test for a Circular Shield Tunnel

The tunnel to be tested is a water storage and drainage tunnel which adopts circular cross section with an inner diameter of 9 m and a segment thickness of 650 mm. The buried depth of this tunnel could reach as large as 50 m–70 m, so the mechanical behaviour of tunnel segment under such buried depth delivers an important contribution to the understanding of tunnel's behaviour which is crucial to the whole project. In order to get comprehensive deformation data of the tunnel segment, laser scanning technique is adopted while the cable-extension displacement sensors are employed to ensure the reliability of the laser scanning data. Figure 10 shows the layout of the full-scale model test system. The device we use is Leica P40 and the accuracy of the scanning profile depends on the distance from the laser scanner to tunnel segment surface. For Leica P40, the systematic scanning error is 1.2 mm + 10 ppm which indicates that within the distance of 10 m, the scanning error is less than 1.2 mm. The furthest distance from laser scanner to segment surface is about 6 m in this test, so the scanning error is around 0.72 mm.

Before the measurements, two restrictions have to be coped with. Firstly, since no obvious vibrations are allowed during laser scanning, a special base needs to be designed so as to prevent the laser scanning device from vibrating. Secondly, the laser scanning device could not be installed at the center of the tunnel lining due to the existence of the central column (Figs. 10 and 11). Consequently, to cover the whole tunnel section, we employ two Leica laser scanning devices mirrored symmetrically on the two sides of the central column. By this way, we could eliminate the impact of the block of laser by the central column because the point cloud data recorded by each device could be unified via commercial software Leica Cyclone (Figs. 13 and 14).

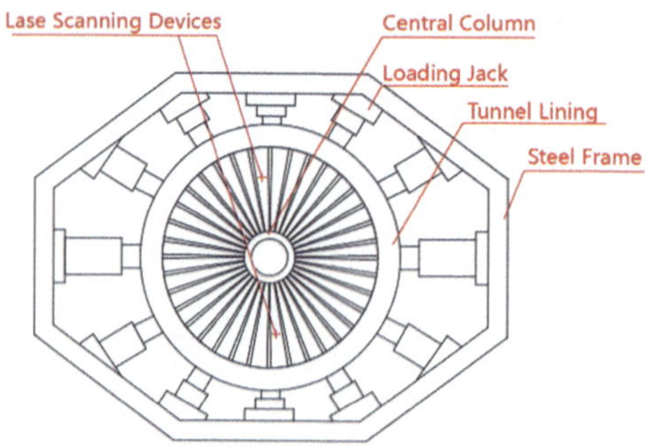

Fig. 10. Layout of the full-scale model test system for circular shield tunnel

Fig. 11. Placement of two Leica laser scanning devices

To allow the registration of the point cloud data from the laser scanning devices into a single point cloud, static black-and-white targets must be attached on the stationary points during the test. We use 6 targets which are attached on the wall of the laboratory (Fig. 12). By identifying the same targets in the point cloud data of the two devices via Cyclone, a universal coordinate system is generated which is shared by the two sets of point cloud data. Then Cyclone could register them with registration error less than 1 mm.

Fig. 12. Targets used in test (No. 4–No.6)

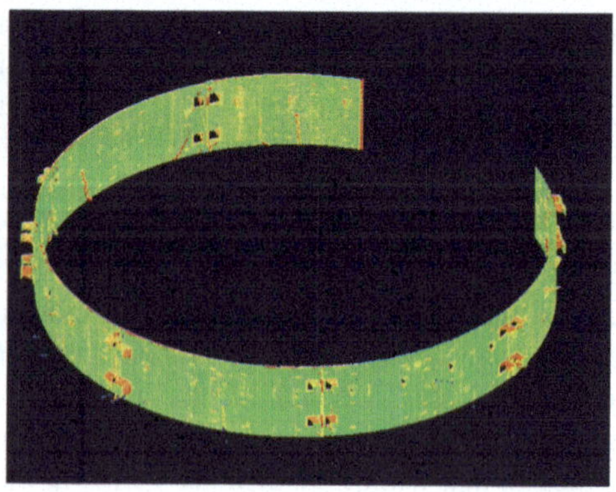

Fig. 13. Tunnel lining laser scanning data of one device before registration

During the test, 2 laser scanning devices controlled by cell phone via Wi-Fi operate at the same time once the jacking loads become stable. The procedure of scanning lasts for 1 min which is fast enough to capture the deformation of tunnel segment without further development. Once the point cloud data have been imported to the computer, we use Cyclone to pre-process the raw data (Fig. 14). After the registration of two series of cloud data, the proposed algorithm could calculate and visualize the deformation automatically within 10 s (Fig. 15).

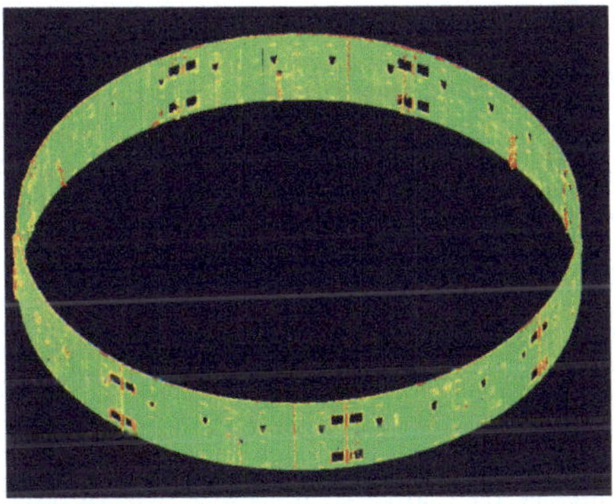

Fig. 14. The laser scanning image of the circular tunnel lining after pre-processing and registration

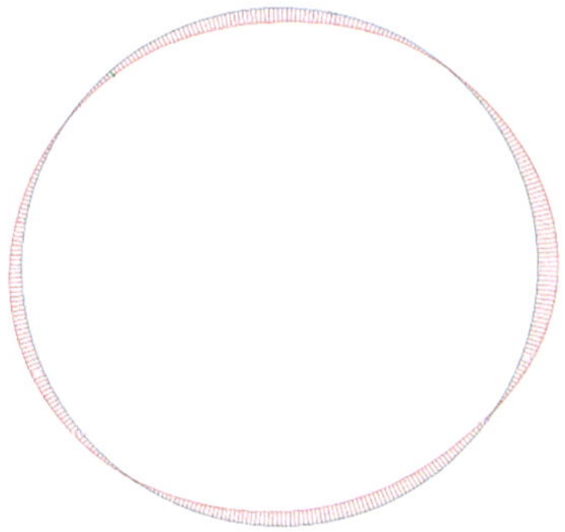

Fig. 15. The deformation contour of the circular tunnel lining

As mentioned above, cable-extension displacement sensors were used to confirm the validity of the laser scanning data. Figure 16 compares the deformation data measured by cable-extension displacement sensors and that determined from laser scanning technique. It is clear that the two sets of data are in good accordance with each other, which proves that both of the laser scanning technique and the proposed data processing method are adequately reliable and accurate.

Even though using cable-extension displacement sensors could also gain the convergence of the tunnel lining, as a "single point measurement", it could only deliver deformation data of dispersed points; whereas, using laser scanning technology, we could acquire the deformation of any point on the tunnel lining by assigning its polar angle through Cloud Viewer software, which could meet the comprehensive data demand of engineering and scientific researches.

4.2 Full-Scale Model Test of a Quasi-rectangular Shield Tunnel

The rectangular shield tunnel has attracted increasing interest of engineers and researchers in recent years for its high space usage efficiency. Compared with the traditional circular shield tunnel, using rectangular shield tunnel could save space by up to 20%, thus, leading to higher economic benefits. However, theoretically the mechanical performance of tunnel with a rectangular cross section is worse than a circular one. Stress concentration and large convergence deformation strongly restrict the development of rectangular shield tunnel. To solve this problem, a new sub-rectangular cross section has

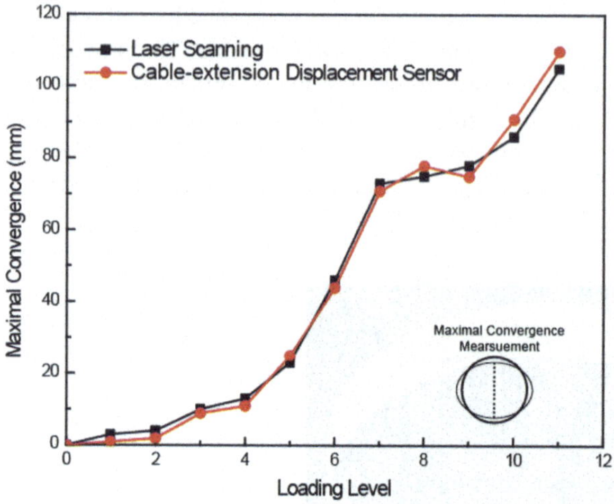

Fig. 16. Comparing the maximal convergence data obtained by laser scanning technique and that monitored by cable-extension displacement sensors

been designed. The new sub-rectangular tunnel is 10.7 m in width and 8.2 m in height with a thickness of 500 mm. To assess the mechanical property of this type of shield tunnel segment, a full-scale model test is conducted which adopts laser scanning technology and the proposed data processing method to acquire tunnel deformation data.

Compared to the traditional test system in which the tunnel lining lies horizontally on a base, an innovative "standing" loading method is developed to evaluate the effect of self-weight of lining on internal force distribution and deformation pattern as shown in Figs. 17 and 18. Since there is no obstacle, only one Leica P40 laser scanning device is used which needs to be fixed at the bottom of the tunnel lining.

After similar procedures described in Sect. 4.1, the pre-processed laser scanning data is shown in Fig. 19 and the deformation contours of tunnel lining under various loading conditions are gained. Figure 20 shows the tunnel deformation contour under self-weight load.

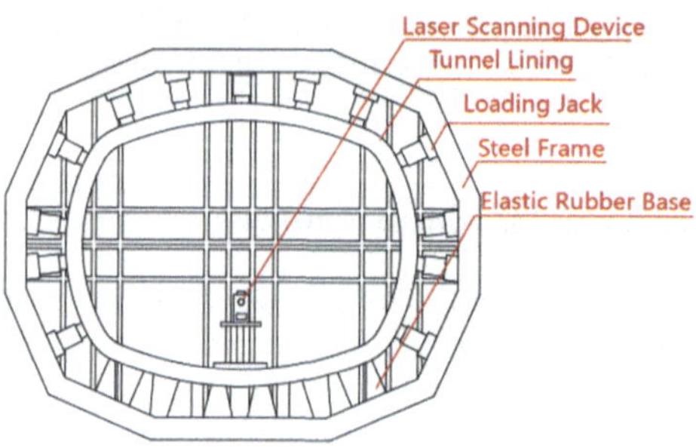

Fig. 17. Layout of full-scale model test system for a sub-rectangular shield tunnel

Fig. 18. A photo of the test system

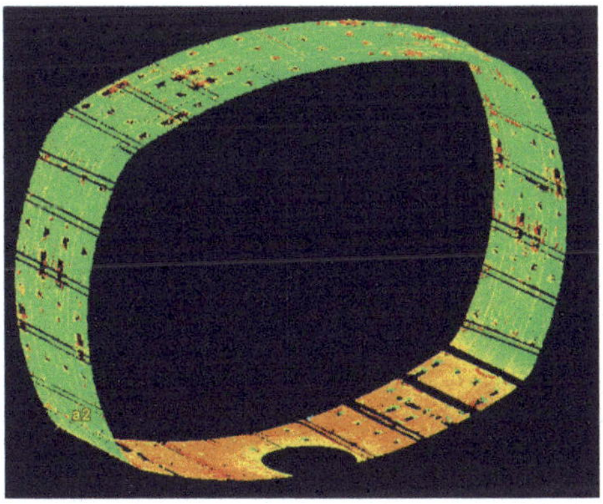

Fig. 19. The laser scanning image of quasi-rectangular tunnel lining after pre-processing

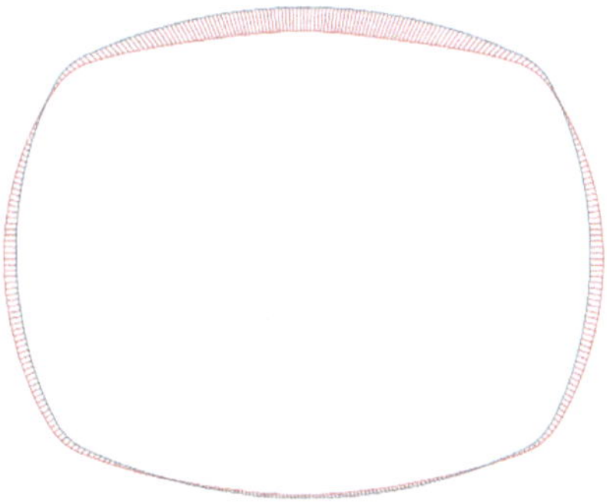

Fig. 20. The deformation contour of sub-rectangular tunnel lining under self-weight load

5 Conclusion

This paper introduces a novel methodology to process laser scanning cloud data with high speed and accuracy. Distinguished from the data processing method available in the literature, this methodology is capable of dealing with different shapes of tunnel cross sections and has the potential to be a universal method to process and visualize the laser scanning data in the field of tunnel engineering.

The proposed methodology starts with pre-generation of a polar coordinate system, followed by projecting the three-dimensional point cloud data onto a two-dimensional plane. A special algorithm known as point discretization and regeneration technique is developed to treat the data missing caused by data pre-processing. The deformation data is acquired by comparing the deformed outline and the design outline. A Graphical User Interface Cloud Viewer is developed to visualize the deformation image with adjustable display factors. One could easily get information data of any point on the tunnel outline by assigning its polar angle in Cloud Viewer.

Full-scale model tests of a circular shield tunnel and a quasi-rectangular shield tunnel adopt the proposed methodology to acquire the deformation data. The results show that (1) both the laser scanning technology and the proposed methodology are valid and accurate, (2) the proposed methodology could process arbitrary tunnel cross sections.

Acknowledgement. The research was financed by the Natural Science Foundation of China (No. 41372276). The technical support from the staff of Shanghai Tunnel Engineering Co. Ltd. are highly appreciated.

References

1. Nuttens, T., Stal, C., De Backer, H., Schotte, K., Van Bogaert, P., De Wulf, A.: Methodology for the ovalization monitoring of newly built circular train tunnels based on laser scanning: Liefkenshoek Rail Link (Belgium). Autom. Constr. **43**, 1–9 (2014)
2. Schueremans, L., Van Genechten, B.: The use of 3D-laser scanning in assessing the safety of masonry vaults—a case study on the church of Saint-Jacobs. Opt. Lasers Eng. **47**, 329–335 (2009)
3. Clarke, T.A.: A Review of Tunnel Profiling Methods (1996)
4. Wang, W., Zhao, W., Huang, L., Vimarlund, V., Wang, Z.: Applications of terrestrial laser scanning for tunnels: a review. J. Traffic Transp. Eng. (Engl. Ed.) **1**(5), 325–337 (2014)
5. Yoon, J.-S., Sagong, M., Lee, J.S., Lee, K.: Feature extraction of a concrete tunnel liner from 3D laser scanning data. NDT&E Int. **42**, 97–105 (2009)
6. Pejić, M.: Design and optimisation of laser scanning for tunnels geometry inspection. Tunn. Undergr. Space Technol. **37**, 199–206 (2013)
7. Fekete, S., Diederichs, M., Lato, M.: Geotechnical and operational applications for 3-dimensional laser scanning in drill and blast tunnels. Tunn. Undergr. Space Technol. **25**, 614–628 (2010)
8. Han, S., Cho, H., Kim, S., Jung, J., Heo, J.: Automated and efficient method for extraction of tunnel cross sections using terrestrial laser scanned data. J. Comput. Civil Eng. **27**(3), 274–281 (2013)

An Experimental Study of Pressuremeter Testing Under Polyaxial Boundary Stress Condition

Lang Liu[1(✉)], Haifeng Fu[2], Rick Chalaturnyk[1], Dingwei Weng[2], Gonzalo Zambrano[1], and Steve Zheng[1]

[1] Department of Civil and Environmental Engineering, University of Alberta, Edmonton, Canada
lliu@ualberta.ca

[2] Research Institute of Petroleum Exploration and Development of PetroChina, Langfang, China

Abstract. The in-situ stress is a critical factor that needs to be assessed in the development of a safe and sustainable oil and gas recovery, but it is also difficult to be accurately measured with the existing methods, especially in the soft sedimentary formations. In this paper, pressuremeter testing, as one of the promising approaches for the in-situ stress determination, is evaluated under a well-controlled polyaxial boundary stress condition. The experiment setup and test procedure are described. The data for the tests under three different stress regimes are presented. The data is corrected for the eccentricity and the movement of the tool during the test to show the non-uniform deformation of borehole in the testing plane. The discrepancy of pressure-expansion curves obtained at different axes is in a good agreement with the theoretical findings showing the effect of the in-situ stress anisotropy on the cavity expansion.

Keywords: Pressuremeter testing · Stress anisotropy

1 Introduction

One of the challenges in the development of the underground resource extraction plan is the uncertainty of in-situ stresses. In the oil and gas industry, an inaccurate estimation of the in-situ stresses would not only lead to the unfavorable design of well completion and but also put the disastrous events, such as the fault activation and the surface subsidence, under poor control (Zoback 2010). In Alberta, the in-situ stress plays a critical role in the assessment of the caprock integrity in the steam stimulation projects, as the minimum in-situ horizontal stress in the caprock is a primary factor to constrain the pressure of the steam injected into the oil sands reservoir.

Methods, such as hydraulic fracturing and overcoring, have been developed and proved successful in determining the stresses in hard rocks (Ljunggren et al. 2003; Schmitt et al. 2012). However, because caprock formations are more ductile and less prone to fracturing, the validity of applying these methods to such the materials should be questioned (ShafieZadeh and Chalaturnyk 2014).

The original version of this chapter was revised: Figure 5 has been corrected. The correction to this chapter is available at https://doi.org/10.1007/978-981-13-0095-0_61

L. Hu et al. (Eds.): GSIC 2018, *Proceedings of GeoShanghai 2018 International Conference: Multi-physics Processes in Soil Mechanics and Advances in Geotechnical Testing*, pp. 449–457, 2018.
https://doi.org/10.1007/978-981-13-0095-0_50

Pressuremeter testing, as one of the commonly used methods in the geotechnical investigation, is considered as a promising method in the evaluation of in-situ stresses in the soft rocks. The response of the borehole at the different orientations under the uniform expansion pressure could provide valuable information to back calculate the possible anisotropy of in-situ stresses in the testing plane.

Figure 1 shows the one dataset obtained from the pressuremeter test in a horizontal borehole drilled into the Boom clay formation at an underground research lab in Mol, Belgium. The test curves at three axes vary as a function of the orientations of strain arms with respect to the maximum and minimum in-plane in-situ stresses.

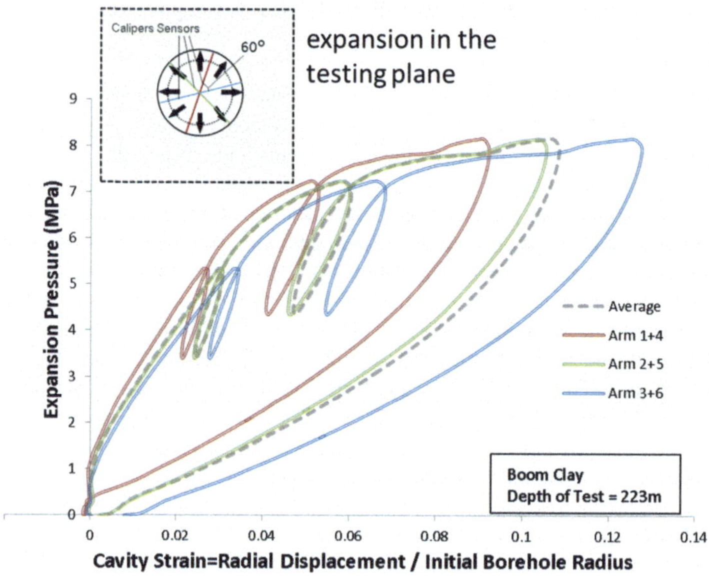

Fig. 1. Pressuremeter test curves measured by independent strain arms at three axes. The test was carried out in the horizontal borehole into the tunnel liner at underground research lab at Mol, Belgium.

Zhou et al. (2016) has explained the "non-circular" effect of the cylindrical cavity during the expansion in a biaxial in-situ stress field with a rigorously derived solution. A numerical study has been carried out by Liu (2015) to address the potential challenges in the interpretation of the pressuremeter tests in the soft sedimentary rock, including the drilling disturbance and the drainage conditions. This work, as part of the assessment of the interpretation technique proposed in Liu et al. (In Preparation)., is to provide the experimental evidence of cavity expansions under the anisotropy boundary stresses. The laboratory configuration of the experiment will be described in the first part of the paper and a preliminary analysis performed upon the tests under three different stress regimes will be presented in the second part.

2 Experiment Setup

2.1 Cement Preparation

In order to duplicate the mechanical properties of the in-situ clayshale materials, white cement (grade 52.5), water and kaolinite were considered as mixing components so that a certain amount of plasticity could be kept without significantly compromising the strength and stiffness. An industry-sized blade mixer was used for cement mixing, three batches of mixing are required to fill a cubic mold with a dimension of 914 mm (H) * 763 mm (W) * 763 mm (L). A small portion of bubble reducer was also added to reduce the voids in the cement slurry. After a preliminary evaluation of the material properties with the small subsamples as well as the large block specimens, the gravimetric mixing proportion of cement, water and kaolinite equal to 1:1:0.4 was determined.

After 9–12 days of curing, the cement block was moved out from the water tank and placed into the loading frame (Fig. 2). A borehole was drilled beforehand with an 83 mm diameter coring bit in the center of the specimen. Due to the swinging effect of the drilling string, the borehole was slightly oversized. Three blocks were prepared for the pressuremeter tests under different boundary stress conditions. Information of these block specimens is listed in Table 1.

Fig. 2. Cement block in the loading frame

2.2 Test Setup

The loading frame consists of four flat jacks for the lateral loadings and eight hydraulic jacks applying vertical loading through a steel platen (Fig. 3). Since the lateral faces of block were not entirely covered by the flat jacks, the actual pressure transmitted into the

Table 1. Basic information of the three block specimens

Specimen No.	Curing days	Initial borehole size (mm)
B1T5	9	86.3
B1T6	12	86.1
B1T7	9	86.2

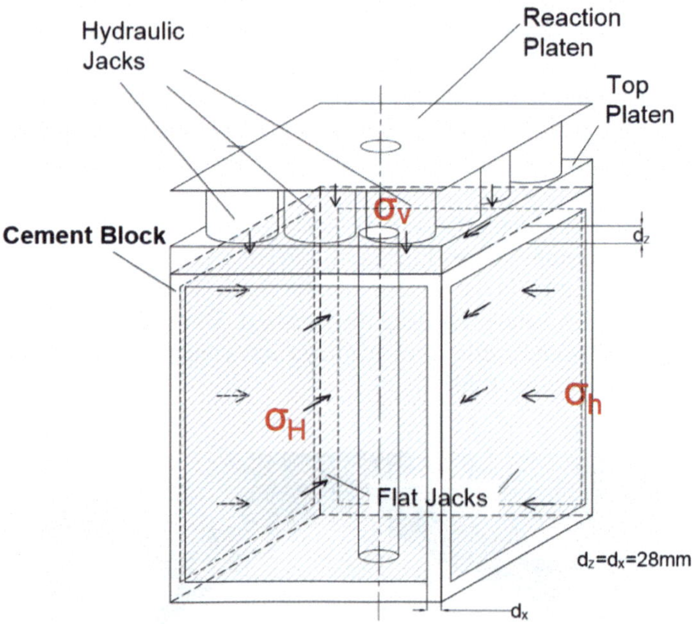

Fig. 3. Schematics of the loading assembly

block was thought to be lower than that applied at the jack. To monitor the deformation of the block under the loading, LVDTs and a single-axis caliper were installed at the outside boundaries and in the borehole respectively.

Three boundary stress conditions were proposed to take into account different stress anisotropies and in-situ stress regimes. The pressures applied at the faces of three blocks after equilibrium are given in Table 2.

Table 2. Boundary stress conditions for three tests

Specimen No.	σ_H (MPa)	σ_h (MPa)	σ_v (MPa)	σ_H/σ_h	Stress regime
B1T5	3.7	3.3	3	1.12	Reverse Fault
B1T6	3.25	2.15	2.6	1.51	Strike-slip
B1T7	3.3	1.7	5	1.94	Normal Fault

The pressuremeter used in the test is a modified version of the High Pressure Dilatometer manufactured by Cambridge In-situ Ltd. The toe of pressuremeter was shortened so that the expandable packer could be fully covered inside of the block (Fig. 4). The outside diameter of the pressuremeter is 73 mm and the total thickness of the packer element including the rubber membrane and the flexible membrane shield is 5.5 mm. Before the test, the stiffness of the membrane and the compliance of the instrument body were carefully calibrated to obtain the factors that would be used in the calculation of the true pressure and displacement at the external surface of the membrane.

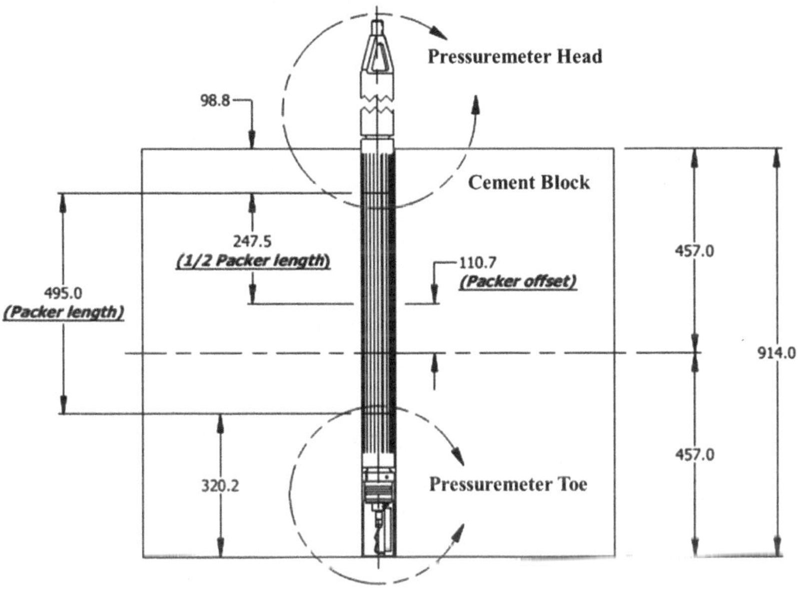

Fig. 4. Dimension of the test setup. Unit: mm

3 Test Results

3.1 Pressuremeter Test Curves

The pressuremeter test was conducted by injecting oil into the probe. The pressure change was maintained at 2 psi/s for both borehole expansion and contraction. The loading rate was thought to be sufficiently low that the interaction between the probe and the borehole would be in a pseudo-static equilibrium at any time of the test. To have extra data to assess the creep potential and the stiffness of the specimen, unload-reload loops and pressure hold tests were also included in the testing procedure.

Figure 5 plots the expansion pressure versus the radial displacement measured in each test. While the displacement data at different azimuths could be averaged to represent the overall deformation of the borehole under the pressure, it remains questionable whether they could be interpreted using the existing analytical solutions which were derived under the assumption of the isotropic boundary stress condition.

3.2 Non-uniform In-Plane Borehole Deformation

Because the pressuremeter was not initially centric and also the borehole itself was not perfectly circular, borehole might not start to expand at the same radius. Also, the probe tended to be centralizing itself in the process of the inflation, the radial displacement captured by each individual arm might not show the actual expansion of the borehole.

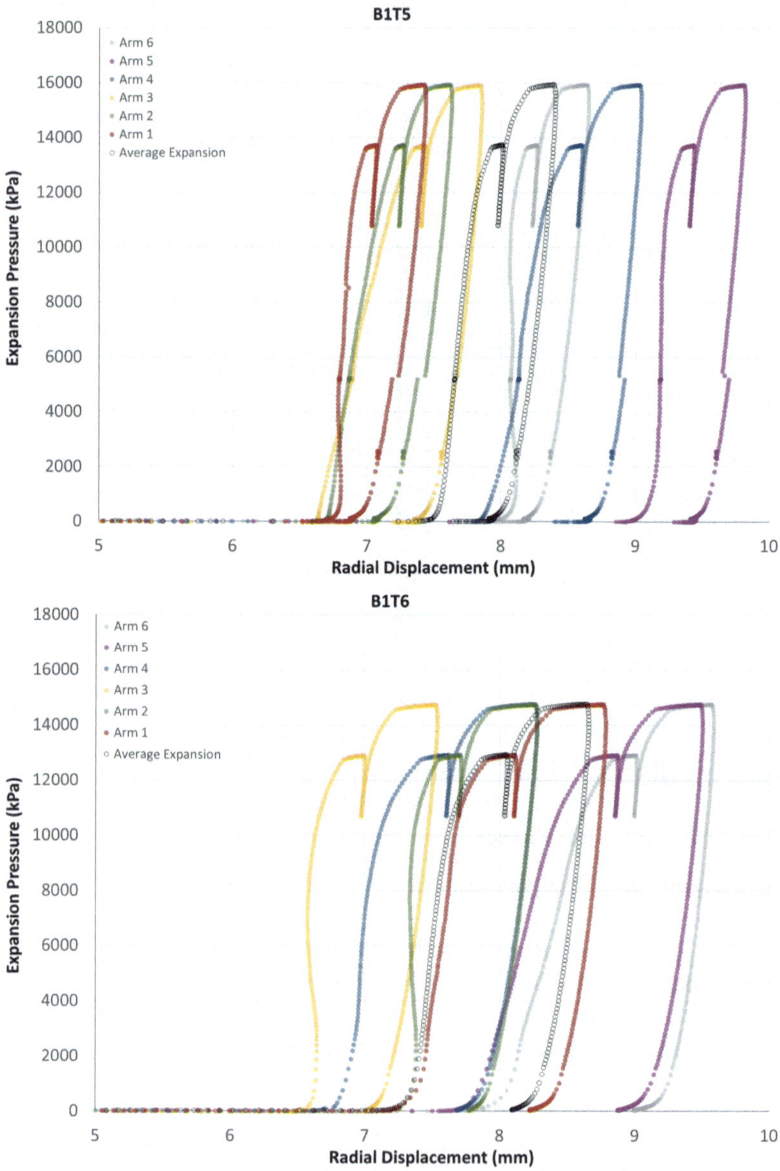

Fig. 5. Pressuremeter test curves under three stress regimes

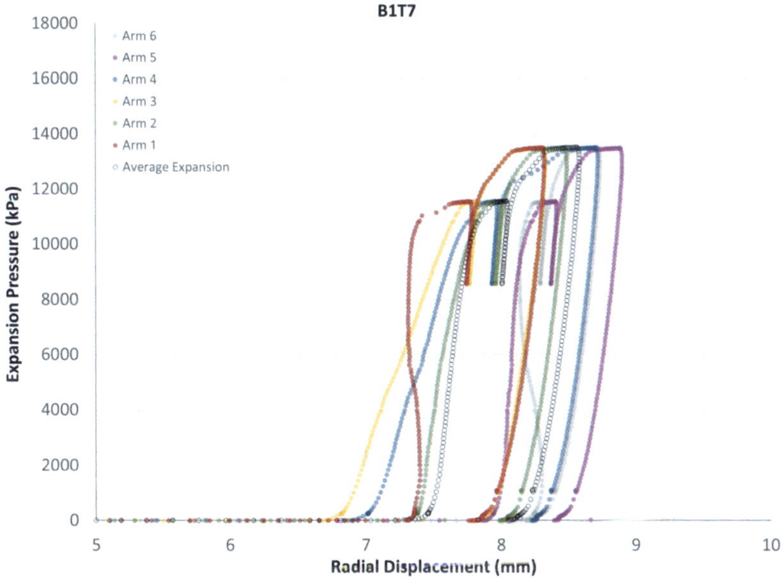

Fig. 5. (*continued*)

To find out the true non-uniform borehole deformation, a correction was made by fitting the displacement data with elliptical function and transforming their coordinates from tool domain to borehole domain. The detail of the correction method was explained by Schwerzmann et al. (2006). With the best fit of the borehole shape, the radial displacement at the maximum horizontal stress direction (0° of the azimuth) and minimum horizontal stress direction (90° of the azimuth) could be calculated. Figure 6 compares the deformations of borehole at two major axes from three tests. Curves were shifted horizontally to have the same origin of the expansion at the pressure level around 3000 kPa.

Under the uniform expansion pressure, borehole tended to have more plastic deformation in the direction of the maximum in-plane boundary stress than in the direction of the minimum in-plane boundary stress. This is in a good agreement with the theoretical findings from Zhou (2016)'s work, addressing the higher shear stress and early borehole yielding at the azimuth aligned with the maximum in-plane in-situ stress. Such the discrepancy of expansions between these two major axes becomes more prominent when the boundary stress ratio increases.

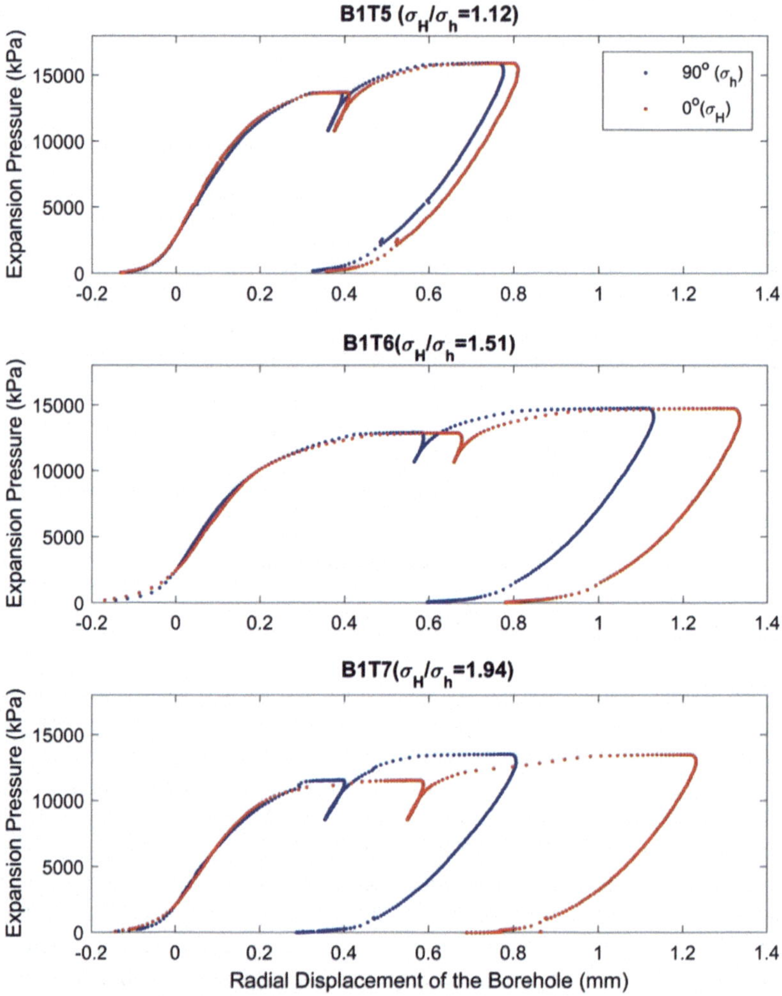

Fig. 6. Anisotropic borehole deformations under three boundary stress conditions

Acknowledgement. This work was financially supported by the Foundation CMG Industrial Research Consortia. The technical support from J. Brandl at the University of Alberta and the lab technicians at RIPED is also gratefully acknowledged.

References

Liu, L.: Numerical Study of Reservoir Geomechanical Pressuremeter Testing under Anisotropic In-situ Stresses. University of Alberta (2015)

Liu, L., Chalaturnyk, R., Zambrano, G.: Numerical investigation and identification of geomechanical parameters from self - bored pressuremeter testing. In Preparation

Ljunggren, C., Chang, Y., Janson, T., Christiansson, R.: An overview of rock stress measurement methods. Int. J. Rock Mech. Min. Sci. **40**, 975–989 (2003)

Zhou, H., Kong, G., Liu, H.: A semi-analytical solution for cylindrical cavity expansion in elastic – perfectly plastic soil under biaxial in situ stress field. Geotechnique **66**(9), 786–788 (2016)

Schmitt, D., Currie, C., Zhang, L.: Crustal stress determination from boreholes and rock cores: fundamental principles. Tectonophysics **580**, 1–26 (2012)

Schwerzmann, A., Funk, M., Blatter, H.: Instruments and methods. Borehole logging with an eight-arm caliper – inclinometer probe. J. Glaciol. **52**, 381–388 (2006)

ShafieZadeh, N., Chalaturnyk, R.: Interpretation challenges for in situ stress from mini-frac tests in soft rocks/hard soils. Shale Energy Eng. 298–309 (2014)

Zoback, M.: Reservoir Geomechanics. Cambridge University Press, Cambridge (2010)

Long-Term Settlement of Subway Tunnel and Prediction of Settlement Trough in Coastal City Shanghai

Zhen-Dong Cui[1,2(✉)], Shan-Shan Hua[1], and Jia-Sen Yan[1]

[1] State Key Laboratory for Geomechanics and Deep Underground Engineering, School of Mechanics and Civil Engineering, China University of Mining and Technology, Xuzhou 221116, Jiangsu, People's Republic of China
czdjiaozuo@163.com

[2] Fujian Research Center for Tunneling and Urban Underground Space Engineering, Huaqiao University, Xiamen 361021, Fujian People's Republic of China

Abstract. Subway tunnels have shown characteristics of severe longitudinal differential settlement since the operation of subways in Shanghai. The differential settlement has an important impact on the internal force of tunnel structure and the waterproofing of joints. According to the in-site measuring data of Subway Line 1, subway tunnel in the soft area is easily prone to developing many settlement troughs with different sizes. The differential settlement in the settlement trough is noticeable, which has a seriously effect on the safety of tunnel structure and subway operation. The differential settlement in the settlement trough relates to the subway vibration load, the engineering activities, the regional land subsidence and the seasonal rain. From the point of view of time and space, the curve fitting and Auto Regressive Integrated Moving Average ARIMA (*p*, *d*, *q*) were combined to predict the long-term settlement in the settlement trough. The Hengshan Road Station and the South Huangpi Road Station were taken as case studies. The results show that the predicted values agree well with the measured data from the perspective of time and space, which provides a new idea for the long-term settlement of subway tunnel.

Keywords: Subway tunnel · Long-term settlement · ARIMA (*p*, *d*, *q*) model

1 Introduction

In order to make rational use of limited urban land and alleviate the urban traffic pressure, subway tunnels have been constructed in more and more cities. Subways play an important role in improving the urban traffic congestion problem and the people's quality of life. In the soft area, the settlement is relatively large during the operation of subway for its special geology conditions. The uneven settlement causes the bolt cracking, which affects the normal operation of subways.

On the one hand, the settlement of subway tunnels occurs during the period of tunnel construction. There are mainly four different approaches to study the settlement in this period, namely empirical methods, analytical methods, finite-element methods

L. Hu et al. (Eds.): GSIC 2018, *Proceedings of GeoShanghai 2018 International Conference: Multi-physics Processes in Soil Mechanics and Advances in Geotechnical Testing*, pp. 458–467, 2018.
https://doi.org/10.1007/978-981-13-0095-0_51

and artificial neural networks. Empirical methods (Attewell and Woodman 1982) were widely used because they predicted the surface deformation effectively during the construction of subway tunnels without considering soil properties. To be more persuasive, analytical methods were used to predict the surface deformation (Loganathan and Poulos 1998; Park 2004; Zhu et al. 2014). Finite-element methods, which could simulate the complicated boundary conditions, were adopted to analyze the excavation of subway tunnel (Ercelebi et al. 2011). The surface deformation during the tunnel excavation is affected by many factors that present complex nonlinear relationships. Artificial neural networks are able to approach any complicated nonlinear relationship and study some unknown systems. Artificial neural networks were applied to predict surface deformation caused by the tunnel construction (Mohammadi et al. 2015).

On the other hand, the subway tunnels experience the settlement during the operation in the soft area. The vertical displacement and the longitudinal stress of the existing tunnel was influenced by the new shield tunnelling (Li et al. 2014; Liang et al. 2016). There existed the significant settlement for the subway tunnel because of foundation pit excavation (Liu et al. 2011). The trend between the subway tunneling settlement and the regional land subsidence were the same (Ye et al. 2007).

Settlement of subway tunnels lasted for a long time. To reduce the economic cost of monitoring, a variety of methods are used to predict the long-term settlement of subway tunnel. Fang et al. (1993) proposed a hyperbolic model to estimate the trend of the maximum ground subsidence over the tunnel curve by analyzing the filed data. Cui and Ren (2014) adopted ARMA (*p*, *q*) model to predict the long-term settlement of subway tunnels. Although the above methods have made some progress in the prediction of tunnel settlement, there are few special methods to predict the long-term settlement of the settlement trough where is more likely to occur settlement. In addition, many prediction models only consider the relationship between the long-term settlement and the time, without considering the influence factor of the space.

This paper analyses long-term settlement behavior according to subway line 1 in Shanghai. The settlement function of the settlement trough is established by the curve fitting method, and the control point of the settlement function is predicted by ARIMA (*p*, *d*, *q*) model. The changes of the settlement with the time and the space are considered by the ARIMA (*p*, *d*, *q*) and the long-term settlement function.

2 Geological Conditions

Shanghai is located on the east of the Yangtze River Delta, with an area of about 6340 km^2. The loose sediments in Yangtze River Delta mainly belong to the Quaternary. The Quaternary deposit is well developed, reaching 300–360 m in the urban area. The subway tunnels in Shanghai are generally situated in the silty clay stratum with characteristics of high water content, high compressibility, low permeability and low shear strength. The disturbed soft clay in Shanghai tends to induce the settlement for a long time. The typical geological section of Shanghai is shown in Fig. 1.

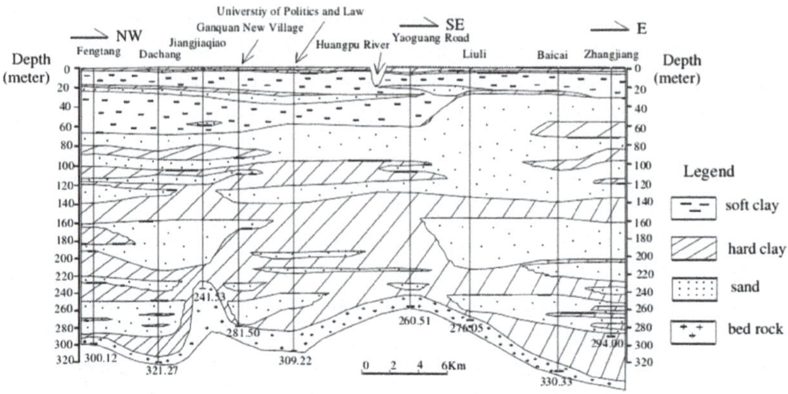

Fig. 1. The typical geological section of Shanghai (Tang et al. 2008)

3 Settlement Behavior of Subway Tunnel

Shanghai Subway Line 1 goes from Xinzhuang Station to Fujin Road Station. As shown in Fig. 2, there has been substantial settlement from 1999 to 2009 and the average accumulated settlement of reaches 111.5 mm at the end of 2009. The maximum settlement in the longitudinal direction is 292.9 mm and the minimum settlement is 0.4 mm. That means the differential settlement is 292.5 mm. The average differential settlement reaches 20.6 mm in the past few years. The differential settlement gradually increases during the long-term settlement of subway tunnel in soft ground. There was a significantly large settlement troughs, located around the Hengshan Road Station.

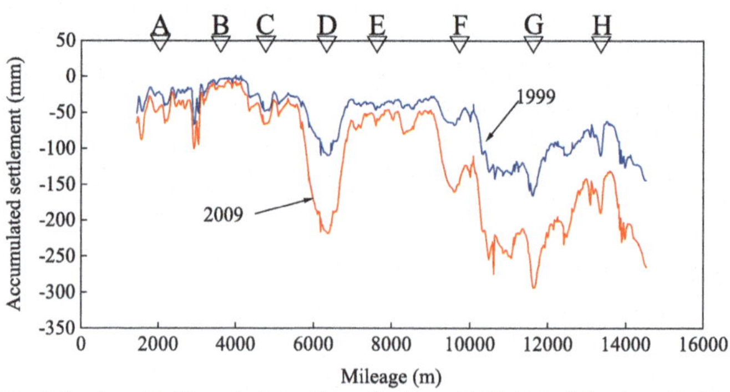

(A-Caobao Road Station; B-Shanghai Stadium Station; C-Xujiahui Station; D-Hengshan Road Station; E-Changshu Road Station; F-South Huangpi Road Station; G-People's Square Station; H-Hanzhong Road Station)

Fig. 2. Accumulated settlement of Subway Line 1

According to the analysis of Subway Lines 1, subway tunnel experienced the significant longitudinal settlement. With the increasing of the operation time, the accumulated settlement increased continuously and subsequently tended to converge.

The differential settlement was relatively large where there was a settlement trough, which would cause subway safety problem.

4 Settlement Trough

The subway tunnels in the soft soil area easily generated the settlement. In order to accurately predict the settlement of settlement trough, improve the safety of subway operation and reduce the cost of long-term monitoring, it is necessary to study settlement regularity about settlement trough in the soft deposit. The Shanghai Metro Line 1, with long operation time and abundant measured data, contributes to analyze the settlement trough in the soft area.

Characteristics of Settlement Trough

The bottom and internal walls of the settlement trough are called the trough bottom and the trough wall respectively. The settlement trough is divided into the left settlement trough and the right settlement trough according to the trough bottom as shown in Fig. 3. Although settlement trough have different size, settlement trough are usually asymmetric. For example, the settlement trough in Hengshan Road Station and People's Square Station have different sizes and both of them are near asymmetric. The whole subway tunnel experienced settlement and the settlement at different mileages is varied by comparing the accumulated settlement in 1999 and 2009. The settlement where there exists a settlement trough is more obvious, at the same time, the accumulated settlement is much larger.

Figure 4 shows the cumulative settlement difference in Hengshan Road Station from 1999 to 2007. The difference at the groove bottom is 112.6 mm and the minimum difference in the trough wall is 31.9 mm, which shows the settlement rate in groove bottom was far greater than groove wall. The accumulated settlement difference presents the downward parabolic and the maximum value is at the bottom of the settlement trough.

Figure 5 illustrates the settlement of the settlement trough in Hengshan Road Station

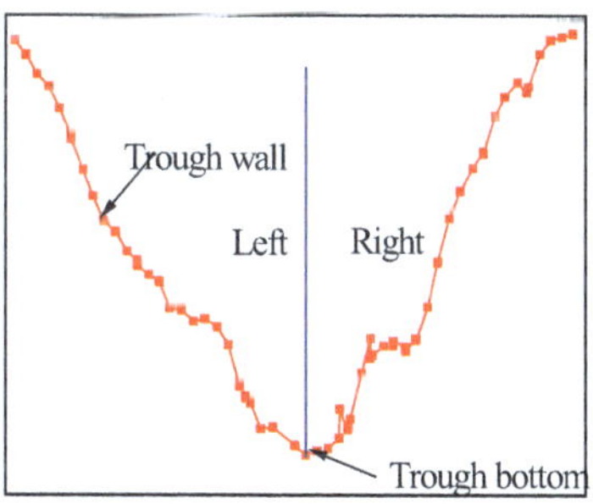

Fig. 3. Schematic diagrams of settlement trough

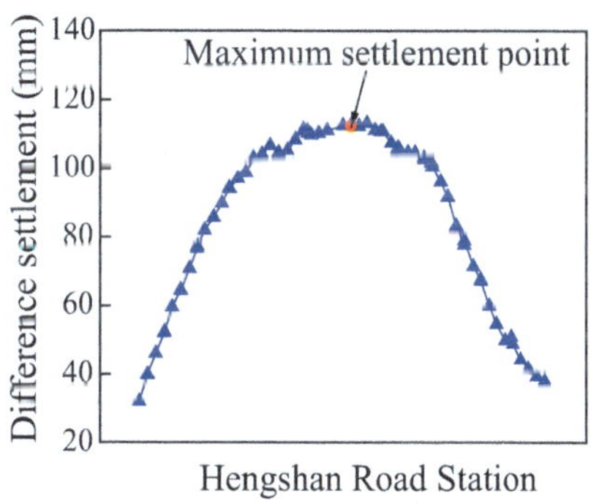

Fig. 4. Accumulated settlement difference from 1999 to 2007

with different years. The settlement trough is overall relatively smooth except some positions. The shape of settlement trough does not change with time. The settlement trend of the settlement trough is positively correlated with the settlement at the trough bottom.

The settlement trough produces settlement when the groove bottom occurs settle-

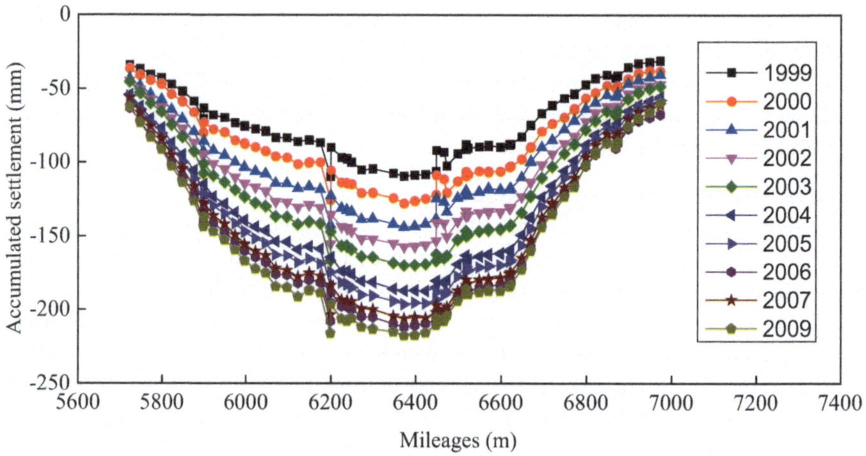

Fig. 5. Accumulated settlement with time at Hengshan road station

ment. At the same time, the settlement trough produces uplift when the groove bottom occurs uplift. The settlement trend of settlement trough may not converge to stable value. The settlement trend of settlement trough is convergent between 1999 and 2003, but the settlement trend of settlement trough is irregular. The settlement trough results from the complex causes including subway vibration load, the engineering activities around the subway tunnel, the regional land subsidence and the changing of underground water during the operations of subways. The soil mass is subjected to different degrees of disturbance in different mileage during the subway construction. The disturbed soil mass during the construction period is prone to settlement in the operation period. The local section of the subway tunnel undergoing engineering activities has larger settlement and the groundwater leakage occurs. The groundwater leakage results in the decrease of pore water pressure, the increase of effective stress and the soil mass consolidation (Shin et al. 2012). The process groundwater seepage is accompanied by the sediment leakage, which aggravates the deformation of subway tunnels (Wu et al. 2014). Because the groundwater leakage is local, the settlement trough is easily formed.

5 Prediction Model and a Case Study

5.1 Model Building

Based on the in-site monitoring data the characteristics of long-term settlement of subway tunnel was analyzed and settlement prediction model of the settlement trough was built.

The shape of the settlement trough is stable, but it is difficult to determine the uplift or settlement of the settlement trough at the next moment. By the in-site monitoring data of the trough settlement, the settlement of settlement trough depends on the bottom of settlement trough. The maximum sedimentation value of the bottom is monitored in the prediction process.

Detailed steps of building the prediction model are as follows.

(1) Data processing

The in-site monitoring data of the left settlement trough and the right settlement trough are carried out the normalization processing, respectively.

(2) Fitting the settlement trough

The curve function for fitting the normalized data of the left settlement trough is called $f_1(x)$, and that of the right settlement trough is called $f_2(x)$, where x is the normalized Euclid distance and $f(x)$ is the normalized settlement.

(3) Determining the control point of the function

According to the number of parameters in the curve function, the least squares algorithm is used to solve the curve parameters. The normalized data is substituted into the function $f_1(x)$ and $f_2(x)$ to get its predictive value, respectively. If the predicted value is different from the monitoring value, the number of input monitoring data is changed to resolve the parameters of the equation, and the error is minimal as soon as possible. The final number input monitoring data is called the control point of the settlement trough curve function.

(4) Predicting the accumulated settlement about the control point at the next time by ARIMA (p, d, q) model

ARIMA (p, d, q), which is used to predict the timed series, is the Autoregressive Integrated Moving Average Model, where p is the order number of AR model and q is the order number of MA model and d is the difference times. ARIMA (p, d, q) model can be used to predict the cumulative settlement of the next moment of the control point. Detailed steps of ARIMA (p, d, q) model analysis are as follows.

(a) To recognize the stationarity of the control point series by unit root test which is calculated by software EVIEWS. The control point series are not stationary in general and need to be treated by difference method. The differential times are the d in ARIMA model.

(b) The p and q in the ARIMA model can be determined by the Autocorrelation Function (ACF) and the partial-autocorrelation function (PACF) table respectively. The ACF and PACF are also calculated by software EVIEWS.
(c) According to the optimal model and its order, moment estimation, maximum likelihood estimation and least square estimation are used to estimating the parameters of the autoregressive part and the moving average part in the model.

By compared the monitoring value and the predicted value of the control point, the forecast results are treated according to the difference between them.

(5) Solving the parameters of curve function

According to the accumulated settlement of the control point at the next moment, the least squares algorithm is used to solve the curve function parameter. The obtained parameter values are substituted into the curve function.

(6) Predicting the settlement of the whole settlement trough

The curve function is used to predict the uneven settlement of the whole settlement trough. The desired predictions are anti-normalization and compared with the actual values.

5.2 Case Studies

The settlement trough in Hengshan Road station and South Huangpi Road Station were used to model verification. According to the prediction model, the settlement curve of the settlement trough can be obtained separately. Formula (1) is the curve function in Hengshan Road station and Formula (2) is the curve function in Huangpi Road station.

$$f(x) = a_1 \sin(b_1 x + c_1) + a_2 \sin(b_2 x + c_2) + a_3 \sin(b_3 x + c_3) + a_4 \sin(b_4 x + c_4) \tag{1}$$

where $a_1, b_1, c_1, a_2, b_2, c_2, a_3, b_3, c_3, a_4, b_4$ and c_4 are the fitting parameters (Tables 1 and 2);

$$f(x) = a_1 \sin(b_1 x + c_1) + a_2 \sin(b_2 x + c_2) + a_3 \sin(b_3 x + c_3) \tag{2}$$

where $a_1, b_1, c_1, a_2, b_2, c_2, a_3, b_3$ and c_3 are the fitting parameters (Table 3);

As shown in Fig. 2, the range of x in Hengshan Road station is from 5721.8 m to 6972.9 m while the range of x in South Huangpi Road Station is from 11501.6 m to 11891.6 m. The fitting parameters can be obtained according to the control point in Hengshan Road station and South Huangpi Road Station separately. The predicted data in Hengshan Road station and South Huangpi Road Station were shown in Figs. 6 and 7.

Table 1. Fitting parameters (from a_1 to c_2)

	a_1	b_1	c_1	a_2	b_2	c_2
$f_1(x)$	1.737	2.188	2.111	0.8412	2.936	0.6177
$f_2(x)$	2.048	2.178	0.2821	1.502	2.622	2.878

Table 2. Fitting parameters (from a_3 to c_4)

	a_3	b_3	c_3	a_4	b_4	c_4
$f_1(x)$	0.027	10.68	0.4724	0.01674	22.22	1.012
$f_2(x)$	0.0347	12.7	2.074	0.01468	20.93	−0.54

Table 3. Fitting parameters (from a_1 to c_3)

	a_1	b_1	c_1	a_2	b_2	c_2	a_3	b_3	c_3
$f_1(x)$	2.184	2.793	−0.1894	1.413	3.297	2.875	0.03753	11.97	1.965
$f_2(x)$	1.128	2.76	−0.4263	5.221	6.131	1.375	4.86	6.34	4.422

By comparing the monitored data with the predicted data, although different settlement troughs have different prediction precision, the range of the prediction precision is almost 0.9–1. At the same time, the prediction performance at the bottom of the settlement trough is superior to that at other locations according to Figs. 8 and 9.

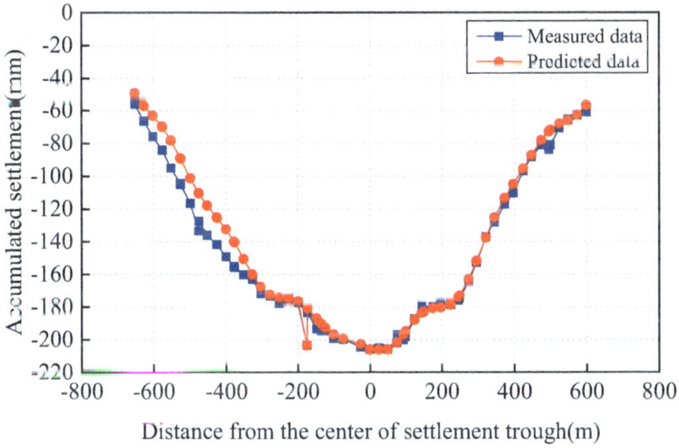

Fig. 6. Comparison of the measured accumulated settlement with the predicted in Hengshan road station

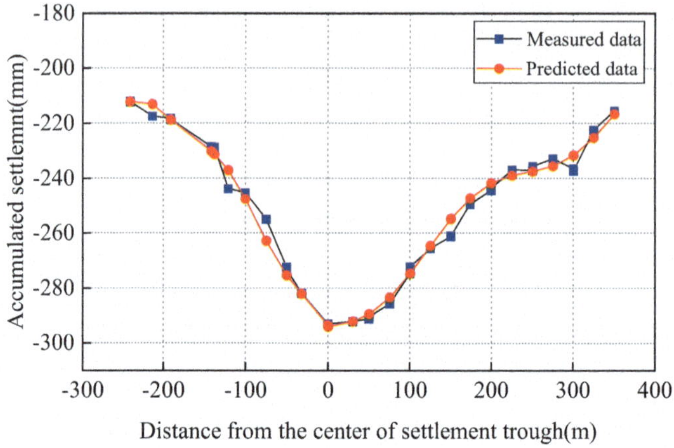

Fig. 7. Comparison of the measured accumulated settlement with the predicted in South Huangpi road station

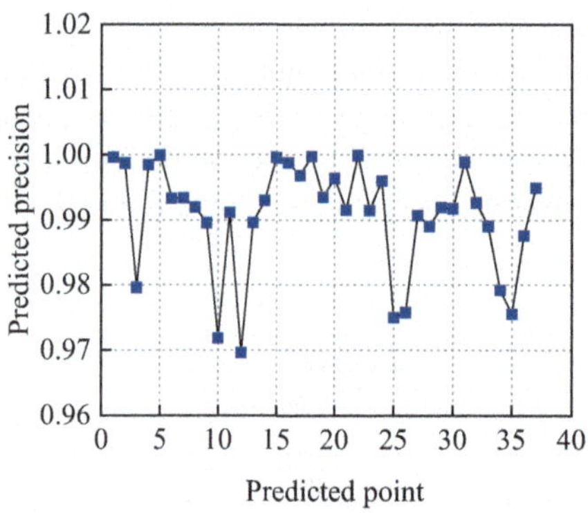

Fig. 9. Prediction precision in South Huangpi road station

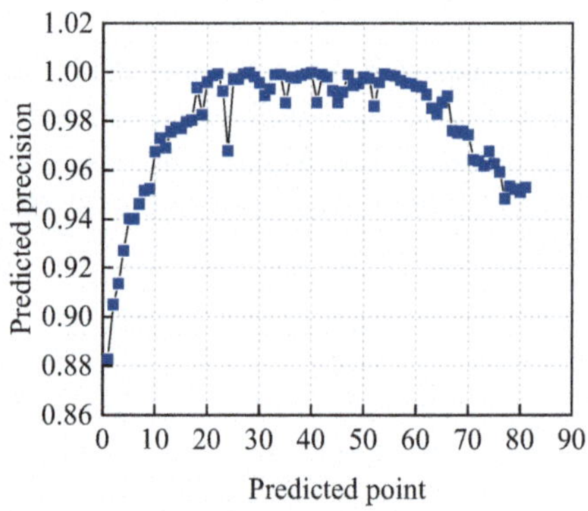

Fig. 8. Prediction precision in Hengshan road station

6 Conclusions

Based on the in-site monitoring data of subway tunnel in Shanghai and the characteristics of the long-term settlement, the settlement troughs were studied and the prediction model for the settlement trough were built in this paper. The conclusions are as follows.

(1) The differential settlement in longitudinal direction is significant where there is a settlement trough;

(2) The curve fitting method and ARIMA (*p, d, q*) model could be used to predict the settlement of the settlement trough. The curve fitting method is able to consider spatial factors, and ARIMA (*p, d, q*) considers the time factor, which effectively improves the prediction accuracy.

Acknowledgements. This work presented in this paper was by National key research and development program (2017YFC1500702) and the research grant (16FTUE03) from Fujian Research Center for Tunneling and Urban Underground Space Engineering (Huaqiao University).

References

Attewell, P.B., Woodman, J.P.: Predicting the dynamics of ground settlement and its derivatives caused by tunnelling in soil. Ground Eng. **15**(8), 13–22 (1982)

Cui, Z.D., Ren, S.X.: Prediction of long-term settlements of subway tunnel in the soft soil area. Nat. Hazards **74**(2), 1007–1020 (2014)

Ercelebi, S.G., Copur, H., Ocak, I.: Surface settlement predictions for Istanbul Metro tunnels excavated by EPB-TBM. Environ. Earth Sci. **62**(2), 357–365 (2011)

Fang, Y.S., Lin, S.J., Lin, J.S.: Time and settlement in EPB shield tunnelling. Tunn. Tunn. **25** (11), 27–28 (1993)

Li, P., Do, S.J., Ma, X.F., et al.: Centrifuge investigation into the effect of new shield tunnelling on an existing underlying large-diameter tunnel. Tunn. Undergr. Space Technol. **42**, 59–66 (2014)

Loganathan, N., Poulos, H.G.: Analytical prediction for tunneling-induced ground movements in clays. J. Geotech. Geoenviron. Eng. **124**(9), 846–856 (1998)

Liu, H.L., Li, P., Liu, J.Y.: Numerical investigation of underlying tunnel heave during a new tunnel construction. Tunn. Undergr. Space Technol. **26**(2), 276–283 (2011)

Liang, R.Z., Xia, T.D., Yi, H.: Effects of above-crossing tunnelling on the existing shield tunnels. Tunn. Undergr. Space Technol. **58**, 159–176 (2016)

Mohammadi, S.D., Naseri, F., Alipoor, S.: Development of artificial neural networks and multiple regression models for the NATM tunnelling-induced settlement in Niayesh subway tunnel, Tehran. Bull. Eng. Geol. Env. **74**(3), 827–843 (2015)

Park, K.H.: Elastic solutions for tunneling-induced ground movements in clays. Int. J. Geomech. **4**(4), 310–318 (2004)

Shin, J.H., Addenbrooke, T.I., Potts, D.M.: A numerical study of the effect of groundwater movement on long-term tunnel behaviour. Geotechnique **52**(6), 391–403 (2012)

Tang, Y.Q., Cui, Z.D., Wang, X.J., et al.: Model test study of land subsidence caused by high-rise building group in Shanghai. Bull. Eng. Geol. Env. **67**(2), 173–179 (2008)

Wu, H.N., Shen, S.L., Chai, J.C., et al.: Evaluation of train-load-induced settlement in metro tunnels. Proc. Inst. Civil Eng. Geotechn. Eng. **168**(5), 396–406 (2014)

Ye, Y.D., Zhu, H.H., Wang, R.L.: Analysis on the current status of metro operating tunnel damage in soft ground and its causes. Chin. J. Undergr. Space Eng. **3**(1), 157–160 (2007). (in Chinese)

Zhu, J.F., Xu, R.Q., Liu, G.B.: Analytical prediction for tunnelling-induced ground movements in sands considering disturbance. Tunn. Undergr. Space Technol. **41**, 165–175 (2014)

Experimental Study on Physical and Mechanical Properties of Phosphorous Tailings

Baoquan Xin[1,2](✉) and Lu Wan[2]

[1] SINOPEC Research Institute of Safety Engineering, Qinigdao 266071, China
xinbq123@163.com
[2] Southwest University of Science and Technology, Mianyang 621010, China

Abstract. It is to study the physical and mechanical properties of phosphate rock and provide data support for the stability control of phosphate tailings dam. A series of laboratory test and a BI-90Plus laser particle size analyzer were used to test the original tailings. 13 basic physical properties and 4 indexes of mechanical properties of tailings are obtained. The results showed that most tailings of the dam are composed of a silty clay with high plasticity, remarkable expansion and contraction behavior. Particle size approximately follows a Gaussian distribution. The results of direct shear tests showed that as the clay content of the tailings increase, the internal friction angle tends to decrease, whereas the cohesion seems to increase gradually. And the difference reaches several times to dozens times. Therefore, the scientific and reasonable tailings discharge in accordance with the grain gradation and physical and mechanical properties of tailings, have an important influence on the stability of the dam. This may be related to the content of clay particles. The calculation results not only provide real data for the tailings dam design and stability, but also provides more accurate data support for engineering mechanical properties, the research of phosphate tailings sorting and storage process optimization.

Keywords: Tailings · Physical and mechanical properties · Shear strength Soil test · Limit equilibrium method · Cohesion

1 Introduction

China's phosphate rock reserves are very rich. About 176 tons of ore resources have been found, ranking second in the world. As a strategic resource of our country, the annual output of phosphate rock is more than 60 million tons. However, the mineral taste is generally low, generally not more than 10%, and the lower is only 2–3%. Therefore, a large number of phosphate ores are stored in tailings reservoirs after tailings are sorted by concentrator [1]. The stability and reliability of tailings dam are very important to the safety of tailings dam. In general, the tailings that has been layered and laminated after discharge are the main materials used to construct the upstream tailings dam. Determining its basic physical and mechanical properties is the basis for studying the stability of the dam body [2].

L. Hu et al. (Eds.): GSIC 2018, *Proceedings of GeoShanghai 2018 International Conference: Multi-physics Processes in Soil Mechanics and Advances in Geotechnical Testing*, pp. 468–476, 2018.
https://doi.org/10.1007/978-981-13-0095-0_52

2 Particle Size Measurement and Gradation Analysis of Tailings

The particle composition and gradation of tailings are important indexes to reflect tailings characteristics comprehensively. Soil mechanics, mineral processing, engineering geology and other fields have shown that grain size is closely related to the nature of soil. In addition, the relative content of the tailings sand grains (usually expressed in terms of particle grading curves) also affects their characteristics. The understanding of the gradation and distribution of tailings is the basis for the study of the physical and mechanical properties of the tailings [3].

The location of the sample is usually the front, middle and the end of the library. In the actual dam construction, the tailings used for dam construction are mostly from the tailings dam. Therefore, the samples in this study were taken from 5 different positions in the front of the library, and then their mineral composition and particle composition were analyzed.

2.1 Mineral Composition

According to the chemical composition analysis, the mineral composition of phosphorus tailings samples is analyzed. The results show that the tailings contain more than 20 chemicals. The contents of CaO, MgO, P_2O_5 and S_iO_2 were the highest, and the total content reached 64.904. The results of the top ten components are shown in Table 1 below.

Table 1. Tailings mineral composition.

Analysis project	CaO	MgO	P_2O_5	SiO_2	SO_3^{2-}	Fe_2O_3	Al_2O_3	K_2O	MnO	BaO
Content (%)	34.707	11.375	10.389	8.433	5.361	1.001	0.914	0.349	0.263	0.179

Content of P_2O_5 in the tailings of this phosphate mine is more than 10%, and the content is still high. Among them, CaO and MgO are easily oxidized and decomposed, which further affect the shear strength parameters and permeability coefficient [4].

2.2 Particle Gradation of Tailings

Tailings particle grading features often size distribution curve. The curve of the logarithm of the particle size as the abscissa (logarithmic), with less than a particle size of the cumulative mass percentage of particles (natural coordinates) as the ordinate. Through the particle grading curve can better understand the composition of tailings particles [5]. In order to accurately analyze the size and composition of the tailings grains, the BI-90Plus laser particle size analyzer was used to analyze the content of the tailings.

The laser particle size analyzer is a fast and convenient instrument for analyzing and testing nanometer particle size. Its detector is PMT with a scattering angle of 15° and 90° with a precision of 1%. Proper amount of tailings was used, and 3 ml samples were allocated according to the requirements of the instrument. 5 groups (group A–E) were tested and analyzed. Tailings particle size distribution shown in Fig. 1. The distribution of particle content (i.e. cumulative content) of tailings smaller than a particle size is shown in Fig. 2.

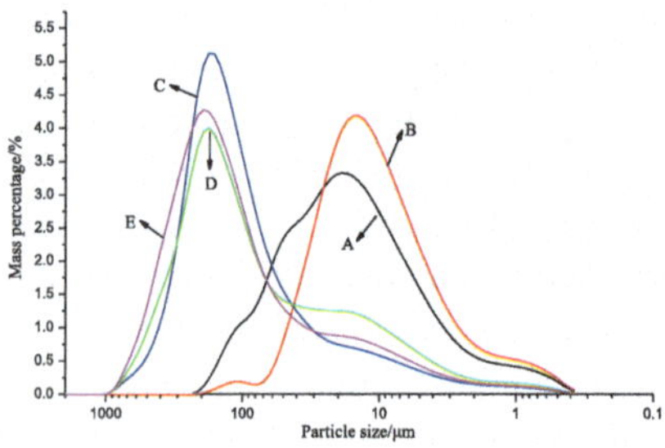

Fig. 1. Tailings particle size distribution curve

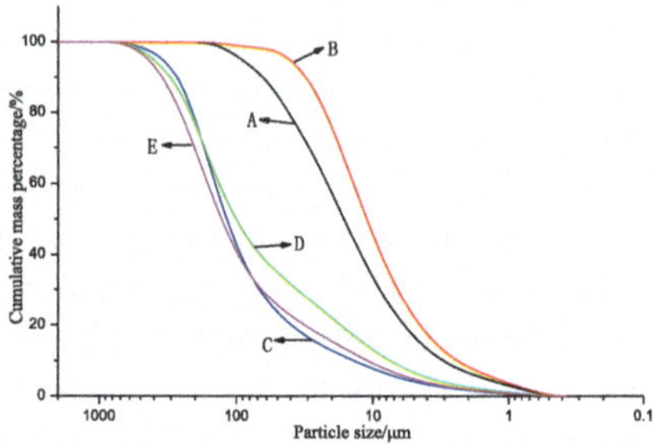

Fig. 2. Tailings particle gradation curves for different groups

It can be seen from Fig. 1 that the mean values of each group were distributed symmetrically, similar to a "clock type" curve, which was in accordance with the Gauss distribution. D_{50} of the 5 groups of samples is less than 0.5 mm, and the particle size is more than 0.25 mm, the quality of the particles is not more than 50%. It is indicating that the tailings particles are finer. Particle size of tailings in the group A and the group B above the size of 0.074 mm did not exceed 50% of the total mass, and the plasticity index was between 10 ~ 17. According to the provisions of [6], it belongs to tail silty clay in viscous tailings. The particle size of groups C, D and E is more than 50% of the total mass, and not more than 85%, which belongs to the tailing powder of sand tailings 0.074 mm.

Figure 2 shows that the particle size of tailings in group A is uneven. The curve B is similar to an inverted *S* shape. Particle size span is not large, which means that the particle distribution of tailings samples is even. Curves of C, D and E are similar in shape and larger in size, indicating that the particles are very heterogeneous.

The main particle size parameters of the 5 tailings are shown in Table 2. Including the equivalent particle size d_{10}, effective particle size d_{30}, median particle size d_{50}, control particle size d_{60}, average particle size d, surface particle size *S.D.* and the volume size *C.V.*

Table 2. Tailings particle analysis test results

Particle size parameter	d_{10}(μm)	d_{30}(μm)	d_{50}(μm)	d_{60}(μm)	d(μm)	*S.D.*(μm)	*C.V.*(%)
Group A	3.3481	8.94027	17.9894	21.9675	29.4298	32.5488	110.598
Group B	2.60665	6.45305	12.2447	14.2560	16.8815	18.4516	109.301
Group C	14.9398	69.5981	129.406	145.973	150.485	125.184	83.1868
Group D	8.00288	36.0142	109.808	133.502	144.651	144.541	99.9235
Group E	11.9139	69.1072	141.602	164.965	174.722	154.489	88.4202

As can be seen from Table 2, the effective particle size d_{10} of group B is the smallest It shows that the tailings of this group have the lowest water permeability, the strongest plasticity and the most expansibility and shrinkage. By Fig. 2, the clay content in group B is about 8.4%, ranging from 5 to 15%, which is suitable for geotechnical tests. The grading characteristics of tailings can be characterized by the inhomogeneous coefficient C_u (d_{60}/d_{10}) and the curvature coefficient C_c ($d_{30}^2/(d_{60}d_{10})$). The results shown in Table 3 below.

Table 3. Quantitative calculation results of tailings gradation characteristics

Group	A	B	C	D	E
C_u	6.561	5.469	9.771	16.682	13.845
C_c	1.087	1.121	2.221	1.214	2.430

In the practical engineering of tailings ponds, it is generally considered that the soil diameters of $C_u \geq 5$ and $1 \leq C_c \leq 3$ are unevenly distributed. The larger size of the pores can be filled with smaller particles, which are well-graded soil. As can be seen from Table 2, the grading of the 5 groups of tailings sand is good, the distribution of continuous grading is even and the anti-seismic liquefaction performance is also better. In addition, the d_{10} and d_{50} of the tailing sand particles increased significantly from tail silty clay (groups A and B) to tail silt (groups C, D and E). This indicates that the content of sand particles (soil particles with a particle size of 0.075 to 2 mm) is increasing, or the content of clay particles is significantly reduced.

In summary, the average particle size of tailings sand particles in group B is the smallest, the clay content is the highest, and the gradation is better. Subsequent experimental studies, if not otherwise specified, the tail ore of group B (tail silty clay) is used as the test material.

3 Basic Physical Properties Test of Tailings

Indicators that characterize the physical properties and status of tailing sand mainly include 3 basic indexes, 4 liquid plasticity indexes and 6 other conversion indexes. The basic indicators include moisture content (w), density (ρ) and particle specific gravity (G_s). According to the test requirements of different indicators [6], respectively, selected a large number of undisturbed sand and sand tailings dam disturbance test.

Liquid plastic indexes include liquid limit (w_L), plastic limit (w_P), plastic index (I_P) and liquid index (I_L). Test is based on FG-III photoelectric liquid plastic limit combined measuring instrument. Cone quality is 76 g, and the cone angle is 30°. When testing, take 250 g tailings sample, all over the 0.5 mm (35 mesh) sieve. 200 g samples were divided into three parts, and 3 soil paste with different water content were modulated by pure water. Experiment was carried out after resting 12 h. Three kinds of tailings samples with different water content are tested. Cone depths of 3 points are controlled at 3–4 mm, 7–9 mm and 15–17 mm respectively. Taking water content as abscissa, cone penetration depth as ordinate, drawing relation curve in double logarithmic coordinate system. The result is shown in Fig. 3.

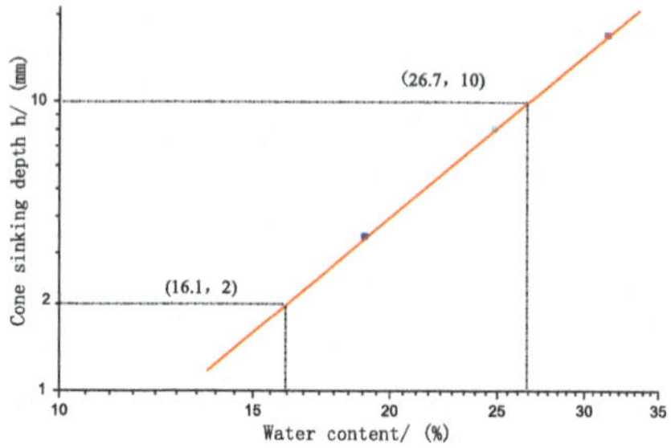

Fig. 3. Curve of cone sinking depth and water cut

In Fig. 3, when the cone depth is 10 mm, the corresponding water content is the liquid limit, and the corresponding water content at the depth of 2 mm is plastic limit. Therefore, the liquid limit w_L of the tailings is 26.7. And the plastic limit w_P is 16.1. The corresponding plasticity index is: $I_P = w_L – w_P = 10.6$, between 10–17, so the sample belongs to tail silty clay. Liquid index is: $I_L = (w – w_P)/(w_L – w_P) = 0.47$. Located in the range of 0.25–0.75, the soil in a plastic state.

In addition, other physical property conversion indexes are void ratio (e), porosity (n), dry density (ρ_d), saturation (S_r), saturation density (ρ_{sat}), and effective density (ρ'). The indicators are summarized in Table 4.

Table 4. Test results of basic physical properties of tailings

Index	Unit	Value	Index	Unit	Value
w	%	21.04	e	——	0.77
S_r	%	75.14	n	——	0.44
ρ_{sat}	g/cm^3	1.98	w_L	%	26.7
ρ	g/cm^3	1.88	w_P	%	16.1
ρ_d	g/cm^3	1.55	I_P	——	10.6
ρ'	g/cm^3	0.98	I_L	——	0.47
G_s	——	2.75		——	——

4 Basic Mechanical Properties Test and Analysis of Tailings

The mechanical property index of tailings is an important basic data for quantitative analysis of slope stability of tailings dam. It mainly includes shear strength characteristic index and seepage characteristic index.

4.1 Shear Strength Characteristics

Shear strength of soil is the key factor to determine the stability of tailings dam. According to the Mohr - Coulomb strength theory, cohesion and internal friction angle are the main indexes affecting the shear strength of the dam. Soil shear strength test indoor mainly direct shear test, triaxial shear (compression) test and unconfined compressive strength test [7]. Tailings dam and its foundation strength index category should be calculated according to the difference of soil strength, according to the test method required in document [6].

(1) **Selection of test methods**

Soil shear strength depends on the effective stress, not the total stress. Therefore, the effective stress method is adopted to calculate the shear strength [8]. Tailings plastic index I_p is greater than 10, which belongs to cohesive soil. Under natural conditions, saturation is less than 80%. Therefore, the direct shear apparatus or the three axle tester should be used to test the shear strength. Normally operated tailings dam will gradually complete drainage consolidation under the action of self-weight. The corresponding test method should be used drainage consolidation method.

For the tailings, such as phosphorus tailings, which contains more clay particles. Direct shear and triaxial shear strength test results are slightly different. However, the direct shear device is simple in structure and easy in operation. It is also suitable for the preparation and drainage of samples. It is the most direct method to test the shear strength of soil samples at present [9].

According to the differences of field drainage conditions, the direct shear test is divided into three types: fast shear (Q), consolidated quick shear (CQ) and slow shear (S). Among them, the slow shear (S) method allows the specimen to be drained and consolidated under vertical pressure, and then rapidly applies the horizontal shear stress to destroy the specimen.

(2) **Process and content of the test**

Equivalent strain-shear device of EDJ-1-type produced by Nanjing Soil Instrument Factory was used in the experiment. According to the requirements of soil test methods, standards and procedures, 4 groups of tailings clay samples were prepared with ring diameter 6.18 cm and high 2 cm ring knife respectively. Then grading pressure is recorded at 100, 200, 300 and 400 kPa respectively to record vertical deformation readings. After consolidation of the sample is stable (the vertical deformation readings per hour do not change beyond 0.005 mm), start the stopwatch and shear at a rate of 0.02 mm/min. Until the dynamometer readings stabilized or significant back, said the specimen has been cut loss. The formula is: $\tau_s = (CR/A_S)*10$. Where: τ_s-Shear stress, KPa; C-Dynamometer calibration coefficient, N/0.01 mm; R-Ergometer reading, 0.01 mm; A_s-Sample area, cm^2; *10*-Unit conversion factor.

(3) **Test results and analysis**

With shear stress as ordinate and shear displacement as abscissa, curves of shearing stress and shearing displacement are plotted under different saturation states, as shown in Fig. 4.

Take the peak point of each curve in Fig. 4 as the shear strength τ. The relationship between τ and vertical pressure P is shown in Fig. 5. Linear fitting results show that the fitting degree R^2 is 0.999, and the test data accuracy is higher. Dip angle of the straight line in the drawing is the internal friction angle of the tailings sample φ_s. Intercept of the line on the ordinate axis is cohesion c_s.

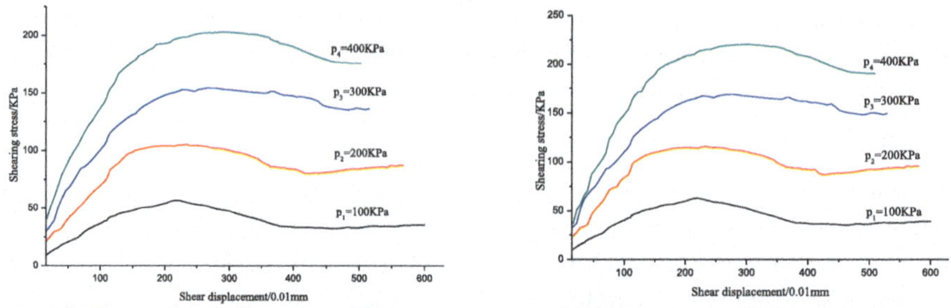

(a) Saturation is less than 80% (b)Saturation of 80% and above

Fig. 4. Curves of shear stress and shear shift under different saturated conditions

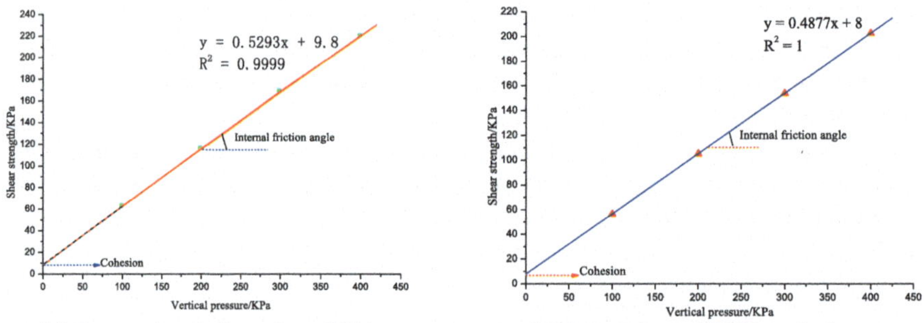

(a) Saturation is less than 80% (b)Saturation of 80% and above

Fig. 5. Relationship between shear strength and vertical pressure at different saturation

Table 5. Test results of shear strength of tailings sand

Tailings name	Saturation is less than 80%		Saturation of 80% and above	
	c_s(KPa)	φ_s(°)	c_s(KPa)	φ_s(°)
Silty clay	9.8	29.0	8.0	27.0
Powder	9.1	31.0	7.2	29.0
Fine sand	7.84	34.0	6.0	32.0

According to the above test method and procedure, the cohesion force and internal friction angle of tail silt and tail sand can be further measured. See Table 5.

Shear strength test results show that as the clay content of the tailings increase, the cohesion tends to increase, whereas the internal friction angle seems to decrease gradually [10]. As clay content (small particle size) of the tailings increase, the structural bond between the particles tends to stronger, and the cohesion seems to greater gradually. Similarly, the particle surface tends to more smooth, that is, the frictional force between particles is small.

4.2 Seepage Characteristics

The permeability of tailings directly affects the seepage rate and soil saturation in dam body. It is closely related to the shear strength and deformation of dam slope soil. Permeability coefficient is the main parameter for characterizing permeability. In general, the smaller the permeability coefficient is, the slower the velocity of the pore pressure dissipates, the more easily the dam will be liquefied.

Tailings sample material (tail silty clay) belongs to fine grained soil (clayey soil and silty soil). So that variable head method is used for seepage test. South -TST55 type penetration instrument was selected as test instrument. Its dimensions are: ϕ 118 mm (except for pipe mouth), height about 155 mm. Specimen size: ϕ 61.8 mm, high 40 mm, sectional area 30 cm^2. According to the requirements of reference [9], tests were made on tailings silty clay, tailings powder and tail fine sand. Each type of tailings are tested 6 times. In the measured results, 4 values within the allowable difference are selected. Then the mean value is taken as the permeability coefficient of the pore ratio, and the result is shown in Table 6.

Table 6. Tailings permeability coefficient

Tailings name	Permeability coefficient(cm/s)				
	A	B	C	D	Average value
Silty clay	1.15	1.05	1.25	1.3	1.19×10^{-4}
Powder	3.5	3.75	3.6	3.7	3.64×10^{-4}
Fine sand	1.35	1.2	1.25	1.2	1.25×10^{-3}

Table 6 shows that as the clay content of the tailings increase, the permeability coefficient tends to decrease. And the difference is much larger than the shear strength parameters, up to several times.

5 Conclusions

(1) Phosphate tailings particles are very fine, particle size and its content was approximately Gauss distribution. Tailings sand particle size distribution is good, continuous distribution evenly. It shows that the anti-earthquake liquefaction performance is good, which is in favor of the long-term stability of the dam body.
(2) The analysis results of tailing particles coincide with the test results of shear strength. The results show that the as the clay content of the tailings increase, the cohesion tends to increase, whereas the internal friction angle and the permeability coefficient seem to decrease gradually. Analysis results show that these are related to the content of clay particles in tailings.
(3) Therefore, the scientific and reasonable tailings discharge in accordance with the grain gradation and physical and mechanical properties of tailings, have an important influence on the stability of the dam.

References

1. Li, J., Tong, X.: Research development of comprehensive utilization of phosphate tailings. Conserv. Utilization Mineral Resour. **8**(05), 57–62 (2015)
2. Yin, G., Yang, Z., Weiz, Z., et al.: Physical and mechanical properties of Yang La-copper's tailings. J. Chongqing Univ. **30**(09), 117–122 (2007)
3. Yao, Z., Zhou, J., Zhang, G., et al.: Experimental study of particle grading impact on piping mechanism. J. Hydraul. Eng. **47**(2), 200–208 (2016)
4. Desisto, S.L.: Subsurface variations in arsenic mineralogy and geochemistry following long-term weathering of gold mine tailing. Appl. Geochem. **73**, 81–97 (2016)
5. Jiajun, X.U., Zuoan, W.E.I., Yulong, C.H.E.N., et al.: Experimental study on the relationship between resistivity and mechanical properties of phosphate tailings. Chin. J. Rock Mech. Eng. **33**(10), 2132–2137 (2014)
6. Ministry of Water Resources of the People's Republic of China. GB/T 50123-1999 Standard test method for geotechnical engineering. China Planning Press, Beijing (2007)
7. Cai, J.: Study on shear strength of undisturbed soil. Rock Soil Mech. **33**(07), 1965–1971 (2012)
8. Gao, D., Chen, Z., Guo, N., et al.: The influence of dry density and matric suction on the deformation and the strength characteristics of the remolded unsaturated loess soils. Chin. J. Rock Mech. Eng. **36**(03), 736–744 (2017)
9. Lu, Y., Tan, X., Guo, X., Li, W.: Methods of rock mass shear strength parameter determination based on borehole testing and their engineering applications. In: Proceedings of the Chinese Academy of rock mechanics and engineering (S2), pp. 4116–4124 (2015)
10. Zhang, Q., Yin, G., Wei, Z., et al.: An experimental study of the mechanical features of layered structures in dam tailings from macroscopic and microscopic points of view. Eng. Geol. **195**, 142–154 (2015)

Comparison of Different Electrical Resistivity Measurement Methods of Soft Marine Clays

Wei Duan[1], Guojun Cai[1(✉)], Songyu Liu[1], Kuikui Li[1], and Anand J. Puppala[2]

[1] Institute of Geotechnical Engineering, Southeast University, Nanjing 210096, China
zbdxdw@163.com, focuscai@163.com, likuikui1993@163.com, liusy@seu.edu.cn

[2] Department of Civil Engineering, The University of Texas at Arlington, Arlington, TX 76019, USA
anand@uta.edu

Abstract. The electrical resistivity of soil is one of the comprehensive indexes of the inherent property of soil, which has important theoretical signification and application value. The measured methods of soil electrical resistivity mainly can be divided into two groups: laboratory tests and in-situ tests. The in situ testing has been widely used in geotechnical site characterization due to its high accuracy and repeatability. Especially the emergence and development of resistivity piezocone penetration test (RCPTU), the RCPTU becomes the main tool of in situ measurement of the electrical resistivity of soil due to its not only include a conventional piezocone penetration test, but also provides a continuous profile of electrical resistivity. The objective of this paper was to compare the different methods for measuring the electrical resistivity of soil. First, the principle of the electrical resistivity of soil was briefly presented. Then, the comparison of different electrical resistivity measurement methods of soft marine clays was made based on the Ningbo marine clay and the advantages and disadvantages of various measured methods were also analyzed and summarized. The results of comparative analysis was verified the reliability of soil electrical resistivity measured by RCPTU.

Keywords: Electrical resistivity · Piezocone penetration test
Site characterization · Soft marine clay

1 Introduction

Electrical resistivity is the basic parameter to describe the conductivity of soil. It is closely linked to other soil physical and mechanical parameters. Soil resistivity can reflect the composition and structure characteristics of soil [1]. It can quickly and accurately obtain the soil structure index, quantitative analysis of soil structure characteristics, and finally establish a reasonable model to determine the soil engineering mechanics characteristic parameters. The value of electrical resistivity depends on soil

L. Hu et al. (Eds.): GSIC 2018, *Proceedings of GeoShanghai 2018 International Conference: Multi-physics Processes in Soil Mechanics and Advances in Geotechnical Testing*, pp. 477–485, 2018.
https://doi.org/10.1007/978-981-13-0095-0_53

porosity, pore shape, pore fluid resistivity, saturation, water content, particle composition, shape, orientation and cementation [2]. With development of a practical application of electrical resistivity theory, resistivity measurement methods have been increasingly and widely used to solve engineering and environmental problems, including soil micromorphology [1, 3], pollution characteristics [4], and soil liquefaction [5], and geotechnical parameters evaluation [6]. Particularly, the advanced resistivity piezocone penetration test (RCPTU) has been widely used in geotechnical site characterization due to its high accuracy and repeatability.

Due to the difficulties in obtaining high quality undisturbed samples in laboratory testing, the in situ measurement methods of electrical resistivity have been became a primary choice. Electrical logging, symmetrical quadrupole vertical electrical method and RCPTU test are the primary and common in situ measurement methods of electrical resistivity and are therefore considered in this study. In the face of many testing methods, different literatures do not unify the method of soil resistivity measurement. The different factors affecting the measurement of soil resistivity are compared with the quadrupole method and the dipole method by Zhou et al. [7]. The two electrical resistivity inversion techniques has been applied for geotechnical site investigation by Wisén et al. [8]. However, the different of electrical resistivity measurement techniques will result in different testing results. In practice, there are many different kinds of electrode arrays or configuration that one could use in the field. The typical electrode arrays are including Wenner, Schlumberger, Pole-pole and Dipole-dipole. Whether the method of electrical resistivity test is correct or not and the accuracy of the test results is very important to the analysis of soil properties.

The purpose of this study was to compare different electrical resistivity measurement methods of soft marine clays. The four kinds of measurement methods of electric logging, symmetrical vertical electrical measurement method, RCPTU, laboratory test were compared and analyzed based on engineering site of Ningbo Metro Line 5 station road. The results can not only provide reference and guidance of the test methods and equipment of soil resistivity, but also eventually provide reliable parameters for the engineering design and construction.

2 Soil Resistivity Theory

Archie (1942) firstly studied the relationship between soil resistivity and its structure, and established the relationship model between resistivity and pore water resistivity [9]. The model is suitable for saturated cohesionless soil and pure sand. Specific equations are as follows:

$$\rho = a\rho_w n^{-m} \tag{1}$$

where ρ is soil resistivity; a is soil parameters; ρ_w is pore water fluid resistivity; n is soil porosity; m is cementation coefficient.

Electrical resistivity of soils is not measured directly, but can be inferred from the measured voltage across an electrode, pair at a constant supplied current (I). The soil resistance, *R*, can be computed by Ohm's law, as follows [10]:

$$R = \frac{V}{I} \tag{2}$$

where *R* = resistance, *I* = electric current, and *V* = electric potential difference.

However, the measured resistance is not an intrinsic property and it depends on the area of the cross section (A) and of the current path length (L). The electrical resistivity, ρ is then a fundamental soil property which can be defined as:

$$\rho = \frac{A}{L}R \tag{3}$$

3 Geological Conditions and Basic Soil Parameters

The site is located in Ningbo Metro Line 5 station road. The Ningbo City is located in the coastal plain and is widely distributed marine sedimentary soft soil. There is a weathered hard crust layer of 1 to 1.5 m. The soft soil layer below the hard crust is mainly composed of clay, muck, mucky clay and mucky silty clay. The mucky soil has typical characteristics of soft soil with high water content, large pore ratio and high compressibility. Table 1 summaries the main physical properties of cohesive soils in investigated sites.

Table 1. Main physical properties of cohesive soils.

Soil layer	γ (kN/m^3)	w (%)	w_L (%)	w_P (%)	I_L
Clay	19.2	30.7	40.4	22.6	0.46
Mucky clay	17.2	50.1	42.4	19.1	1.41
Mucky silty clay	16.9	49.2	36.5	15.2	1.76
Clay	18.9	34.4	40.0	20.8	0.64
Silty sand	19.4	27.5			

4 Testing Method

4.1 Electrical Logging

The test equipment is adopted JDC-1 type resistivity logging instrument and its supporting device of JDX-1 soft electrode. The instrument connecting underground ground electrode system, and supporting the use of PC, can measure the apparent resistivity and natural potential parameters. The field test is shown in Fig. 1.

Fig. 1. Field test of electric logging method

4.2 Quadrupole Vertical Electrical Method

This study adopts the type of WDDS-1 digital resistivity instrument. The way it works: the electrical resistivity can be calculated by the measurement of the potential difference (V) of each electrode and supply current (I) using the formula (2). The method provides underground the artificial electric field and changes the power supply electrode pitch based on the electrical difference of stratum. Thus it can be obtained the apparent resistivity curve of each measuring depth. The ρ values of different depths can be calculated using the weighted average method based on the values of thickness and resistivity values of each electrical layer.

$$\rho = \frac{\rho_1 \cdot h_1 + \rho_2 \cdot h_2 + \cdots + \rho_i \cdot h_i}{h_1 + h_2 + \cdots h_i} \; (\Omega \cdot \mathrm{m}) \tag{4}$$

Where ρ_i = soil resistivity of ith layer $(\Omega \cdot \mathrm{m})$; h_i = thickness of ith layer (m).

4.3 RCPTU Testing

RCPTU field tests were conducted using g a lightweight truck with a 20 ton capacity hydraulic system, as per American Society for Testing and Materials ASTM D5778 (ASTM 2012). The RCPTU system was consist of a hydraulic pushing and leveling system, 1-m length segmental rods, cone penetrometers and a data acquisition system. The dimensions of the probe are: diameter 35.7 mm, conical tip area 10 cm^2, and friction sleeve area 150 cm^2, and the data was collected every 5 cm. The cone tip resistance (q_c), sleeve friction (f_s), penetration pore pressures (u_2) behind the tip at the shoulder, and ρ were be simultaneously measured in the process of penetration. The field RCPTU test with the resistivity piezocone probe with a four-electrode array are shown in the Fig. 2.

4.4 Soil Sampling and Laboratory Testing

High quality soil samples were taken at different depths for laboratory testing. The soil samples were collected by averages of a stationary piston sampler 76 mm in diameter at 1.0 m interval from ground level to the penetration depth, When the Shelby tube sampler was withdrawn from the borehole, the soil sample at the end of the tube was excavated with waxing sealing at both ends.

Fig. 2. Schematic diagram of field RCPTU test

Laboratory test was tested by Wenner equidistant quadrupole method (Fig. 3). The small electrodes are buried in the four small holes arranged on the measured surface soil. The current I flows into the outer two electrode, while the potential difference V between the two inner electrodes can be measured by a potential difference meter or a high resistance voltmeter. Soil resistivity is calculated as follows:

$$\rho = \frac{4\pi aR}{\left(1 + \frac{2a}{\sqrt{a^2 + 4b^2}} - \frac{a}{\sqrt{a^2 + b^2}}\right)} \tag{5}$$

where R = measured resistance; a = straight line spacing; b = buried depth of electrode.

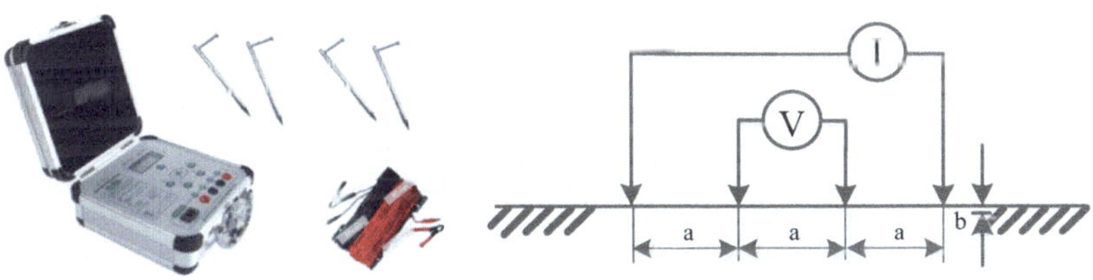

Fig. 3. Schematic diagram of resistivity test device

5 Analysis of Test Results and Discussion

5.1 Test Results of Electrical Logging

According to the test results of electrical logging, the variation curves of electrical resistivity with depth are obtained. The test results are shown in Fig. 4.

It can be seen from Fig. 4 that there is a 1.5 m weathered crust layer, commonly referred to plain fill; underlain by silt clay of 5.5 m, the average value of ρ is 5.02 $\Omega \cdot$ m; Mucky silty clay, clay and silt sand are also distributed in the lower part. The ρ of soil changes obviously with the changing of soil layer. The ρ value is

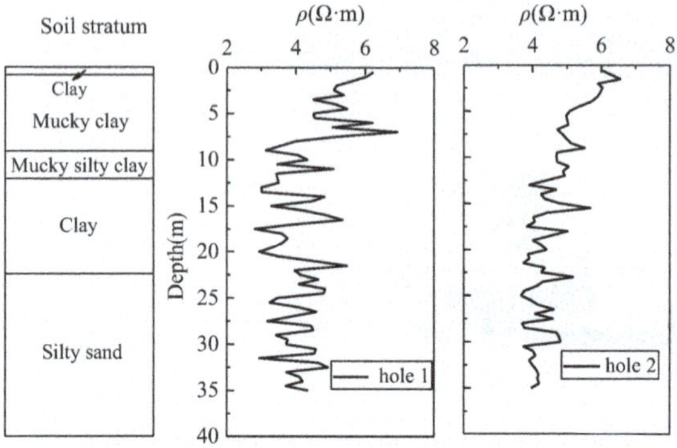

Fig. 4. Resistivity curve of electrical logging

larger in surface clay and mucky silty clay, and decreases with the depth. The ρ value of surface clay is very high, the average value can reach 5.62 $\Omega \cdot$ m. The ρ value of silty sand layer is the lowest with average value of 4.09 $\Omega \cdot$ m. The results indicate that the soil electrical resistivity decreases with the increase of porosity.

5.2 Test Results of Symmetrical Quadrupole Vertical Electrical Method

The test result of resistivity profile is shown in Fig. 5. Taking the thickness of each electrical layer as the weight to weight average and obtaining the soil electrical resistivity values of 1 m, 2 m, 5 m and 10 m in each electrical sounding (Table 2).

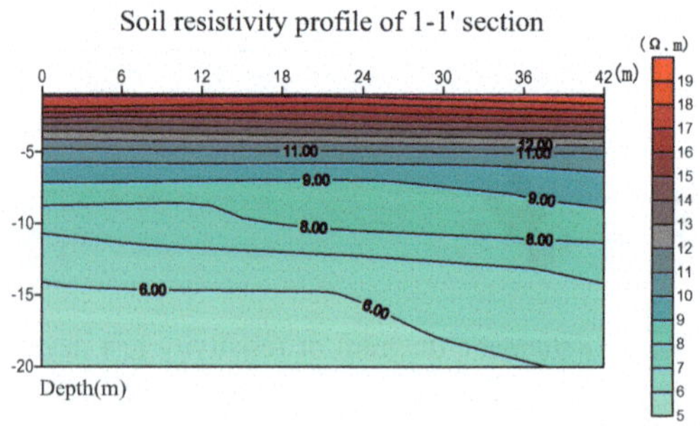

Fig. 5. Test results of soil electrical resistivity

Table 2. Test results of soil resistivity in shallow soil

Physical point	Soil resistivity ($\Omega \cdot$ m)			
	1 m	2 m	5 m	10 m
D1	23.9	16.2	10.6	7.1
D4	19.6	20.8	13.0	8.0

It can be seen from the Table 2 that the range of ρ value in 1 m is 4.7 to 23.9 $\Omega \cdot$ m with an average value of 11.7 $\Omega \cdot$ m; the range of ρ value in 2 m is 6.4 to 20.8 $\Omega \cdot$ m with an average value of 10.9 $\Omega \cdot$ m; the range of ρ value in 5 m is 6.4 to 13.0 $\Omega \cdot$ m with an average value of 8.7 $\Omega \cdot$ m; the range of ρ value in 10 m is 4.6 to 8.0 $\Omega \cdot$ m with an average value of 5.7 $\Omega \cdot$ m. The maximum test depth is 10 m and the electrical resistivity decreases with the increasing of depth, which is closely related to soil layering and soil properties. But compared with electrical logging method, its soil electrical resistivity values are greater.

5.3 Test Results of Resistivity Piezocone Penetration Test

According to RCPTU test results, the curves of electrical resistivity along depth are shown in Fig. 6. As can be seen from Fig. 6, the ρ changes obviously with the change of soil layer. The ρ values are larger in mucky clay and decreases with the increasing depth. In silty sand, the varied range of are great and the range of ρ values are 7.05 to 12.55 $\Omega \cdot$ m. The ρ values of surface clay are very high and the average value is up to 10.55 $\Omega \cdot$ m. The electrical resistivity of mucky clay below surface clay is the highest. The ρ values of silty sand are low with average value of 7.05 $\Omega \cdot$ m. The electrical resistivity is closely related to properties of soil layer.

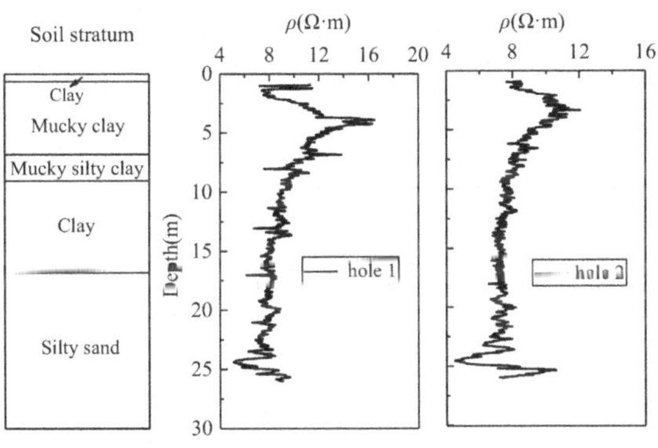

Fig. 6. Resistivity curve of RCPTU

5.4 Comparison and Discussion

Combined with laboratory test, the results are plotted in Fig. 7. It can be noted that the trends of ρ along the depth between schlumberger vertical electrical test and electric logging method is basically the same. The trend of RCPTU results and laboratory test results are high consistency. By contrast, the deviation of electrical logging method from laboratory test results is large, and the vertical electric method deviates from laboratory test results with small amplitude. The results show that the RCPTU is closer to the laboratory tests and verify the RCPTU is a promising in situ method to measure ρ of soil.

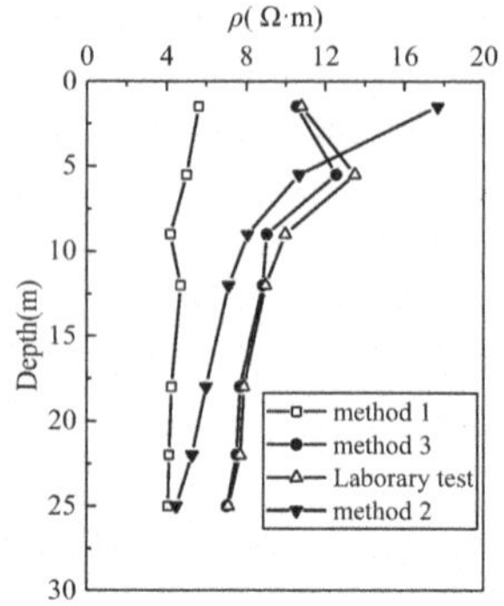

Fig. 7. Comparison of different methods of soil electrical resistivity

6 Conclusion

The three kinds of in-situ test of electric logging, schlumberger vertical electrical measurement method and resistivity piezocone penetration test were performed in the site of the Ningbo City Metro Line 5 project. The in situ results were compared with laboratory tests and the following conclusions are obtained:

(1) The electrical resistivity changes with the change of soil properties, and the soil resistivity decreases gradually from ground surface to bottom. It can be seen that the electrical resistivity reflects the change of soil structure and strength to a certain extent. The change of soil electrical resistivity can be used as a reference parameter for soil classification and soil layer stratification.
(2) The soil electrical resistivity is closely related to the structure and strength. It can reflect the basic physical and mechanical properties, such as soil water content, saturation, porosity, compaction, consolidation and permeability characteristics. The electrical resistivity is an important parameter in the analysis of geotechnical testing.
(3) The electrical resistivity measured by RCPTU test is between the other in situ methods and is basically close to laboratory test. Therefore, the RCPTU can be used for predicting geotechnical parameters. With the development of theoretical study on soil electrical resistivity, RCPTU test technology will be widely used in geotechnical situ investigation due to its rapid, in-situ and high accuracy.

References

1. Kim, J.H., Yoon, H.K., Lee, J.S.: Void ratio estimation of soft soils using electrical resistivity cone probe. J. Geotech. Geoenviron. Eng. **137**(1), 86–93 (2010)
2. Fukue, M., Minato, T., Horibe, H., Taya, N.: The micro-structures of clay given by resistivity measurements. Eng. Geol. **54**(1), 43–53 (1999)
3. Arulanandan, K.: Dielectric method for prediction of porosity of saturated soil. J. Geotech. Eng. Div. **117**(2), 319–330 (1991)

4. Arulanandan, K., Muraleetharan, K.K.: Level ground soil-liquefaction analysis using in situ properties: I. J. Geotech. Eng. **114**(7), 753–770 (1988)
5. Long, M., Donohue, S., L'Heureux, J.S., Solberg, I.L., Rønning, J.S., Limacher, R., Lecomte, I.: Relationship between electrical resistivity and basic geotechnical parameters for marine clays. Can. Geotech. J. **49**(10), 1158–1168 (2012)
6. Zhou, M., Wang, J., Huang, S., et al.: Experimental investigation on influencing factors in soil resistivity measurement. Rock Soil Mech. **11**, 3269–3275 (2011)
7. Wisén, R., Christiansen, A.V., Dahlin, T., Auken, E.: Experience from two resistivity inversion techniques applied in three cases of geotechnical site investigation. J. Geotech. Geoenviron. Eng. **134**(12), 1730–1742 (2008)
8. Archie, G.E.: The electrical resistivity log as an aid in determining some reservoir characteristics. Trans. AIME **146**(01), 54–62 (1942)
9. Lunne, T., Robertson, P.K., Powell, J.J.M.: Cone Penetration Testing in Geotechnical Practice. Chapman & Hall, London (1997)
10. ASTM D: Standard Test Method for Performing Electronic Friction Cone and Piezocone Penetration Testing of Soils. ASTM D-5778 (2000)

A Penetration Processing Study of Piezocone Penetration Test in Cutoff Wall

Xuepeng Li[1], Guojun Cai[1(✉)], Songyu Liu[1], and Yuchao Li[2,3]

[1] Institute of Geotechnical Engineering,
Southeast University, Nanjing 210096, China
focuscai@163.com
[2] MOE Key Laboratory of Soft Soils and Geoenvironmental Engineering,
Zhejiang University, Hangzhou 310058, China
[3] College of Civil Engineering and Architecture,
Zhejiang University, Hongzhou 310058, China

Abstract. The penetration processing of piezocone penetration test (CPTU) is very important for the analysis of stress-strain of soil bentonite cutoff wall (SBCW) in landfill site. In this paper, CPTU is conducted in the soil bentonite cutoff wall, which is located in Jiangsu province, China. A finite-element procedure is also used to analyze the processing of penetration. The Modified Cam Clay model was adopted to represent the stress-strain response of the soil-bentonite backfills. In this model, a tube was added to the bottom of the cone. This study shows that the largest value of radius stress is in the shoulder of cone. And the influenced domain of PEMAG is ten times of the cone radius.

Keywords: Cutoff wall · CPTU · Penetration · Bentonite · Landfill site

1 Introduction

Soil-bentonite backfills have been widely employed in Geotechnology engineering applications to control the contamination for landfill site [1–3]. The soil-bentonite is becoming more and more common for these backfills are typically cheaper that treatment systems and cause less risk of contaminant exposure during the construction [4]. The hydraulic conductivity, compressibility and strength are very important considerations for the soil-bentonite backfills [5, 6]. Chai et al. [7] predicted the coefficient of consolidation from non-standard piezocone dissipation curves by finite-element procedure. Baligh [8] assumed that the soil deformation caused by CPT may be determined with reasonable accuracy purely from an incompressibility requirement. Most of the studies were conducted using laboratory technology to evaluate the above parameters [9–11].

Piezocone penetration test (CPTU) is gaining wide acceptance due to its reliability, accuracy and continuous data acquisition [12–14]. The cone is firstly penetrated into soil and then pause a lot for the dissipation of pore water pressure. However, the penetration processing of cone is also very important, because the stress-stain behavior is changed by the penetration [15, 16].

This paper presents a finite-element approach for the analysis of the penetration processing of piezocone penetration test in soil bentonite cutoff wall. The main purpose

L. Hu et al. (Eds.): GSIC 2018, *Proceedings of GeoShanghai 2018 International Conference: Multi-physics Processes in Soil Mechanics and Advances in Geotechnical Testing*, pp. 486–492, 2018.
https://doi.org/10.1007/978-981-13-0095-0_54

of the present study is to investigate the stress-strain behavior of the SBCW. A tube with the radius of 1 mm added to the bottom of cone is been used in this model [17]. This paper shows the largest value of radius stress is in the shoulder of cone. And the influenced domain of PEMAG is ten times of the cone radius.

2 Materials and Methods

2.1 Constituent Materials

The backfills were comprised of 95% in-situ excavated soil (dry weight) and 5% bentonite (dry weight). The water content of the soil bentonite backfills was 50% corresponding to 125 mm slump according to the laboratory tests using a standard 300 mm slump cone. The particle-size distributions of the constituent materials are shown in Fig. 1. The soil bentonite cutoff wall has the properties with 10 m depth, 0.6 m width, and 15 m length.

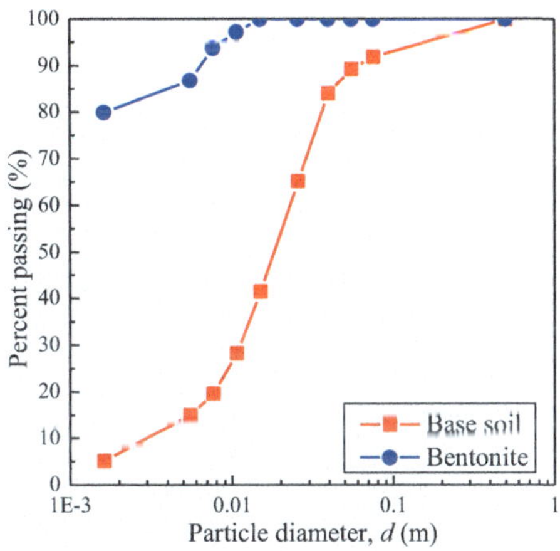

Fig. 1. Particle-size distributions for the constituent soils

2.2 CPTU Test

The piezocone penetration tests were performed using a lightweight truck with 20-ton capacity hydraulic system. The equipment was operated with a rate of penetration of 20 mm/s. The cone tip resistance (q_c), sleeve friction (f_s), and pore water pressure at the should of the cone (u_2) can be obtained by CPTU technology [18, 19].

2.3 Finite Element Technology

The Modified Cam Clay (MCC) model was adopted to represent the stress-strain response of the soil bentonite backfills. The corresponding model parameters of the soil are listed in Table 1. γ, λ, κ, M and v represent the value of total unit weight, slope of consolidation line in e - lnp' plot (p' is the effective mean stress), slope of rebound line in in e - lnp' plot, stress ratio at failure and Poisson's ratio respectively. e_0 represent the initial void ratio (the value of top or bottom surface of soil is 0.96 and 0.54).

Table 1. Corresponding model parameters for the study

γ (kN/m^3)	k (cm/s)	λ	κ	M	e_0	v
17.7	4.1E-8	0.1	0.01	1.2	0.96, 0.54	0.33

The penetration speed simulated in the modeling was assumed to be constant at 20 mm/s as same as the speed of CPTU in suit test. The overall penetration depth of the piezocone was 5 m. Figure 2 shows the typical deformed mesh of the model with different penetration depth. The tube with a radius of 1 mm is added at the bottom of the cone. The model is corrected and can be used in the next study.

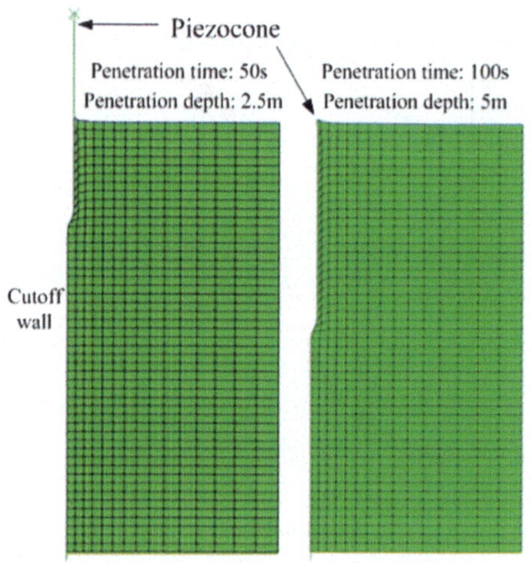

Fig. 2. The typical deformed mesh of model

3 Results and Discussion

3.1 The Radial Displacement

Figure 3 gives the radial displacement of soil bentonite backfills with different penetration depths. The data are obtained from the contours of radial displacement at the end of penetration (t = 100 s, depth is 5 m). In this Figure, D represents the distance from the periphery of the cone. D = r_0 or 2 r_0 represent the distance of one or two times the radius of the cone from the periphery of the cone. The maximum radial displacement is 0.0052 m, 0.0023 m when the distance is r_0 and 2 r_0 respectively. And the maximum value is located in the shoulder of the cone.

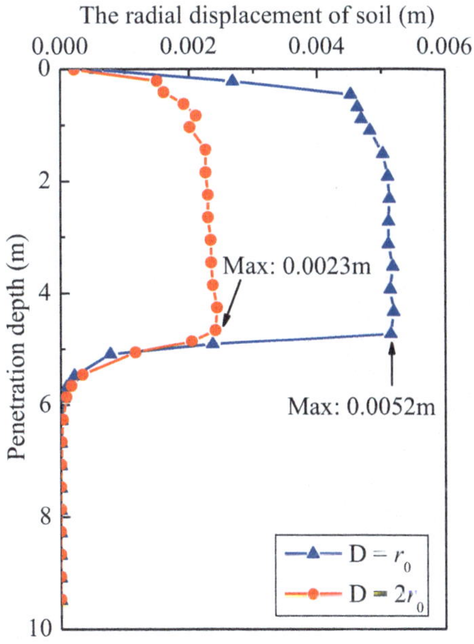

Fig. 3. The radial displacement with different penetration depths

3.2 The Radial Stress

Figure 4 shows the contours of radial stress of soils at the end of penetration with the penetration time 100 s. Figure 5 gives the radial stress of soils with different penetration depths. The maximum radial stress is 432 kPa, 244 kPa when the distance is r_0 and 2 r_0 respectively. And the maximum value is also located in the shoulder of the cone. D = r_0 or 2 r_0 represent the distance of one or two times the radius of the cone from the periphery of the cone.

The radial stress with different distance from the periphery of cone is shown in Fig. 6. Four different locations (Location A, Middle location, cone shoulder and cone tip) are listed in the Figure respectively. It notes that Location A is located at a distance with one time the radius of the cone from cone shoulder along the positive direction of the Y axis. Middle Location is located at the middle between the cone shoulder and cone tip. The results show that the radial stress decreases as the distance increases and the maximum value is at the cone shoulder.

3.3 Plastic Strain Magnitude (PEMAG)

Figure 7 gives the contours of plastic strain magnitude (PEMAG) at the end of penetration (t = 100 s, depth = 5 m). The maximum value is 1.141 from the contours. Figure 8 shows the PEMAG with different distance from the periphery of cone. When the distance from the periphery of cone is 0.18 m, the value of PEMAG is only 0.02.

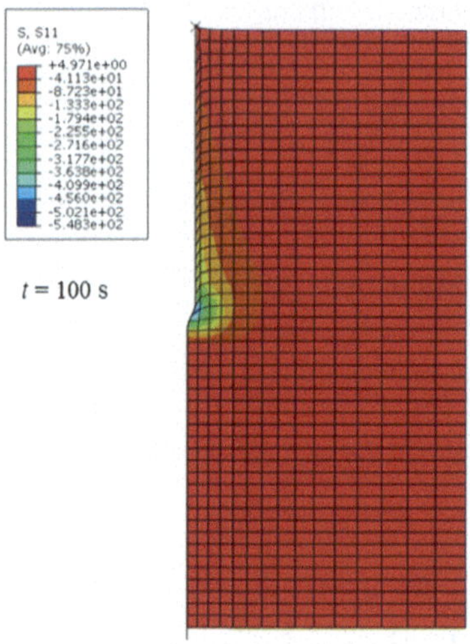

Fig. 4. Contours of radial stress

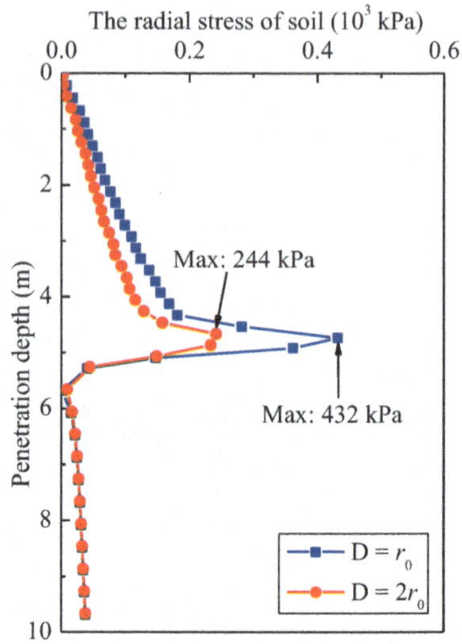

Fig. 5. The radial stress with different depths

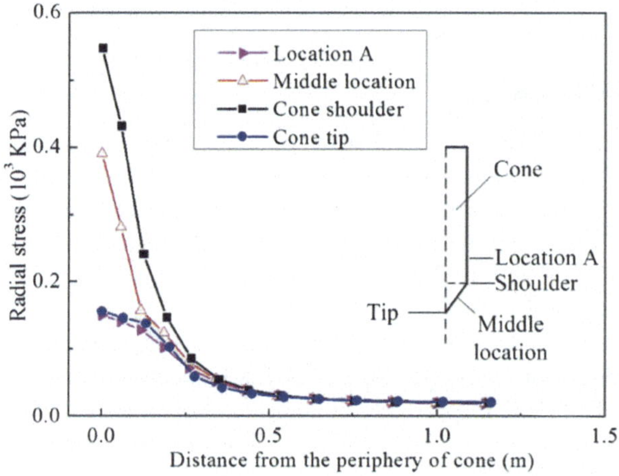

Fig. 6. The radial stress with different distance from the periphery of cone

The results present the uninfluenced domain in this model is 0.18 m. The PEMAG value in this study shows the influenced domain is about ten times radius of cone. And the value of PEMAG in cone shoulder is also larger than value of cone tip as the same as the previous conclusion.

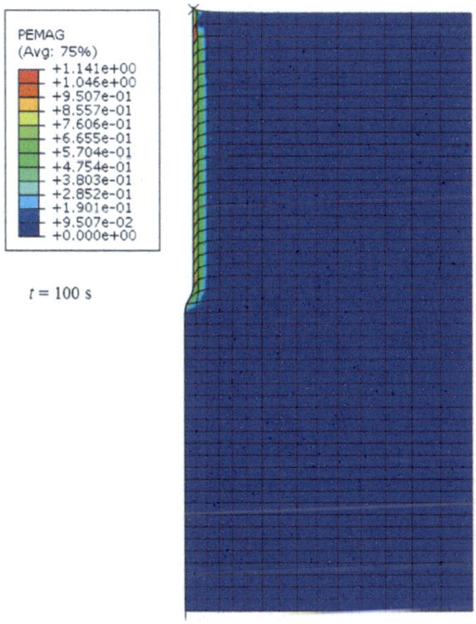

Fig. 7. Contours of PEMAG at the end of penetration

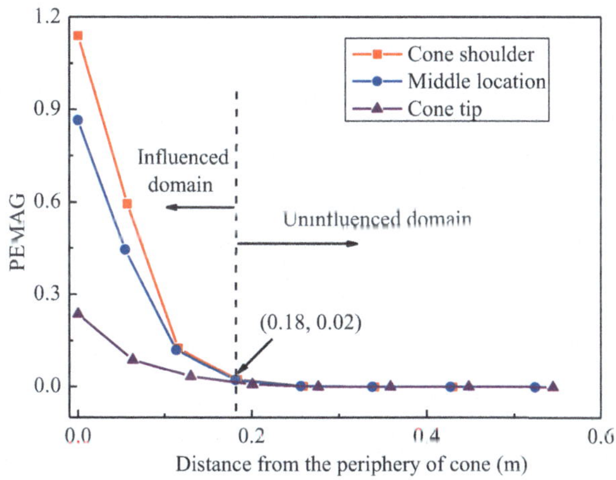

Fig. 8. PEMAG with different distance from the periphery of cone

4 Conclusion

A finite-element procedure can be effectively used to simulate the CPTU penetration processing in soil bentonite cutoff wall. A tube with a radius of 1 mm was added to the bottom of the cone in this model. The paper shows that the radial displacement, radial stress and PEMAG at cone shoulder are larger than those located at cone tip and Middle location. And the paper also shows that the influenced domain of PEMAG is about ten times of cone radius.

References

1. Evans, J.C.: Vertical cutoff walls. In: Daniel, D.E. (ed.) Geotechnical Practice for Waste Disposal. Chapman & Hall, London (1993)
2. LaGrega, M.D., Buckingham, P.L., Evans, J.C.: Hazardous Waste Management. McGraw Hill, Boston (2001)
3. Yeo, S.S., Shackelford, C.D., Evans, J.C.: Consolidation and hydraulic conductivity of nine model soil-bentonite backfills. J. Geotech. Geoenviron. **131**(10), 1189–1198 (2005)
4. Shackelford, C.D., Jefferis, S.A.: Geoenvironmental engineering for in situ remediation. In: Proceedings of International Conference on Geotechnical and Geoenvironmental Engineering (GeoEng2000), Technomic, Lan-caster, PA, vol. 1, pp. 121–185 (2000)
5. D'Appolonia, D.J.: Soil-bentonite slurry trench cutoffs. J. Geotech. Geoenviron. **106**(4), 399–417 (1980)
6. Bohnhoff, G.L., Shackelford, C.D.: Consolidation behavior of polymerized bentonite-amended backfills. J. Geotech. Geoenviron. **140**(5), 04013055 (2014)
7. Chai, J.C., Sheng, D., Carter, J.P., et al.: Coefficient of consolidation from non-standard piezocone dissipation curves. Comput. Geotech. **41**, 13–22 (2012)
8. Baligh, M.M.: Strain path method. J. Geotech. Eng. **111**(9), 1108–1136 (1985)
9. Gleason, M.H., Daniel, D.E., Eykholt, G.R.: Calcium and sodium bentonite for hydraulic containment applications. J. Geotech. Geoenviron. **123**(5), 438–445 (1997)
10. Malusis, M.A., Barben, E.J., Evans, J.C.: Hydraulic conductivity and compressibility of soil-bentonite backfill amended with activated carbon. J. Geotech. Geoenviron. **135**(5), 664–672 (2009)
11. Castelbaum, D., Shackelford, C.: Hydraulic conductivity of bentonite slurry mixed sands. J. Geotech. Geoenviron. **135**(12), 1941–1956 (2009)
12. Baligh, M.M., Vivatrat, V., Ladd, C.C.: Cone penetration in soil profiling. J. Geotech. Geoenviron. **112**(7), 727–745 (1980)
13. Lunne, T., Robertson, P.K., Powell, J.J.M.: Cone Penetration Testing in Geotechnical Practice. Blackie Academic and Professional, London (1997)
14. Cai, G.J., Liu, S.Y., Puppala, A.J.: Comparison of CPT charts for soil classification using PCPT data: Example from clay deposits in Jiangsu Province. China. Eng. Geol. **121**(1), 89–96 (2011)
15. Yu, H.S., Herrmann, L.R., Boulanger, R.W.: Analysis of steady cone penetration in clay. J. Geotech. Geoenviron. **126**(7), 594–605 (2000)
16. Chai, J.C., Agung, P.M.A., Hino, T., et al.: Estimating hydraulic conductivity from piezocone soundings. Geotechnique **61**(8), 699–708 (2011)
17. Fei, K., Zhang, J.W.: Application of ABAQUS software in geotechnical engineering. China Water and Power Press, Beijing (2010). (in Chinese)
18. Cai, G.J., Liu, S.Y., Puppala, A.J.: Reliability assessment of CPTU-based pile capacity predictions in soft clay deposits. Eng. Geol. **141**, 84–91 (2012)
19. Campanella, R.G., Robertson, P.K., Gillespie, D.: Cone penetration testing in deltaic soils. Can. Geotech. J. **20**(1), 23–35 (1983)

T-bar Penetration Tests in Fujian Cohesive Soil

P. Peng, G. J. Cai(✉), and S. Y. Liu

Institute of Geotechnical Engineering, Southeast University,
Nanjing 210096, China
ocptpeng@163.com, focuscai@163.com, liusy@seu.edu.cn

Abstract. The effect of cyclic loading on the strength of the soft soils is considerable in in-situ tests of offshore engineering. Parameters obtained by T-bar cyclic penetration test are crucial to evaluate the undrained shear strength of original and disturbed soil, due to solving the error caused by the very low resistance and the correction of overburden pressure in piezocone test (CPTU). Results of a series of tests which consist of T-bar penetration tests, vane shear tests and CPTU at Fujian hydraulic reclamation sites are presented in this research. The determination of the undrained shear strength (s_u) based on T-bar penetration tests is validated by the data from vane shear tests. It is shown that the bearing capacity factor ($N_{\text{T-bar}}$) is limited from of 9 to 10, which is identical with the theoretical value. The T-bar penetration test is shown to be effective comparing with CPTU.

Keywords: T-bar penetration test · CPTU · Hydraulic reclamation
Undrained shear strength

1 Introduction

With the focus on marine development over the last few decades, offshore engineering gradually developed into the deep sea, at the same time the scale of marine engineering is also increasing. This puts forward higher requirements for the investigation of soil bearing capacity, especially the effect of cyclic loading on marine engineering design [1]. In general, deep-sea sedimentary clay have the characteristics of low shear strength, high sensitivity, low surface strength (<20 kPa) and shear strength increasing with depth (1–2 kPa/m) [2], which makes it difficult and costly to obtain high-quality undisturbed soil samples. These factors make the determination of design parameters more and more dependent on the in-situ testing, and one of the most widely used in-situ testing is piezocone penetration testing (CPTU) [3].

CPTU technology has the features of speed, economy and feasibility. Not only can be used for accurate soil classification, CPTU can also be used to estimate undrained shear strength, over consolidation ratio, sensitivity, compression modulus, shear modulus, consolidation coefficient permeability, coefficient and other characteristics of soils. Hence the widely use of CPTU in soft soils engineering [4]. The in-situ test equipment has been developed with the improvement of marine engineering, in order to meet the deep sea operations. In 1994, Randolph proposed full-flow penetrometers based on full-flow theory, in order to solve the accuracy decreased in offshore CPTU [5].

L. Hu et al. (Eds.): GSIC 2018, *Proceedings of GeoShanghai 2018 International Conference: Multi-physics Processes in Soil Mechanics and Advances in Geotechnical Testing*, pp. 493–500, 2018.
https://doi.org/10.1007/978-981-13-0095-0_55

This paper performed T-bar penetration tests in Fujian hydraulic reclamation sites, and studied in detail feasibility of the tests for determining undrained shear strength and sensitivity.

2 Full-Flow Penetrometer

Vane shear test and CPTU were two of the in-situ testing techniques used in offshore engineering. The vane shear test has significant disturbances on soft soils to affect the feasibility of the test. And it shows that the accuracy of the data obtained by CPTU is reduced as the water depth increases, from a large number of CPTU data. Reasons for this perturbation are pointed out by Randolph [2]:

(1) The penetration resistance is too low leading to lower measurement accuracy, compared to the high pressure environment under the sea.
(2) The contribution of verburden stress to the cone resistance.

To solve the problems, Randolph et al. proposed a new type of full-flow penetrometers. Kinds of full-flow penetrometers are shown in Fig. 1, including T-bar, ball and plate [6–8].

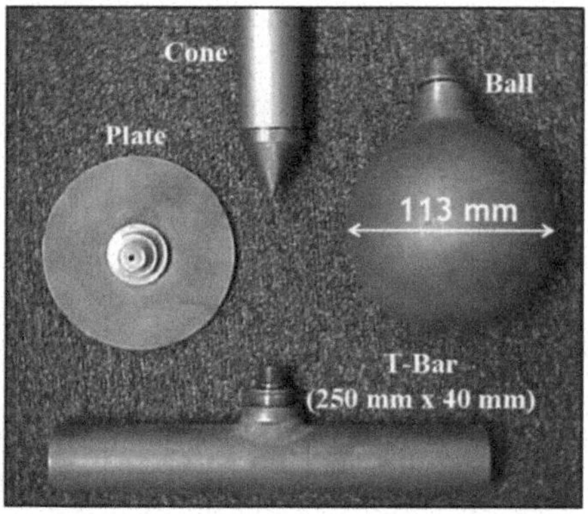

Fig. 1. T-bar, ball and plate full-flow penetrometer

At present, the standard mainly used of T-bar penetration tests is NORSOK G-001 (Standards Norway 2004). It Stipulates that a T-bar penetrometer should have the shape of 40 mm diameter and 250 mm length, which gives a projection area of 10 000 mm^2.

3 Parameter from T-bar Penetration Tests

3.1 Undrained Shear Strength

The correction of net resistance ($q_{\text{T-bar}}$) against the overburden pressure is shown as Eq. 1 [9]:

$$q_{\text{T-bar}} = q_{\text{m}} - \frac{[\sigma_{v0} - u_0(1-\alpha)]A_{\text{s}}}{A_{\text{p}}} \tag{1}$$

where: q_m is the penetration resistance; u_0 is the hydrostatic pore pressure; α is the interface friction ratio; A_s is the cross sectional area of the shaft; A_p is the projected area of the penetrometer.

Undrained shear strength (s_u) can be estimated as Eq. 2, by using net resistance ($q_{\text{T-bar}}$) and T-bar resistance factor ($N_{\text{T-bar}}$).

$$s_{\text{u}} = \frac{q_{\text{T-bar}}}{N_{\text{T-bar}}} \tag{2}$$

where: $q_{\text{T-bar}}$ is the net resistance after the correction; $N_{\text{T-bar}}$ is the bearing capacity factors for the T-bar.

Results of resistance factor based on upper and lower bound theory are shown in Fig. 2, including CPTU, T-bar and Ball [10]. The resistance factor ($N = q_u/s_u$) is a function of the interface friction ratio (α). Obviously, the resistance factors of T-bar from both upper and lower bound theory are practically the same.

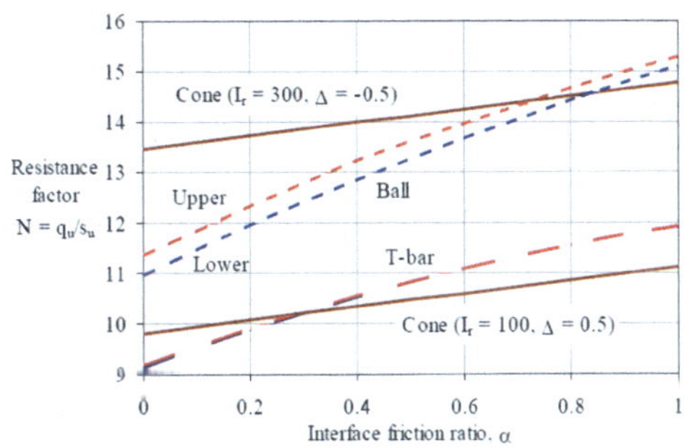

Fig. 2. Theoretical factors for cone, T-bar and ball

Martin and Randolph et al. originally studied the $N_{\text{T-bar}}$ [11]. The quadratic and linear fit of resistance factor of T-bar ($N_{\text{T-bar}}$) to interface friction ratio (α) based on the upper bound theory are obtained:

$$N_{\text{Tbar-ideal}} \sim 9.14 + 4.14\alpha - 1.34\alpha^2 \tag{3}$$

$$N_{\text{Tbar-ideal}} \sim 9 + 3\alpha \tag{4}$$

Zhou and Randolph et al. argue that the soft soil will produce a relatively high strain rate and strain softening will occur when the soft soil flows through the surface of T-bar [12]. Therefore, the effects of strain rate and strain softening should be taken into account when determining the bearing capacity.

Studies by DeJong et al. (2011) have been carried out on determining bearing capacity factors for various sites and comparing them to the theoretical values and to results from the direct simple shear (DSS) and the vane share test (VST) [13]. Results from Australia, Norway, west Africa and Canada was reported, and it was shown that the bearing capacity factors derived from laboratory tests and in-situ tests were larger than the theoretical values. For the particular engineering condition, theoretical values are unreliable, and it should be derived with empirical methods. The laboratory tests on high quality samples and other in-situ tests should be carried out for the correcting of bearing capacity factors.

3.2 Sensitivity

Resistance can be measured in the process of pulling up in T-bar penetration tests, therefore a cyclic penetration test can be carried out in a certain depth range by repeating the penetration and pull-out and recording the resistance value. With the progress of the cyclic tests, the disturbance of soft soils gradually accumulates, and the strength stabilizes at a particular value after about 10 cycles. Studies of data from Australia and Norway offshore sites are reported by Einav and Randolph in Fig. 3 [14].

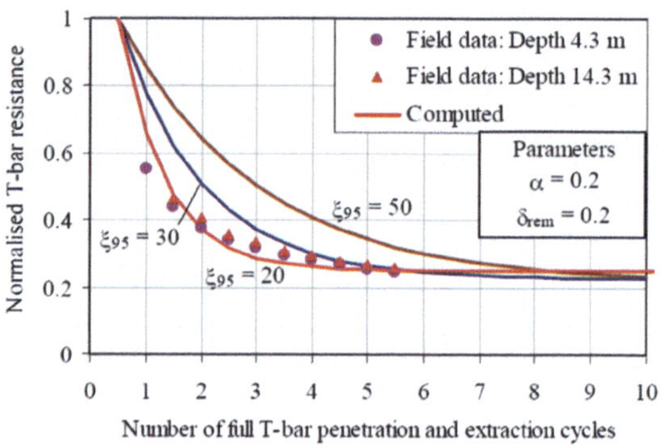

Fig. 3. Calibration of degradation model from cyclic T-bar tests

The sensitivity of the soft soils can be measured by cyclic penetration tests as Eq. 2:

$$S_T = \frac{s_u}{s_{ur}} = \frac{q_{in}}{q_{rem}} \tag{5}$$

where: s_u is the undrained shear strength of undisturbed soil; s_{ur} is the undrained shear strength of remolded soil; q_{in} is the penetration resistance for the initial penetration; q_{rem} is the penetration resistance at the end of the cycle.

4 Field Test

4.1 Sites

The tests were carried out in a sea reclamation project located on the Lianjiang County, Fuzhou City shown in Fig. 4. The topography of the site was originally coastal beach, while the hydraulic reclamation project was commenced in 2014. During the project, the ground was pre-filled till 8 m depth by organic soil which had a characteristic of very low share strength, high compressibility, water content and plasticity index. The sites was covered with sand (1.5 m–2.0 m thickness of fill). The tests were initiated after two years' self-weight consolidation, and no ground improvements were used under these fields during this period. Soil types from top were filled sand, miscellaneous fill, organic soil, silty clay, silt, round gravel, gravel layer.

Fig. 4. Location of site in Fujian

4.2 Equipments

Piezocone and T-bar penetrometer were used in this study. The equipment used was developed by institute of geotechnical engineering, Southeast University shown in

(a) (b)

Fig. 5. (a) Equipments, (b) Pre-embedded pipe

Fig. 5(a). The T-bar probe was constructed by hardened stainless steel with a diameter of 40 mm, a length of 250 mm, and a projected area of 100 cm^2.

As the ground was covered by sand layer, the T-bar penetration tests had to be carried out in a pre-embedded pipe, in order to ensure that the penetration process proceeded smoothly as Fig. 5(b).

5 Typical Resaults from T-Bar Penetration Tests

The study was conducted including T-bar penetration tests, CPTU and VST GB50021-2001. Some examples of the tests are shown in Fig. 6 respectively.

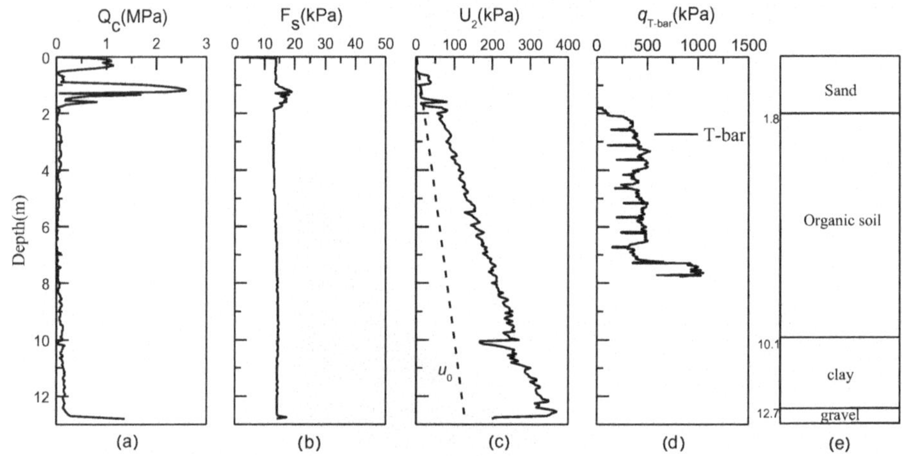

Fig. 6. Data from CPTU tests and T-bar penetration tests (a, b, c) Penetration resistance, friction resistance and pore pressure from CPTU tests, (d) Penetration resistance from T-bar penetration tests, (e) Soil classfication.

According to the CPTU data, the ground can be divided into sand, organic soil, clay and gravel. The values of q_c are very low and close to zero in soft organic soil. Relatively, the $q_{\text{T-bar}}$ values can obviously reflect the intensity changes. It shows that the T-bar penetration tests are more reliable than CPTUs in very soft soil.

Bearing capacity factors can be determined from the $q_{\text{T-bar}}$ values and the results of VST using Eq. 2. To support the determination of $N_{\text{T-bar}}$ values, VSTs were set at a distance of 1 m from each T-bar penetration test. The calculated data for $N_{\text{T-bar}}$ of the spot determined by VST data is 9.5. The results of bearing capacity factors versus are distributed within the range of 9 to 12, which is consistent with the study of Lunne [15].

A typical cyclic test result is shown in Fig. 7. The cycles were conducted at 3.8 m and 6.8 m with a depth of 1 m. Process of penetration and extraction cycle leads to significant loss in strength. Soil softens to a fully remoulded state after 5–10 cycles, while the strength tends to be stable. The share strength in each cycle calculated with $N_{\text{T-bar}}$ is shown in Fig. 7b.

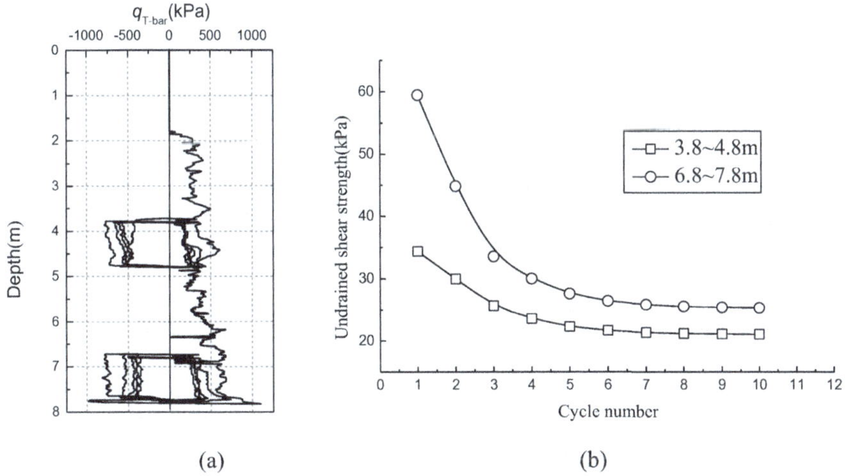

Fig. 7. Example for cyclic test results (a) Resistance, (b) Strength degradation curves

The sensitivity S_T can be determined from cyclic T-bar penetration tests as Eq. 5. The S_T of this sample spot is 1.51 in 3.8–4.8 m, and 2.12 in 6.8–7.8 m, similar to VST result, which is 1.50 in 3.8–4.8 m, and 2.04 in 6.8–7.8 m. Generally, the difference between S_T results measured by T-bar penetration test and VST is within 10%. There shows a high feasibility of T-bar penetrometer for estimating the sensitivity of very soft organic soil.

6 Conclusions

This paper presents the results of T-bar penetration tests on soft organic soil of hydraulic reclamation sites in Fujian. Some results can be obtained as follows:

(1) It makes a great contribution for T-bar penetration test to compute the undrained shear strength of very soft organic soil, since conventional CPTU results remain very low penetration resistance. The bearing capacity factor $N_{T\text{-bar}}$ estimated in the sites is within 9 to 12, conform to the theorctical value. It seems to be promised in the use of the T-bar penetrometer in the calculation of undrained shear strength and the sensitivity of very soft organic soil for hydraulic reclamation sites.

(2) For further study, more extensive tests were required to be conducted to provide the experience for determination of $N_{T\text{-bar}}$ by empirical methods.

References

1. Randolph, M.F., Cassidy, M., Gourvenec, S., Erbrich, C.: Challenges of offshore geotechnical engineering. In: The 16th International Conference of Soil Mechanics and Geotechnology Engineering, Osaka, Japan, vol. 1, pp. 123–176. Taylor & Francis, London (2005)

2. Lunne, T., Andersen, K.H., Eng low, H.: Guidelines for offshore in situ testing and interpretation in deepwater soft soils. Can. Geotech. J. **48**(4), 543–556 (2011)
3. Cai, G.J., Liu, S.Y., Puppala, A.J.: Comparison of CPT charts for soil classification using PCPT data: example from clay deposits in Jiangsu Province, China. Eng. Geol. **121**, 89–96 (2011)
4. Guo, S.Z., Liu, R.: Application of cone penetration test in offshore engineering. Chin. J. Geotech. Eng. **37**(supp 1), 207–211 (2015)
5. Stewart, D.P., Randolph, M.F.: T-bar penetration testing in soft soils. J. Geotech. Eng. **120** (12), 2230–2235 (1994)
6. Randolph, M.F., Hefer, P.A., Geise, J.M, Watson, P.G.: Improved seabed strength profiling using T-bar penetrometer. In: International Conference on Offshore Site Investigation and Foundation Behaviour –"New Frontiers", London, pp. 221–235. Society for Underwater Technology, London (1998)
7. Kelleher, P.J., Randolph, M.F.: Seabed geotechnical characterisation with the portable remotely operated drill. In: International Symposium on Frontiers in Offshore Geotechnics, Perth, Australia, pp. 365–371. Taylor & Francis, London (2005)
8. Peuchen, J., Adrichem, J., Hefer, P.A.: Practice notes on push-in penetrometers for offshore geotechnical investigation. In: International Symposium on Frontiers in Offshore Geotechnics, Perth, Australia, pp. 973–979. Taylor & Francis, London (2005)
9. Chung, S.F., Randolph, M.F.: Penetration resistances in soft clay for different shaped penetrometers. In: 2nd International Conference on Geotechnical and Geophysical Site Characterisation, ISC-2, pp. 671–678. Millpress, Porto (2004)
10. Randolph, M.F.: Characterisation of soft sediments for offshore applications. In: Proceedings of the 2nd International Conference on Geotechnical and Geophysical Site Characterization, ISC-2, pp. 209–232. Millpress, Porto (2004)
11. Martin, C.M., Randolph, M.F.: Upper bound analysis of lateral pile capacity in cohesive soil. Geotechnique **56**(2), 141–145 (2015)
12. Zhou, H., Randolph, M.F.: Resistance of full-flow penetrometers in rate-dependent and strain-softening clay. Geotechnique **59**(2), 79–86 (2009)
13. Dejong, J.T., Yafrate, N.J., De Groot, D.J.: Evaluation of undrained shear strength using full-flow penetrometers. J. Geotech. Geoenviron. Eng. **137**(1), 14–26 (2011)
14. Einav, I., Randolph, M.F.: Combining upper bound and strain path methods for evaluating penetration resistance. Int. J. Numer. Methods Eng. **63**(14), 1991–2016 (2005)
15. Lunne, T., Low, H.E., Randolph, M.F.: Estimation of intact and remoulded undrained shear strengths from penetration tests in soft clays. Geotechnique **60**(11), 843–859 (2010)

Field Test of Multi-anchored-Plating Cantilever Retaining Wall with Pre-stressed Force

Yingyong Li[1], Hongbo Zhang[2(✉)], Xiuguang Song[2], and Liang Lu[2]

[1] Department of Transportation of Shandong Province, Jinan 250002, China
[2] School of Civil Engineering, Shandong University, Jinan 250061, China
zhanghongbo@sdu.edu.cn

Abstract. Traditional anchored-plating walls have been used for tens of years to bear large lateral earth pressure in high embankment, slopes, etc. However, because of its own deficiency, such as complex construction, difficult compaction and larger deformation, the application of anchored-plating wall has been limited. Therefore, by combining features of cantilever retaining wall with anchored plates, a new type of retaining wall, multi-anchored-plating wall, was put forward to overcome these questions. Integrating all cables along the height of wall in one section into one anchorage and applying prestressing force along cables would be main features of the type of retaining wall. In order to discuss the difference between cantilever retaining wall, anchored plating wall and multi-anchored-plating wall, field tests have been conducted to investigate the performance. During construction, the lateral earth pressure behind wall, foundation pressure and cable force had been monitored on site. Test results showed that the distribution of lateral earth pressure would be changed greatly after applying prestressed force. The main feature was the stress concentration around the anchorage. Also, lateral earth pressure of retaining wall except anchorage point lied between the K_0 line and K_a line. It is observed that there was a serious stress loss for pre-stressed cables because of soil compaction and cable relaxing. And the final remained cable force was about 35–45% of design load.

Keywords: Lateral earth pressure · Lateral displacement · Surcharge Pre-stressed force · Multi-anchored-plate retaining wall

1 Introduction

Reinforced soil walls can be broadly classified into metallic, geosynthetic and multi-anchor categories (Miyata et al. 2009). Thereinto, the third category are multi-anchor walls (MAWS) constructed with multiple steel plate anchors bolted to round bar sections that are attached at the opposite end to the wall facing. Traditionally, multi-anchored systems typically comprise a facing of some sort which is connected by the anchor tendon to a fixed length, grouted in natural soil, or an anchor block or deadman, placed in backfill (Abdelouhab et al. 2011). The new type of Multi-anchor-plates retaining wall mentioned in this paper combined traditional cantilever retaining wall, anchored plates and cables. Because cables were placed at the

L. Hu et al. (Eds.): GSIC 2018, *Proceedings of GeoShanghai 2018 International Conference: Multi-physics Processes in Soil Mechanics and Advances in Geotechnical Testing*, pp. 501–508, 2018.
https://doi.org/10.1007/978-981-13-0095-0_56

same height in one section connected with wall face and anchored plates separately with different length, the structure also could be named as dispersed pressure retaining wall. By adopting this method, larger earth pressure could be divided into smaller parts and exerted on anchor plates (Yoo and Kim 2008). Therefore, construction procedure could be simplified greatly. Also, the lateral deformation could be decreased even eliminated by applying prestressed force.

Because of the complexity of multi-anchor-plates retaining wall system, a series of influencing factors, such as the placement of anchor plate and tensioning of anchorage cable et al., would disturb the distribution of soil pressure greatly on the retaining wall. Test results by Rowe (1952) showed that the stress concentration occurs at the anchor level and that the pressure distribution was not triangular behind and in front of the wall. (Goeland Patra 2008) also stated that the Rankine and Coulomb theories underestimated the height of the center of the lateral earth pressure. This would affect the design because of the changed moment arm when the moment equilibrium is considered to determine the wall penetration depth, anchor force, and maximum bending moments. (Yoo and Jung 2006) presents the results of field instrumentation on a 7-m high anchored segmental retaining wall. The measured and the inferred horizontal earth pressures showed that the horizontal earth pressures exerting on the wall facing are approaching the at-rest state.

Despite the large number of multi-anchor walls in the world, multi-anchor-plates cantilever retaining wall has not been reported in the open literature. One of the critical design issues argued by designers when the system was first introduced was the design earth pressure for use in internal stability calculations. This paper presents the monitoring results of the wall, including site conditions, wall construction, instrumentation, and observed performance to reveal a type of reasonable distribution of lateral earth pressure on the multi-anchored-plates cantilever retaining wall.

2 Field Monitoring of Multi-anchored-Plate Cantilever Retaining Wall

2.1 Components

As shown in Fig. 1, the multi-anchored-plates retaining wall comprises three main components such as wall facing units, anchor bar-block systems, and backfill (Fig. 1). The retaining wall is 6.5 m height and 0.4 m width on the top with 1:0.1 slope ratio. The width of wall heel and wall toe equals to 2.5 m and 1.5 m separately. Each anchorage pier is connected with four pre-stressed cables with different length. And each rod is attached to a reinforced concrete anchored plate. The standard anchor plates are 450 mm height 1080 mm width and 300 mm thickness. The epoxy-coated pre-stressed cable is also cemented to double-protect any possible corrosion during its design life. Typical cable diameter is 15.2 mm. The yield strength and modulus of the cable is 223 N/mm^2 and 1860 MP are respectively. By using this type of anchorage system, the total force can be applied in step via different cables.

The multi-anchored-plate cantilever retaining wall is designed so as to satisfy four general modes of failure: external stability, internal stability, local facing stability, and

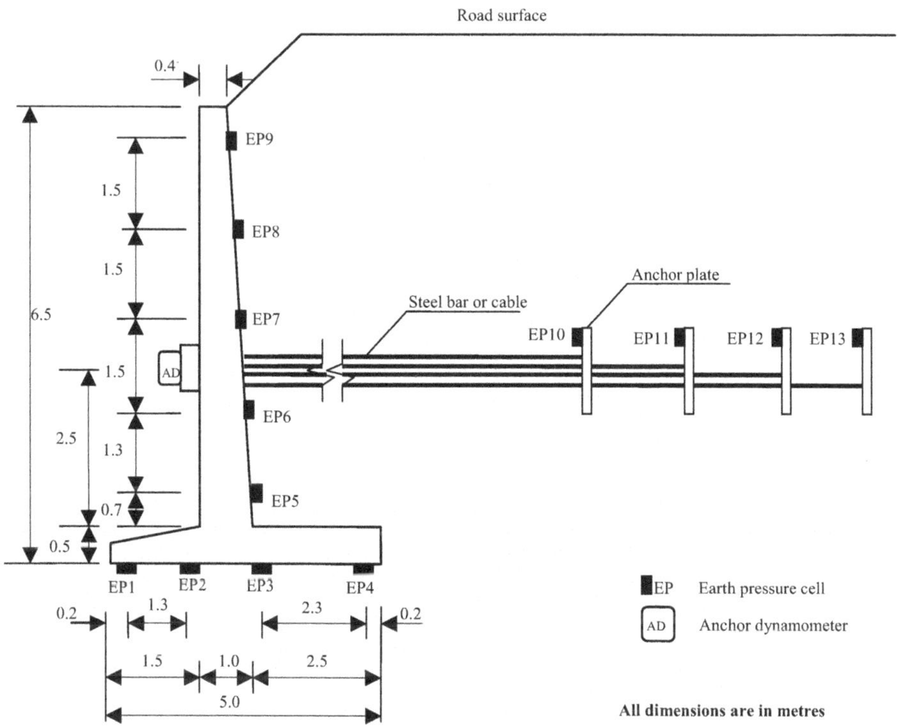

Fig. 1. The structure and instrument locations of multi-anchored-plates retaining wall

global/overall stability within the framework of limit equilibrium together with the tied-back wedge analysis (RECO 1998). The tied-back wedge analysis refers to the lateral stability provided by the horizontal layers of anchor bar block system to anchor the active failure wedge. Among the important design assumptions, the manufacturer of the wall system recommends the internal stability analysis be performed assuming that an active earth force acting on the wall facing is supported by the pullout resistance mobilized by the anchor blocks. It should, however, be pointed out that a higher state of stress is usually assumed in the current design approach (FHWA 1997) for metallic-reinforced soil systems and in anchored systems than for walls with extensible reinforcements. The main focus of the instrumentation program was to check the validity of the design approach that assumes an active state of stress in the internal stability analysis.

2.2 Instrumented Wall

The instrumented wall was located at a construction site for a Highway in Qingzhou, Shandong Province, China. The entire wall section at the project site was approximately 145 m long, and a relatively straight portion was selected for instrumentation. The instrumented wall section has an exposed height of 4.5 m with an embedment depth of approximately 2 m. The uniform length of one section of retaining wall is

10 m. Four anchor plates were installed at a horizontal distance from wall facing were 5 m, 9 m, 13 m, 16 m respectively. Construction of the wall followed the general procedure for that of a typical anchored-plate retaining wall except multi-anchored-plate installation and applying pre-stressed force.

As shown in Fig. 2, when the filling height was above 0.5 m on the location of anchorage system, tendons and anchor plates were placed and casted into the a trench excavated along the surface of compacted soil. Then covered cables with PVC bellows and assembled anchorage system. Finally, backfilled and compacted the trench with manual tamping machine. When the filling height reached the design value, apply tension loads one by one according to the technique requirement. Finally, the pipe was grouted with cement slurry for cables anticorrosion.

(a)Filling embankment (b) Excavating trench and casting anchor plate (c) Tensioning cables

Fig. 2. The construction process of multi-plates retaining wall

In-situ weathered granite soil available at the site was used as the backfill material. Laboratory tests performed on the backfill material revealed that the percent passing the No. 200 sieve was approximately 9% with the coefficients of uniformity and a curvature of $C_u = 12$ and $C_c = 1.4$, respectively, and that the maximum dry unit weight was $\gamma_{\mathrm{d,max}} = 18\ \mathrm{kN/m^3}$ with the optimum water content of $w_{opt} = 14\%$ according to the standard Proctor test (ASTM D 698). The estimated internal friction angle and cohesion at a density corresponding to the as-compacted state was approximately 35° and 10 kPa.

Based on the Unified Soil Classification System (USCS) the soil was classified as SW-SM. The gravel was also extended to a distance of 300 mm behind the facing column to create a drainage layer. Construction specification required that the backfill material be compacted to a minimum of 95% of standard Proctor. Walk-behind vibrating drum rollers were used to compact the backfill material located within 1 m of the back of the wall facing, while a 10-ton drum roller was used elsewhere.

An extensive monitoring program was devised prior to construction to evaluate the mechanical behavior of the wall. Monitored items include foundation pressure, horizontal earth pressures exerted on the wall facing units and the anchor blocks and total pre-stressed force in the anchorage, as shown in Fig. 1. All measurements were obtained over a seven-month period of April 10, 2011–November 25, 2011.

3 Test Results and Discussion

3.1 Foundation Pressure of Retaining Wall

As shown in Fig. 3, before tensioned cables, the foundation pressure on point EP3 is the maximum, then point EP1, EP2 and EP4. With the filling height increased, the pressure increment on point EP3 has been the most obvious. But after applied the pre-stressed force, the pressure on point EP3 became bigger while pressure on other points reduced. It showed that after applied pre-stressed force, the retaining wall rotated inward around the bottom of retaining wall unlike before applied pre-stressed force.

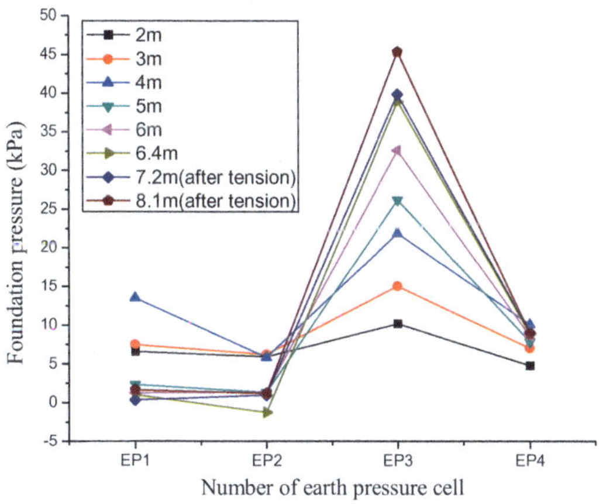

Fig. 3. Distribution of Foundation pressure

3.2 Lateral Earth Pressure Exerted on Wall Facing

As shown in Fig. 4, the variation of distribution of lateral earth pressure during construction time can be divided into three phases;

Phase I, before applied the first stage pre-stressing force, the distribution of lateral earth pressure is more like an irregular trapezoid. From EP9 to EP8, the lateral earth pressure of retaining wall increased quickly. From EP8 to EP6, the pressure increase gradually. But from EP6 to EP5, the pressure decreased quickly.

Phase II, when the filling height reached 7.2 m, applied the first stage pre-stressed force, 60% of total force. The lateral earth pressure around anchor position increased more obviously, such as EP6–EP8. But for other points, the variation of pressure was minus. What's more, by compared with K_a and K_0 line, before applied pre-stressed force, most of lateral earth pressure lied between both lines. However, after applied pre-stressed force, earth pressure around anchorage was over K_0 line. This variation could certificate the limitation effect of multi-plate retaining wall.

Phase III, when the filling height reached 8.1 m, applied the second stage pre-stressing force, 320 kN. After applied the second force, the peak value of lateral earth pressure got another increment. And the distribution of lateral earth pressure

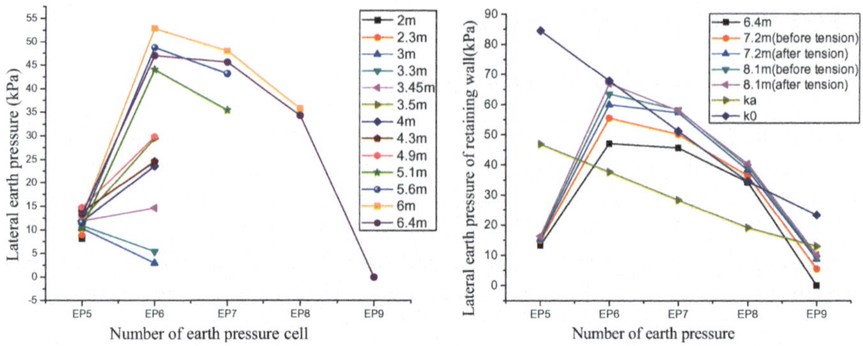

(a)Before applied pre-stressing force (b)After applied pre-stressing force

Fig. 4. Variation of lateral earth pressure of retaining wall

becomes more like triangular shape. At the same time, most of the lateral earth pressure except the top and bottom could exceed the K_0 line.

3.3 Tensile Force of Cables

Although the tension load is 320 kN, monitoring data showed that the final average locked load of cables was just about 55% design load. It was mainly because soil compression, cables relaxing and anchorage lock loss. As shown in Fig. 5, after locked, in the first 2 months, the average drop amplitude of pre-stressing force was 25%–30%. However, with the time extending, tendons force increased again but were lower than locked values. Compare with the first 2 months, the average increase amplitude of pre-stressing force was about 18% in next 2 months. And after that, the monitored force was only about 35–45% of design pre-stressed force and fluctuated within a narrow range.

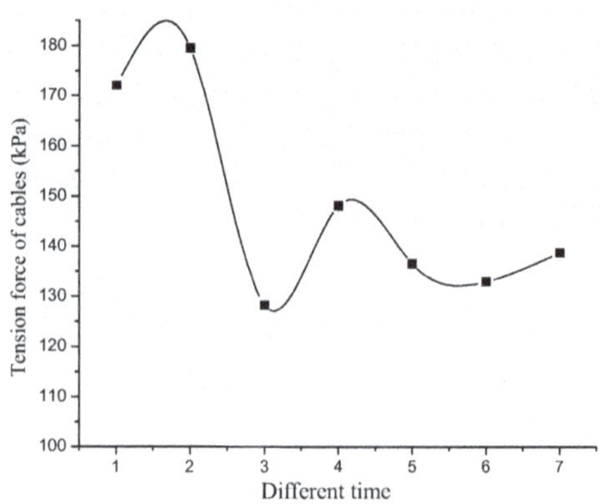

Fig. 5. Variation of cable force at different conditions. **Note:** 1–2 means the first and second stage of tension, respectively. 3–7 means service time of highway was 2, 4, 8, 12, 18 months, respectively.

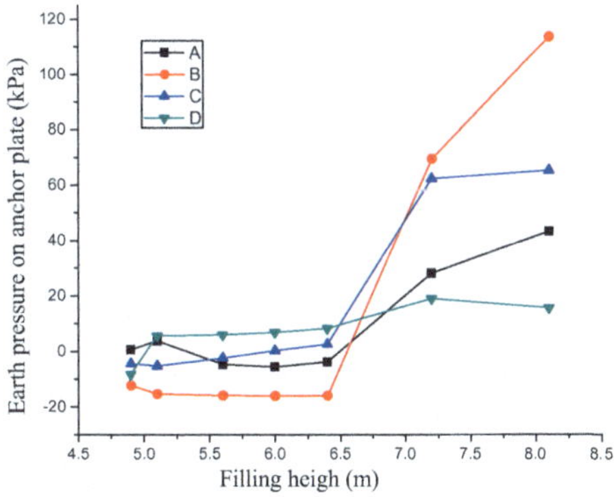

Fig. 6. Pressure of anchor plate versus filling height. **Note:** Mark A, B, C and D represent distance of anchor plates to retaining wall, that is 5, 9, 13 and 16 m, respectively.

3.4 Lateral Earth Pressure on Anchor Plate

As shown in Fig. 6, before applied the pre-stressed force, lateral earth pressure on anchor plates increased gradually with the increment of filling height. But after applied the pre-stressed force, lateral earth pressure on anchor plates increased sharply. And with the increment of filling height, the earth pressure on anchor plates also increased gradually. Another important phenomenon is that earth pressure exerted on these four anchor plates was different, as shown in Fig. 7. And the sequence B, C, A and D sorted by earth pressure value can be got from Fig. 7. This is because for the farthest anchor plate, the relaxation of pre-stressed cable is more serious, which caused the bigger loss of pre-stressed force. However, for the nearest anchor plate, the embankment compactness near retaining wall could not been guaranteed well. Therefore, the loss of pre-stressed force was also bigger.

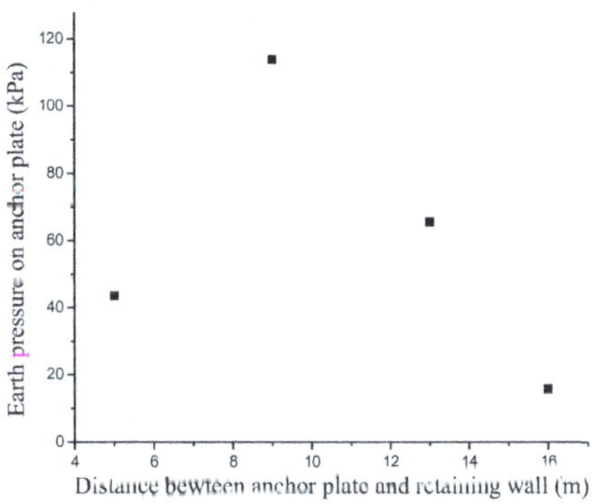

Fig. 7. Lateral earth pressure on anchor plates vs. length of cables

4 Conclusions

This paper presents the results of field instrumentation on a 6.5-m high multi-anchored cantilever retaining wall. Field monitoring datum showed that by applying pre-stressed force, most of lateral earth pressure on wall facing was near even over K_0 line. Therefore, the stress state could be adjusted well and the lateral displacement was reduced under lateral earth pressure. However, there was a larger stress loss because of soil compaction and cable relaxing. Therefore, although applied the same pre-stressed force for 4 different cables, the remained force of nearest and farthest cable was smaller than other cables. And after about 8 months service, the pre-stressed force could reach a normal stable state. And remained force was only about 35–45% of design pre-stressed force.

Acknowledgement. This research work was conducted when the first author visited the University of Kansas as a visiting scholar, which was sponsored by the China Scholarship 610 Council. This research work was also supported by Grant No. 51208284 from the Natural Science Foundation of China. Their financial support is gratefully acknowledged.

References

Abdelouhab, A., Dias, D., Freitag, N.: Numerical analysis of the behaviour of mechanically stabilized earth walls reinforced with different types of strips. Geotext. Geomembr. **29**(2), 116–129 (2011)

Yoo, C., Kim, S.B.: Performance of a two-tier geosynthetic reinforced segmental retaining wall under a surcharge load: full-scale load test and 3D finite element analysis. Geotext. Geomembr. **26**(6), 460–472 (2008)

Miyata, Y., Bathurst, R.J., Konami, T.: Measured and predicted loads in multi-anchor reinforced soil walls in Japan. Soils Found. **49**(1), 1–10 (2009)

Rowe, P.W.: Anchored sheet-pile walls. Ice Proc. **1**(1), 27–70 (1952)

Goeland, S., Patra, N.R.: Effect of arching on active earth pressure for rigid re-taining walls considering translation mode. Int. J. Geomech. **8**(2), 123–133 (2008)

Yoo, C. Jung, H-Y.: Case history of geosynthetic reinforced segmental retaining wall failure. J. Geotech. Geoenviron. Eng. **132**(12), 1538–1548 (2006)

Robust Linear Regression for Undrained Shear Strength Data

Jun Lin[1], Guojun Cai[1(✉)], Songyu Liu[1], and Anand J. Puppala[2]

[1] Institute of Geotechnical Engineering, Southeast University, Nanjing, Jiangsu, China
focuscai@163.com

[2] Department of Civil Engineering, The University of Texas at Arlington, Arlington, TX, USA

Abstract. Outlier data has attracted considerable interesting geotechnical data. When doing classical linear least squares regression, if the regression data satisfied certain regression weights, the ordinary least squares regression is considered as the best method. However, the estimating and regression results may be inaccurate in case of these data not meeting given assumptions. Particularly in least squares regression analysis, there is some data (outliers) violating the assumption of normally distributed residuals. Under situation of regression data blending to outliers, robust regression is the best fit method. It can discriminate outliers and offer robust results when the regression data exists outliers. The purpose of this study is to make use of robust regression method to trend regression in geotechnical data analysis. Without defining absolute outliers from geotechnical testing data, outlier data of undrained shear strength is detected based on robust regression result.

Keywords: Undrained shear strength · Robust regression · Outlier data

1 Introduction

Geotechnical engineers face a number of uncertainties [1, 2]. Soil materials formed from geological weathering processes, and by physical means to deliver the soil to the current position [3]. In the forming process, the soil is influenced by various stress, pore fluid, and physical and chemical changes. Therefore, it is not surprising that there are always some outliers in geotechnical data. When dealing with geotechnical problems, empirical correlations between in situ or laboratory test results and geotechnical parameters are often used in geotechnical design. When establishing such empirical correlations, mostly adopted method is regression analysis, including linear or non-linear regression [4].

Linear least squares regression (LLR) is a modeling approach by far the most widely used. When people say they use "regression", "linear regression" or "least squares" to adapt their data, they usually mean doing LLR. LLR is not only the most widely used method of modeling, but also have adapted to a variety of circumstances, beyond its immediate scope [5].

L. Hu et al. (Eds.): GSIC 2018, *Proceedings of GeoShanghai 2018 International Conference: Multi-physics Processes in Soil Mechanics and Advances in Geotechnical Testing*, pp. 509–515, 2018.
https://doi.org/10.1007/978-981-13-0095-0_57

A mathematical method that finds the best-fit curve for a given set of points is to minimizing the sum of the squares of the distances of regression data deviating from the curve. The sum of squares of the offset distances is used instead of the absolute values of the offset distances because this allows the residuals to be treated as a continuously differentiable quantity. Whereas, because of the use of the square of the offset, peripheral points may have a disproportionate effect on fit. Whether the results are desirable or not, it depends on the issue of question [6].

The statistical observations of outliers are significantly different from the other sample values. Least-squares regression is obviously the best option if errors are normally distributed. Then, other means is eagerly required if these errors are not normally distributed. One particular distribution is the long tail error distribution of great concern. One solution is still to use the LLR method after removing the largest remaining value as outliers. However, this solution may be infeasible if several larger residual values exist by reason that the poor nature of the outlier tests. In addition, outlier testing is an acceptance/rejection process. The testing processes are neither smooth or statistically efficient. Robust regression (RR) is another option for least-squares regression in the case of the data contaminated with outliers. It can also be used to detect influential observations when the data is exposed to outliers [7].

It is difficult to define absolute outliers from geotechnical testing data, but it is possible to indicate the least predictable or relatively outlying data points using statistical tools. The objective of this paper is to demonstrate the advantages of RR analysis used in geotechnical data comparing with least square regression analysis. The procedure of RR is discussed shortly, based on LLR method. And then, Regression analysis is operated on undrained shear strength (s_u) data derived from CPTU test with both RR and LLR. Comparing regression result, the outlier of s_u data can be detected based on RR method.

2 s_u Data from CPTU

Unlike traditional sampling and laboratory tests, the piezocone penetration (CPTU) tests overcome the sampling disturbances with ease. In addition, comparing with conventional sampling and laboratory tests, CPTU tests can define an profile of s_u along with depth with remarkable less time and effort. Besides, the profile of s_u results is nearly continuous with depth, instead of at relatively few points of sampling and testing.

Various methods were proposed to determine s_u results from CPTU data. Generally, these methods can be divided into theoretical and empirical relationships. The cone penetration into soils is a complex process. Because of the limitation of theoretical methods applied to simulating soil behavior during cone penetration process, empirical relationships are more favored to determine s_u data in this study. These empirical relationships mainly include the direct or indirect correlations between cone penetration resistance and s_u. Another reason for adopting empirical determining relationships is to avoid too much on site and lab work.

A regular practice to determine to s_u is to establish a relationship between s_u and a net cone resistance. The net cone resistance is defined as $q_t - \sigma_{v0}$, where σ_{v0} is the in-situ total overburden stress. The equation links s_u to the net cone resistance is given as:

$$s_u = \frac{q_t - \sigma_{v0}}{N_{kt}} \tag{1}$$

where N_{kt} is an constant quantity. Numerous studies have been conducted to determine the proper values of N_{kt}. In this research, N_{kt} is chosen to be 12.

3 Robust Linear Regression

3.1 Procedure of Robust Linear Regression

When the error distribution is not a normal distribution, the linear least squares estimation is not suitable, especially when the error has a heavy tail characteristic. The usual approach is to remove these relatively large weight data from the observed data in the least squares regression process. Another approach, so-called "robust regression," uses a more sophisticated approach that makes the method insensitive to outlier data. The most extensively used robust regression method is m-estimate. Such estimates can be viewed as a generalization of maximum likelihood estimates and is therefore called "m-estimate."

Considering the most generally linear regression model,

$$\begin{aligned} y_i &= \alpha + \beta_1 x_{i1} + \beta_2 x_{i2} + \ldots + \beta_k x_{ik} + \varepsilon_i \\ &= x_i' \beta + \varepsilon_i \end{aligned} \tag{2}$$

for the i th of n observations.

To estimate b for β, the linear regression form is

$$\hat{y}_i = \alpha + b_1 x_{i1} + b_2 x_{i2} + \ldots + b_k x_{ik} + \varepsilon_i \tag{3}$$

and the residuals are given by

$$e_i = y_i - \hat{y}_i \tag{4}$$

In M-estimation method, the estimator b is inferred by minimizing a specific objective function over all b,

$$\sum_{i=1}^{n} \rho(e_i) = \sum_{i=1}^{n} \rho(y_i - x_i' b) \tag{5}$$

where the specific function ρ defines the weight of each residual in the specific function.

Let $\varphi = \rho'$ be the derivative of ρ. φ is called the influence curve. Deriving the partial derivative of the objective function, combining with the coefficients b and setting the equation to zero and, then yields a coefficient estimation equation with $k+1$ rank:

$$\sum_{i=1}^{n} \varphi(y_i - x_i' b)x_i' = 0 \tag{6}$$

Define the weight function

$$\omega(e) = \frac{\varphi(e)}{e} \tag{7}$$

and let

$$\omega_i = \omega(e_i) \tag{8}$$

The equation of the estimated coefficient can be rewritten as

$$\sum_{i=1}^{n} \omega_i(y_i - x_i' b)x_i' = 0 \tag{9}$$

To solve these estimating equations is equivalent to a weighted least-squares regression, finding min $\sum \omega_i^2 e_i^2$.

However, the weights depend on the residuals, the residuals depend on the estimated coefficients, and the estimated coefficients depend on the weights. Therefore, the iterative weighting least squares (IRLS) is used to solve this problem: A solution (called iteratively reweighted least-squares), is therefore required:

Set least-squares estimates as initial estimates $b^{(0)}$.

In every iteration t, residuals $e_i^{(t-1)}$ are determined, and corresponding weights $\omega_i^{(t-1)}$ from the former iteration are also calculated.

To solve new weighted-least-squares estimates

$$b^{(t)} = \left[X' W^{(t-1)} X\right]^{-1} X' W^{(t-1)} y \tag{10}$$

where X is the model matrix, with x_i' as its i th row, and $W^{(t-1)} = diag\left\{\omega_i^{(t-1)}\right\}$ is the current weight matrix.

Repeating step 2 and step 3 until the estimated coefficients tend to converge.

The asymptotic covariance matrix of b is

$$v(b) = \frac{E(\varphi^2)}{[E(\varphi')^2]}(XX')^{-1} \tag{11}$$

Using $\sum [\varphi(e_i)]^2$ to estimate $E(\varphi^2)$, and $[\sum \varphi'(e_i)/n]^2$ to estimate $[E(\varphi')^2]$ produces the estimated asymptotic covariance matrix, $\hat{v}(b)$ (which is not reliable in small samples).

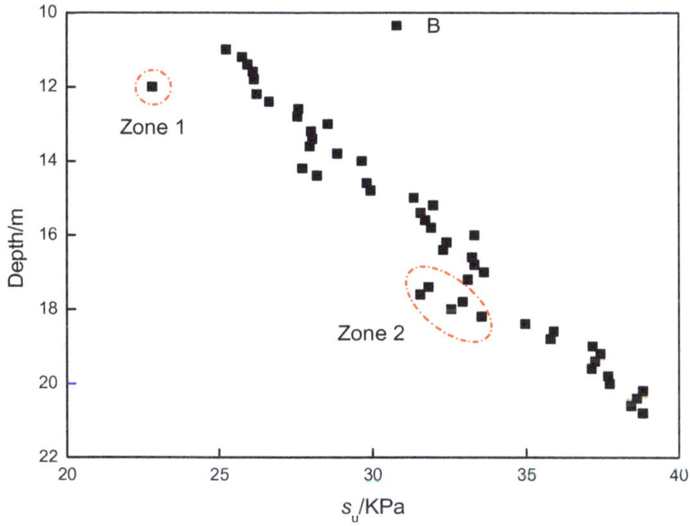

Fig. 1. s_u data from CPTU test

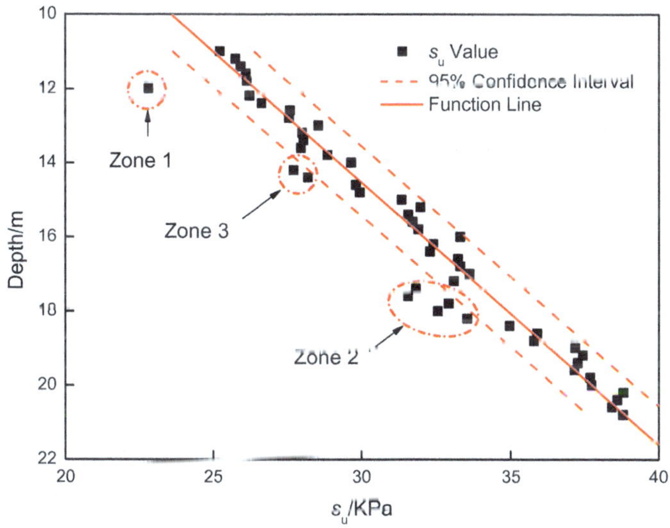

Fig. 2. RR result of s_u data

3.2 Results of RR Regression and LLR Regression

A case study demonstrates the RR method on the s_u data form CPTU test. The number of s_u data used for regression is about 50. Visual inspection on the s_u data, the outlier data is likely located in Zone 1 and Zone 2, shown in Fig. 1. To detect the outlier data and get a better function to describe the $s_u s$ data trend with depth, RR method and LLR method are used to regression analysis.

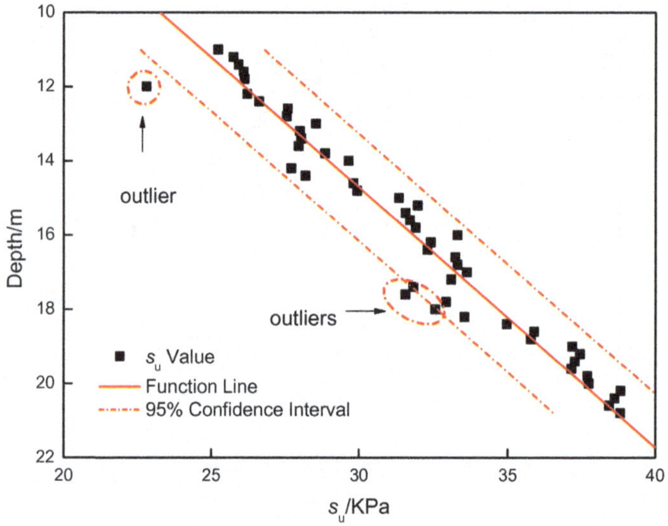

Fig. 3. LLR result of s_u data

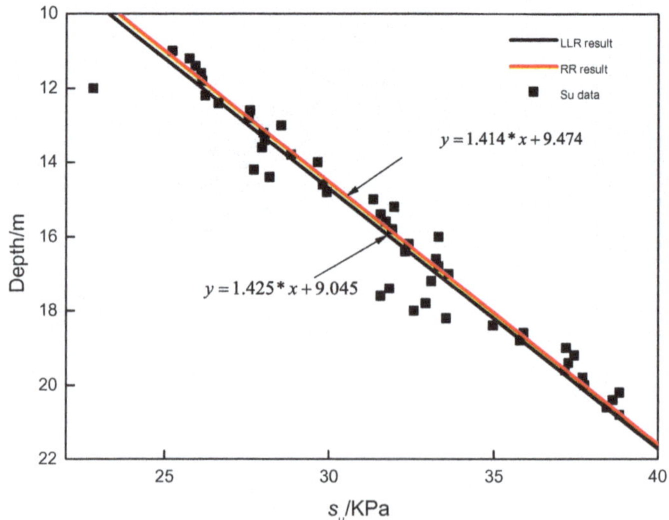

Fig. 4. Comparison of all results

The RR result is shown in Fig. 2 and LLR result in Fig. 3. Both results are provided with regression functions with 95% confidence interval. Take an inspection on the results in Figs. 2 and 3. It is obliviously that the RR regression function has a narrower 95% confidence interval (Fig. 2) than the LLR regression function (Fig. 3). And the RR regression function lies in the middle of s_u data in the trend direction. The s_u data scatters in both side of the RR regression function in equally chance. As a contrast, the LLR regression function has a wider 95% confidence interval. And the LLR crosses less s_u data than RR regression in the trend direction. Most important is that the s_u data scatters more in upper 95% confidence interval than in lower 95% confidence interval, as shown in Fig. 3. It means that the LLR is bias in such situation. As shown in Fig. 2, data in Zone 1 and Zone 2 and Zone 3 is more likely to be outlier data (Fig. 4).

4 Conclusion

This paper demonstrates the RR regression analysis and LLR analysis in the case that outlier data existed in the s_u data. The regression analysis results show that RR method can deal with the outlier data in s_u data very well. The RR regression function can give a more desirable result than the LLR function. Usually, the RR regression function has a narrower confidence interval than the LLR regression function. It is highly recommended that RR regression analysis should be adopted in the case that there is some outlier data existing in the geotechnical data.

References

1. Zhang, L.M., Tang, W.H., Zhang, L.L., Zheng, J.G.: Reducing uncertainty of prediction from empirical correlations. J. Geotech. Geoenviron. Eng. **130**(5), 526–534 (2004)
2. Ching, J., Phoon, K.: Characterizing uncertain site-specific trend function by sparse Bayesian learning. J. Eng. Mech. **143**(7), 4017028 (2017)
3. Baecher, G.B., Christian, J.T.: Reliability and Statistics in Geotechnical Engineering. Wiley, New York (2003)
4. Gillins, D.T., Bartlett, S.F.: Multilinear regression equations for predicting lateral spread displacement from soil type and cone penetration test data. J. Geotech. Geoenviron. Eng. **141** (4), 04013047 (2015)
5. Yuen, K.V., Ortiz, G.A.: Outlier detection and robust regression for correlated data. Comput. Methods Appl. Mech. Eng. **313**, 632–646 (2016)
6. Gürünlü Alma, Ö.: Comparison of robust regression methods in linear regression. Int. J. Contemp. Math. Sci. **6**(9–12), 409–421 (2011)
7. Davies, P.L.: Aspects of robust linear regression. Ann. Stat. **21**(4), 1843–1899 (1993)

Correlation Analysis of CPT Test Results and the Compaction Index for Calcareous Sand

Xiaocong Liang[1,2(✉)], Xiaobing Guo[3], Xin Wang[1,2], Sheng Chen[1,2], and Deyong Wang[1,2]

[1] CCCC Fourth Harbor Engineering Institute Co., Ltd., Guangzhou 51000, China
mariobeatrice@sina.com
[2] Key Laboratory of Environment Protection & Safety of Transportation Foundation Engineer of CCCC, Guangzhou 51000, China
[3] CCCC Guangzhou Dredging Co., Ltd., Guangzhou 51000, China

Abstract. In previous studies, the standard penetration test (SPT) has been used to determine the compaction of sand and develop resulting correlations. However, correlations are rarely based on the cone penetration Test (CPT), especially those correlating with relative compaction values. Several drilling boreholes were selected to collect samples for laboratory testing as part of the Kuwait LNGI project. The test results showed that the relative density and fines content exhibit a linear correlation at a specified level of relative compaction and the samples with high fines contents yielded larger maximum density values based on compaction tests compared with those based on relative density tests. Moreover, a comparative analysis of the correlation between q_c and the relative density was conducted based on three methods: Jamiolkowski's method, Baldi's method and Jamiolkowski's method with a carbonate content correction. The analysis results revealed that Jamiolkowski's method yielded higher q_c values than Baldi's method at a 90% compaction level, and the corrected Jamiolkowski method yielded low q_c values. Furthermore, in accordance with the above analysis, it was more reasonable to establish acceptance criteria based on Baldi's method for the compaction of sand with a carbonate content less than 45%. However, the corrected Jamiolkowski method should be applied when the carbonate content is greater than 45%.

Keywords: Relative compaction · Relative density · Correlation analysis Calcareous sand · Acceptance criteria

1 Introduction

The compaction index is very important in the design process of ground improvement for granular materials and can directly reflect the ground improvement compaction degree. To obtain the compaction degree of ground improvement, the in situ density of granular materials must be determined. However, since the sampling of granular materials is easily disturbed, the laboratory determination of the density is often

L. Hu et al. (Eds.): GSIC 2018, *Proceedings of GeoShanghai 2018 International Conference: Multi-physics Processes in Soil Mechanics and Advances in Geotechnical Testing*, pp. 516–524, 2018.
https://doi.org/10.1007/978-981-13-0095-0_58

unrepresentative of the actual value and cannot reflect the actual in situ compaction status. With the development of electric cone penetration Test (CPT) equipment, the use of CPT testing in the inspection of ground improvement has become increasingly popular due to its reliability, repeatability and stable performance. To investigate the compaction degree of granular soils, studies of the correlation between CPT test values and the compaction index have been conducted by some researchers [1–3].

However, most of the CPT test results were correlated with the relative density values, and the correlation analyses rarely included the relative compaction values. In addition, the correlations were mostly based on silica sand, and for calcareous sand, correlations should be corrected due to its crushability and compressibility. Therefore, there is a need to analyze the correlation between the relative density and the compaction degree and to investigate the correlation for calcareous sand based on CPT test values. This approach will provide a new acceptance criteria method for ground improvement over large areas.

2 Correlation Analysis

2.1 Correlation Analysis for the Compaction Index

The compaction index is typically expressed as the relative density D_r or relative compaction degree R_c [4]. The formulas of these two indexes are presented as following Eqs. (1) and (2).

$$D_r = \frac{\frac{1}{\rho_{\mathrm{dmax}}} - \frac{1}{\rho_{\mathrm{d}}}}{\frac{1}{\rho_{\mathrm{dmim}}} - \frac{1}{\rho_{\mathrm{dmax}}}} \tag{1}$$

where ρ_{d}, ρ_{dmax}, and ρ_{dmin} are the dry density, maximum density and minimum density, respectively.

$$R_c = \rho_d / \rho'_{\mathrm{dmax}} \tag{2}$$

where ρ'_{dmax} is the maximum density obtained from the compaction test, which is different from ρ_{dmax} obtained from the maximum density test.

It is worth noting that the two maximum dry densities, ρ_{dmax} and ρ'_{dmax}, mentioned above were obtained from different tests. ρ_{dmax} refers to standard ASTM D-4254 and is based on the vibration compaction method. ρ'_{dmax} refers to standard ASTM D-1557 [5, 6], in which the layered sample is compacted in a standard cylindrical mold.

2.2 Correlation Analysis for the Compaction Index

The earliest study regarding the correlation of these two compaction indexes based on laboratory tests was performed by Lee and Singh [7]. The correlation was performed based on 47 selected samples. The maximum densities ρ_{dmax} and ρ'_{dmax} were deemed equal by the authors.

$$R_c(\%) = 80 + 0.2D_r(D_r > 40\%) \tag{3}$$

In addition, Youssef and Gihan [8] proposed an empirical expression based on 20 selected samples from the Fayoum area, as shown in Eq. (4). The maximum densities ρ_{dmax} and ρ'_{dmax} were distinguished when performing the analysis.

$$D_r = 5.5R_c - 4.47(0.85 < R_c < 1) \tag{4}$$

This formula was converted as follows.

$$R_c(\%) = 81.27 + 0.18D_r(D_r > 20.5\%) \tag{5}$$

Although the maximum densities ρ_{dmax} and ρ'_{dmax} are distinguished in Eq. (4), a similar linear correlation is still obtained using both equations. Furthermore, it can be concluded from reference [9] that the higher the fine content of the sand is, the lower the value of the maximum dry density based on ASTM D-4254 and the higher the value of the maximum dry density based on ASTM D-1557.

3 Correlation Analysis Between CPT and the Relative Density

3.1 Baldi's Method

The method proposed by Baldi [10] in 1986 is one of the most widely applied methods. The sample used to build the correlation was uncemented silica sand from the Ticino and Hokksund district. The formula is as follows:

$$D_r = \frac{1}{C_2}\ln\left(\frac{q_c}{C_O(\sigma')^{C_1}}\right) \tag{6}$$

where C_0, C_1, and C_2 are experimental coefficients; σ' is the overburden effective stress; and q_c is the cone resistance. The suggested C_0, C_1, and C_2 values for normal consolidated sand were 157, 0.55, and 2.41, respectively, and those for overconsolidated sand were 181, 0.55, and 2.61, respectively.

3.2 Jamiolkowski's Method

With reference to the literature [11], the previous correlation method was mainly suitable for sands with fines contents lower than 15%. Moreover, the previous method did not account for the boundary effects that reduce q_c. Therefore, Jamiolkowski conducted several chamber tests in 2001 and proposed a correlation considering boundary effects and the effect of sand compressibility. The material used in the correlation analysis was normal consolidated quartz silica sand. The empirical formula is shown below, and the correlation is shown in Fig. 1.

$$D_r = 100\left[0.268.\ln\left(\frac{q_c/\sigma_{atm}}{\sigma'/\sigma_{atm}^{0.5}}\right) - 0.675\right] \tag{7}$$

where σ_{atm} is 1 atmospheric pressure, which is equal to 100 kPa.

4 The Al-Zour LNGI Project in Kuwait- Case Study

4.1 Introduction of the Al-Zour LNGI Project

The Al-Zour LNGI project is located in the Al-Zour region of Kuwait. The project has a reclamation area of 50,000 m^2 and an average reclamation depth of 20 m. The designed filling material is sand with a fines content lower than 15%, and the hydraulic reclamation method was adopted. The strata in the borrow area include clay or silt, caprock, loose to medium dense silty sand, and dense to very dense sand with silt. The relative compaction was required to reach 95% above the water level and greater than 90% below the water level. Considering such design requirements, three locations in the borrow area were adopted to drill boreholes for test samples. Eight samples were collected to perform a series of tests, including a sieve analysis, relative density test, compaction test and specific gravity test. The reference standards were ASTM D-422, ASTM D-2487, ASTM D-4254 and ASTM D-1557.

4.2 Correlation Analysis

(1) Correlation analysis between the maximum densities ρ_{dmax} and ρ'_{dmax}

Based on the collected test results, the correlation regarding the maximum densities ρ_{dmax} and ρ'_{dmax} can be obtained, as shown in Fig. 1. Furthermore, the ratio of $\rho_{\text{dmax}}/\rho'_{\text{dmax}}$ for different soil types which named by ASTM D-2487 [12] is shown in Fig. 2.

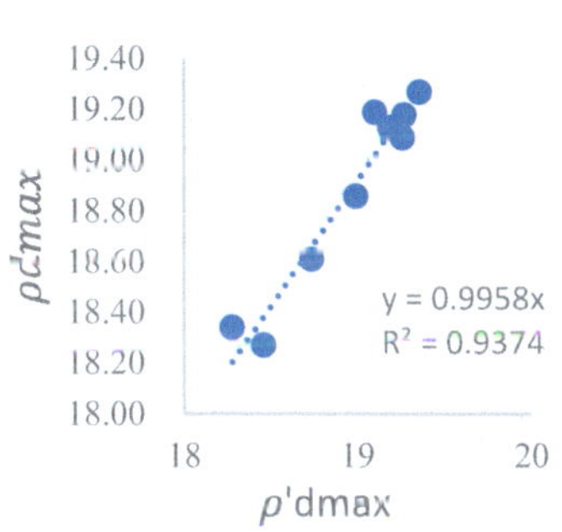

Fig. 1. ρ_{dmax} and ρ'_{dmax} Correlation

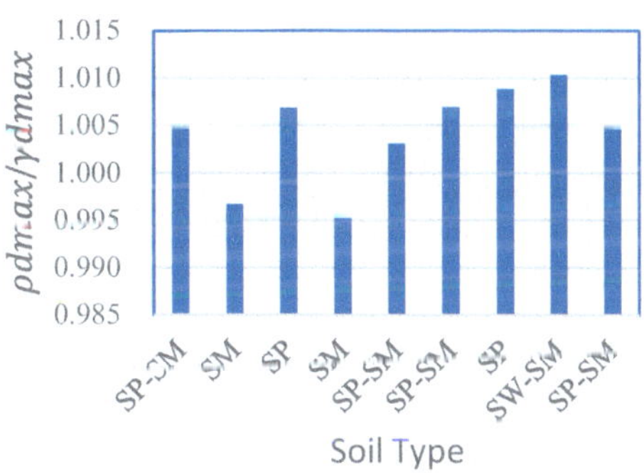

Fig. 2. Ratio of $\rho_{\text{dmax}}/\rho'_{\text{dmax}}$ for different soil types

Figures 1 and 2 show that the two maximum densities ρ_{dmax} and ρ'_{dmax} are linearly correlated. However, their ratio varies slightly for different soil types. The ratios obtained from the relative density test method for soil types SP, SP-SM and SW-SM with a fines content lower than 12% are higher than those from the compaction test. Soil type SM, which has a fines content larger than 12%, exhibits the opposite trend. Thus, the results display a similar consistency with the reference [9], which has concluded with the fine contents increasing, the compaction test tend to yield a higher value than relative density test.

(2) Correlation between relative density and relative compaction

The relative density is not constant at the same relative compaction for different samples and varies within a certain range. Based on the collected test result, the relative density varies from 43.96 to 69.21 at 90% relative compaction and varies from 71.28 to 85.91 at 95% relative compaction. Further analysis suggests that the relative density is correlated with the fine content under constant relative compaction conditions. The correlation is shown in Fig. 3.

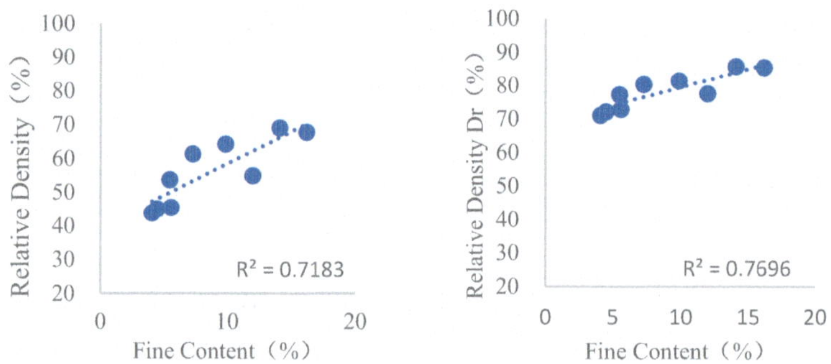

(a) 90% of relative compaction (b) 95% of relative compaction

Fig. 3. Correlation between the relative density and fines content

Notably, the relative density is linearly correlated with the fine content, as shown in Fig. 4. The coefficient of determination (R^2) values are 0.7183 and 0.7696, which reflect good correlation performance. A high fines content results in a large relative density at constant relative compaction. Furthermore, the correlation can be expressed as follows:

$$D_r = 1.90\text{FC} + 39.60(R_c = 90\%) \quad (8)$$

$$D_r = 1.08\text{FC} + 68.91(R_c = 95\%) \quad (9)$$

where FC is the fine content (%).

The above empirical equation was based on filling material with a fine content less than 20%. Since the fines content of the filling material was required to be less than

Table 1. Comparative results of the relative density

Relative compaction	Laboratory test		Lee and Singh method (1971)	Youssef and Gihan method (2007)
	Fine content 5%	Fine content 10%		
90%	49.1	58.6	50	48.5
95%	74.4	79.7	75	76.3

15%, two controlling fines content values, 5% and 10%, were adopted for comparison based on Eqs. (3) and (5). The comparative results are shown in Table 1.

The results in Table 1 indicate that Lee and Singh's method and Youssef and Gihan's method exhibit good consistency with the laboratory test for a filling material with a 5% fines content, which indicated that proposed specified correlation based on Al-Zour LNGI project is reasonable and applicable.

4.3 Correction Based on the Carbonate Content

With reference to Lunne [13], most of the correlations between CPT and the relative density have been established based on silica sand. However, a high carbonate content leads to greater compressibility and crushability for a granular material. Therefore, the associated effect on the correlation cannot be neglected at high carbonate contents. In accordance with the geotechnical investigation of the Al-Zour project, the carbonate content of the borrow area is less than 30%. Based on the classification method proposed by PIANC [14], the material below the designed dredging level is mainly slightly calcareous sand to carbonate sand.

Moreover, Wehr [15] suggested that only sand with a carbonate content higher than 50% should be considered for correction and proposed the following carbonate correction factor (CF) for carbonate sand.

$$\mathrm{CF} = 1.36 + 0.0046D_r \tag{10}$$

The above formula exhibits a positive correlation with the relative density, which indicates that the correction factor will increase as the relative density increases. Lunne [16] suggested a less laborious method of establishing the correction factor based on oedometer tests. If the compressibility $C_c/(1 + e0)$ of carbonate sand is less than 0.05, the correction factor does not need to be considered. If the compressibility of $C_c/(1 + e0)$ is higher than 0.05, a correction factor of 1.4 should be adopted for the relative density within the range of 40% to 60%, and a correction factor of 1.5 should be adopted for relative density values higher than 60%.

To obtain the compressibility of carbonate sand, five samples were selected for carbonate content and oedometer tests. Notably, the first four samples were collected in a high carbonate area. The fifth sample was artificially created sand with an assumed maximum carbonate content that existed in the borrowing area. The test results can be seen in Table 2.

Table 2. Compressive test results

S.N.	Sample ID	Carbonate content (%)	Compressive index C_c	Initial void ratio e_0	$C_c/(1 + e_0)$
1	M3	20.73	0.0025	0.4473	0.0017
2	L7	21.7	0.0018	0.4587	0.0012
3	K8	24.47	0.0019	0.528	0.0012
4	K7	26.97	0.0018	0.4678	0.0012
5	T1	45.77	0.0034	0.3537	0.0025

Based on the test results presented in the above table, the compressibility $C_c/(1 + e0)$ of the first four samples ranges from 0.0012–0.0017, with a carbonate content of 20.73–26.97%. These compressibility values are less than the 0.05 value. Even the compressibility of the last sample, which has a maximum carbonate content of 45.77%, is only 0.0025, which is still less than 0.05. Hence, it is not necessary to apply the correction factor when the carbonate content is less than 45.77, as supported by the findings of Lunne.

Furthermore, Hyde [17] found that carbonate particles undergo considerable breakage if the carbonate content surpasses 45%. Overall, based on the suggested correction methods and test results above, the effect of breakage is minor for most reclaimed sands and can be neglected when the carbonate content is less than 45%. However, if the carbonate contest surpasses 45%, the correction factor should be applied and the breakage of carbonate particles should be considered in determining the CPT results.

4.4 Correlation Analysis Between CPT and the Compaction Index

In accordance with the sieve analysis results of samples taken from the reclamation area, the fines content of most of the sand sample was approximately 5%. Referring to Table 1, the relative density at 90% relative compaction is 49.1. Substituting the obtained relative compaction into Eq. (10), a correction factor of 1.58 can be obtained. Jamiolkowski's method includes the effect of the compressibility of sand and provides a boundary constraint for the correlation. Thus, when sand with a high carbonate content is encountered, the CF should be based on Jamiabolkowski's method.

To obtain a suitable correlation, the correlations based on Baldi's method, Jamiolkowski's method and the corrected Jamiolkowski method were compared at the same relative density of 49.1. The comparative curves are shown in Fig. 4.

To achieve a relative compaction level of 90%, Jamiolkowski's method requires a larger q_c value than Baldi's method. However, after determining the correction factor based on Jamiolkowski's method, a lower q_c value is obtained than that based on Baldi's method. This observation is mainly due to the low compressibility of the material used in Baldi's proposed correlation. Further analysis shows that Baldi's method yields a CF nearly equal to that of Jamiolkowski's method at 1.38.

Therefore, as mentioned in Sect. 4.3, if the carbonate content is less than 45%, it is not necessary to consider the effects of carbonate particle compressibility. The Baldi

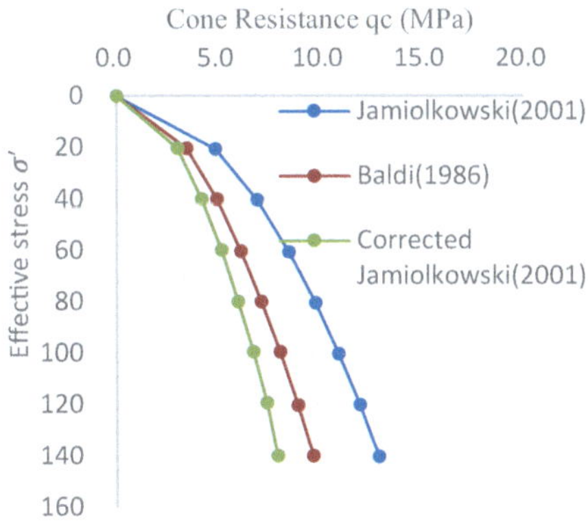

Fig. 4. Comparative curves for different correlation methods

method should be applied to establish the acceptance criteria between the CPT results and relative compaction. However, if the carbonate content is larger than 45%, Jamiolkowski's method with a correction factor is suggested.

5 Conclusion

The correlation between the relative density and relative compaction was developed, and the differences between the two maximum densities obtained from tests based on ASTM D-4254 and ASTM D-1557 were investigated. Furthermore, different correlations between the CPT test results and the relative density considering a carbonate content correction were compared. The following conclusions were drawn from the above analysis results and the results of previous studies

(1) The maximum density values obtained from the relative density test, referred to ASTM D-4254 for soil types SP, SP-SM and SW-SM with a fines content lower than 12%, were higher than those from the compaction test, referred to ASTM D-1557. However, soil type SM, which had a fine content higher than 12%, displayed the opposite trend.
(2) The fine content displayed a linear correlation with the relative density. The relative density increased when the fines content increased at a specified and constant level of relative compaction.
(3) Jamiolkowski's method produced a larger q_c value than Baldi's method at the designed compaction degree of 90%. However, Jamiolkowski's method with a correction factor yielded a lower q_c value than Baldi's method.

If the carbonate content is less than 45%, it is not necessary to apply a carbonate content correction. In this case, Baldi's method should be used to establish the acceptance criteria. However, if the carbonate content exceeds 45%, Jamiolkowski's method with a correction factor is suggested.

References

1. Villet, W.C.B., Mitchell, J.K.: Cone resistance, relative density and friction angle, cone penetration testing and experience. ASCE (1981)
2. Schmertmann, J.H.: Static cone to compute static settlement over sand. J. Soil Mech. Found. Div. **96**, 1011–1043 (1970)
3. Schmertmann, J.H.: Guidelines for cone penetration test, performance and design. Report No. FHWA-TS-78-209, U.S. Department of Transportation, Washington, D.C. (1978)
4. Lambe, T.W., Whitman, R.V.: Soil Mechanics, SI Version. Wiley, New York (1979)
5. ASTM D 4254: Standard test methods for minimum index density and unit weight of soils and calculation of relative density. ASTM Special Technical Publications (2016)
6. ASTM D1557: Standard test methods for laboratory compaction characteristics of soil using modified effort. ASTM Special Technical Publications (2012)
7. Lee, K.L., Singh, A.: Relative density and relative compaction. J. Soil Mech. Found. Div. **97**(7), 1049–1052 (1971)
8. Gomaa, Y., Gihan, A.: Correlation between relative density and compaction parameters. In: Twelfth International Colloquium on Structural and Geotechnical Engineer, 12th ICSGE, Cairo (2007)
9. Arcement, B.J., Wright, S.G.: Evaluation of Laboratory Compaction Procedures for Specification of Densities for Compacting Fine Sands. Earth Walls (2001)
10. Baldi, G., Bellotti, V.N., Ghionna, N., Jamiolkowski, M., Pasqualini, E.: Interpretation of CPT's and CPTU's – 2nd part: drained penetration of sands. In: Proceedings of the 4th International Geotechnical Seminar Field Instrumentation and In-Situ Measurements, Nanyang Technological Institute, Singapore, 25–27 November 1986, pp. 143–156 (1986)
11. Jamiolkowski, M., LoPresti, D.C.F., Manassero, M.: Evaluation of relative density and shear strength of sands from cone penetration test and flat dilatometer test. In: Soil Behavior and Soft Ground Construction (GSP119), pp. 201–238. American Society of Civil Engineers, Reston (2001)
12. ASTM D2487: Standard Practice for Classification of Soils for Engineering Purpose (Unified Soil Classification System). ASTM Special Technical Publications (2016)
13. Lunne, T., Robertson, P.K., Powell, J.J.M.: Cone Penetration Testing in Geotechnical Practice. Blackie Academic and Professional, London (1997)
14. PIANC: Classification of soils and rocks for the maritime dredging process report of marcom working group 144 (2014)
15. Wehr, W.J.: Influence of the carbonate content of sand on vibro compaction. In: 6th International Conference on Ground Improvement Techniques, Coimbra, Portugal (2005)
16. Lunne, T.: Guidelines for use and interpretation of CPT in hydraulically constructed fill. Report No. 20041367-3, NGI (2006)
17. Hyde, A.F.L., Kam, M.W., et al.: The effect of silica content on the propertied of carbonate sand. In: Proceedings of the XIII, ICSMFE, New Delhi, pp. 267–270 (1994)

Sonar Detection for Karst Cavities and Fracture Under Bored Pile by Generalized S-Transform

Zhen-Ming Shi[1,2], Liu Liu[1,2], Ming Peng[1,2(✉)], Cheng-Cheng Liu[3], Feng-Juan Tao[1,2], and Chun-Sheng Liu[4]

[1] Key Laboratory of Geotechnical and Underground Engineering of Ministry of Education, Department of Geotechnical Engineering, Tongji University, Shanghai 200092, China
pengming@tongji.edu.cn

[2] Department of Geotechnical Engineering, College of Civil Engineering, Tongji University, Shanghai 200092, China

[3] Department of Physical Science and Engineering, Institute of Acoustics, Tongji University, Shanghai 200092, China

[4] Wuhan Changsheng Engineering Exploration Technology Development Co., Ltd., Wuhan 430312, Hubei, China

Abstract. The existing of karst cavities under the bored-pile bottom is of great danger to the quality and stability of the project. Sonar detection is a new technology to ensure the quality of bedrock in construction. Because of the complexity of pile hole materials, detecting the fractures and cavities is still a technical challenge which needs to be studied further. The generalized S-transform was adopted to extract the arrival times of the reflection from the karst cavities at a certain receiver, which can be used to calculate the depth and size of karst cave. Seven models with karst cavities and fractures with different depths were used to verify the feasibility and the effect of the depth of fracture to the detecting karst cavity with the generalized S-Transform. A high-orders staggered-grid finite-difference method was adopted in wave field modelling to obtain the synthetic test signal. The results show that it is feasible and effective to detect karst cavities and fracture under bored pile using the generalized S-transform. The changes caused by the reflected wave significantly appear as peaks in the time-frequency spectrum calculated by generalized S-Transform, which has great advantage in identification for weak reflections.

Keywords: Sonar detection · Karst cavities · Pile · Generalized S-transform FDTD

1 Introduction

Karst distribute in about one-third of total land area of China (Yan et al. 2009). Bored piles are commonly applied in construction in karst area for its varied adaptability. However, the instability of karst cavity roof under pile bottom is a potential geological disaster threatening construction (Zhao et al. 2007). It is dangerous when karst cavity under pile is not clearly detected.

L. Hu et al. (Eds.): GSIC 2018, *Proceedings of GeoShanghai 2018 International Conference: Multi-physics Processes in Soil Mechanics and Advances in Geotechnical Testing*, pp. 525–533, 2018.
https://doi.org/10.1007/978-981-13-0095-0_59

Despite attempts at traditional geophysics methods (El-Qady et al. 2005; Zajc et al. 2015), slurry in the pile hole make them difficult to be used in pile hole. The resolution and detection range of surface geophysics methods are hard to meet the requirement of pile construction. In practical, advance drillings are required for each pile. But false negative still occur for the limited size of the drillings. Sonar detection is a new technology developed by Shi et al. (2016), which makes full use of slurry to couple air and bedrock. Sonar detection is being applied gradually in projects for its the advantages of low cost and high speed. However, the explanation of sonar detection results largely depends on the experience. Better signal processing methods are urgently needed to identify the arrival time of reflection from the karst cavity.

The generalized S-transform which was developed in 1994 for analyzing geophysics data (Stockwell and Mansinha 1996) provides frequency dependent resolution while maintaining a direct relationship with the Fourier spectrum. The time–frequency distribution of a weak signal can be finely depicted by the generalized S-transform in time–frequency domain (Mcfadden et al. 1999; Pinnegar and Mansinha 2004). In this study, we take advantage of this characteristic of the generalized S-transform to identify the arrival times of the reflected wave in the test signal. The location of the karst cavity or fracture can be calculated according to the arrival times of the reflected wave. Seven models with a karst cavity and a fracture of different depths were used to verify the feasibility of detecting karst cavity using the generalized S-Transform.

2 Sonar Detection for Karst Cavity Under Bored Pile

The sonar detection is a new technology to detect karst cavities under bored pile. As shown in Fig. 1, the sonar detector consists of sonar transducers, a computer and connected cables. The sonar transducers consist of a sonar transmitter in the central and four sonar receivers. The sonar detection makes full use of drilling slurry in the pile hole, which can couple the impedance between air and bedrock so that more elastic wave can transmit into bedrock. The sonar elastic wave is transmitted as shown in Fig. 2. The sonar elastic wave will transmit and reflect at different impedance surfaces.

The offset of the detection is rather small compared to the depth of geological reflector D (the roof or floor of karst cavity or fracture). Thus, the depth of geological reflector D can be calculated as below:

$$D = \frac{1}{2} C \times \Delta t = \frac{1}{2} C \times (t_r - t_d) \quad (1)$$

where C is the sound-wave velocity; Δt is the travel time of the sound wave, which is the difference between the arrival time of the reflected wave t_r and the arrival time of the directive wave t_d. Thus, the key of the detection is the identification of the arrival time of reflected wave t_r.

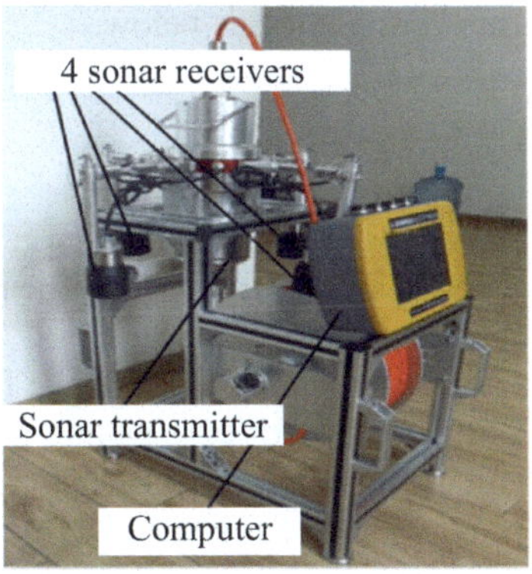

Fig. 1. Sonar detector for karst cavity under pile

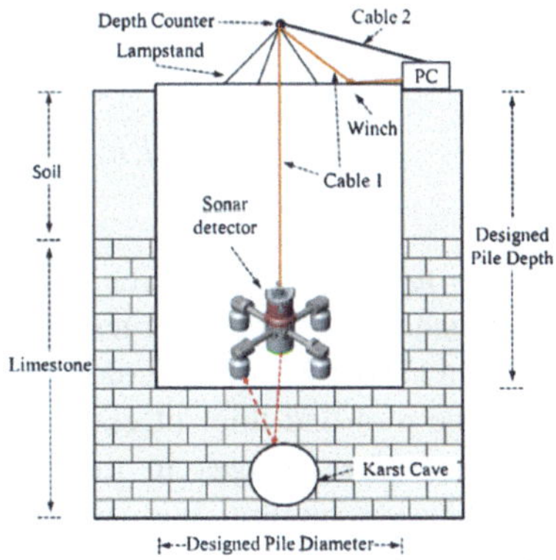

Fig. 2. Schematic diagram of sonar detection

3 The Generalized S-Transform

The generalized S-transform of the signal $x(t)$ is defined as (Pinnegar and Mansinha 2004):

$$S(\tau,f) = \int_{-\infty}^{+\infty} x(t)\omega(\tau - t,f)e^{-i2\pi ft}dt \quad (2)$$

$$\omega(\tau,f) = \frac{|f|^p}{k\sqrt{2\pi}}\exp[-\frac{f^{2p}\tau^2}{2k^2}] \quad (3)$$

$$\sigma(f) = \frac{k}{|f|^p} \quad (4)$$

where, τ and t denote time. f is frequency. ω is the Gauss's window function of Generalized S-Transform. $\sigma(f)$ is scale factor of Gauss's window function, p is a regulatory factor, k is a regulatory parameter.

In order to choose time-frequency resolution, a regulatory factor p and parameter k are introduced in the $\sigma(f)$. The time window will be widened with the increase of k and the decrease of p. The regulatory factor p has larger effect on the time window's shape (width and height) compared to regulatory parameter k. Thus, a wider time window can be adopted at low frequency to get a higher frequency resolution. A narrower time window can be used to obtain a higher time resolution in high frequencies.

According to the Fourier Transform, Convolution theorem and formulas above, the generalized s-transform can be expressed as below:

$$GST(\tau,f) = \int_{-\infty}^{+\infty} X(\alpha+f)e^{-\frac{2\pi^2\alpha^2k^2}{f^{2p}}}e^{-i2\pi\tau t}d\alpha \quad (5)$$

where, $X(\alpha + f)$ is the calculated from $x(t)$ using Fourier Transform and shifting.

We can derive the discrete time S-transform from formula (5) as below:

$$GST[j,n] = \sum_{m=0}^{N-1} X\left[\frac{n+m}{NT}\right] e^{-\frac{2n^2k^2m^2}{n^{2p}}} e^{\frac{i2\pi mj}{N}} \quad (6)$$

Where, $f = n/NT$, $\tau = jT$, T is the sampling interval and sampling frequency. Then the time–frequency distribution of test signal can be calculated by Formula (6).

When a karst cavity or fracture present under pile bottom, the reflected wave from fracture, the roof and floor of cavity will change the wave pattern propagating at the pile bottom. The changes in the test signal significantly appear as peaks in the time-frequency spectrum. We can identify those peaks to detect the karst cavity and fracture.

4 Modelling Results Analysis

4.1 Numerical Models

A model without karst cavity and fracture, a model with a karst cavity and five models with a karst cavity and a fracture with different depths were used to verify the feasibility of the Generalized S-Transform. The diameters of pile hole in the six models are 1.6 m as shown in Fig. 3. The geometric parameters of models are shown in Table 1:

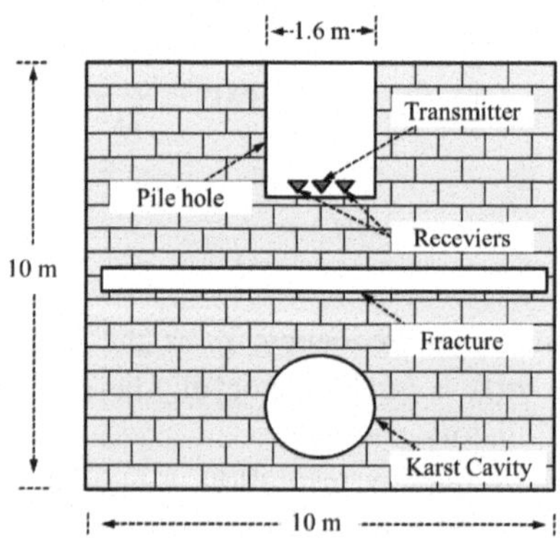

Fig. 3. Numerical model of detection for karst cavity and fracture under bored pile

Table 1. The geometric parameters of models

The Number of Models	1	2	3	4	5	6	7
The depth of fracture (m)	**Null**	**Null**	**1.0**	**1.5**	**2.0**	**2.5**	**3.0**
The thickness of fracture (m)	Null	Null	0.5	0.5	0.5	0.5	0.5
The depth of cavity (m)	Null	4.0	4.0	4.0	4.0	4.0	4.0
The radius of cavity (m)	Null	1.0	1.0	1.0	1.0	1.0	1.0

The calculating section is 10 m × 10 m in the all models. High-orders staggered grid finite-difference method was used in the modelling with a cell size of 0.05 m and a time step of 2×10^{-6} s. In this paper, the second-order accuracy was adopted in time domain, and tenth-order in space domain. Perfectly matched layers (PML) were used to eliminate the reflections from the edges of the computation domain (Collino and Tsogka 2001). The test signal of the receiver with an offset of 0.4 m was recorded (i.e. the offset is the half of the radius of the pile hole). The length of the test signal is 7×10^{-3} s. The material properties of different media are shown in Table 2. Different damping coefficient of P-wave and S-wave were adopted in the modelling.

Table 2. Material properties of media in numerical models

Material properties	v_p (m/s)	v_s (m/s)	ρ (kg/m^3)	ε_L (s^{-1})	ε_T (s^{-1})
Bedrock	4000	2300	2300	10	50
Slurry in hole	2000	0	1500	1500	2500
Filling in karst cavity	2000	1100	1500	1000	1500
Filling in fracture	2000	1100	1500	1000	1500

4.2 Modelling Results

Figure 4(a) is the modelling test signal of model 1 without karst cavity and fracture. All the reflected wave in the test signal are the reflections from the wall of pile hole. Figure 4 (b) is the corresponding generalized S-Transform amplitude spectrum, in which there are five energy peaks point A at 0.472 ms, B at 0.960 ms, C at 1.382 ms, D at 1.862 ms and E at 2.288 ms. The interval between the arrival times of multiple reflections is around 0.450 ms. The errors might be caused by the grid in the shape of square, while the pile hole is a circular. It shows that the energy of multiple reflection gradually weaken for the attenuation in slurry. It verifies that the generalized S-Transform can both identify the arrival time of the reflections and the energy of the reflections.

Figure 4(c) and (d) are the analysis of model 2 with only a karst cavity. The reflection from the roof and floor of karst cavity can be obviously extracted at Point A and B in Fig. 4(c). It finds that the energy of Point E in Fig. 4(d) is much stronger than Point E in Fig. 4(b). The reason is that the reflection from cavity roof interfere the 5th times reflection from hole wall. The arrival time of reflections from roof and floor can be identified by point E and F respectively.

Figure 4(e) to (n) are the modelling test signal of the models with the fractures of different depths and the same karst cavity and their time-frequency spectrums by generalized S-Transform. The identification for the fractures are easily influenced by the multiple reflections from the wall of hole in the time domain. But in the time frequency spectrum, the reflection will cause a significant enhancement of energy (see the red section in Fig. 4(f), (h) and (j) and the light part D in Fig. 4(l)). An obvious vacancy can be found before the point E in Fig. 4(n), which is caused by the inverse phase interference of the reflections from the fracture and the hole wall. Points G in these figure are caused the reflected S wave from the fracture.

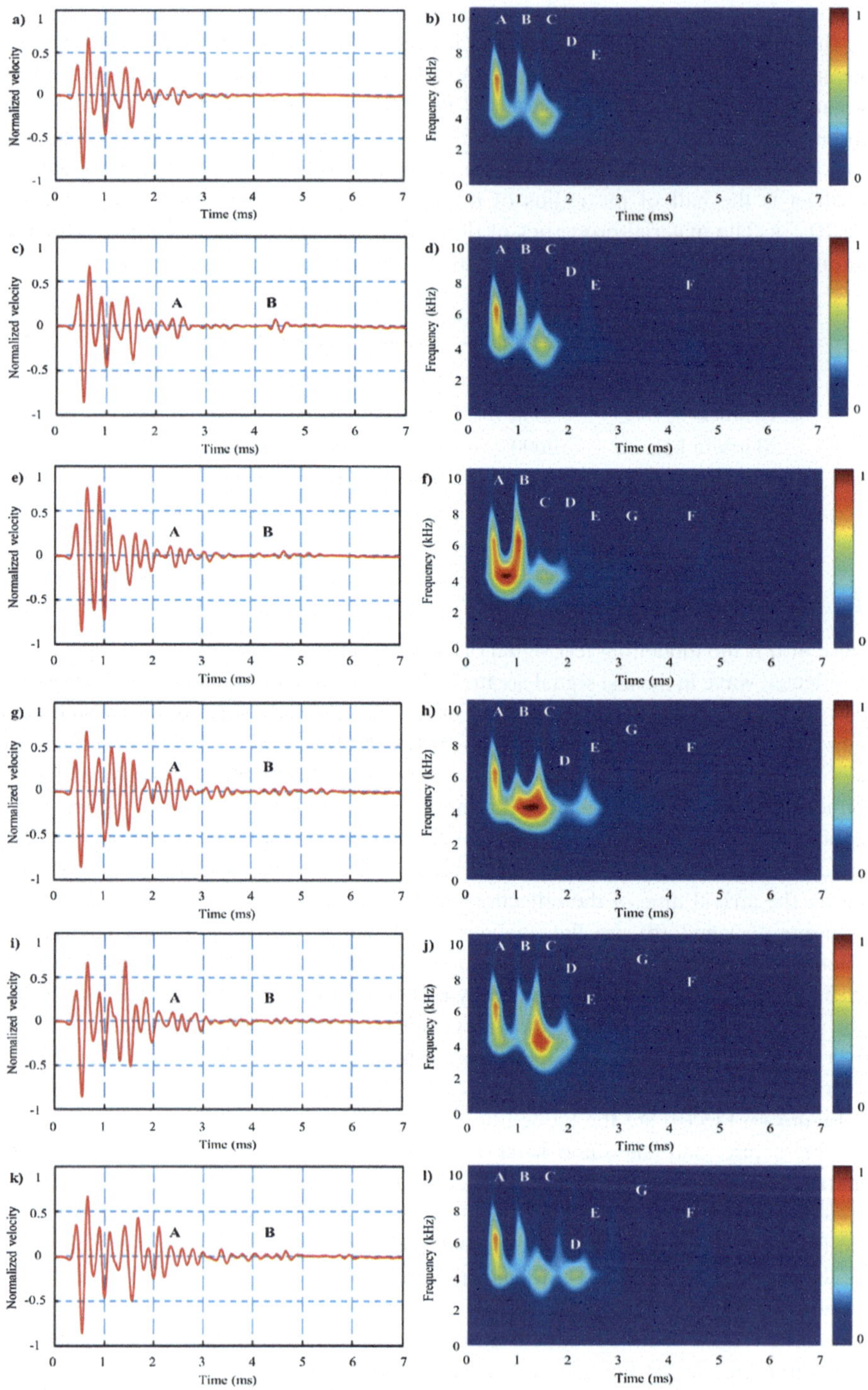

Fig. 4. Test signal and its corresponding amplitude of generalized S-Transform; (a) and (b) model 1; (c) and (d) model 2; (e) and (f) model 3; (g) and (h) model 4; (i) and (j) model 5; (k) and (l) model 6; (m) and (n) model 7.

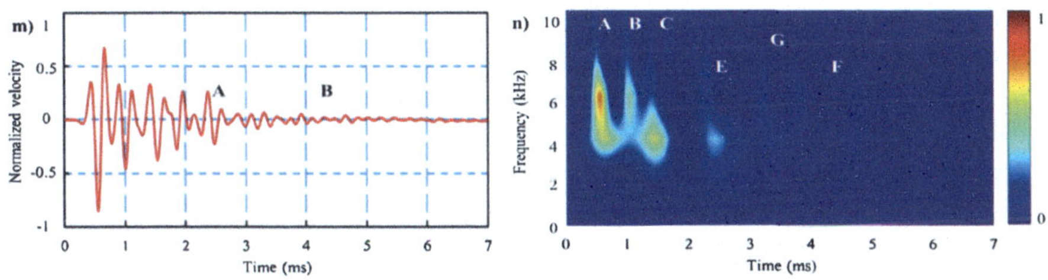

Fig. 4. *(continued)*

Table 3. Calculation errors for the arrival time of reflection from cavity in the models

The depth of fracture (m)	Arrival time of point E t_E (ms)	Arrival time of point F t_F (ms)	Absolute error of t_E (ms)	Absolute error of t_F (ms)
1.0	2.300	4.686	0.072	0.026
1.5	2.324	4.686	0.048	0.026
2.0	2.254	4.694	0.118	0.034
2.5	2.342	4.682	0.030	0.022
3.0	2.292	4.702	0.080	0.042

Points E and F in Fig. 4(f), (h), (j), (i) and (n) are the peaks caused by the reflections from the roof and floor of karst cavity. The reflections from the cavity floor (point B in Fig. 4(e), (g), (i), (k) and (m)) are difficult to be identified in the original test signal. But there is a significant energy peak in the time-frequency spectrum by generalized S-Transform. The arrival time calculated by the points E and points F and their errors are shown in Table 3. The true values of the arrival time of the reflections are obtained by the subtraction between the test signals of the models with cavity and without cavity. It shows that the absolute errors for the most of the identifications for the arrival time are not greater than 0.1 ms. The relatively larger error in the model with a fracture at the depth of 2 m might be caused by the interference of the multiple reflections from fracture. The absolute error of the arrival time for the reflections from the floor are smaller than that from roof. The reason is that the identification for the reflection from the cavity floor is less affected by the multiple reflections and reflected S-wave.

5 Conclusions

This paper provides an analysis method for the sonar detection to identify the karst cavities and fracture under pile based on generalized S-Transform. Seven numerical models with different depths of fractures were conducted to verify this method. The following conclusions could be drawn:

1. It is feasible and effective to detect the karst cavities and fracture using generalized S-transform. The changes caused by the reflected wave significantly appear as peaks in the time-frequency spectrum, which has great advantage in identification for weak reflection.
2. The reflection from the fracture will cause a significant enhancement of energy or a vacancy in the generalized S-Transform amplitude spectrum. Multiple reflections from the fracture make the error of the identification for karst cavity larger.
3. The multiple reflections from the wall of pile hole has large effect on the identification for the karst cavities and fractures. The decay of the multiple reflections from hole wall is easily identified by generalized S-transform. The reflection from the geological structure under pile can be observed if there is another enhancement of energy.
4. Only different depths of fracture are considered in this study for the limited length of this paper. The angle, thickness, filling of the fracture and many other factors will be studied in the future. The test signal of field experiment will be introduced to study the method as well.

Acknowledgments. The authors appreciate the financial support provided by the Natural Science Foundation of China (No. 41731283) and the fundamental research funds for the central universities. Special thanks to the Wuhan Changsheng Engineering Exploration Technology Development Co., Ltd. for the joint development of sonar detector.

References

Yan, E., Cheng, J., Liu, L.: Stability analysis and evaluation of soil cave foundation under the role of groundwater in karst area. Int. J. Intell. Syst. Appl. **1**(1), 60–67 (2009)

Zhao, M.H., Jiang, C., Cao, W.G.: Study on bearing capacity of rock-socked piles and safe thickness of cave roofs in karst region. Chin. J. Geotech. Eng. **29**(11), 1618–1622 (2007)

El-Qady, G., Hafez, M., Abdalla, M.A., et al.: Imaging subsurface cavities using geoelectric tomography and ground-penetrating radar. J. Cave Karst Stud. Natl. Speleol. Soc. Bull. **67**(3), 174–181 (2005)

Zajc, M., Celarc, B., Gosar, A.: Structural–geological and karst feature investigations of the limestone–flysch thrust-fault contact using low-frequency ground penetrating radar (Adria–Dinarides thrust zone, SW Slovenia). Environ. Earth Sci. **73**(12), 1–13 (2015)

Shi, Z., Liu, L., Ming, P., et al.: A sonar detection technology for karst cavities under bored piles and its application. Chin. J. Rock Mech. Eng. **35**, 177–186 (2016)

Stockwell, R.G., Mansinha, L.: Localization of the complex spectrum: the S transform. IEEE Trans. Signal Process. **44**(4), 998–1001 (1996)

Mcfadden, P.D., Cook, J.G., Forster, L.M.: Decomposition of gear vibration signals by the generalized S-transform. Mech. Syst. Signal Process. **13**(5), 691–707 (1999)

Pinnegar, C.R., Mansinha, L.: Time-local Fourier analysis with a scalable, phase modulated analyzing function: the S-transform with a complex window. Signal Process. **84**(7), 1167–1176 (2004)

Collino, F., Tsogka, C.: Application of the perfectly matched absorbing layer model to the linear elastodynamic problem in anisotropic heterogeneous media. Geophysics **66**(1), 294 304 (2001)

Calculation of the Lateral Cyclic Response of Monopiles in Sand

Feng Yu[1,2], Chen-rong Zhang[1,2(✉)], and Mao-song Huang[1,2]

[1] Department of Geotechnical Engineering, Tongji University, Shanghai 200092, China
zcrong33@tongji.edu.cn
[2] Key Laboratory of Geotechnical and Underground Engineering of Ministry of Education, Tongji University, Shanghai 200092, China

Abstract. Monopiles are widely used to support the upper structures of wind turbines in offshore engineering. Due to the harsh environment conditions, they are subjected to long-term lateral cyclic loading from waves and winds through the design life. In this paper, a robust and practical beam on Winkler foundation model is presented for the behavior of monopiles under cyclic lateral load, in which the stiffness degradation of sand is incorporated based on the accumulated soil strain from results of drained cyclic triaxial tests. The reasonability of the model is proved by comparisons with a published centrifuge model test.

Keywords: Monopiles · Axial strain · Sand · Winkler foundation model Calculating method · Cyclic load

1 Introduction

Monopiles for offshore wind turbines are designed to sustain long-term lateral cyclic loads from winds and waves. Besides the maximum static capacity, the accumulations of the pile deformation and rotation under long-term lateral cyclic load are of importance for the maintenance of offshore wind turbines to avoid exceeding the serviceability limit during the design life time.

The p-y curves are widely used in the practice of laterally loaded piles, in which a lot of work is given about the complex cyclic pile-soil interaction in sand. Reese et al. [1] constructed a series of cyclic p-y curves based on a few full-scale tests to consider the cyclic-loading deterioration, which are adopted by API [2]. Based on experiment results, Little and Briaud [3] and Long and Vanneste [4] advised the power function of number of cycles N to simulate the cyclic soil reaction along the pile. As the empirical cyclic p-y curves are fitted based on the results of slender small piles with limited cyclic numbers, they are not suitable for monopiles with long-term cyclic load. Achmus et al. [5] considered the reduction in soil stiffness in the FEM, based on the accumulated strain from cyclic triaxial tests. Considering the numerical errors of sophisticated soil constitutive model of a large number of cycles, the semi-empirical approach by Achmus et al. [5] is a better choice for the calculation of lateral deformation responses of a pile with respect to the number of long-term cyclic loads.

L. Hu et al. (Eds.): GSIC 2018, *Proceedings of GeoShanghai 2018 International Conference: Multi-physics Processes in Soil Mechanics and Advances in Geotechnical Testing*, pp. 534–544, 2018.
https://doi.org/10.1007/978-981-13-0095-0_60

In this study, a robust and practical beam on Winkler foundation model is presented, in which both the cyclic stiffness of soil in reloading and unloading phases are derived from the hysteric loop of sand under the cyclic triaxial test and the explicit cumulative strain model of soil. The proposed method to consider the response of a laterally loaded monopile is used to calculate the results of a published centrifuge test, which shows a good agreement.

2 Simplified Method of Monopile Under Cyclic Lateral Load

2.1 Governing Equation for a Laterally Loaded Pile Under a Winkler Foundation

The governing equation of a laterally loaded pile is:

$$E_pI_p\frac{d^4u}{dz^4} + k_hu = 0 \tag{1}$$

where E_pI_p is the bending rigidity of the pile, k_h is the subgrade reaction modulus, u is the horizontal displacement of the pile.

According to Vesic's equation [6], the subgrade reaction modulus k_h is calculated as:

$$k_h = \frac{0.65E}{1-\nu^2}\left(\frac{Ed_p^4}{E_pI_p}\right)^{1/12} \tag{2}$$

where E *and* ν are the elasticity modulus and Poisson's ratio of soil.

Dividing the pile into n elements, Eq. (1) can be solved by finite difference method as:

$$[K_L]\{u\} = \{F_L\} \tag{3}$$

$$\{u\} = [u_0\ u_1\ \cdots\ u_i\ \cdots\ u_{n-1}\ u_n] \tag{4}$$

$$\{F_L\} = \left[2h^3\frac{H_0}{E_pI_p} + 2h^2\frac{M_0}{E_pI_p} - h^2\frac{M_0}{E_pI_p}\ \cdots\ 0\right] \tag{5}$$

where $[K_L]$ is the stiffness matrix, H_0 *and* M_0 are the horizontal force and bending moment at pile head, $h = L/n$.

To investigate the cyclic responses of a laterally loaded pile, the subgrade reaction modulus of the pile in Eq. (2) is assumed to change with the cyclic stiffness of sand, the derivation of which is shown below.

2.2 Cyclic Stiffness Model of Sand

The degradation stiffness model of sand to account for the cyclic effect is elucidated in Fig. 1. For one-way cyclic load, in a complete cycle, the secant modulus of the sand includes two aspects: the modulus in a reloading phase E_{lsN}, and the modulus in an unloading phase E_{usN}, where N represents the cyclic number. To obtain the values of

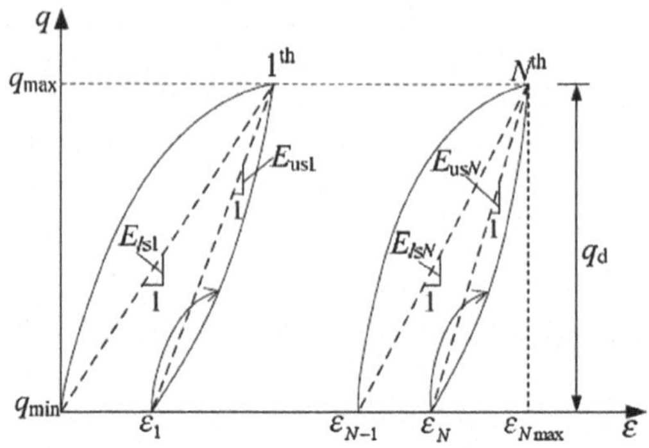

Fig. 1. Calculating diagram of the secant modulus.

the above two moduli, which are inputs for the beam on Winkler foundation model, the development of the hysteretic curve and the expression of the accumulated plastic strain in each cycle are needed. The former part is shown below and the latter part is introduced in Sect. 2.3.

In a cyclic triaxial test (Fig. 1), $q_{\max}$ and $q_{\min}$ are the maximum and minimum deviatoric stresses, and the amplitude of a cyclic load is $q_d = q_{\max} - q_{\min}$. ε_N and $\varepsilon_{N\max}$ represent the accumulated residual strain and peak strain in the *N*th cycle. E_{lsN} and E_{usN} are the reloading and unloading secant moduli of sand in the *N*th cycle, respectively. Which are expressed as:

$$E_{lsN} = \frac{q_d}{\varepsilon_{N\max} - \varepsilon_{N-1}} \tag{6}$$

$$E_{usN} = \frac{q_d}{\varepsilon_{N\max} - \varepsilon_N} \tag{7}$$

The hyperbolic stress-strain curve is widely used as the backbone curve of sand under triaxial test as:

$$\varepsilon = \frac{q}{E_o} / (1 - \frac{q}{q_{ult}}) \tag{8}$$

in which, q and q_{ult} are the deviatoric stress and peak strength of sand. E_O is the tangent modulus.

With Masing's rule [7], the hysteretic loop of sand is given as:

$$(\varepsilon - \varepsilon_r) = \frac{q - q_r}{E_{ON}} / (1 - \left|\frac{q - q_r}{\xi q_{ult}}\right|) \tag{9}$$

in which, q_r is the reference stress, ε_r is the reference strain and ξ is an coefficient. They are related to the reloading and unloading phases. For the reloading curve, $q_r = q_{\min}$, $\varepsilon_r = \varepsilon_{\min}$, $\xi = 1$. For the unloading curve, $q_r = q_{\max}$, $\varepsilon_r = \varepsilon_{\max}$, $\xi = 2$. E_{ON} is the tangent modulus of sand for the *N*th cycle.

In a single cycle, under the assumption of an unchanged E_{ON} both for the loading and unloading phases (Fig. 2), the accumulated maximum strain for the *N*th cycle is:

$$\varepsilon_{N\max} = \frac{q_d}{E_{ON}}/(1 - \frac{q_d}{q_{ult}}) + \varepsilon_{N-1} \tag{10}$$

The accumulated residual strain for the *N*th cycle is:

$$\varepsilon_N = -\frac{q_d}{E_{ON}}/(1 - \frac{q_d}{2q_{ult}}) + \varepsilon_{N\max} \tag{11}$$

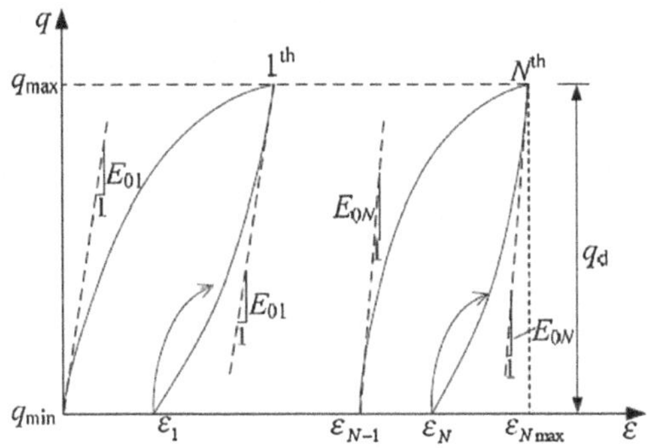

Fig. 2. Calculating diagram of the initial equivalent modulus.

Substituting Eq. (10) into Eq. (11), we can get:

$$\varepsilon_N = -\frac{q_d}{E_{ON}}/(1 - \frac{q_d}{2q_{ult}}) + \frac{q_d}{E_{ON}}/(1 - \frac{q_d}{q_{ult}}) + \varepsilon_{N-1} \tag{12}$$

The tangent modulus of sand in the *N*th cycle E_{ON} is obtained from Eq. (12), which is shown as below:

$$E_{ON} = \frac{q_d}{\Delta\varepsilon_N}(\frac{1}{1 - \frac{q_d}{q_{ult}}} - \frac{1}{1 - \frac{q_d}{2q_{ult}}}) = \frac{q_{eq}}{\Delta\varepsilon_N} \tag{13-1}$$

where $\Delta\varepsilon_N = \varepsilon_N - \varepsilon_{N-1}$ is the plastic strain increment in the *N*th cycle, q_{eq} is defined as the equivalent cyclic load, which is the function of the cyclic load amplitude and peak strength of sand:

$$q_{eq} = q_d(\frac{1}{1 - \frac{q_d}{q_{ult}}} - \frac{1}{1 - \frac{q_d}{2q_{ult}}}) \tag{13-2}$$

Considering Eqs. (10)–(13-2), the secant modulus of sand for the reloading and unloading curves in Eqs. (6) and (7) can be written as:

$$E_{lsN} = E_{ON}(1 - \frac{q_d}{q_{ult}}) = \frac{q_{eq}}{\Delta\varepsilon_N}(1 - \frac{q_d}{q_{ult}}) \tag{14}$$

$$E_{usN} = E_{ON}(1 - \frac{q_d}{2q_{ult}}) = \frac{q_{eq}}{\Delta\varepsilon_N}(1 - \frac{q_d}{2q_{ult}}) \tag{15}$$

It can be seen from Eqs. (14) and (15) that the two moduli of sand are functions of the residual strain increment $\Delta\varepsilon_N$ in each cycle. As $\Delta\varepsilon_N$ approaches infinitesimal at large cyclic number, it will result in infinite modulus of sand when the cycle number is large enough, which means an irrational accumulation of deformation for the laterally loaded pile. Therefore, a revision on the secant modulus of sand is given as below:

$$E^*_{\mathrm{ls}N} = q_{eq}(1 - \frac{q_d}{q_{ult}}) \cdot (\frac{1}{3A\Delta\varepsilon_N + \alpha B}), \quad \alpha = \frac{1 - \frac{q_d}{q_{ult}}}{1 - \frac{q_d}{2q_{ult}}} \tag{16}$$

$$E^*_{\mathrm{us}N} = q_{eq}(1 - \frac{q_d}{2q_{ult}}) \cdot (\frac{1}{A\Delta\varepsilon_N + B}) \tag{17}$$

where the dimensionless stress level α makes that the two moduli equal to elastic modulus of sand when the accumulated strain $\Delta\varepsilon_N$ is 0. B is a parameter as:

$$B = \frac{q_{eq}}{E_e}(1 - \frac{q_d}{2q_{ult}}) = \frac{q_d}{E_e}(\frac{1}{\alpha} - 1) \tag{18}$$

in which E_e is the elastic modulus of sand. The parameter A is used to control the development of modulus for the reloading and unloading curves.

As shown in Fig. 3, when the cycle numbers are large enough, the secant modulus of soil in the reloading phase equals to that in the unloading phase. Hence, no accumulated soil strain occurs.

2.3 Accumulated Axial Strain of Sand Under Cyclic Triaxial Test

An explicit model for the accumulation of axial residual stain of clay due to long-term cyclic loading is presented by Huang et al. [8, 9], based on the triaxial undrained test data of Shanghai clay, in which the influence of cycle numbers, the initial mean stress, the initial stress ratio and the cyclic stress amplitude are considered. Furthermore, with the triaxial undrained test results of Fujian standard sand, Wang et al. [10] proved that the above empirical model can also be applied to predict the axial accumulated strain of sand. It is used in this paper to calculate modulus of sand in each cycle in Eqs. (16) and (17).

In the $p - q$ plane as shown in Fig. 4, the soil stress states before cyclic load are the static mean stress p_s and static deviatoric stress q_s. For sand sample subjected to drained one-way cyclic load, the confining pressure is kept constant and the axial stress component is cyclically varied, with an amplitude of the deviatoric stress as q_d.

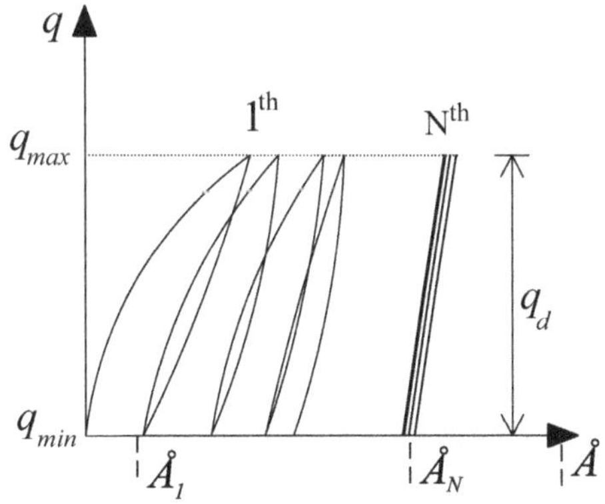

Fig. 3. Schematic diagram of the revised soil secant stiffness.

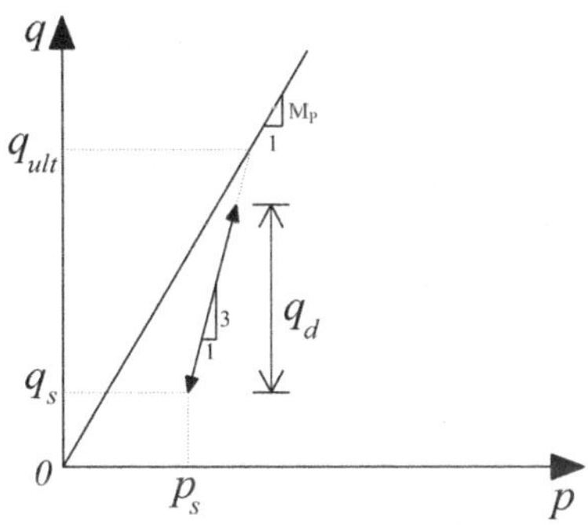

Fig. 4. Cyclic triaxial drainage shear test load path.

The explicit model assumes that the accumulated axial strain of sand increases exponentially with cyclic number, which is shown as below:

$$\varepsilon_N = a \cdot D^{*m} \left(\frac{p_s}{p_a} \right)^c N^b \tag{19}$$

in which a, m, c, b are material parameters obtained from test results. It should be noted that the initial mean stress is normalized by a reference confining pressure p_a (taken as 101 kPa). Furthermore, a normalized size of the deviatoric stress amplitude D^* is defined as:

$$D^* = \frac{D_p - D_s}{D_{\max} - D_s} \tag{20}$$

in which the cyclic deviatoric stress amplitude q_d is related to the static deviatoric stress q_s and the peak deviatoric stress of sand q_{ult} by normalizations as $D_p = (q_d + q_s)/q_{ult}$, $D_s = q_s/q_{ult}$ and $D_{\max} = q_{ult}/q_{ult} = 1$. The calculation of peak strength of the sand q_{ult} is given below.

Denoting the peak friction angle of sand as φ_p, the inclination of the sand at peak is:

$$M_p = \frac{6 \sin \varphi_p}{3 - \sin \varphi_p} \tag{21}$$

Since the inclination of the stress path for the drained triaxial test is 3, q_{ult} can be related to the initial stress state as:

$$q_{ult} = \frac{M_p}{3 - M_p}(3p_s - q_s) \tag{22}$$

Bolton [11] conducted a comprehensive review of experimental data on the shear strength of sand and considering the dilatancy of sand, he suggested an empirical correlation of the peak frictional angle φ_p and critical state friction angle φ_c as:

$$\varphi_p = \varphi_c + 3D_r(10 - \ln p_f) - 3 \tag{23}$$

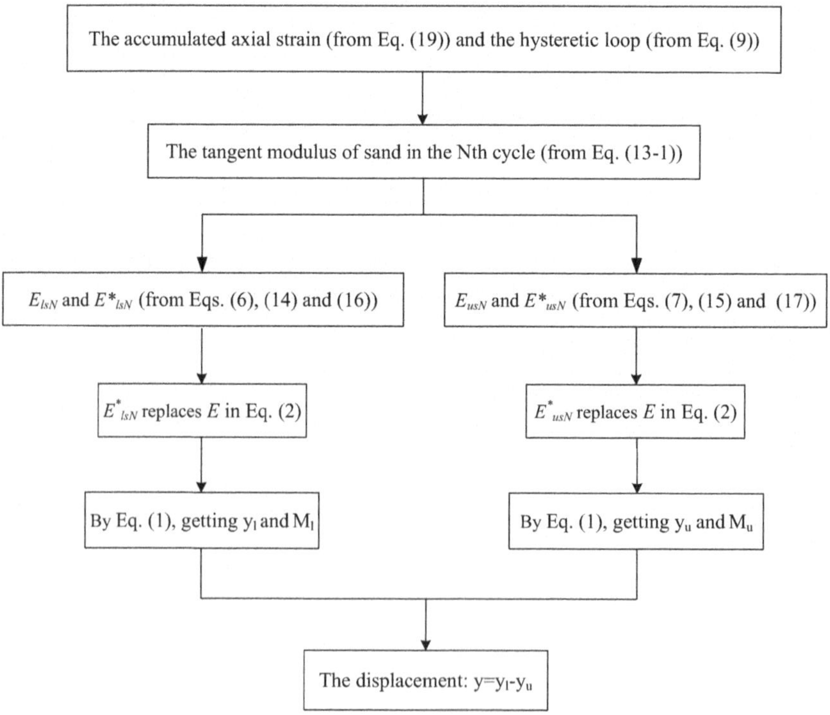

Fig. 5. Flow chart for computing the displacement and bending moments of the pile.

where D_r is the relative density of sand, p_f is the mean stress at peak.

Figure 5 shows the flow chart for computing the displacement and bending moments of the pile, the computation steps are as follows:

(1) Combining the accumulated axial strain formula (from Eq. (19)) and the hysterctic loop formula (from Eq. (9)), we can get the tangent modulus of sand in the *N*th cycle (from Eq. (13-1)).
(2) Based on the tangent modulus of sand, the secant modulus of sand for the reloading and unloading curves in Eqs. (6) and (7) can be written as Eqs. (14) and (15). Then Eqs. (14) and (15) are revised as Eqs. (16) and (17) respectively.
(3) The reloading secant modulus E^*_{lsN} and unloading secant modulus E^*_{usN} replace the elasticity modulus E in Eq. (2), the subgrade reaction modulus k_h can be gotten.
(4) Putting k_h into Eq. (1), we can get the displacement and bending moments of the pile. y_l and M_l are the displacement and bending moments in a reloading phase, y_u and M_u are the displacement and bending moments in an unloading phase, the residual displacement of pile is $y = y_l - y_u$.

3 Verification

In the following part, the proposed beam on Winkler foundation method, with a stiffness degradation model to consider the cyclic interaction between pile and sand, is used to calculate the responses of a lateral cyclic loaded pile in sand in a centrifuge test by Zhu et al. [12].

The pile is driven into sand at 1 g, with an embedment depth of 50 m and the load is applied 2.7 D above the ground surface (D is the diameter of the pile). The model pile is made of aluminum alloy pipe, having an elasticity modulus of 68.9 GPa and a Poisson's ratio of 0.33. The properties of the test pile in model and prototype scales are shown in Table 1. In the test, the bending moments of the test pile are measured through 15 pairs of strain gauges placed on the pile shaft and the lateral displacements at the pile head are measured by two laser displacement transducers at the pile head. The lateral load is applied by a motor-controlled loading device and the load pattern is a one-way lateral cyclic load with a load amplitude as 2100 kN and the maximum cycle number as 995.

Table 1. Properties of the test pile in model and prototype scales

Type of pile	Diameter/m	Wall thickness/m	Embedment depth/m	Flexural rigidity
Model pile	0.03	0.002	0.6	1193 $N \cdot m^2$
Prototype pile	2.5	0.045	50	56.66 $GN \cdot m^2$

The test soil is Fujian standard sand, having a critical state friction angle of $\varphi_c = 35°$. The minimum and maximum void ratios are $e_{\min} = 0.607$ and $e_{\max} = 0.952$, respectively. Using the raining method to prepare the sand sample, the dry sand is measured having the relative density of 65%, the void ratio of 0.727 and the dry density of 1.524 g/cm. At a centrifuge acceleration level of 83 g, the average peak friction of the sand is calculated as $\varphi_p = 39°$. Based on the one-way cyclic drained triaxial test results by Wang et al. [10], the parameters to describe the accumulated cyclic strains of the Fujian standard sand in Eq. (19) are advised as a = 1.382, b = 0.0386, c = 0.737, m = 1.494. The parameters in respect to the reloading and unloading moduli of sand in Eqs. (16) and (17) are $A = 1.11e^5$, $B = 0.008$.

An elastic finite element analysis of the responses of the test pile under two static lateral loads (corresponding to the minimum and maximum loads in a load cycle) is performed to obtain the stress state of the sand around the pile before and during cyclic load. Then the static mean stress p_s, the static deviatoric stress q_s and the cyclic deviatoric stress q_d of the sand can be obtained, which are inputs for the calculation of the soil secant modulus during the cyclic load.

Figure 6 gives the comparisons of the calculated and measured accumulated displacements at pile head for the centrifuge test. Figures 7 and 8 give the distributions of the calculated and measured results of the lateral deflections and bending moments at the first and last cycles, respectively. In Fig. 6, it is noted that the large residual displacement accumulation occurs in the first 100 cycles and decreases rapidly as the cyclic numbers increase. After 300 load cycles, the accumulation of the displacement is small, and the rate of which reaches an approximately constant value. In Figs. 7 and 8, both the calculated deflections near the pile head and the maximum moments of the laterally loaded pile are smaller than the measured values, although a good agreement of the location of the maximum moments exists. Based on the analysis above, the prediction of the general trend of the accumulated displacements with cycle numbers is satisfactory. However, the further investigation in the underestimation of the responses of the laterally loaded pile in the deflection and moments along the pile is needed.

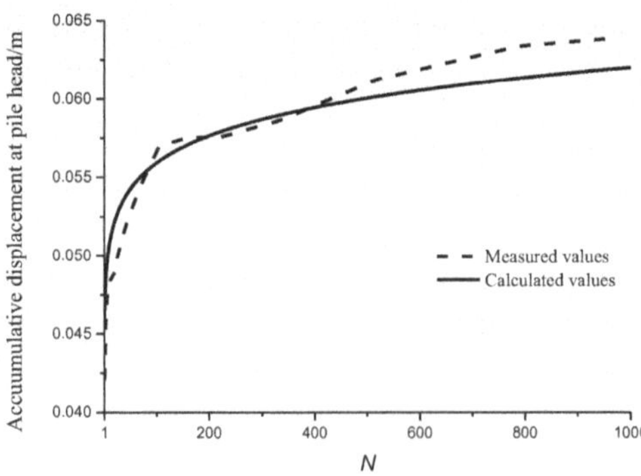

Fig. 6. Comparisons of the calculated and measured cumulative displacements at the pile head.

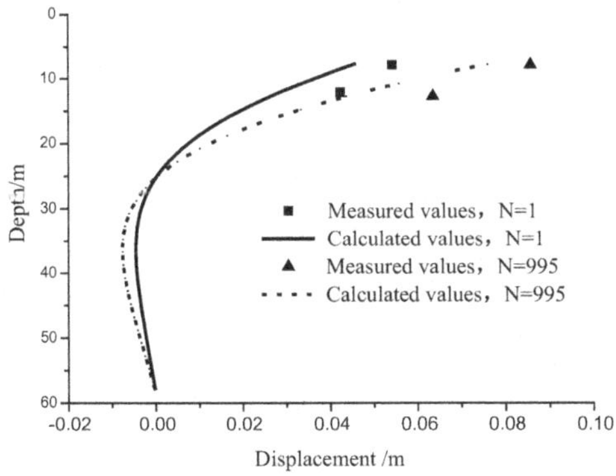

Fig. 7. Comparisons of the calculated and the measured deflections along the pile.

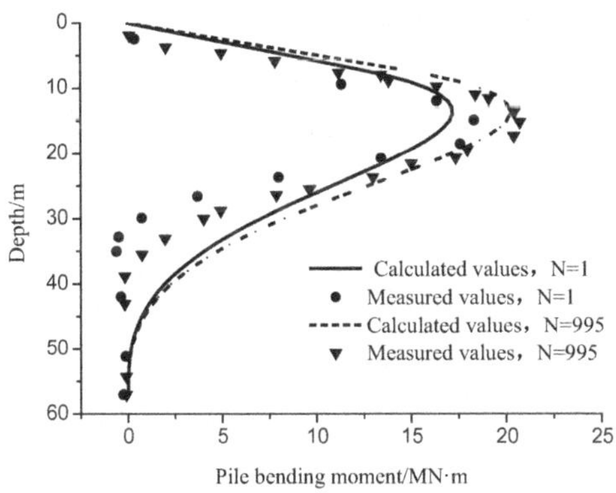

Fig. 8. Comparisons of the calculated and the measured bending moments along the pile.

4 Conclusions

Based on the analysis above, the following conclusions can be obtained:

(1) The stiffness model of sand under the cyclic load is given to consider the cyclic pile-soil interaction for a Winkler foundation, expressed as a reloading modulus and an unloading soil modulus, in which the hysteretic curve of sand and the accumulated soil strain under cyclic load are incorporated. The parameters in the model are determined by the results of cyclic triaxial tests of sand under drained conditions.

(2) The beam on a Winkler foundation method is proposed in this paper to calculate the cyclic responses of a laterally loaded pile in sand. A comparison with a centrifuge model test is given to prove the feasibility of the proposed method. The prediction of the general trend of the accumulated displacements with cycle numbers is satisfactory, while an underestimation of the responses of the laterally loaded pile in the deflections and moments along the pile is observed.

Acknowledgement. This work is financially supported by the National Natural Science Foundation of China (Grant Nos. 51779175 and 51579177).

References

1. Reese, L.C., Cox, W.R., Koop, F.D.: Analysis of laterally loaded piles in sand. In: Proceedings of the 6th Annual Offshore Technology Conference (OTC 2080), Houston, Texas, vol. 2, pp. 473–485 (1974)
2. American Petroleum Institute: Recommended practice for planning, designing and constructing fixed offshore platforms – working stress design. In: API Recommended Practice 2A-WSD (RP2A-WSD), 21st edn. API, Dallas (2000)
3. Little, R.L., Briaud, J.L.: Full scale cyclic lateral load tests on six single piles in sand. In: Miscellaneous Paper GL-88-27, Geotechnical Div Texas A&M Univ, College Station (1988)
4. Long, J., Vannestev, G.: Effects of cyclic lateral loads on piles in sand. J. Geotech. Eng. **120**(1), 225–244 (1994)
5. Achmus, M., Kuo, Y.S., Abdel-Rahman, K.: Behavior of monopile foundations under cyclic lateral load. Comput. Geotech. **36**(1), 725–735 (2009)
6. Vesic, A.B.: Bending of beam resting on isotropic elastic solid. J. Eng. Mech. Div. **87**(2), 35–53 (1961)
7. Ramberg, W., Osgood, W.R.: Description of stress-strain curves by three parameters. Technical Note 902, National Advisory Committee for Aeronautics, Washington, D.C. (1943)
8. Huang, M.S., Li, J.J., Li, X.Z.: Cumulative deformation behaviour of soft clay in cyclic undrained tests. Chin. J. Geotech. Eng. **28**(7), 891–895 (2006). (in Chinese)
9. Huang, M.S., Yao, Z.M.: Explicit model for cumulative strain of saturated clay subjected to cyclic loading. Chin. J. Geotech. Eng. **33**(3), 325–331 (2011). (in Chinese)
10. Wang, L., Zhu, B., Lai, X.H.: Cyclic accumulated deformation of sand and its explicit model. Chin. J. Geotech. Eng. **37**(11), 2024–2029 (2015). (in Chinese)
11. Bolton, M.D.: The strength and dilatancy of sands. Geotechnique **36**(1), 65–78 (1986)
12. Zhu, B., Xiong, G., Liu, J.C., Sun, Y.X., Chen, R.P.: Centrifuge modelling of a large-diameter single pile under lateral loads in sand. Chin. J. Geotech. Eng. **35**(10), 1807–1815 (2013). (in Chinese)

Correction to: Proceedings of GeoShanghai 2018 International Conference: Multi-physics Processes in Soil Mechanics and Advances in Geotechnical Testing

Liangbo Hu, Xiaoqiang Gu, Junliang Tao, and Annan Zhou

Correction to:
L. Hu et al. (Eds.): ***Proceedings of GeoShanghai 2018 International Conference: Multi-physics Processes in Soil Mechanics and Advances in Geotechnical Testing*****,, GSIC 2018, https://doi.org/10.1007/978-981-13-0095-0**

In the original version of the book, the incorrect second author name "We-Wei Du" has to be corrected as "Wei-Wei Du" in the chapter "Model Test of Seepage Erosion Around Defective Buried Pipelines", frontmatter and backmatter. The erratum book has been updated with the change.

In the original version of the book, an incorrect image was labelled as Figure 5 in Chapter 50, which has now been corrected. The book and the chapter have been updated with the change.

The updated version of these chapters can be found at
https://doi.org/10.1007/978-981-13-0095-0_30
https://doi.org/10.1007/978-981-13-0095-0_50

L. Hu et al. (Eds.): GSIC 2018, *Proceedings of GeoShanghai 2018 International Conference: Multi-physics Processes in Soil Mechanics and Advances in Geotechnical Testing*, p. C1, 2021.
https://doi.org/10.1007/978-981-13-0095-0_61

Author Index

L. Hu et al. (Eds.): GSIC 2018, *Proceedings of GeoShanghai 2018 International Conference: Multi-physics Processes in Soil Mechanics and Advances in Geotechnical Testing*, pp. 545–547, 2018.
https://doi.org/10.1007/978-981-13-0095-0

The manufacturer's authorised representative in the EU is Springer Nature Customer Service Centre GmbH, Europaplatz 3, 69115 Heidelberg, Germany. If you have any concerns regarding our products, please contact ProductSafety@springernature.com

Printed and bound by CPI Group (UK) Ltd, Croydon, CR0 4YY
25/09/2026
02222884-0016